Multi-Agent Systems for Traffic and Transportation Engineering

Ana L. C. Bazzan
Instituto de Informática, UFRGS, Brazil

Franziska Klügl
Örebro University, Sweden

Information Science REFERENCE

INFORMATION SCIENCE REFERENCE

Hershey · New York

Director of Editorial Content:	Kristin Klinger
Senior Managing Editor:	Jamie Snavely
Managing Editor:	Jeff Ash
Assistant Managing Editor:	Carole Coulson
Typesetter:	Jeff Ash
Cover Design:	Lisa Tosheff
Printed at:	Yurchak Printing Inc.

Published in the United States of America by
Information Science Reference (an imprint of IGI Global)
701 E. Chocolate Avenue
Hershey PA 17033
Tel: 717-533-8845
Fax: 717-533-8661
E-mail: cust@igi-global.com
Web site: http://www.igi-global.com/reference

and in the United Kingdom by
Information Science Reference (an imprint of IGI Global)
3 Henrietta Street
Covent Garden
London WC2E 8LU
Tel: 44 20 7240 0856
Fax: 44 20 7379 0609
Web site: http://www.eurospanbookstore.com

Library of Congress Cataloging-in-Publication Data

Multi-agent systems for traffic and transportation engineering / Ana L.C. Bazzan and Franziska Klugl, editor.
p. cm.
Includes bibliographical references and index.

Summary: "This book aims at giving a complete panorama of the active and promising crossing area between traffic engineering and multi-agent system addressing both current status and challenging new ideas"--Provided by publisher.

ISBN 978-1-60566-226-8 (hbk.) -- ISBN 978-1-60566-227-5 (ebook)

1. Transportation--Data processing. 2. Transportation engineering--Data processing. 3. Traffic engineering--Data processing. 4. Intelligent agents (Computer software) I. Bazzan, Ana L. C. II. Klügl, Franziska.
TA1230.H36 2009
388.3'10285--dc22

2008042399

British Cataloguing in Publication Data
A Cataloguing in Publication record for this book is available from the British Library.

All work contributed to this book is new, previously-unpublished material. The views expressed in this book are those of the authors, but not necessarily of the publisher.

List of Reviewers

Adrian Agogino, *University of California, NASA Ames Research Center, USA*
Gustavo Kuhn Andriotti, *Universität Würzburg, Germany*
Michael Balmer, *IVT, ETH Zürich, Suisse*
Brahim Chaib-Draa, *Laval University, Canada*
Michel Ferreira, *Universidade do Porto, Portugal*
Leonardo Garrido, *Centro de Computación Inteligente y Robótica, Tecnológico de Monterrey, México*
Dominik Grether, *Institut für Verkehrssystemplanung und Verkehrstelematik, TU Berlin, Germany*
Qi Han, *Eindhoven University of Technology, The Netherlands*
Florian Harder, *Institut für Geographie, Universität Würzburg, Germany*
Rainer Herrler, *SSI Schäfer Noell GmbH, Germany*
Peter Jarvis, *NASA Ames Research Center, USA*
David J. Kaup, *University of Central Florida, USA*
Hubert Klüpfel, *TraffGoHT GmbH, Germany*
Kai Nagel, *Institut für Verkehrssystemplanung und Verkehrstelematik, TU Berlin, Germany*
Gregor Lämmel, *Institut für Verkehrssystemplanung und Verkehrstelematik, TU Berlin, Germany*
Shoichiro Nakayama, *Kanayawa University, Japan*
Rex Oleson, *University of Central Florida, USA*
Denise de Oliveira, *Universidade Federal do Rio Grande do Sul, Brazil*
Guido Rindsfüser, *Emch+Berger AG Bern, Suisse*
Roseli F. Romero, *ICMC-USP, Brazil*
Rosaldo Rossetti, *University of Porto (FEUP), Portugal*
Andreas Schadschneider, *Universität zu Köln, Germany*
Heiko Schepperle, *Universität Karlsruhe (TH), Germany*
Takeshi Takama, *Stockholm Environment Institute, Sweden, University of Oxford, UK*
Harry Timmermans, *Eindhoven University of Technology, The Netherlands*
Sabine Timpf, *University of Augsburg, Germany*
Kagan Tumer, *Oregon State University, USA*
Karl Tuyls, *Eindhoven University of Technology, The Netherlands*
Matteo Vasirani, *University Rey Juan Carlos, Madrid, Spain*
Li Weigang, *UnB, Brazil*
Marco Wiering, *Universiteit Utrecht, The Netherlands*
Shawn Wolfe, *NASA Ames Research Center, USA*
Tomoshisa Yamashita, *National Institute of Advanced Industrial Science and Technology (AIST), Japan*

Table of Contents

Detailed Table of Contents

This chapter tackles micro-simulation by discussing design and implementation issues of MATSim, as well as an experiment in which this simulator is used to study daily traffic in Switzerland.

Continuing the discussion around microscopic simulation, in this chapter, the TRASS simulation framework, a multi-layer architecture, is presented and evaluated in the context of several application scenarios.

In this chapter, a multi-agent model is proposed aiming to cope with the complexity associated with microscopic traffic simulation modelling. Using a prototype with some of the features introduced, the authors discuss scenarios using car-following and lane-changing behaviours.

The authors of this chapter investigate the behaviour of pedestrians and human crowds, focussing on aspects related to physical movement. It thus starts with a review of methods and approaches, and continue with a discussion around validation issues, aiming at reducing the gap between the multi-agent and pedestrian dynamics communities.

This chapter discusses the "Social Potential" model for implementing multi-agent movement in simulations by representing behaviors, goals, and motivations as artificial social forces.

In this chapter, Sabine Timpf presents a vision for simulating human navigation within the context of public, multi-modal transport, showing that cognitive agents require the provision of a rich spatial environment. She introduces spatial representations and wayfinding as key components in the model. She illustrates her vision by a case study that deals with multi-modal public transport.

This chapter presents two extensions of a system for managing autonomous vehicles at intersections. In the first, it is demonstrated that for intersections with moderate to low amounts of traffic, a completely decentralized, peer-to-peer intersection management system can reap many of the benefits of a centralized system without the need for special infrastructure at the intersection. In the second extension, it is shown that the proposed intersection control mechanism can mitigate the effects of catastrophic physical malfunctions in autonomous vehicles.

Providing services and infrastructure for autonomous vehicles at intersections is also the topic of this chapter in which the authors describe an agent-based valuation-aware traffic control system for intersections. Their approach combines valuation-aware intersection-control mechanisms with driver-assistance features such as adaptive cruise and crossing control.

Collaborative driving is the focus of this chapter. The authors describe an agent-based cooperative architecture that aims at controlling and coordinating vehicles, also showing that reinforcement learning can be used for this purpose.

The authors of this chapter tackle the issue of how road users can learn to coordinate their actions with those of other agents in a scenario without communication. Further, the authors explore the impacts of agent reward functions on two traffic related problems (selection of departure time and selection of lane).

In this chapter, the authors discuss multiagent learning in the context of a coordination mechanism where teams of agents coordinate their velocities when approaching the intersection in a decentralized way, improving the intersection efficiency.

Preface

The increasing demand for mobility in the 21st century poses a challenge to researchers from several fields to devise more efficient traffic and transportation systems designs, including control devices, techniques to optimize the existing network, and also information systems. More than ever, interdisciplinary approaches are necessary. A successful experience has been the cross-fertilization between traffic, transportation, and artificial intelligence that dates at least from the 1980s and 1990s, when expert systems were built to help traffic experts control traffic lights. Also, information on how to combine parking and public transportation can be provided by intelligent systems, and transportation and logistics have also benefited from artificial intelligence techniques, especially those tied to optimization.

During the last decade, there has been a tremendous progress in traffic engineering based on agent technology. However, given the increasing complexity of those systems, a product of the modern way of life and new means of transportation, the individual choices must be better understood if the whole system is to become more efficient. Thus, it is not surprising that there is a growing debate about how to model transportation systems at both the individual (micro) and the society (macro) level. This may raise technical problems, as transportation systems can contain thousands of autonomous, intelligent entities that need to be simulated and/or controlled. Therefore, traffic and transportation scenarios are extraordinarily appealing for (multi-)agent technology.

Additionally, traffic scenarios became very prominent as test beds for coordination or adaptation mechanisms in multi-agent systems. Many examples of successful deployments of tools and system exist.

This book is a collection of contributions addressing topics that arose from a cross fertilization between traffic engineering and multi-agent system. Hence, this book summarizes innovative ideas for applications of different agent technologies on traffic and transportation related problems.

CHALLENGES AND APPROACHES

The second half of the last century has seen the beginning of the phenomenon of traffic congestion. This arose due to the fact that the demand for mobility in our society has increased constantly. Traffic congestion is a phenomenon caused by too many vehicles trying to use the same infrastructure at the same time. The consequences are well-known: delays, air pollution, decrease in speed, and risky manoeuvres thus reducing safety for pedestrians as well as for other drivers.

The increase in transportation demand can be met by providing additional capacity. However, this may no longer be economically or socially attainable or feasible. Thus, the emphasis has shifted to improving the existing infrastructure without increasing the overall nominal capacity, by means of an optimal utilization of the available capacity. Two complementary measures can be taken: improving the management systems by use of recent developments in the areas of communication and information technology, and improving the management via control techniques. The set of all these measures is framed as Intelligent Transportation Systems (ITS).

Artificial intelligence and multi-agent techniques have been used in many stages of these processes. During the last decade, there has been a tremendous progress in traffic engineering based on agent technology. The approaches can be classified into three levels: integration of heterogeneous traffic management systems, traffic guidance, and traffic flow control.

The first of these levels is discussed in several papers, for example the platform called Multi-Agent Environment for Constructing Cooperative Applications - MECCA/UTS – (Haugeneder & Steiner, 1993), as well as in Ossowski et al. (2005), in Rossetti and Liu (2005), and in van Katwijk et al. (2005).

Regarding traffic guidance, it is generally believed that information-based ITS strategies are among the most cost-effective investments that a transportation agency can make. These strategies, also called Advanced Traveler Information Systems (ATIS), include highway information, broadcast via radio, variable message systems, telephone information services, Web/Internet sites, kiosks with traveler information, and personal data assistant and in-vehicle devices. Many other new technologies are available now to assist people with their travel decisions. Multi-agent techniques have been used for modeling and simulation of the effects of the use of these technologies, as well as the modeling of behavioural aspects of the drivers and their reaction to information. Details can be found in Balmer et al. (2004), Bazzan and Klügl (2005), Bazzan et al. (1999), Burmeister et al. (1997), Elhadouaj et al. (2000), Klügl and Bazzan (2004), Klügl et al. (2003), Paruchuri et al. (2002), Rigolli and Brady (2005), Rossetti et al. (2002), Tumer et al. (2008), and Wahle et al. (2002).

Regarding the third level mentioned above – traffic control – a traffic control loop was proposed by Papageorgiou (2003). It applies to any kind of traffic network if one is able to measure traffic as the number of vehicles passing on a link in a given period of time. With the current developments in communication and hardware, computer-based control is now a reality. The main goals of Advanced Transportation Management Systems (ATMS) are: to maximize the overall capacity of the network; to maximize the capacity of critical routes and intersections which represent the bottlenecks; to minimize the negative impacts of traffic on the environment and on energy consumption; to minimize travel times; and to increase traffic safety. In order to achieve these goals, devices to control the flow of vehicles (e.g. traffic lights) can be used. However other forms of control are also possible. For classical approaches please see: TRANSYT (Robertson, 1969; TRANSYT-7F, 1988), SCOOT (*Split Cycle and Offset Optimization Technique*) (Hunt et al., 1981), SCATS (*Sydney Coordinated Adaptive Traffic System*) (Lowrie, 1982), and TUC (*Traffic-responsive Urban Traffc Control*) (Diakaki et al., 2002). Regarding the use of multiagent systems, some work in this area can be found in Bazzan (2005), Bazzan et al. (2008), Camponogara and Kraus (2003), Dresner and Stone (2004), France and Ghorbani (2003), Nunes and Oliveira (2004), Oliveira et al. (2004), Oliveira et al. (2005), Rochner et al. (2006), Silva et al. (2006), Steingrover et al. (2005), Wiering (2000).

ORGANIZATION OF THE BOOK

The book is organized into three parts. The first is a collection of chapters that focus on agent-based simulation of transportation and traffic scenarios for traffic reproduction, both for vehicular traffic and pedestrian traffic. A second section is a compilation about traffic control and management, mainly using traffic lights. A third part deals with agent-based approaches for related themes such as air traffic management and logistics.

A brief description of each of the chapters follows, starting with those in Section I.

In Chapter I, Takama discusses congestion and adaptation by means of a multi-agent system (MAS) aiming to analyse real transport and traffic problems. The chapter contribution is both a methodological discussion and an empirical case study. The latter is based on real stated-preference data to analyse the effect of a real road-user charge policy and a complimentary park and ride scheme at the Upper Derwent Valley in the Peak District National Park, England.

Han and colleagues (Chapter II) discuss an agent-based modeling approach focusing on the dynamic formation of (location) choice sets. Individual travellers learn through their experiences with the transport systems, changes in the environments and from their social network, based on reinforcement learning, Bayesian learning, and social comparison theories.

Chapter III tackles micro-simulation by discussing design and implementation issues of MATSim, as well as an experiment in which this simulator is used to study daily traffic in Switzerland.

Continuing the discussion around microscopic simulation, in Chapter IV, the TRASS simulation framework, a multi-layer architecture, is presented and evaluated in the context of several application scenarios.

In Chapter V, a multi-agent model is proposed aiming to cope with the complexity associated with microscopic traffic simulation modelling. Using a prototype with some of the features introduced, the authors discuss scenarios using car-following and lane-changing behaviours.

Schadschneider and colleagues investigate the behaviour of pedestrians and human crowds, focussing on aspects related to physical movement. Chapter VI thus starts with a review of methods and approaches, and continue with a discussion around validation issues, aiming at reducing the gap between the multi-agent and pedestrian dynamics communities.

Chapter VII discusses the "Social Potential" model for implementing multi-agent movement in simulations by representing behaviours, goals, and motivations as artificial social forces.

In Chapter VIII, Sabine Timpf presents a vision for simulating human navigation within the context of public, multi-modal transport, showing that cognitive agents require the provision of a rich spatial environment. She introduces spatial representations and the basics of wayfinding as key components in the model. She illustrates her ideas by a case study that deals with multi-modal public transport.

Chapters IX to XV compose Section II of this book and have in common the focus on traffic control.

Chapter IX presents two extensions of a system for managing autonomous vehicles at intersections. In the first, it is demonstrated that for intersections with moderate to low amounts of traffic, a completely decentralized, peer-to-peer intersection management system can reap many of the benefits of a centralized system without the need for special infrastructure at the intersection. In the second extension, it is shown that the proposed intersection control mechanism can mitigate the effects of catastrophic physical malfunctions in autonomous vehicles.

Providing services and infrastructure for autonomous vehicles at intersections is also the topic of Chapter X in which the authors describe an agent-based valuation-aware traffic control system for intersections. Their approach combines valuation-aware intersection-control mechanisms with driver-assistance features such as adaptive cruise and crossing control.

Collaborative driving is the focus of Chapter XI. The authors describe an agent-based cooperative architecture that aims at controlling and coordinating vehicles, also showing that reinforcement learning can be used for this purpose.

Tumer, Welch, and Agogino (Chapter XII) tackle the issue of how road users can learn to coordinate their actions with those of other agents in a scenario without communication. Further, the authors explore the impacts of agent reward functions on two traffic related problems (selection of departure time and selection of lane).

In Chapter XIII, the authors discuss multi-agent learning in the context of a coordination mechanism where teams of agents coordinate their velocities when approaching the intersection in a decentralised way, improving the intersection efficiency.

Yamashita and Kuramatami (Chapter XIV) propose a cooperative car navigation system with route information sharing, based on multi-agent simulation. They use a scenario from Tokyo in which drivers can share information about their route choices. Results have confirmed that the mechanism has reduced the average travel time of drivers sharing information and that the network structure influenced the effectiveness of the mechanism.

Exchange of information is also tackled in the Chapter XV, this time by traffic signal agents. The authors show that these agents can learn better than independent ones, by sharing information about their environment.

Section III (chapters XVI to XVIII) of the book brings a collection of topics that are related to transportation and focus on different agent technologies such as agent-based simulation.

In Chapter XVI the authors apply agent-based solutions to handle job arrival uncertainty in a real-world scenario. This approach is compared to an on-line optimization approach across four scenarios, with the results indicating that the agent-based approach is competitive.

Chapter XVII deals with the use of agent-based simulation for modelling the organisational structure and mechanisms in the context of regional transport corridors.

Collaborative air traffic flow management is the topic of Chapter XVIII. This chapter describes the design and methodology of a multi-agent simulation for this problem. This is then used to evaluate several policies for the management of air traffic flow.

REFERENCES

Balmer, M., Cetin, N., Nagel, K., & Raney, B. (2004). Towards truly agent-based traffic and mobility simulations. In Jennings, N., Sierra, C., Sonenberg, L., & Tambe, M., (Eds.), *Proceedings of the 3rd International Joint Conference on Autonomous Agents and Multi Agent Systems, AAMAS*, volume 1, (pp. 60–67), New York, USA. New York, IEEE Computer Society.

Bazzan, A. L. C. (2005). A distributed approach for coordination of traffic signal agents. *Autonomous Agents and Multiagent Systems*, *10*(1), 131–164.

Bazzan, A. L. C., de Oliveira, D., Klügl, F., & Nagel, K. (2008). Adapt or not to adapt – consequences of adapting driver and traffic light agents. In Tuyls, K., Nowe, A., Guessoum, Z., and Kudenko, D., editors, *Adaptive Agents and Multi-Agent Systems III*, volume 4865 of *Lecture Notes in Artificial Intelligence*, (pp. 1–14). Springer-Verlag.

Bazzan, A. L. C., & Klügl, F. (2005). Case studies on the Braess paradox: simulating route recommendation and learning in abstract and microscopic models. *Transportation Research C*, *13*(4),299—319.

Bazzan, A. L. C., Wahle, J., & Klügl, F. (1999). Agents in traffic modelling - from reactive to social behavior. In *Advances in Artificial Intelligence*, number 1701 in Lecture Notes in Artificial Intelligence, (pp. 303–306), Berlin/Heidelberg. Springer. Extended version appeared in Proc. of the U.K. Special Interest Group on Multi-Agent Systems (UKMAS), Bristol, UK.

Burmeister, B., Doormann, J., & Matylis, G. (1997). Agent-oriented traffic simulation. *Transactions Society for Computer Simulation*, *14*, 79–86.

Camponogara, E., & Kraus Jr., W. (2003). Distributed learning agents in urban traffic control. In Moura-Pires, F. & Abreu, S., (Eds.), *EPIA*, (pp. 324–335).

Diakaki, C., Papageorgiou, M., & Aboudolas, K. (2002). A multivariable regulator approach to traffic-responsive network-wide signal control. *Control Engineering Practice*, *10*(2), 183–195.

Dresner, K., & Stone, P. (2004). Multiagent traffic management: A reservation-based intersection control mechanism. In Jennings, N., Sierra, C., Sonenberg, L., & Tambe, M., (Eds.), *The Third International Joint Conference on Autonomous Agents and Multiagent Systems*, (pp. 530–537), New York, USA. New York: IEEE Computer Society.

Elhadouaj, S., Drogoul, A., & Espié, S. (2000). How to combine reactivity and anticipation: the case of conflicts resolution in a simulated road traffic. In *Proceedings of the Multiagent Based Simulation (MABS)*, (pp. 82 – 96). Springer-Verlag New York.

France, J., & Ghorbani, A. A. (2003). A multiagent system for optimizing urban traffic. In *Proceedings of the IEEE/WIC International Conference on Intelligent Agent Technology*, (pp. 411–414), Washington, DC, USA. IEEE Computer Society.

Haugeneder, H., & Steiner, D. (1993). MECCA/UTS: A multi-agent scenario for cooperation in urban traffic. In *Proc. of the Special Interest Group on Cooperating Knowledge Based Systems*.

Hunt, P. B., Robertson, D. I., Bretherton, R. D., & Winton, R. I. (1981). *SCOOT - a traffic responsive method of coordinating signals*. TRRL Lab. Report 1014, Transport and Road Research Laboratory, Berkshire.

Klügl, F., & Bazzan, A. L. C. (2004). Simulated route decision behaviour: Simple heuristics and adaptation. In Selten, R. & Schreckenberg, M., (Eds.), *Human Behaviour and Traffic Networks*, (pp. 285–304). Springer.

Klügl, F., Bazzan, A. L. C., & Wahle, J. (2003). Selection of information types based on personal utility - a testbed for traffic information markets. In *Proceedings of the Second International Joint Conference on Autonomous Agents and Multi-Agent Systems (AAMAS)*, (pp. 377–384), Melbourne, Australia. ACM Press.

Kosonen, I. (2003). Multi-agent fuzzy signal control based on real-time simulation. *Transportation Research C*, *11*(5), 389–403.

Lowrie, P. (1982). The Sydney coordinate adaptive traffic system - principles, methodology, algorithms. In *Proceedings of the International Conference on Road Traffic Signalling*, Sydney, Australia.

Morgan, J. T., & Little, J. D. C. (1964). Synchronizing traffic signals for maximal bandwidth. *Operations Research*, *12*, 897–912.

Nunes, L., & Oliveira, E. C. (2004). Learning from multiple sources. In Jennings, N., Sierra, C., Sonenberg, L., & Tambe, M., editors, *Proceedings of the 3rd International Joint Conference on Autonomous Agents and Multi Agent Systems, AAMAS*, volume 3, (pp. 1106–1113), New York, USA. New York, IEEE Computer Society.

Oliveira, D., Bazzan, A. L. C., & Lesser, V. (2005). Using cooperative mediation to coordinate traffic lights: a case study. In *Proceedings of the 4th International Joint Conference on Autonomous Agents and Multi Agent Systems (AAMAS)*, (pp. 463–470). New York, IEEE Computer Society.

Oliveira, D., Ferreira, P., Bazzan, A. L. C., & Klügl, F. (2004). A swarm-based approach for selection of signal plans in urban scenarios. In *Proceedings of Fourth International Workshop on Ant Colony Optimization and Swarm Intelligence - ANTS 2004*, volume 3172 of *Lecture Notes in Computer Science*, (pp. 416–417), Berlin, Germany.

Ossowski, S., Fernández, A., Serrano, J. M., Pérez-de-la-Cruz, J. L., Belmonte, M. V., Hernández, J. Z., García-Serrano, A. M., & Maseda, J. M. (2005). Designing multiagent decision support systems for traffic management. In Klügl, F., Bazzan, A. L. C., & Ossowski, S., (Eds.), *Applications of Agent Technology in Traffic and Transportation*, Whitestein Series in Software Agent Technologies and Autonomic Computing, (pp. 51–67). Birkhäuser, Basel.

Papageorgiou, M. (2003). Traffic control. In R. W. Hall, (Ed.), *Handbook of Transportation Science*, chapter 8, (pp. 243–277). Kluwer Academic Pub.

Paruchuri, P., Pullalarevu, A. R., & Karlapalem, K. (2002). Multi agent simulation of unorganized traffic. In *Proceedings of the First International Joint Conference on Autonomous Agents and Multi-Agent Systems (AAMAS)*, volume 1, (pp. 176–183), Bologna, Italy. ACM Press.

Rigolli, M., & Brady, M. (2005). Towards a behavioural traffic monitoring system. In Dignum, F., Dignum, V., Koenig, S., Kraus, S., Singh, M. P., & Wooldridge, M., editors, *Proceedings of the fourth international joint conference on Autonomous agents and multiagent systems*, (pp. 449–454), New York, NY, USA: ACM Press.

Robertson (1969). TRANSYT: A traffic network study tool. Rep. LR 253, Road Res. Lab., London.

Rochner, F., Prothmann, H., Branke, J., Müller-Schloer, C., & Schmeck, H. (2006). An organic architecture for traffic light controllers. In Hochberger, C. & Liskowsky, R., (Eds.), *Proc. of the Informatik 2006 - Informatik für Menschen*, number P-93 in Lecture Notes in Informatics, (pp. 120–127). Köllen Verlag.

Rossetti, R. & Liu, R. (2005). A dynamic network simulation model based on multi-agent systems. In Klügl, F., Bazzan, A. L. C., & Ossowski, S., (Eds.), *Applications of Agent Technology in Traffic and Transportation*, Whitestein Series in Software Agent Technologies and Autonomic Computing, (pp. 181–192). Birkhäuser, Basel.

Rossetti, R. J. F., Bordini, R. H., Bazzan, A. L. C., Bampi, S., Liu, R., & Van Vliet, D. (2002). Using BDI agents to improve driver modelling in a commuter scenario. *Transportation Research Part C: Emerging Technologies*, *10*(5–6), 47–72.

Silva, B. C. d., Basso, E. W., Bazzan, A. L. C., & Engel, P. M. (2006). Dealing with non-stationary environments using context detection. In Cohen, W. W. & Moore, A., (Eds.), *Proceedings of the 23rd International Conference on Machine Learning ICML*, (pp. 217–224). New York: ACM Press.

Steingrover, M., Schouten, R., Peelen, S., Nijhuis, E., & Bakker, B. (2005). Reinforcement learning of traffic light controllers adapting to traffic congestion. In Verbeeck, K., Tuyls, K., Nowé, A., Manderick, B., & Kuijpers, B., (Eds.), *Proceedings of the Seventeenth Belgium-Netherlands Conference on Artificial Intelligence (BNAIC 2005)*, (pp. 216–223), Brussels, Belgium. Koninklijke Vlaamse Academie van Belie voor Wetenschappen en Kunsten.

TRANSYT-7F (1988). *TRANSYT-7F User's Manual*. Transportation Research Center, University of Florida.

Tumer, K., Welch, Z. T., & Agogino, A. (2008). Aligning social welfare and agent preferences to alleviate traffic congestion. In Padgham, L., Parkes, D., Müller, J., & Parsons, S., (Eds.), *Proceedings of the 7th Int. Conference on Autonomous Agents and Multiagent Systems*, (pp. 655–662), Estoril. IFAAMAS.

van Katwijk, R. T., van Koningsbruggen, P., Schutter, B. D., & Hellendoorn, J. (2005). A test bed for multi-agent control systems in road traffic management. In Klügl, F., Bazzan, A. L. C., & Ossowski, S., (Eds.), *Applications of Agent Technology in Traffic and Transportation*, Whitestein Series in Software Agent Technologies and Autonomic Computing, (pp. 113–131). Birkhäuser, Basel.

Wahle, J., Bazzan, A. L. C., & Kluegl, F. (2002). The impact of real time information in a two route scenario using agent based simulation. *Transportation Research Part C: Emerging Technologies*, *10*(5–6),73–91.

Wiering, M. (2000). Multi-agent reinforcement learning for traffic light control. In *Proceedings of the Seventeenth International Conference on Machine Learning (ICML 2000)*, (pp. 1151–1158).

Ana L. C. Bazzan
Instituto de Informática, UFRGS, Brazil

Franziska Klügl
Örebro University, Sweden

Acknowledgment

We would like to thank all authors and referees for their work. We are also grateful to Cornelia Triebig for the valuable work on the edition of the final version of this book. Finally we would like to thank the Alexander von Humboldt Foundation for the fellowship that has allowed the joint work of both editors in Würzburg, Germany, during the period of 2006-2007.

Ana L. C. Bazzan
Franziska Klügl
The Editors

Section I
Reproducing Traffic

Chapter I
Adaptation and Congestion in a Multi-Agent System to Analyse Empirical Traffic Problems:
Concepts and a Case Study of the Road User Charging Scheme at the Upper Derwent Valley, Peak District National Park

Takeshi Takama
University of Oxford, Stockholm Environment Institute, UK

ABSTRACT

This study discusses adaptation effects and congestion in a multi-agent system (MAS) to analyse real transport and traffic problems. Both methodological discussion and an empirical case study are presented in this chapter. The main focus is on the comparison of an analysis of a MAS simulation analysis and an analysis that solely uses discrete choice modelling. This study explains and discusses some important concepts in design empirical MAS in traffic and transportation, including validation Minority Game and adaptation effects. This study develops an empirical MAS simulation model based on real stated-preference data to analyse the effect of a real road-user charge policy and a complimentary park and ride scheme at the Upper Derwent Valley in the Peak District National Park, England. The simulation model integrates a transport mode choice model, Markov queue model, and Minority Game to overcome the disadvantages of a conventional approach. The results of the simulation model show that the conventional analysis overestimates the effect of the transportation and environment policy due to the lack of adaptation affects of agents and congestion. The MAS comprehensively analysed the mode choices, congestion levels, and the user utility of visitors while including the adaptability of agents. The MAS also called as agent-based simulation successfully integrates models from different disciplinary backgrounds, and shows interesting effects of adaptation and congestion at the level of an individual agent.

INTRODUCTION

Traffic congestion and associated air pollution are considered the most significant threat to the UK tourism industry, as they leave a negative impression on visitors. In particular, tourists to National Parks are heavily dependent on their private cars. According to underlying economic theory, Road-User Charging is a suitable tool to ensure that road users (i.e. car drivers) pay for the external costs generated from their travel (Hensher & Puckett, 2005; Steiner & Bristow, 2000). Currently, one of the major objectives of installing Road-User Charging is to reduce traffic congestion levels. It is likely that a Road User Charging scheme around the Upper Derwent Valley (the Valley) in the Peak District National Park (Figure 1) will be considered a viable option for reducing traffic levels. At the same time, it is important to examine to the extent to which visitors feel uncomfortable about the scheme.

This study develops a multi-agent system (MAS) simulation including a discrete choice model to analyse the effect of the Road User Charging at the Valley on congestion levels at parking areas and the mode choice of visitors. The focus of this study is the comparison of an analysis of MAS simulation modelling and an analysis that solely uses discrete choice modelling.

Figure 1. The Upper Derwent Valley is located between two large cities, Manchester and Sheffield. The entrance to the Upper Derwent Valley by car is only from the A57 and only through Derwent Lane, which comes to a dead-end. There are four parking areas on the Derwent Lane.

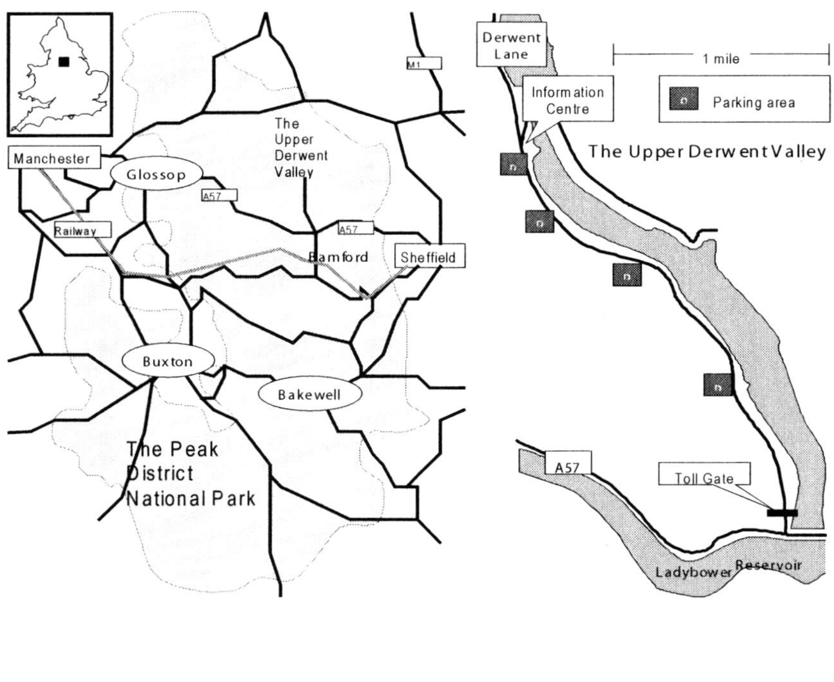

CASE STUDY SITE DESCRIPTION

The Upper Derwent Valley is located *between* two large cities, Manchester and Sheffield (Figure 1). Access to the Valley by private cars is easy, not only from local towns but also from these nearby cities through the A57. The entrance to the Upper Derwent Valley by car is only from the A57 and only through Derwent Lane, which comes to a dead-end. There are four parking areas on Derwent Lane. The approximate parking capacity of each parking area is 134, 77, 58, and 18 vehicles respectively from the Information Centre. Only the first parking area requires a parking ticket, which costs £2.50 for one-day parking or 50 pence per hour. Tourists try to park as close to the parking area of the Upper Derwent Information Centre as possible since the Information Centre is the final parking area to visit the scenic area of the Valley. However, the Information Centre charges a parking fee, so tourists who are not willing to pay a parking fee will instead choose the Derwent Overlook (the second parking area). It is important to underline that even on the busiest days congestion on the roads such as the A57 and Derwent Lane is minimal, but severe congestion occurs around the Information Centre and the second parking area.

A bus service is also planned in this area as a complementary policy tool of the Road User Charging scheme. Overall, 700 questionnaires were distributed in the Valley during the summer of 2003, and 323 were returned (i.e. a return rate of 46.1%) to collect information about decision making processes of agents with the stated preference approach and the arriving rates of vehicles. Several key person interviews including parking officers and local authorities were also carried out.

The age distribution of visitors is highly skewed, and two modes at '**35-44**' and '**55-64**' are present in the distribution ('**Age**' in Figure 2). This age distribution matches observations made during the survey. In addition, the income distribution ('**Income**') also supports this trend. Some 20% of visitors to the Valley are non-workers, and most of these visitors are assumed to be elderly people, since the proportion of students is nominal (i.e. 5%). Some 27.0% and 12.8% of visitors from local areas and other areas come to the Upper Derwent Valley at least every other month.

Figure 2. Proportions of visitors' characteristics. The age and income distribution shows that visitors are largely family and elderly members. These visitors come from local villages as well as two large cities and visit as much as once a week, but not more.

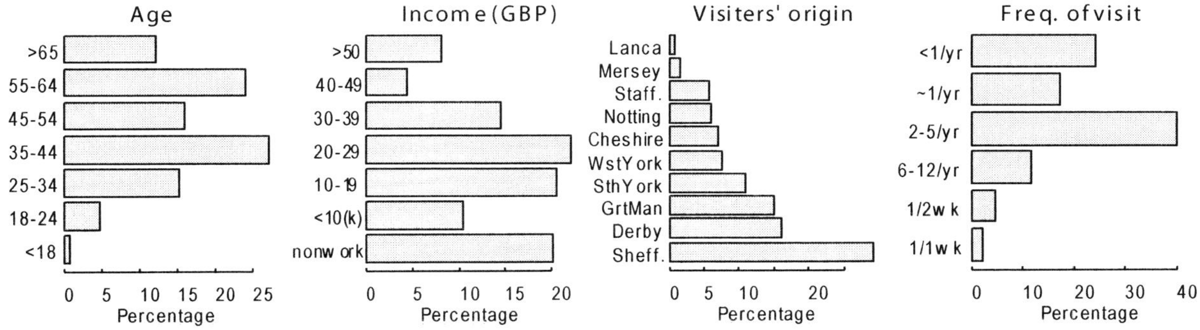

APPROACH OF THIS STUDY

Structure of Conventional Analysis Based Solely on Equation-Based Models

Before explaining the simulation model used in this study, the advantages and disadvantages in the conventional analysis are briefly examined. Two types of equations describing travel behaviour, mode choice and parking location choice, are used in this study. They are the multinomial model and logit model as shown in Figure 3.

The multinomial discrete choice model on mode choice is based on travel information such as "parking fee" and "searching time for parking space". The detailed explanation and its applications to the real world are found in previous studies (Ben-Akiva & Lerman, 1985; Greene, 2003). The other probabilistic equation model on parking location is based on characteristics of travellers such as age, and the frequency of visit. The major advantage of the analysis using probabilistic equation-based models is that they simplify problems of the real world, so the approach does not require complex input data compared with a MAS simulation model. Also, the equations clearly represent the individual human behaviour. However, when the problems are interrelated with one another, these advantages over-simplify the problem. It is important to underline that the characteristics of tourists are not directly modelled in the equation of the mode choice. For example, a tourist who used to park at the Information Centre pays a parking fee and spends nominal time on searching and walking. Therefore, the tourist is more likely to change travel mode from car to bus than other tourists who arrive directly at the second park area. Hence, we can still link up the mode choice between tourists with different characteristics, but it is difficult to know how mode choices affect decision-making regarding the parking location. In other words, this analysis connects unidirectionally two equations only with the imagination of researchers.

Moreover, in the case of a Road User Charging scheme, this research is concerned with how the scheme reduces congestion in the area, together with other relevant factors such as search time for parking space. From this analysis, there is no clear indication about the effect of a Road User Charge on the congestion level at parking areas in the Upper Derwent Valley. For example, a possible scenario about the effect of a toll fee on congestion is that "*if the probability of a tourist going by car is reduced by 10% due to a toll fee, the congestion level at parking areas in the Valley would be reduced by 10%*". However, this scenario is likely to overestimate the reduction of the congestion since an unblocked parking area will attract other potential car travellers.

Figure 3. Structure of equation-based analysis - Two types of equations describing travel behaviour, mode choice and parking location choice, are used in this study and the equations are connected unidirectionally only by the imagination of researchers.

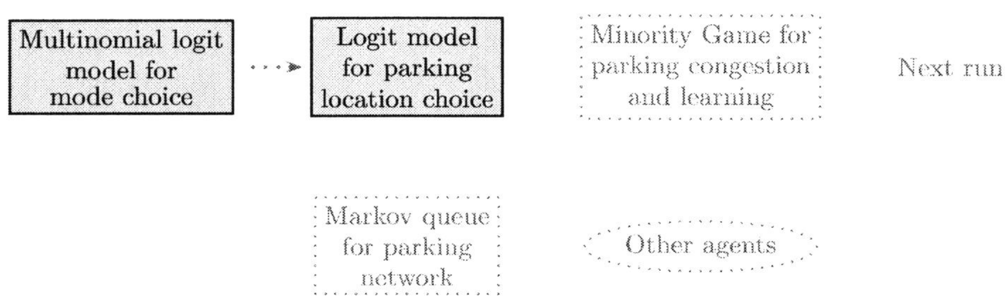

Also, any model of the parking network in the Valley is absent in the analysis, which solely uses a discrete choice model. For example, in the model it is assumed that a tourist can definitely park at the Information Centre if she or he decides to park there. This is because these models cannot formulate the concept of adaptation effects and congestion, which requires dynamic links amongst the tourists – an example of "*over simplification*" (Parunak et al., 1998). In conclusion, although the analysis solely using probabilistic equation-based models clearly presents the probabilities of travellers' choices independently, it could be dangerous to infer congestion levels at parking areas and tourists' mode choices based on these equations.

The Structure of MAS Simulation Modelling

The MAS model in this study is the integration of four modules (Figure 4). The two equation-based models in the previous section stay as main modules of the agents' decision-making at a micro level. However, an adaptive learning process is added to the Multinomial discrete choice model with the strategies' success scores in a Minority Game, which is explained later. In addition, this simulation model includes a statistical model of a Markov queue model based on real arrival rate and departure rate of vehicles from/to each parking area. The Markov queue model connects the two equation models, bidirectionally. The outputs of the Markov queue model are inter-linked with the Minority Game through the interaction of various agents and the whole system proceeds to the next time step. The Markov queue model and Minority Game simulate the movement of vehicle therefore these modules eventually determine the travel time and the experience of finding a parking space. Then, these outputs become the input variables of the equation-based models in the following time steps. Therefore, this model is dynamic and includes the concept of adaptation and congestion.

One more clear difference in the structures of the two analyses is that the calculation in Figure 3 occurs once at a system level, so there is only one set of calculations in the analysis solely with an equation-base model. In contrast, in the MAS simulation model, the whole set of calculations in Figure 4 is carried out for each agent and at each time step, and the system level results are the aggregation of these many small calculations. This is very useful concept since the user utility and behaviour of each individual agent is analysed by tracking individual calculations.

Figure 4. Structure of MAS simulation - The two logit models stay as main modules at a micro level. However, an adaptive learning process is added with a Minority Game. A Markov queue model simulates the movement of vehicles.

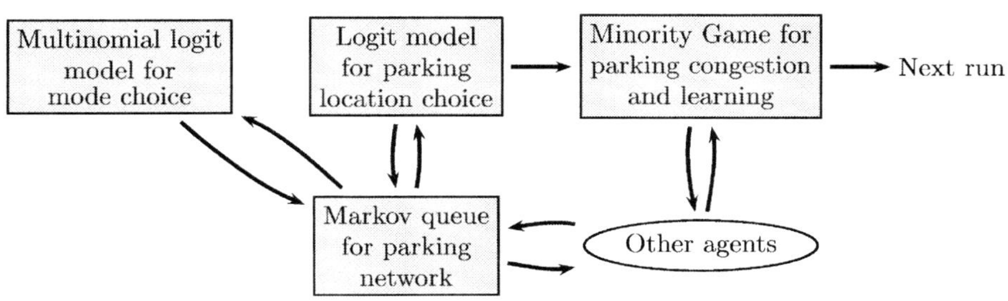

FOUR MODULES OF THE MAS SIMULATION MODEL

This section explains the development and validation of the four modules in MAS introduced above.

Discrete Choice Analysis of Parking Location

The first module is a logit model of parking location choice. A parking fee is only charged at the first parking area at the Information centre. In contrast, if the drivers do not park at the Information Centre, they have to walk to the Information Centre, but they do not have to pay the parking fees. Therefore, the distribution of parking costs is rather categorical, 'Park' or 'Not park' at the Information centre, and a binary logic model was used with BIOGEME[1] to analysis the choice (Bierlaire, 2003). The three significant factors in the logit model are visitors' age, visiting frequency per year, and the Willingness To Pay (WTP) to the road user charging (Table 1). The age of a trip leader is a categorical variable and the categorical age variable is commonly used in transport modelling (Bierlaire, 2001). The observed utility functions for the two choices are:

1st Parking area:	V_i^I	=	β^{often} (No. of visit) + β^{tage} (Age) + β^{tWTP} (WTP)
Other parking areas:	V_i^O	=	α^{Other}

The logistic form of the fitted model used to estimate the probability of parking at the 1st parking area is below:

$$P(1st)=\frac{\exp(U_i^1)}{\exp(U_i^1)+\exp(U_i^O)}$$

U_i^* is unobserved utility including an error term, i.e. $U_i^* = V_i^* + \varepsilon_i$. According to this model, the probability of parking at the Information Centre rises as age increases at a given WTP and frequency of visit level. In contrast, the probability declines as a visitor travels to the Valley more often. From

Table 1. Parameters for binary logit model for parking location choice

Coefficient	Estimate	Robust Std. Error	Robust t-value
α(Other)	1.240	0.518	2.395
β(often)	-0.033	0.015	-2.251
β(age)	0.214	0.104	2.051
β(WTP)	0.229	0.084	2.737

Number of observations = 268

L(0) = -185.763

L(^β) = -144.046

$\bar{\rho} = 0.203$

the modelling estimation, an equity issue is clearly presented with the current monetary policy tool, a parking fee. Elderly visitors are more willing to pay the parking fees to park at the Information Centre. In other words, elderly visitors are more disadvantaged when required to pay parking fees. The results of the logit model were validated as they show a similar trend as the parking cost and visitors' characteristics from the survey.

Multinomial Mixed Logit Models of a Mode Choice

This section explains a mode choice model, which is implemented as a module of MAS. The model is base on the survey data and respondents were asked how they would travel to the Valley if the road user charging and park & ride schemes were put into effect in the Valley in *ceteris paribus* conditions (e.g. with same trip members). The visitors to the Upper Derwent Valley were expected to respond to the schemes in one of "Auto", "Bus" and "Cancel" modes:

'Auto' option: Pay a toll for road use and drive into Derwent Lane to get to the Information Centre.
'Bus' option: Come near the Valley with any travel mode, and then use the complementary park & ride service to get to the Upper Derwent Information Centre.
'Cancel' option: Cancel the trip to the Valley and instead go somewhere else or stay at home.

After reviewing previous research (Fowkes, 2000; Ortúzar & Willumsen, 2001, p.283; Steiner & Bristow, 2000) and interview local authorities, four attributes of the mode choices on travel time and costs were chosen, and four different levels were selected for each attribute: road user charging (£), park & ride fare (£), frequency of bus service (minutes), searching a car park & walking time to the information centre (minutes), and Parking fee (£).

The correlations among three alternatives were insignificant, so that the nested logit (McFadden, 1981) and the error component model of the mixed logit model[2] were also insignificant. The reason for the insignificant correlations among alternatives could be due to a simultaneous decision making process since destination (Trip | Cancel) and mode (Bus | Auto) are likely to affect the processes simultaneously in this situation (Steiner & Bristow, 2000, p.99).

The heteroscedastic taste of time and cost with the multinomial mixed logit model are significant, but no socio-economic factors are significant. Possibly, socio-economic factors are efficiently captured by the taste variation of the mixed logit model. The insignificant group size can be explained by the discussion of the marginal or average road pricing principle (Nash, 2003; Rothengatter, 2003). In this case, road user charging seems to be as effective as the marginal pricing principle, so that additional trip members are not as important as the first member to calculate the cost of travel, i.e. the toll is not simply divided by the number of trip members. The best-fitted utility functions with the multinomial mixed logit model are:

$$\text{Auto:} \quad V_i^A = \alpha^{Auto} + \beta^{time}(\text{Toll} + \text{Parking fee}) + \beta^{time}(\text{Search \& walk})$$

$$\text{Bus:} \quad V_i^B = \beta^{cost}(\text{Bus fare} + \text{Parking fee}) + \beta^{time}(\text{Headway})$$

$$\text{Cancel:} \quad V_i^C = \alpha^{cancel}$$

The log-likelihood ratio test (McFadden, 1974), which compares the model fitness of a generic model with that of a specific model, showed that the parameters for costs and time are generic. Thus, no alternative specific coefficient is present in these utility functions. Also, the test showed the multinomial mixed logit model significantly improves the model fitness compared with the conventional multinomial logit model. These functions do not show a mean and standard error for the coefficients of cost, time, and lagged dependent variable, but these are expressed in the summary statistics for the estimates of the utility functions (Table 2). The cost and time attributes are presented in pounds and minutes, respectively.

For example, the logistic form of the fitted model for choosing the Auto option is:

$$P(Auto) = \frac{exp\ (U_i^A)}{exp\ (U_i^A) + exp\ (U_i^B) + exp\ (U_i^C)}$$

The other logistic forms for the other two options are similar to the one above. All six coefficients have no significant correlation with one another as calculated by the robust t-test. The standard errors express that the probabilities of negative coefficients are >99.99% for β^{time} and 97.93% for β^{cost}. Therefore, the problem of a positive coefficient is negligible[3]. In this case, the value of time was 7.24 pence per minutes. This is close to the non-commuting values of time in the report from the Department for Transport, i.e. 7.55 pence per minute[4] (Department for Transport, 2004). The parameters are reasonably validated.

After the implementation of schemes with a hypothetical £3 toll fee, the probabilities of travel mode are the ones shown in Table 3. More than half of visitors, who used to park at the Information Centre (labelled as "Centre" in Table 3), are likely to keep using their own cars to visit the Valley. In contrast,

Table 2. Parameters for multinomial mixed logit model with normal distributed taste and panel data structure for mode choice. m and s represent the mean and a standard error of a coefficient, respectively.

Coefficient		Estimate	Robust Std. Error	Robust t-value
α(Cancel)		-4.627	0.299	-15.497
α(Auto)		1.873	0.141	13.277
β(cost)	m	-0.704	0.040	-17.463
	σ	0.089	0.043	2.058
β(time)	m	-0.051	0.003	-15.019
	σ	0.025	0.004	7.120
lagged	m	0	-	-
	σ	3.070	0.222	13.819

Number of observations = 3840

L(0) = -4218.67

L(^β) = -2730.05

$\bar{\rho} = 0.351$

more than two third of the visitors, who used to park at the other parking areas, are likely to keep using their own cars to visit the Valley. This result shows that the effect from the road user charging scheme is not equal for all types of visitors. Elderly visitors are more likely to be affected by the road user charging scheme as they have a strong preference for parking at Information Centre (Eckton, 2003). On the other hand, the purpose of the road user charging scheme is to reduce the congestion level around Derwent Lane, and it is, consequently, effective in achieving this policy aim from the model results.

Markov Queue Parking Network Module

The Markov queue module simulates the parking network of the Upper Derwent Valley. The main objective of this module is to determine the searching time of parking spaces and walking time between the parking space and the primary destination assumed to be the Information Centre. The discrete choice models explained above cannot explain the mechanics of a parking network system in the Valley, which is the necessary component for modelling parking congestion and the interaction of agents. Also, it is no feasible to collect dynamic searching time and walking time from surveys. From a pilot survey, car drivers were found not to remember the exact searching time and walking time, or only answered the approximate time, such as 5 or 10 minutes.

In contrast, arrival rates into the parking network system are easy to measure at the input point of the system. Parking time is also obtained easily by questionnaires since car drivers have a good memory about the arriving time at a parking area and departure time from a parking area. With the Markov queue theory (Chernick, 1999; Hinkley, 1988), departure rates are calculated from parking hours, i.e. the departure rate of the parking network system is the inverse of parking time. Simultaneously, the congestion level of a network system can be estimated with the Markov queue model. The major factor, which determines searching time, is a congestion level in a parking area, so searching time can also be estimated with the Markov queue model. A parking location is determined after a car driver finds a parking space and consequently walking distance and time are approximated though this process.

Previous studies have shown that it is difficult to implement the mathematical Markov queue model to solve real world problems, thus simulation approaches had been recommended (Arnott & Rowse, 1999; Norris, 1997). Additionally, time-driven and event-driven concepts are added to the Markov queue simulation model.

Data on Arrival and Departure Rates

In the Markov processes, the distribution of the arrival and departure rates are usually considered as coming from the Poisson distribution. The distributions of rates were checked to see whether they come

Table 3. Expected probabilities of each mode choice between parking locations. W + S stands for search and walking minutes and Parking means parking fee for the Auto option.

Park at	Toll	W + S	Parking	Probability		
	(pounds)	(mins)	(pounds)	Auto	Bus	Cancel
Centre	3	0	2.5	0.54	0.42	0.04
Other	3	20	0.0	0.71	0.27	0.02

from the underlying distributions by the bootstrap method (Davison & Hinkley, 1997; Efron, 1979) with R[5] (Ihaka and Gentleman, 1996).

The departure rates at each parking area were collected from parking beat surveys, which were undertaken over three days, namely the 23rd, 26th, and 27th of August 2001, by the Transport Office of Derbyshire County Council. The 23rd of August 2001 was a normal summer weekday, in contrast, the 26th and 27th of August 2001 were a Sunday and the summer bank holiday, which were usually the busiest days in the Valley. All parked cars were recorded, so the data set acted as a population. In total, 1961 cars were recorded. The rest of the data was collected during August, 2003 for 10 days. From the observations, the steps of the parking network were determined as below (Figure 5).

There are two input points (i.e. Information Centre and Derwent Overlook) and four output points (i.e. all four parking areas). The state transition in the system at a macro level is represented as Figure 6.

The arrival rate and the departure rate of cars per minute are symbolised as λ and μ, respectively in this study. The number in the circle is the number of cars in the parking network system and N is the overall parking capacity, i.e. 134 + 77 + 58 + 18 = 287. The arrival rate per minute was counted at the Information Centre with 30-minute intervals (30λ) between 10:00 and 15:00 hrs.

Arrival Rate to the Upper Derwent Valley Parking Areas

The arrival rate during holidays and weekend displays a trend related to the time of day. The arrival rate peaks around noon and gradually decreases afterward. A commonly assumed distribution for counted data is the Poisson distribution (Pfeiffer & Schum, 1973, p.200), and if this assumption is valid, the Fano factor should be one, i.e. $\phi = \sigma^2 / m = 1$ (Stevens & Zador, 1998, p.213). The bootstrap simulation method (Chernick, 1999; Hinkley, 1988) was used to estimate the Fano factors. Eight out of eleven bootstrapped ϕ for each observed time of day have confidence intervals containing '1' by the percentile method[6]. Therefore, the evidence shows that the distributions of arrival rates during holidays come from the Poisson distribution. A triangular function fits with the time dependency of arrival rates, $30\lambda_t$. The function for $30\lambda_t$ is:

Figure 5. Markov queue network. The numbers in circles are parking capacities.

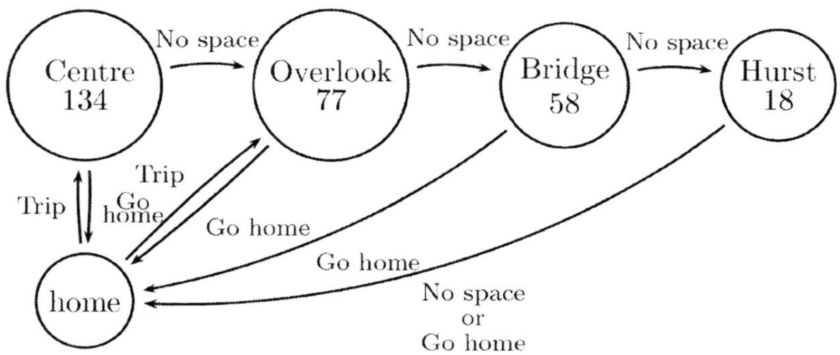

Figure 6. State transition diagram of the system

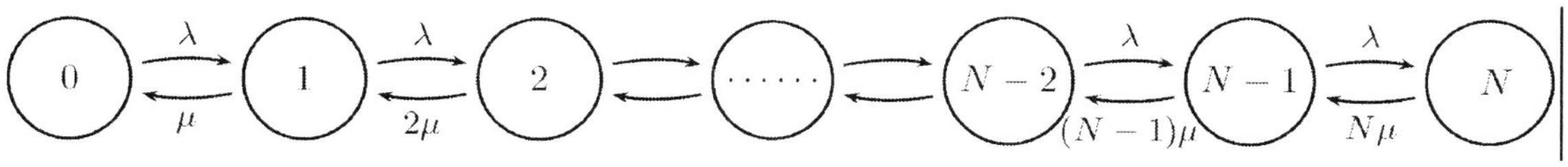

$$30\lambda_{hour} = \begin{cases} \alpha^1 + \beta^1 \times hour, & \text{If } 10{:}00 \leq hour \leq 12{:}30 \\ \alpha^2 + \beta^2 \times hour, & \text{otherwise} \end{cases}$$

Where: hour is time of day and its interval is [10:00, 15:00]

Since the distribution is heteroscedastic, the weighted least squares method (i.e. dividing each observation by the variance of the error term for that observation) is used to fit the linear models. The estimation of all parameters in equation is significant with more than 99% of confidence (Table 4). The p-value for the model fitness is also significant at the 99% confidence level and the model fitness of R^2 is high, i.e. 0.92 for $10{:}00 \leq t \leq 12{:}30$ and 0.86 for $12{:}00 < t$. In addition, the large standard deviation around noon is explained by the property of the Poisson distribution. The standard deviation of the Poisson distribution is positively related with mean, so the larger the mean a distribution has, the larger the standard deviation that distribution has.

From the results, we can reasonably validate that the distribution of arrival rates come from a time dependent Poisson distribution. Therefore, the equation was used to produce the arrival rate in the Markov queue model of the parking network system.

Departure Rate from the Upper Derwent Valley Parking Areas

The departure rate from the Upper Derwent Valley is defined as the expected number of cars leaving the Valley per hour. It is also defined as the inverse of parking hour. The distributions of the parking hours between dates and parking areas were checked to see if they were different. Overall, the distribution of the parking hour is skewed to the right, i.e. longer parking hours. Since the expected departure rate per hour (60μ) is the inverse of the parking hours, it is 1/2.736 = 0.367 in this case. The bias from the non-linear transformation was found to be very small (< 0.00001) from a bootstrap simulation, so a correction for the parameter estimation is not necessary. The departure rate per hour from the Upper Derwent Valley was determined as 0.367.

Table 4. Significance of triangular function of arrival rates, $30\lambda t$

	α		β		**Overall**	
t	Est.	p-val.	Est.	p-val.	p-val.	R^2
$10{:}00 \leq t \leq 12{:}30$	-129.18	0.008	15.36	0.002	0.002	0.92
otherwise	225.95	0.003	-12.98	0.008	0.008	0.86

Figure 7. Means of time dependent upon arrival rate and a triangular function with the peak at 12:30

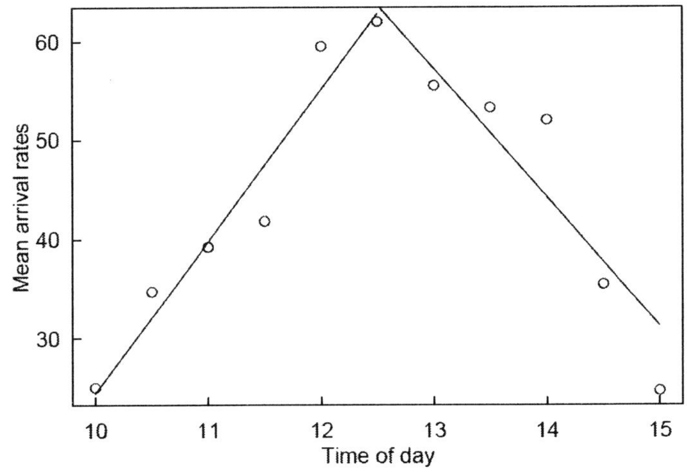

Moreover, in the Markov queue process, the service time (in this case, parking hours) is distributed exponentially (Norris, 1997, p.182). The exponential distribution with a rate of 0.367, is fairly close to the density function of the parking hour (Figure 8). Therefore, the departure rate is also reasonably validated as Markov and consequently it comes from the Poisson distribution.

Markov Queue Network Simulation Model

The distributions of arrival and departure rates satisfied and validated the requirements for the Markov queue network model (Vose, 2000, p.235). Therefore, the Markov queue network model was developed with the RePast toolkit[7] (Collier et al., 2003; Ross, 1997, pp.88-89). There is no reason to assume that a car, which comes to the Upper Derwent Valley, must leave after other cars, which have come earlier, i.e. First In First Out (FIFO). Therefore, the departure from the Valley in this simulation is System In Random Order (SIRO).

This model is the combination of event-driven[8] and time-driven approaches (Febbraro & Sacco, 2004; Jain & Neal, 2004; Peterson, 1981). A macroscopic timing determines the arrival and departure of cars, so these events are time-driven events for the system (Cheng, 1998). Therefore, the arrival and departure are not controlled by each car in this simulation. At the same time, each car acts according to the micro level events they encounter between two macro level events. The micro level events are to enter and exit parking areas, and these events change the driving speed of cars, i.e. event-driven events for each car. This approach becomes more beneficial in a successive study with a MAS model of the Upper Derwent Valley, in which the number of cars going to the Valley changes according to the number of private car visitors at every time step. In this case, only the arrival rate (λ) needs to change according to the change in car numbers. Figure 9 shows the pseudo-code for the main loop of the macroscopic timing in a simulation day (a representative day in a given week).

To verify and validate the overall simulation performance, the Markov simulation was run based on observed data, i.e. without agents' mode choice. The overall car numbers in the simulated valley was 774.818 and its confidence interval [777.2413, 772.3947] captured the expected car numbers in the

Figure 8. Exponential distribution of rate 0.367 and density of parking hour

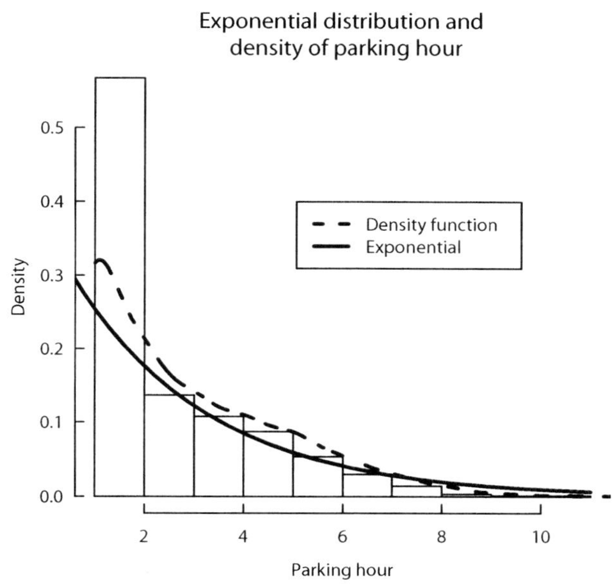

real valley. Also, the number of cars in each parking area was similar to the actual data. Therefore, the model is reasonably verified and validated.

Minority Game in the MAS Simulation of Upper Derwent Valley

This section explains the Minority Game, which is used as an additional decision making module for the MAS (Arthur, 1994; Challet et al., 2004; Challet & Zhang, 1997; Edmonds, 1999), to assess the congestion level of the four parking areas. In general, tourists should not come to the Valley by Auto when they cannot park in their target parking area. Also, tourists arriving by bus will be glad that they chose the Bus option if they find out there are no empty spaces in the parking areas, where the buses

Figure 9. Pseudo-code for Main loop of car movements

```
Start the Markov queue trip of cars
      Initialise the current time and the next event time
      Initialise the arrays of cars in the valley
      Set initial state as the initial car number
      while the current time is before the end time, repeat
            Calculate the next event time based on state
            if the next event time is before the end time
                  Move cars in the Valley and update state
                  Next event occurs and update state
            End 'if' condition
            Set the current time to the next event time
      End 'while' loop
End the Markov queue trip of cars
```

pass and stop. These two situations indicate that the tourists are indirectly playing a Minority Game. The Minority Game seems well suited to study the problem of congestion in transportation sectors (Bazzan et al., 2000; Dia, 2002; Lee et al., 2001; Klügl & Bazzan, 2004; Peeta et al., 2005).

Unlike conventional Minority Games, the winners and losers are not determined by a fixed proportion of agents in this model (e.g. 51%). Rather, the winners and losers are determined by the gap between the forecasted utility and actual return, i.e. if the return from a choice is less than expected, the tourist made a wrong choice[9]. However, the winning side is more personally determined and neither side is required to be the overall wining side. Also, the utility and actual return is not made up from a theory or some imaginary threshold, but calculated by the utility functions of the mixed logit discrete choice model. For example, if too many visitors choose Auto, more people are likely to underestimate their decision since the searching time to find a parking space and walking time to the Information Centre are on average longer in the congested condition. However, some of these tourists are still able to park where they want to if they are lucky. Therefore, within the same choice, there are both winners and losers in this Minority Game. Put differently, tourist agents in this model use personal experiences rather than centralised information to make their decisions; therefore the agents also have personalised results in the Minority Game (de Cara et al., 2000).

Adaptability and Strategies of Agents in the Game

Although the former sections use the phrase: 'visitors make a choice', strictly speaking, strategies determine a choice instead of agents in the MAS. Agents choose the best strategy and follow the choice according to that strategy. The reason for this two-step decision making is due to imperfect information. It is impossible to obtain perfect information to win in the Minority Game since you need to know the decision making of many other agents. So, at least, the assumption of perfect information fails in this situation. Therefore, even if the mechanism of the decision making process was correct, the output may be wrong.

According to the Minority Game, the following two thought patterns can be suggested. For example, if parking areas are severely congested at the time of travel, searching time and walking time tend to be longer for visitors with the Auto option. These visitors may think: 1) the parking area will be congested so "I will not go to the Valley by car next time", or 2) many visitors will be discouraged to go to the Valley by car so parking areas will be empty; therefore, they think that "I will go to the Valley by car next time". Thus, searching time and walking time can have a negative affect as well as a positive one on the derived utility of Auto. From the description above, three thought patterns were considered for this simulation.

The three thought patterns of visitors depend on which mode takes the congestion related utility:

Thought pattern 1: visitor believes that the parking area will be congested again next time, so discourages a visitor from going to the Valley by car, i.e. add β^{time} (Search & walk) into U^A

Thought pattern 2: visitor believes that the parking area will be less congested next time, so discourages a visitor from going to the Valley by bus, i.e. add β^{time} (Search & walk) into U^B

Thought pattern 3: visitor believes that the parking area will be less congested next time, so discourages a visitor from cancelling the trip. , i.e. add β^{time} (Search & walk) into U^C

Thought pattern 1 is the same as the result from the multinomial mixed logit model mechanism. Thought pattern 2 and 3 try to cut the ground from under the feet of other agents. In other words, thought patterns 2 and 3 are the second thoughts from the result of the multinomial discrete choice models. These thought patterns are sceptical about the result of the multinomial discrete choice like Hume's evaluative scepticism (Clark, 1998) and agents are making decisions even in the uncertain situation without previous experience by applying an 'act-then-learn' framework rather then the 'learn-then-act' one, which is used in adaptation planning (Beltratti, 1996, p.119;).

Also, each of the three strategy types is sub-categorised into five strategies according to the experience they use for the mode choice. Namely, agents could use any of the last five experiences to calculate the choice. From the survey carried out in the summer of 2003, it was unlikely that travellers remembered any detailed trip information more than for the last five trips. This means that some agents decide on the travel model based on the last trip experience while other agents may use the fifth oldest trip experience.

The best strategy with the maximum success score was chosen before each trip. Also, this minority game used the horizon of strategy successfulness. The horizon is related with the "adaptability" of agents, since a long horizon makes agents consider too much historical information, which may not be relevant to the current situation (Liu et al., 2004, pp.347-351). The length of the horizon is a parameter *H*, which represents the horizon for which each strategy scores. Therefore, the success score of each strategy is a virtual point in the last *H* steps an agent experienced:

$$\theta_t^s = \sum_{i=t-1-H}^{t-1} R_i^{x_i^s} / H$$

where

x = The selected choice by strategy s at i
R^x = Return from the selected choice at i

As shown in the equation above, success score θ of any given strategy s at a time step t is the moving average of the return from a selected choice by the strategy within the scope of horizon *H*. The choice made by a strategy is not relevant with the choice used by an agent, which possesses the strategy. All strategy-scores θ are calculated whether or not the strategies were chosen by the agent. Similarly, although *H* was set to five in this model, the length of the horizon is also irrelevant with the length of experience remembered. The max length of horizon is set to five because of survey results, i.e. most people do not remember trips older than the last five trips.

In conclusion, there are five adaptation elements in this MAS simulation of Minority Game:

1. Adaptation happens in **uncertain situations with an act-then-learn framework**,
2. Adaptation is the change of a **decision making process at a strategic level** in the uncertain situation,
3. An agent does not have all strategies available in the world and **adaptability is restricted by the number of strategies** the agent possesses,

4. Adaptation processes **happen quicker if agents do not continue to use old information**, which may be irreverent to the current situation,
5. Adaptation is **necessary to succeed in fluctuating environments.**

DECISION MAKING PROCESS OF AGENTS

This section uses an example to illustrate of the decision making process. At a given time step, i.e. week, the probability of each choice is calculated by each strategy according to the multinomial mixed logit model. Nevertheless, the thought patterns 2 and 3 swap the utility of searching & walking time from U^A according to their rules. Then, there are five memories, so that a set of 15 possible strategies and the sets of probabilities associated with the mode choice of an agent can be like the one in Table 5. These 15 strategies are possible strategies, but in reality, there are only a maximum of five strategies for each agent according to the calibrated memory distribution, i.e. some agents may have only one strategy. For example, a subset of five strategies can be like the one in Table 6.

Next, this agent needs to find the best strategy to make a mode choice. The set of strategies in Table 7 is the same set of strategies as in Table 6, but now with five horizon values. The choice in the table is the choice in each strategy made in each experienced time step. R is the return from the predicted choice and actual parking condition in the Valley. Then, θ is the moving average of the five returns. It is important to mention that the returns, R, are not necessarily the same among the strategies, even if the choice is the same at any given horizon, since the game is based on localised experience, but not the centralised information. R is expected to be negative according to microeconomic theory, i.e. cost and travel time is expected to affect the utility of visitors negatively (Hess et al., 2005).

In this example, thought pattern 2 with memory 1 has the highest success score, so this is the current best strategy. However, this best strategy may change in the future since it is a moving average. In the strategy of thought pattern 2 with memory 1, Auto has the probability of 0.6 (Table 6), so this option is likely to be chosen by this agent, but this is still determined by the probabilities. Although this decision making process involves guessing and baffling other agents' calculations, the basis is still the multinomial discrete choice model. Thus, this process is not just throwing a dice, but the result is still connected with the situation of the Upper Derwent Valley. The complete agent decision marking process throughout all four modules explained above is shown as a flowchart in Figure 10.

Table 5. Example set of 15 possible strategies in an agent. The numbers represent probabilities associated with the memories of mode choices.

Prob. of chance	**Memory 1**			**Memory 2**			**Memory 3**			**Memory 4**			Memory 5		
	Auto	Bus	Cancel	Auto	Bus	Cancel	Auto	Bus	Cancel	Auto	Bus	Cancel	Auto	Bus	Cancel
Thought pattern 1	0.50	0.40	0.10	0.40	0.40	0.20	0.65	0.35	0.10	0.40	0.45	0.15	0.45	0.30	0.25
Thought pattern 2	0.60	0.20	0.20	0.45	0.30	0.25	0.70	0.25	0.15	0.45	0.35	0.02	0.55	0.10	0.35
Thought pattern 3	0.55	0.45	0.00	0.45	0.45	0.10	0.70	0.30	0.00	0.45	0.50	0.05	0.55	0.40	0.05

Table 6. Example set of five assigned strategies in an agent

Prob. of chance	**Memory 1**			**Memory 2**			**Memory 3**			**Memory 4**			Memory 5		
	Auto	Bus	Cancel	Auto	Bus	Cancel	Auto	Bus	Cancel	Auto	Bus	Cancel	Auto	Bus	Cancel
Thought pattern 1	0.50	0.40	0.10				0.65	0.35	0.10						
Thought pattern 2	0.60	0.20	0.20	0.45	0.30	0.25									
Thought pattern 3	0.55	0.45	0.00												

Table 7. Finding the best strategy in an example strategy set. 'TP' stands for thought pattern, 'M' stands for memory, and R is return from the selected choice.

Success score	θ	**Horizon 1**		**Horizon 2**		**Horizon 3**		**Horizon 4**		Horizon 5	
		Choice	*R*	Choice	*R*	Choice	*R*	Choice	*R*	Choice	*R*
TP1M1	**-7.4**	Auto	-2.4	Auto	-0.5	Auto	-2.2	Auto	-2.3	Bus	0.0
TP1M3	**-6.2**	Bus	0.0	Auto	-5.3	Cancel	-0.9	Bus	0.0	Bus	0.0
TP2M1	**-2.8***	Auto	0.4	Bus	0.0	Auto	-0.7	Auto	-2.3	Auto	-0.2
TP2M2	**-6.4**	Auto	-2.4	Auto	-0.4	Auto	-0.4	Bus	0.0	Auto	-3.2
TP3M1	**-5.2**	Cancel	-0.9	Auto	-1.0	Auto	-1.5	Cancel	-0.9	Cancel	-0.9

Figure 10. Flowchart of agents' decision making

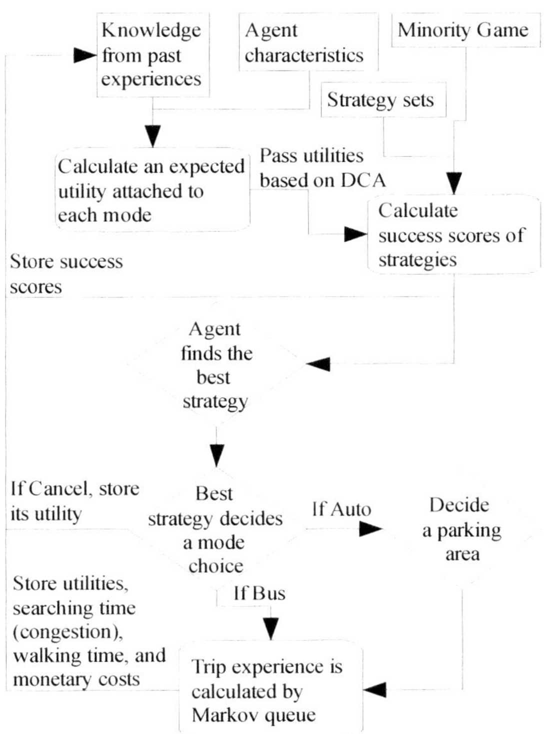

SIMULATION RESULTS

Simulation Setting

This MAS model uses numerous parameters and is rich in local rules, so only important settings are explained here. Unless specified otherwise, the values of parameters are the same throughout this paper. First of all, the agent population size or travel group size is 3000. The agent size does not affect the behaviours of vehicles. Since this is considered a sample from a larger agent population, vehicle number is automatically reflected to the ratio between the population and the agent size. The parameters for the decision making process of agents are based on the stated preference survey, including the ones explained in the previous sections.

"Real" bus fares are 50 pence per person, the parking fee for the Bus option is 50 pence per vehicle, and the toll is £3. These costs come from interviews with the local authority (Derbyshire County Council, 2003, per. com.). Agents do not know these travel related pieces of information before they experience them, therefore each agent picks up "believed" values randomly from possible ranges at the beginning of each simulation (Table 8). The one time step is at a representative day in a week and possibly one of weekends, i.e. the stepsize is one week. The first 520 time steps, which are 10 years in simulation time, are treated as the initial transient period, so the outputs for that the period are discarded from the analysis. This long transition period is necessary in this model, as agents need experiences before the proper simulation starts. As explained above, although visitors come to the Valley frequently, according to the survey data, it is unlikely that people visit the Valley more than once a week. The frequencies of visits are based on the survey data, so approximately 75% of agents are expected to go to the Valley five times during the initial transient period. Therefore, all five memory spaces are filled by the end of this period[10], i.e. these agents are likely to gather enough real experience from the simulation. The remaining agents use believed values even after the initial transient period. After the transit period, the simulation was run for 150 weeks or 150 time steps.

Walking speed: Walking speed is set according to agents' age between 4.2 and 3.0 feet per second, i.e. the older the slower. The difference between the walking speeds of the old (the top three older categories) and the young (the bottom four youngest categories) is 0.7 feet per second. Previous research on walking speed is in urban areas and is not recreational walking (Knoblauch et al., 1996), so these findings are used only as a rough standard.

Table 8. Range of believed values

Variable	Range	Justification
Toll fee	[0, 5] (£)	Up to the R.U.C. in London.
Bus fare	{0, 0.1, 0.2, 0.5, 1} (£)	"One-coin value" from local authority
Searching times	[0, 18.41] (minute)	Based on Markov queue model
Walk distance	[0, 4792.09] (metre)	Based on Markov queue model
Parking fee for Auto	{0, 0.5, 1.5, 2.0, 2.5} (£)	The real range in the Valley
Parking fee for Bus	{0, 0.1, 0.2, 0.5, 1} (£)	"One-coin value" from local authority
Headway	{15, 30, 45, 60}(minute)	From interview and current situation

Trip group size: As mentioned above, an agent is defined as a trip leader in a group in a vehicle and so the number of members is treated as a part of the agent's characteristics. The bigger trip party has to pay more bus fares, or can share the costs of the toll and parking fees. Therefore, this factor is important for this study. The empirically observed distribution is used for the estimation. A group size between two and four contribute to the 80% of the total distribution. The maximum group size is assumed to be ten because such a large group in a vehicle was actually observed even though proportionally it is nominal.

Frequency of visit: Visiting frequency is probabilistic; therefore, "visit once every other week" does not guarantee that an agent visits the Valley this time if it did not visit on the last time step. Instead, this concept says that this agent is likely to visit the Valley, on average, 36 times a year. Moreover, although the frequency is assumed fixed, agents may change their visiting tendency, e.g. if an agent learnt that it is not good to visit the national park any more, its real visiting frequency becomes zero. The frequency of visit is based on the survey results, and it depends on the travel origin.

Seasonality in traffic demand: The traffic flow data of 2003 on the A57 are used to estimate the seasonal demands in the Valley. The final week of August and the first week of September are set as the busiest weeks (the vertical dotted lines in Figure 11). June to September demands are set as 90% of the demand of the busiest weeks, i.e. high season (the darkest background in Figure 11), April, May, and October are set as 80% of the busiest weeks, i.e. intermediate season (the intermediate background), and the rest of periods are set as 60% of the busiest weeks, i.e. low season (the no background colour). The seasonality affects the frequency of agents' visits if their visiting frequency is less than 'once every other week'. For example, an agent expected to come once a year still comes once a year, but is more likely to come during high season than low season.

Figure 11. The left figure shows mode choices before the Road User Charging, and the right figure shows mode choices after the Road User Charging. The background contrasts show the traffic seasonality in the Valley.

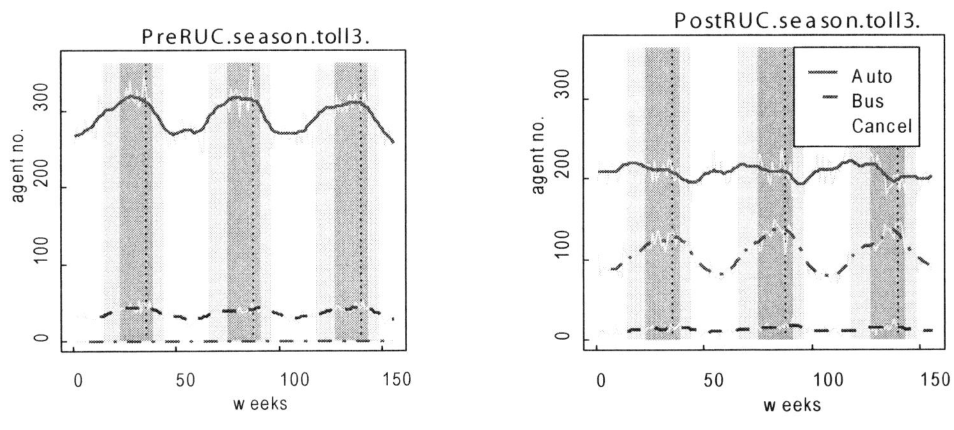

Road User Charge and Park and Ride Schemes

Traffic Pattern

The demand for Auto is reduced after the implementation of the Road User Charging as the demand shifts to Bus and Cancel. (Figure 11) The lines in bold are smoothed using the LOWESS method with value 0.1 (Cleveland, 1981), and raw data are shown as the pale lines. The amount of shift is greater in the high seasons since the trend of Auto is relatively more flattened after the implementation. This is because the preference of agents (or strategies strictly) in this model is logarithmic and not linear with given parameters. This means that an extra 10 vehicles in the parking areas in an extremely congested situation puts off agents coming by car more than the same extra 10 vehicles in a less congested situation. This phenomenon is consistent even at localised viewpoints.

Congestion at a Parking Area

Figure 12 shows the proportion of time the first parking area is congested. Generally, congestion levels are reduced after the Road User Charging is implemented. However, medium congested periods increases after the implementation, i.e. the darkest and the bottom band areas (100% full) shrink while the second bottom areas (75% full) stretch in Figure 12.

The level of 100% full congestion gets significantly lowered while that of 75% full congestion remains at relatively the same level after Road User Charging is introduced, i.e. more severe congestion is reduced. Therefore, the model in this section shows that the Road User Charging scheme reduces the demand of Auto effectively in more realistic conditions and the reduction in the demand corresponds with the reduction in the congestion level in the parking areas. Since the scheme reduces demand and congestion more efficiently at extreme conditions, the scheme solves the severe congestion problem at parking areas, which is reported by many visitors, whilst the scheme can still attract visitors in less congested conditions.

Figure 12. Congestion levels in the first parking areas, i.e. Information Centre. The left graphs shows Congestion levels before the Road User Charging, and the right graph shows after the Road User Charging.

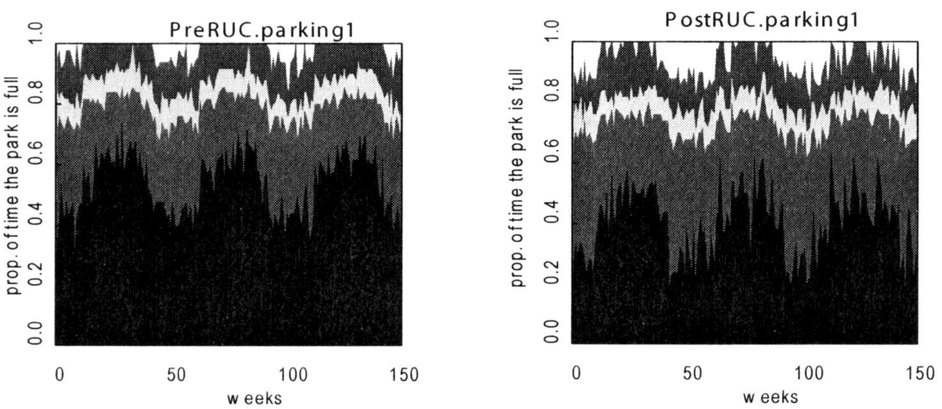

Simulation with Elderly Exemption

This section focuses more on the elderly visitors. From observation, a large proportion of visitors to the Valley are elderly people. The result from the questionnaire shows that 24% of visitors are aged between 55 and 64 and 12% of visitors are aged over 65, so overall more than a third of visitors are retirement aged people. Many visitors to the Valley carry a large amount equipment to have a picnic and it will be difficult for elderly people to carry this equipment without a vehicle. Although this model possibly takes into account this difficulty within an existing factor 'walking speed', this internalisation is likely to be underestimated so that special attention should be given to this fact. There are five to six disables parking spaces in front of the Information Centre, which is a primary destination, but these spaces should strictly speaking be used by visitors with true disabilities. Also, this space is not enough during the high season and will never satisfy the majority of elderly visitors even if the space is doubled.

One possible solution is to give the elderly a discount on the toll fee. Elderly visitors, who are eligible to receive the exemption, were defined by age, from 55 or 65, in this section. While all other settings remain the same from previous results, this simulation introduces an elderly exemption. When the discount was only £1, the overall demand for Auto did not increased significantly (Middle graph in Figure 13). In contrast, when all visitors older than 55 received a full exemption from the toll fee, the trend of Auto use rose vertically rose and that of Bus use fell (Right end graph). With 100 sets of this situation[11], the percentage rose in overall Auto demand by 12.71% with a standard deviation of 1.07%.

DISCUSSION

Comparison with the Analysis that Solely Uses Equation-Based Models

When a researcher conducts the same analysis solely with the multinomial discrete choice model, the result could be similar but the contents of the result have to be examined carefully.

Figure 13. Mode choices with elderly exemption. Elderly visitors (aged from 55 years) are eligible for a discounted toll fee. Results are shown for a discount of between £0 and £3, respectively from left to right.

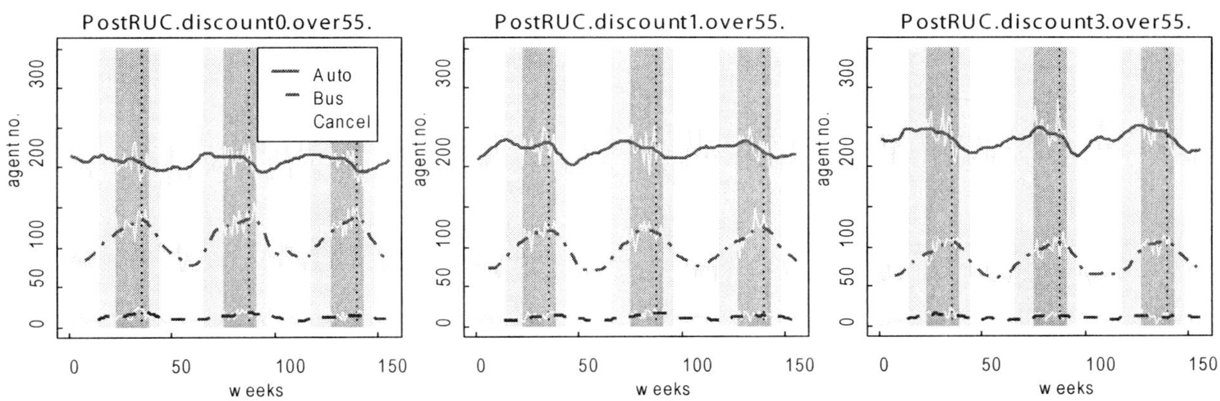

The multinomial discrete choice model cannot calculate searching time, walking time, and a parking fee for Auto, so this approach is as sophisticated as the MAS model. Leaving this question aside, assuming that these values are the central point of possible values and the other parameters are the same as those of the current agent simulation, the overall percentage increase in Auto by the discrete choice model is 15.09%. The two approaches produce very similar outputs, but the difference is obvious when the breakdown by age categories is examined.

The Auto demand rises only in the top two elderly age categories and those in the rest of the age categories stay the same with the analysis using an equation-based model (Right graph in Figure 14). In contrast, the demand of Auto is higher for the two elderly age categories in the MAS analysis (Left graph), while the demand declines in younger age categories due to the side effect from the congestion in parking areas. As more elderly visitors come to the Valley by car, the parking areas are more congested, and consequently this situation discourages other visitors from coming to the Valley by car. Moreover, this discouragement reduces the parking congestion more than expected, so that this possibly encourages other visitors, namely the elderly visitors, to come to the Valley by car, simultaneously. Hence, the proportional rises of elderly visitors are more prominent and the demand for Auto by some agents is reduced in a MAS analysis. There are also variations in the trend of Auto numbers in MAS model. This is an important issue, which is discussed further in the next section.

As shown in Figure 3, the conventional transport analysis does not have the direct feedback mechanism with parking congestion or in other words, there is no concept of congestion in the analysis. Therefore, the analysis solely with equation based modelling ignores the fact that some visitors may still visit the Valley by car as the parking areas will be less congested because the toll fee suppresses the others' private car usage (Stopher, 2004). As a result of this, the conventional transport analysis solely with the discrete choice model underestimates the Auto demand of elderly visitors and overestimates Auto demand of younger visitors compared with the MAS model. This is due to underrating the side effect of parking congestion.

Figure 14. Proportional change in Auto by age categories with elderly exemption from age 55 and over. The left graph is the result from MAS simulation analysis. The right graph is from the conventional equation-based analysis.

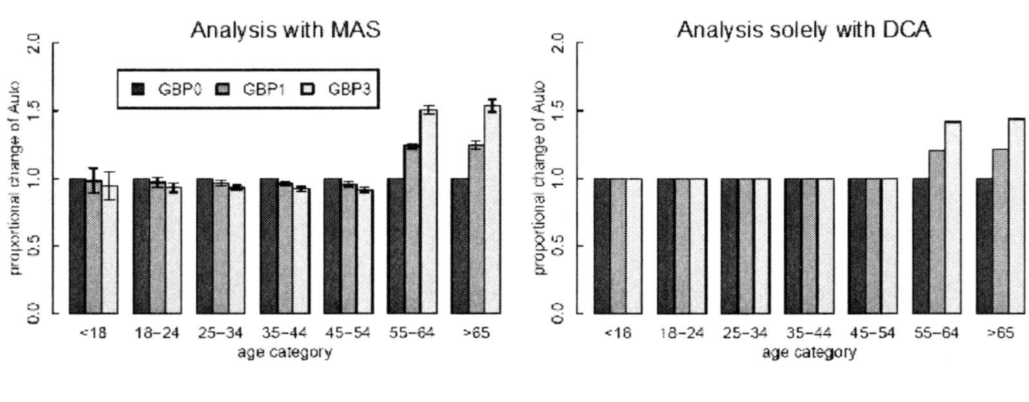

Variation in the Model and Unpredictability in the Minority Game

The change in the Auto number was varied in the MAS model. Especially, the inverse trend is sometimes observed in the age class younger than 18 in Figure 14. This is evidence of the variation within agents' decision making. Each agent behaves individually, so it is not necessarily the case that two agents make exactly the same choice when inputs and characteristics are exactly the same. The mean value of the sum of individual decisions could be the same as the result of a system level analysis, such as the analysis only using a discrete choice model, but a deviation is associated with the former, but not with the latter. The stability of the result is strongly affected by sample size. The proportion of agents in an age category below 18 years is only 0.6% or only 18 or 19 out of 3,000 agents, and the variances are very large in this model. With the small agent size and wide variation, the result from the youngest age category is very sensitive.

It should be emphasised again that, the reason for the wide variation is partly due to the frequency of visits, but more importantly, it is partly because of unpredictability in the Minority Game. It is impossible to achieve perfect rationality in the Minority Game since this kind of rationality requires that an agent is aware of decisions from many other agents. Due to the cognitive limitations of individuals, this type of information is usually inaccessible in real life (Klügl & Bazzan, 2004), especially in the case of parking congestion. This study uses sceptical rational strategies to formulate agents' decision making, and similarly, no strategy can globally be a best strategy in this dynamic situation (Thompson & Richardson, 1998). When many agents find the same best strategy, the strategy is no longer the best strategy since these agents move in the same direction and this is no longer the minority side.

In the end, these agents are making a decision, but at the same time, partly throwing a die to select a choice at every time step. This causes a wide variety in the decision making process of agents. For example, the previous section shows that the Auto demand by elderly visitors increases when the exemption is given to them, but the change is not uniform even within the same age category. Figure 15 shows the cumulative number of Auto chosen by each elderly agent aged over 65 with the same visiting frequency of between 2 and 5 times a year. These graphs show a cumulative sum with time, so a hori-

Figure 15. Cumulative numbers of Auto chosen by each elderly agent age over 65 without and with full elderly exemption from age over 55, respectively from left to right graphs. Visiting frequency is the same between the graphs, namely between 2 and 5 times a year.

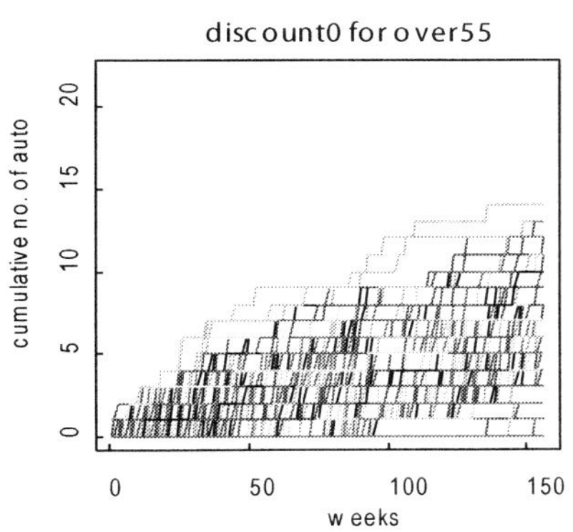

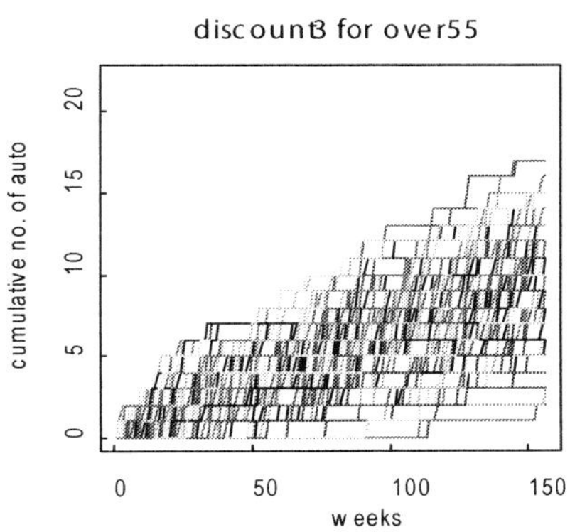

zontal line at the bottom means the agent never chooses Auto. Without an exemption, an elderly agent goes to the Valley by Auto between 14 and 0 time by the end of a simulation run. In contrast, with an exemption of GBP3, the range is between 17 and 1 times.

The lines are widely spread in both graphs, although these agents are in the same age and frequency category. The number of times Auto was chosen is distributed at a lower level without a toll exemption. In contrast, more lines are distributed at a higher level with the exemption of £3. The difference indicates that many elderly agents, who cannot afford to visit the Upper Derwent Valley by car, are supported by the exemption. However, some other elderly agents still prefer other choices such as Bus or Cancel. This is partially because extremely bad experiences discourage the agents from choosing Auto and cause them to adapt. On the other hand, this could be because an internal preference assigned with other characteristics determines alternative choices. Therefore, analysis of the agent user utility is necessary to justify the comfort of elderly visitors.

User Utility Distribution Amongst Agents

User utilities shown in Figure 16 were calculated from the utility functions of the mixed logit model, their units are utility. Therefore, these values are meaningful only in comparison, not in absolute terms. In other words, the negative trends in the user utility do not mean that the agents in this category are worse off, but only that the relative difference between the two plots provides some explanations. The plots show the improvement of the agents' user utility after the implementation of the elderly exemption, in the same age and frequency category.

Without the exemption, the user utility ranged between -59.01 and 4.68 with a mean of -25.45. With the full exemption, the user utility ranged between -47.87 and 32.870 with a mean of 9.202. There are still equity problems within the same agent category since some agents keep increasing their user utility while some other agents struggle to raise their user utility. This is the nature of the Minority Game, and could be a real phenomenon in many competitive societies.

Figure 16. Cumulative user utility of each elderly agent aged over 65 without and with full elderly exemption from age over 55, respectively from left to right graphs. Visiting frequency is the same between the graphs, namely between 2 and 5 times a year.

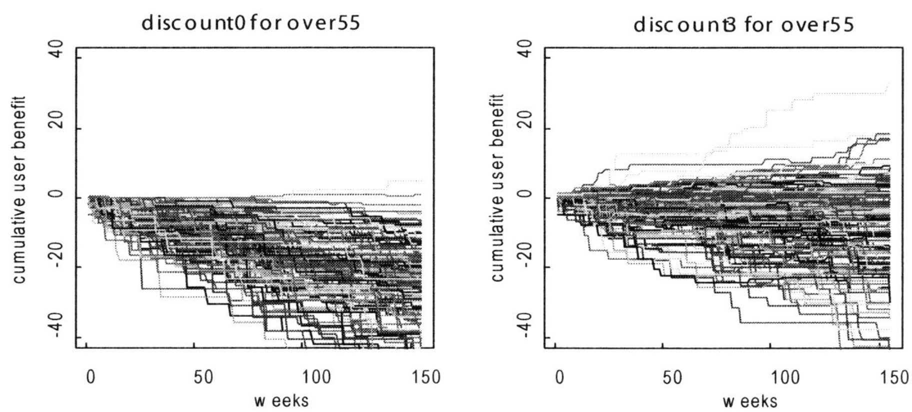

Lastly, the cumulative utility is examined in the case of elderly exemption for those aged over 55. As discussed before, the variances of user utility are very large at the end of the simulation, also the user utility of agents are improved as the exemption rises in the top two elderly age categories (Figure 17). These are the expected phenomenon and the trend is similar to that of the bar graph about the relationship between the number of Auto uses and the exemption (left graph in Figure 14), except one important issue.

In the bar graph of the Auto number, the number chosen by the younger agents decreases as the exemption increases due to the congestion in the parking areas caused by the increase in elderly drivers. In contrast, the user utility of younger agents does not decrease much, if at all. The reason is simply the adaptation of agents. As parking areas are congested, it is more likely that one has a bad experience and so younger agents learn that there is not much point to in going by car in this situation. Then, younger agents stop using Auto and switch mode to Bus or even Cancel, but older agents still try to go to the Valley by car since the bad experience of the congestion is substituted by the exemption. For younger agents, alternative choices are better options even though their initial motivation is Auto. Congestion is a relatively more important factor. Because of this adaptation, the user utility of younger agents does not decrease much. Therefore, the exemption for elderly visitors at any level as presented in this study is a possible and suitable scheme to reduce the difficulty specific to the elderly visitors, which can be underestimated in the simulation model.

In addition, the overall Auto demand increases with the exemption in some cases. This means that the exemption partially ruins the chance to reduce the congestion level in the parking areas, which is one of the main purposes of implementing the road user charging scheme. Therefore, this fact should also be considered when the exemption is implemented. Moreover, this study uses the exemption for elderly visitors, but the idea could be applicable to other social groups as well.

CONCLUSION AND SUGGESTION

This simulation model produced comprehensive outcomes including mode choices, congestion levels, the user utility, and adaptation of visitors. The study focus is mainly on concept and process. Having

Figure 17. Boxplots of agent user utility at the end of the simulation in the situations of elderly exemption from age over 55. The level of discount is none, £1, and £3. All agents have the visiting frequency category between 2 and 5 per year.

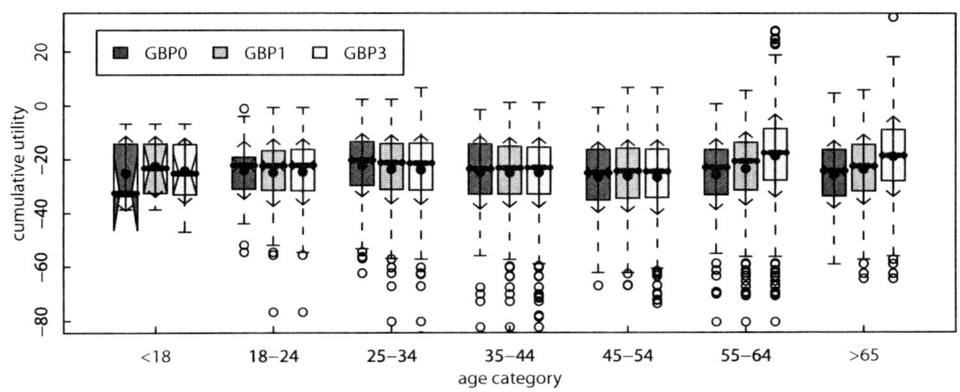

said that, the results showed that the road user charging scheme would reduce the Auto demand in the Upper Derwent Valley and proved that the reduction eased the congestion in the parking areas. The reduction in Auto demand and parking congestion was effective especially when overcrowding occurred, for example during the August Bank Holiday. Although a further study should be conducted to finalise the possibility of exemption for the elderly, the model showed the exemption improves the comfort of elderly visitors without sacrificing that of younger visitors significantly.

This model showed the oversimplification in the conventional equation-based analysis, which gave significant biases when real world problems were analysed by ignoring adaptation effects. In the case of the Upper Derwent Valley, the over simplification not focusing on the parking network and consequently the concept of congestion and adaptation effects, which required the dynamic modelling of the linkages amongst tourists. MAS models have the advantage of dynamic modelling, connecting between modules in the model, and incorporating adaptation concepts. Therefore the MAS model simulates the situation of the Upper Derwent Valley more realistically.

This research focused on the improvement of transport choice problems by MAS so that the decision making parts are still based on equations namely discrete choice analysis. In the future research, as well as collecting further empirical data, it will be interesting to check how the model produces different outcomes when these equations are replaced by rules-based logics based on further empirical evidence and fewer assumptions. In conclusion, this project established a multi agent simulation model to examine a road user charging scheme with the role of congestion in mind.

ACKNOWLEDGMENT

This work is based on the author's doctoral thesis at the Transport Studies Unit/Oxford University Centre for the Environment (OUCE). His work was funded primarily by the Oxford Kobe scholarship and in part by Light senior scholarship from St. Catherine's College. The author wishes to acknowledge the priceless support and supervision by Prof. John Preston, the data collection by Mr. Nikolaos Thomopoulos, and final advise and check by Dr. Richard Taylor, Dr. Sukaina Bharwani, and Ms. Anna Taylor.

REFERENCES

Arnott, R., & Rowse, J. (1999). Modeling parking. *Journal of Urban Economics, 45*(1), 97-124.

Arthur, W. B. (1994). Inductive reasoning and bounded rationality. *American Economic Review, 84*, 406–411.

Axelrod, R. (1997). *The complexity of cooperation*. Princeton, NJ: Princeton University Press.

Axtell, R., Axelrod, R., Epstein, J., & Cohen, M. (1997). Replication of agent-based models, aligning simulation models: A case study and results. *The Complexity of Cooperation,* (pp. 183-205).

Bazzan, A. L., Bordini, R. H., Andrioti, G. K., & Vicari, R. M. (2000). Wayward agents in a commuting scenario - personalities in the minority game. *In Proceedings of the Fourth International Conference on MultiAgent Systems (ICMAS-2000)*, 55, Washington DC. IEEE Computer Society.

Beltratti, A. (1996). *Models of economic growth with environmental assets* Norwell, MA: Kluwer Academic Publishers.

Ben-Akiva, M., & Lerman, S. R. (1985). *Discrete choice analysis: theory and application to travel demand.* Cambridge: MIT Press.

Bierlaire, M. (2001). The acceptance of modal innovation: The case of swissmetro. In *1st Swiss Transport Research Conference*, Monte Verità.

Bierlaire, M. (2003). Biogeme: a free package for the estimation of discrete choice models. In *Proceedings of the 3rd Swiss Transport Research Conference*, Monte Verita, Ascona.

Challet, D., Marsili, M., & Zhang, Y.-C. (2004). *Minority Games Interacting agents in financial markets.* Oxford , UK: Oxford University Press.

Challet, D., & Zhang, Y.-C. (1997). Emergence of cooperation and organization in an evolutionary game. *Physica A, 246*, 407–415.

Cheng, Y. (1998). Hybrid simulation for resolving resource conflicts in train traffic rescheduling. *Computers in Industry, 35*(3), 233-246.

Chernick, M. R. (1999). *Bootstrap Methods A Practitioner's Guide.* New York: John Wiley & Sons Inc.

Cleveland, W. S. (1981). LOWESS: A program for smoothing scatterplots by robust locally weighted regression. *The American Statistician*, 35, 54.

Collier, N., Howe, T., & North, M. (2003). Onward and upward: The transition to repast 2.0. In *Proceedings of the First Annual North American Association for Computational Social and Organizational Science Conference*, page 5, Pittsburgh, PA USA. June, Electronic Proceedings.

Davison, A., & Hinkley, D. (1997). *Bootstrap Methods and their Application.* Cambridge Series in Statistical and Probabilistic Mathematics. Cambridge, UK: Cambridge University Press.

de Cara, M. A. R., Pla, O., & Guinea, F. (2000). Learning, competition and cooperation in simple games. *The European Physical Journal B, 13*, 413–416.

Department for Transport (2004). Values of time and operating costs. *Technical report, Transport Analysis Guidance Unit.* Retrieved August 4, 2008 from http://www.webtag.org.uk/webdocuments/3_Expert/5_Economy_Objective/3.5.6.htm#1_2b.

Derbyshire County Council (2003). Re: toll and bus fare for road user charge at the Upper Derwent valley. *e-mail and interview through phone, 2nd – 6th, June. Currently (Nov., 2003).*

Dia, H. (2002). An agent-based approach to modelling driver route choice behaviour under the influence of real-time information. *Transportation Research Part C: Emerging Technologies, 10*(5-6), 331–349.

Eckton, G. D. C. (2003). Road-user charging and the Lake District National Park. *Journal of Transport Geography, 11*(4), 307-317.

Edmonds, B. (1999). Modelling socially intelligent agents. *Applied Artificial Intelligence, 12*, 677–699. http://www.cpm.mmu.ac.uk/cpmrep26.html.

Edmonds, B., & Moss, S. J. (2005). *From KISS to KIDS: An "anti-simplistic" modeling approach* Manchester Metropolitan University Business School.

Efron, B. (1979). Bootstrap methods: Another look at the jackknife. *The Annals of Statistics, 7*, 1-2.

Epstein, J., & Axtell, R. (1996). *Growing Artificial Societies: Social Science from the Bottom Up*. Brookings Institution Press.

Fagiolo, G., Windrum, P., & Moneta, A. (2006). *Empirical validation of agent-based models: A critical survey*. LEM Working Paper Series.

Febbraro, A. D., & Sacco, N. (2004). On modelling urban transportation networks via hybrid petri nets. *Control Engineering Practice, 12*(10), 1225-1239.

Frenken, K. (2006). Technological innovation and complexity theory. *Economics of Innovation and New Technology, 15*(2), 137-155.

Fowkes, A. S. (2000). Recent developments in stated preference techniques in transport research. In Ortézar, J., (Ed.), *Stated Preference Modelling Techniques, volume 4 of PTRC Perspectives*, (pp. 37–52). PTRC Education and Research Services Ltd, London.

Gilbert, N. (2004). Open problems in using agent-based models in industrial and labor dynamics. In R. Leombruni, & M. Richiardi (Eds.), *Industry and labor dynamics: The agent-based computational approach* (pp. 401-405). Sigapore: World Scientific.

Gilbert, N., & Troitzsch, K. G. (1999). *Simulation for the Social Scientist*. Buckingham: Open University Press.

Clark, G. L. (1998). Stylized facts and close dialogue: Methodology in economic geography. *Annals of the Association of American Geographers, 88*(1), 73-87.

Greene, W. H. (2003). *Econometric analysis*. London: Prentice Hall International.

Hensher, D., & Puckett, S. (2005). Road user charging: The global relevance of recent developments in the United Kingdom. *Transport Policy, 12*(5), 377–383.

Hess, S., Bierlaire, M., & Polak, J. W. (2005). Estimation of value of travel-time savings using mixed logit models. *Transportation Research Part A: Policy and Practice, 39*(2-3), 221-236.

Hinkley, D. V. (1988). Bootstrap methods. *Journal of the Royal Statistical Society. Series B, 50*(3), 321-337.

Ihaka, R. & Gentleman, R. (1996). A language for data analysis and graphics. *Journal of Computational and Graphical Statistics, 5*, 299-314.

Jain, S., & Neal, R. M. (2004). A Split-Merge Markov Chain Monte Carlo Procedure for the Dirichlet Process Mixture Model, *Journal of Computational and Graphical Statistics, 13*, 158-182.

Klügl, F., & Bazzan, A. L. C. (2004). Route decision behaviour in a commuting scenario: Simple heuristics adaptation and effect of traffic forecast. *Journal of Artificial Societies and Social Simulation, 7*(1).

Knoblauch, R. L., Pietrucha, M. T., & Nitzburg, M. (1996). Field studies of pedestrian walking speed and start-up time. *Transportation Research Record*. No. 1538 Pedestrian and Bicycle Research.

Lee, K., Hui, P. M., Wang, B.-H., & Johnson, N. F. (2001). Effects of announcing global information in a two-route traffic flow model. *Journal of the Physical Society of Japan*, *70*(12), 3507–3510.

Lindner, C. C., & Rodger, C. A. (1997). *Design Theory*. Boca Raton, FL: CRC Press.

Liu, X., Liang, X., & Tang, B. (2004). Minority game and anomalies in financial markets. *Physica A: Statistical and Theoretical Physics*, *333*, 343–352.

McFadden, D. (1974). Conditional logit analysis of qualitative choice behaviour. In Zarembka, P., (Ed.), *Frontiers in Econometrics*. New York: Academic Press.

Mokhtarian, P. L., & Salomon, I. (2001). How derived is the demand for travel? some conceptual and measurement considerations. *Transportation Research Part A: Policy and Practice*, *35*(8), 695-719.

Nash, C. (2003). Marginal cost pricing and other pricing principles for user charging in transport: a comment. *Transport Policy*, *10*, 345–348.

Norris, J. (1997). *Markov chains*. Cambridge: Cambridge University Press.

Oota, J. (1995). Introduction to petri net. *Technical report, Faculty of Information Science and Technology*, Aichi prefectural university.

Ortúzar, J. d. D., & Willumsen, L. G. (2001). *Modelling transport*. Chichester: John Wiley.

Parunak, V., Savit, R., & Riolo, R. (1998). Agent-based modeling vs. equation-based modeling: A case study and users guide. *Proceedings of Workshop on Multi-agent systems and Agent-based Simulation (MABS'98)*, 10–25.

Peeta, S., Zhang, P., & Zhou, W. (2005). Behavior-based analysis of freeway car-truck interactions and related mitigation strategies. *Transportation Research Part B: Methodological*, *39*(5), 417–451.

Peterson, J. L. (1981). *Petri net theory and the modeling of systems*. Englewood Cliffs, N.J: Prentice-Hall.

Pfeiffer, P. E., & Schum, D. A. (1973). *Introduction to Applied Probability Theory*. New York: Academic Press.

Redmond, L. S., & Mokhtarian, P. L. (2001). The positive utility of the commute: modeling ideal commute time and relative desired commute amount. *Transportation*, *28*(2), 179-205.

Resnik, M. D. (1987). *Choices: An introduction to decision theory* Minnesota: University of Minnesota Press.

Repast Organization for Architecture and Development (2005). Repast 3.0. Retrieved August 4, 2008 from http://repast.sourceforge.net.

Ripley, B. D. (1987). *Stochastic simulation*. New York: John Wiley & Sons Inc.

Ross, S. M. (1997). *Simulation*. New York: Academic Press.

Rothengatter, W. (2003). How good is first best? marginal cost and other pricing principles for user charging in transport. *Transport Policy, 10*, 121–130.

Steiner, T. J., & Bristow, A. L. (2000). Road pricing in national parks: a case study in the Yorkshire Dales National Parks. *Transport Policy, 7*, 93–103.

Stevens, C. F., & Zador, A. M. (1998). Input synchrony and the irregular firing of cortical neurons. *Nature Neuroscience*, 1, 210-217.

Stopher, P. R. (2004). Reducing road congestion: a reality check. *Transport Policy, 11*(2), 117–131.

Thompson, R. G., & Richardson, A. J. (1998). A parking search model. *Transportation Research Part A: Policy and Practice, 32*(3), 159–170.

Tsamboulas, D. A. (2001). Parking fare thresholds: a policy tool. *Transport Policy, 8*(2), 115-124.

Vose, D. (2000). *Risk analysis: a quantitative guide*. Chichester: John Wiley.

FURTHER READINGS

General MAS Issues: Agent-Based Simulation

Aumann, C. (2007). A methodology for developing simulation models of complex systems. *Ecological Modelling*, 202(3-4), 385–396.

Comment: General discussion on developing complex simulation models.

Axelrod, R. (1997). *The Complexity of Cooperation*. Princeton, NJ: Princeton University Press.

Comment: General text book on the social simulation of MAS

Axtell, R., Axelrod R., J.M. Epstein & M.D. Cohen. (1996) Aligning Simulation Models: A Case Study and Results, *Computational and Mathematical Organization Theory*, 1(2), 123-141.

Comment: Issue on comparing multiple MAS – docking

Bonabeau, E. (2002). Agent-based modeling: Methods and techniques for simulating human systems. *In Proceedings of the National Academy of Sciences*, 99, 7280–7287. National Acad Sciences.

Comment: General introductory paper on ABM.

Barreteau, O. (2003). Our companion modelling approach. *Journal of Artificial Societies and Social Simulation*, 6(2).

Comment: A new empirical modelling approach with validation in mind

Bonabeau, E., Dorigo, M., & Theraulaz, G. (1999). *Swarm Intelligence: From Natural to Artificial Systems*. Oxford, UK: Oxford University Press.

Comment: An alternative of RePast – A foumous MAS/ABM platform developed at Santa Fe Institute. The book covers some transportation.

Challet, D., Marsili, M., & Zhang, Y. (2005). *Minority Games*. Oxford, UK: Oxford University Press.

Comment: Good introductory book about the social aspect of MAS

Epstein, J. & Axtell, R. (1996). *Growing Artificial Societies: Social Science from the Bottom Up*. Brookings Institution Press.

Comment: Good introductory text book on the social simulation of MAS

Gilbert, N. & Troitzsch, K. G. (1999). *Simulation for the Social Scientist*. Buckingham: Open University Press.

Comment: General introductory text book about computation simulation approach.

Grimm, V., Revilla, E., Berger, U., Jeltsch, F., Mooij, W. M., Railsback, S. F., Thulke, H.-H., Weiner, J., Wiegand, T., & DeAngelis, D. L. (2005). Pattern-oriented modeling of agent-based complex systems: Lessons from ecology. *Science*, 310(5750), 987–991.

Comment: A modelling approach between theory driven and data drive approaches.

Janssen, M. A. & Ostrom, E. (2006). Empirically based, agent-based models. *Ecology and Society*, 11(2), 37–49.

Comment: Overview on empirical studies with MAS.

Richiardi, M., Leombruni, R., Sonnessa, M., & Saam, N. (2006). A common protocol for agent-based social simulation. *Journal of Artificial Societies and Social Simulation*, 9(1).

Comment: Discussion on the standardisation of MAS approach.

Wooldridge, M. (2002). *An introduction to multiagent systems*. Chichester: John Wiley and Sones.

Comment: A general textbook on MAS from a viewpoint of computer science.

Comparision and Integration with Econometrics/Stochastic Approaches

Ben-Akiva, M. & Lerman, S. R. (1985). *Discrete choice analysis : theory and application to travel demand*. Cambridge: MIT Press.

Comment: One of the key textbooks on discrete choice analysis.

Friedman, M. (1953). *The methodology of positive economics*. Essays in Positive Economics, 3–43.

Comment: Rational of econometric approach as a forecasting tool

Gatti, D. D., Gaffeo, E., Gallegati, M., Giulioni, G., Kirman, A., Palestrini, A., & Russo, A. (2007). Complex dynamics and empirical evidence. *Information Sciences*, 177(5), 1204–1221.

Comment: Discussion between econometric approach and MAS

Greene, W. H. (2003). *Econometric analysis*. London: Prentice Hall International.

Comment: Economitrics text book, which has special focus on discrete choice analysis.

Leombruni, R. & Richiardi, M. (2005). Why are economists sceptical about agent-based simulations? *Physica A: Statistical Mechanics and its Applications*, 355(1), 103–109.

Comment: Discussion between economic model and MAS.

McNally, M. G. (2000). The four step model. In Hensher, D. A. & Button, K. J., editors, *Handbook of Transport Modelling*, chapter 3, 35–52. Oxford, UK: Elsevier.

Comment: General explanation on transportation four stage models.

Oi, W. & Shuldiner, P. (1962). *An Analysis of Urban Travel Demands*. Evanston, IL: Northwestern University Press.

Comment: One of the early examples of transportation four stage models.

Train, K. (2003). *Discrete Choice Methods with Simulation*. Cambridge: Cambridge University Press.

Comment: A text book on discrete choice modells.

Congestion and Minority Game

Adler, J. & Blue, V. (2002). A cooperative multi-agent transportation management and route guidance system. *Transportation Research Part C: Emerging Technologies*, 10(5), 433–454.

Comment: MAS for a congestion problem.

Fletcher, M. & Deen, S. M. (2001). *Multi agent design issues in congestion management*. Retrieved August 4, 2008 from http://citeseer.ist.psu.edu/380820.html.

Comment: An interesting MAS working paper on congestion.

Jianping, S., Chunxiao, Y., & Zhaosheng, Y. (2003). Multi-agent urban expressway control system based on generalized knowledge-based model. *In Intelligent Transportation Systems, 2003. Proceedings. 2003 IEEE*, 2, 1759–1763.

Comment: MAS for a congestion problem.

Microsimulation Model

Norris, J. (1997). *Markov chains*. Cambridge: Cambridge University Press.

Comment: general Markov chain theory

Arnott, R. & Rowse, J. (1999). Modeling parking. *Journal of Urban Economics*, 45(1), 97-124.

Comment: parking model

Hopcroft, J. E. & Ullman, J. D. (1979). *Introduction to Automata Theory, Languages, and Computation*. Reading, MA: Addison Wesley.

Comment: Networked queue model and cellular automata

Ben-Akiva, M. E. (2005). *DynaMIT: Real Time Traffic Estimation and Prediction System.* Massachusetts Institute of Technology, Cambridge.

Comment: One of the most well-known software tools for microsimulation in transportation research

Lee, K., Hui, P. M., Wang, B.-H., & Johnson, N. F. (2001). Effects of announcing global information in a two-route traffic flow model. *Journal of the Physical Society of Japan*, 70(12), 3507-3510.

Comment: Cellular automata approach

Febbraro, A. D. & Sacco, N. (2004). On modelling urban transportation networks via hybrid petri nets. *Control Engineering Practice*, 12(10), 1225-1239.

Comment: Microsimulation and petri net model

Validation and Verfication

Balci, O. (1994). Validation, verification, and testing techniques throughout the life cycle of a simulation study. Annals of Operations Research, 53(1), 121–173.

Comment: Life cycle of validation with validation and verification of simulation model.

Fagiolo, G., Windrum, P., & Moneta, A. (2006). Empirical validation of agent-based models: A critical survey. note: LEM Working Paper Series, Retrieved August 4, 2008 from http://www.econ.iastate.edu/tesfatsi/EmpValid.Fagiolo. 2006-14.pdf.

Comment: Working paper on empirical validation in MAS.

D'Aquino, P., Page, C. L., Bousquet, F., & Bah, A. (2003). Using self-designed role-playing games and a multi-agent system to empower a local decision-making process for land use management: The selfcormas experiment in senegal. *Journal of Artificial Societies and Social Simulation*, 6(3).

Comment: An empirical modelling approach for MAS with validation in mind

Gilbert, N. (2004). Open problems in using agent-based models in industrial and labor dynamics. In Leombruni, R. & Richiardi, M., editors, *Industry and Labor Dynamics: the agent-based computational approach,* 401–405. Sigapore: World Scientific.

Comment: Discussion on the difficulties of empirical MAS

Leombruni, R. & Richiardi, M. (2005). Why are economists sceptical about agent-based simulations? *Physica A: Statistical Mechanics and its Applications*, 355(1), 103–109.

Comment: A critical review on agent-based simulation model

Oreskes, N., Shrader-Frechette, K., & Belitz, K. (1994). Verification, validation, and confirmation of numerical models in the earth sciences. *Science*, 263(5147), 641.

Comment: Discussion on validation and verification in empirical studies.

Richiardi, M., Leombruni, R., Sonnessa, M., & Saam, N. (2006). A common protocol for agent-based social simulation. *Journal of Artificial Societies and Social Simulation*, 9(1).

Comment: Protocol suggesting validation and verification of MAS

Sargent, R. G. (2004). Validation and verification of simulation models. *In Proceedings of the 36th conference on Winter simulation*, 17–28.

Comment: Validation and verification on simulation models especially on MAS.

Windrum, P., Fagiolo, G., & Moneta, A. (2007). Empirical validation of agent-based models: Alternatives and prospects. *Journal of Artificial Societies and Social Simulation*, 10(2), 8.

Comment: Discussion on empirical validation on MAS

Other General Transportation MAS Examples

Abkin, M., Bea, R., Corker, K., Gilgur, A., Jadhav, A., Lee, S., Pritchett, A., & Verma, S. (2002). Examining air transportation safety issues through agent-based simulation incorporating human performance models. *In Proceedings of The 21st Digital Avionics Systems Conference*, 2, 7A5–1– 7A5–13.

Comment: Transportation MAS to model the bahaviour of pilots and air traffic controllers.

Au, E., Chiu, D., Leung, H.-f., Lee, O., & Wong, M. (2005). A multi-modal agent based mobile route advisory system for public transport network. *In System Sciences, 2005. HICSS '05. Proceedings of the 38th Annual Hawaii International Conference*, 92b.

Comment: MAS in route and mode choice in Hongkong.

Gjerdrum, J., Shah, N., & Papageorgiou, L. G. (2001). A combined optimization and agent-based approach to supply chain modelling and performance assessment. *Production Planning and Control,* 12, 81–88.

Comment: An example in stochastic approach with MAS in the field of transportation.

Kavicka, A. (2007). Simulations of transportation logistic systems utilising agent-based architecture. *International Journal of Simulation Modelling.*

Comment: A MAS example on logistic system.

Laichour, H., Mandiau, R., & Maouche, S. (2001). Traffic control assistance in connection nodes: multi-agent applications in urban transport systems. *In Intelligent Data Acquisition and Advanced Computing Systems: Technology and Applications, International Workshop on, 2001.*, 133–137.

Comment: Transport network model with MAS.

Raney, B., Cetin, N., Vollmy, A., Vrtic, M., & Axhausen, K. (2003). An agent-based microsimulation model of swiss travel: First results. *Networks and Spatial Economics*, 3(1), 23–41.

Comment: A large scale empirical transportation MAS.

Raney, B. & Nagel, K. (2003). Truly agent-based strategy selection for transportation simulations. *In 82nd Annual Meeting of the Transportation Research Board*, Washington, DC.

Comment: MAS in transportation with a modular approach.

Zeng, C.-h. & Li, Q.-s. (2004). Route selecting in the light of the theory of multimode transportation based on multi-agent. *In Systems, Man and Cybernetics, 2004 IEEE International Conference on*, 4, 4023–4027.

Comment: MAS on mode and route choice based on transport cost and time

Zutt, J., Aronson, L., van der Krogt, R., Roos, N., & Witteveen, C. (2002). Multi-Agent Transport Planning. *In Proceedings of the Fourteenth Belgium-Netherlands Artificial Intelligence Conference (BNAIC'02),* 387–394.

Comment: MAS application to transport planning.

ENDNOTES

1 http://transp-or.epfl.ch/page63023.html

2 The error component model tries to capture the correlation between alternatives, which share unravelled attributes. Therefore, the idea is similar to that of the nested logit model.

3 The positive time coefficient can be explained by the pleasure of walking and driving (Redmond and Mokhtarian, 2001; Mokhtarian and Salomon, 2001).

4 £4.46 (non-working hour in 2002 price) / 60 minutes ×1.0158 (non-work value of time growth from 2002 to 2003) = 7.55 pence per minute

5 http://www.r-project.org/

6 Here the 100(1 - α)% confidence interval is simply given by the $\alpha/2$ and 1 - $\alpha/2$

7 http://repast.sourceforge.net/

8 Petri net is one of the discrete event system models and applicable in this situation. In this model, an event is in a discrete state (e.g. `On', `Off') and each event occurs at anytime (asynchronous) without the influence of other events (concurrency) (Peterson, 1981,Oota, 1995).

9 This gap-utility approach is similar to the *regret* approach in operational research (e.g. Resnik, 1987, pp. 28-30).

10 As explained in Minority Game section, the survey data shows that travellers to the valley do not remember any trip related information older than last five trips.

11 Each set has exactly the same setting including the seed of random number generator except for the level of elderly exemption.

Chapter II
A Multi-Agent Modeling Approach to Simulate Dynamic Activity-Travel Patterns

Qi Han
Eindhoven University of Technology, The Netherlands

Theo Arentze
Eindhoven University of Technology, The Netherlands

Harry Timmermans
Eindhoven University of Technology, The Netherlands

Davy Janssens
Hasselt University, Belgium

Geert Wets
Hasselt University, Belgium

ABSTRACT

Contributing to the recent interest in the dynamics of activity-travel patterns, this chapter discusses a framework of an agent-based modeling approach focusing on the dynamic formation of (location) choice sets. Individual travelers are represented as agents, each with their cognition of the environment, habits, and activity-travel patterns. Agents learn through their experiences with the transport systems, changes in the environments and from their social network. Conceptually, agents are assumed to have an aspiration level associated with choice sets that in combination with evaluation results determine whether the agent will start exploring or persist in habitual behavior; an activation level of each (location) alternative that determines whether or not the alternative is included in the choice set in the next time step, and an expected (utility) function to evaluate each (location) alternative given current beliefs. Each of these elements is dynamic. Based on principles of reinforcement learning, Bayesian learning, and social comparison theories, the framework specifies functions for experience-based learning, extended and integrated with social learning.

INTRODUCTION

So-called activity-based models have rapidly gained interest in the transportation research community. These models predicts and simulate in a coherent fashion multiple facets of activity-travel behavior including which activities are conducted, when, where, for how long, with whom, and the transport mode involved. To the extent that these models have been actively implemented, some type of simulation is used to implement predicted activity-travel in time and space. The majority of these simulations are based on Monte Carlo simulations; others use agents such as Albatross (Arentze and Timmermans, 2001, 2003a, 2004a) and Aurora (e.g., Joh, et al. 2006). An overview of these developments is given in Timmermans, et al. (2002). In addition to these comprehensive models, agent-based simulations have also been suggested for particular facets of activity-travel choice (e.g., Charypar and Nagel, 2003; Balmer, et al. 2004; Rosetti and Liu, 2004; Hertkort and Wagner, 2005; Rindsfuser and Klügl, 2005).

Although the theoretical underpinnings of these agent-based models differ, they have in common the assumption that individuals will choose within their choice sets the alternative they prefer, sometimes subject to a set of constraints (Ben-Akiva and Boccara, 1995; Pellegrini, et al., 1997, Cascetta and Papola, 2001). In most of these models, however, the construction and composition of individual choice sets is not explicitly modeled. Choice sets are typically assumed given or derived on the basis of some arbitrary rule (Swait and Ben-Akiva, 1987; Thill and Horowitz, 1997; Swait, 2001). The delineation of choice sets is particularly important in large-scale micro-simulation systems, which are receiving increasing attention in activity-based travel-demand modeling and integrated land-use – transportation systems. As expected, knowing the choice set from which an alternative is selected significantly decreases the complexity and may improve the performance of these large-scale systems (Shocker, et al., 1991). In this context, the choice set refers to the set of discrete alternatives known by the individual, which is a subset of the universal choice set that consists of all alternatives available to the decision maker. Known means that the individual knows the attributes that are potentially relevant for evaluation under specific contextual conditions in the activity-travel decision-making process. Note that this definition differs from commonly used terminology in marketing, where a distinction is made between awareness set, evoked set, consideration set and choice set (Timmermans and Golledge, 1990). We can refine our framework along these lines, but that is beyond the goal of the present chapter.

As a part of the FEATHERS model (Arentze, et al., 2006; Janssens, et al., 2006), an extension and elaboration of Aurora (Joh et al., 2006), an agent-based system which incorporates different types of dynamics and learning discussed in Arentze and Timmermans (2003) is developed. This chapter discusses the conceptual framework that addresses one type of dynamic: the formation and dissolution of personal choice sets, which lays the foundation for the longer-term dynamics of the FEATHERS models. It should be stated from the outset that the discussion below mainly concerns location choice set, but the basic mechanism can be applied as building blocks for multiple facets of activity-travel patterns, including person choice set, mode choice set, and so on.

We assume that individuals conduct activities to satisfy specific needs and try to organize their activities and travel in time and space in some satisfactory way, influenced by their cognition of the environment. If the environment is stationary, one might assume that as a result of repeated trials some Pareto optimum or steady state will be established: activity-travel patterns are stabilized and become habitual. However, in reality, the space-time environment is non-stationary and individuals' needs may change as well, as a result of, for example, changes in socio-demographics. Furthermore, critical incidents may imply that individuals are triggered to change their behavior. Under these circumstances, the actual performance of the transport and land-use system for an individual may decrease below some critical level – the aspiration level of the individual, leading him/her to search for alternatives such that the expectations regarding his/her activity-travel pattern can be achieved. In addition, an individual's cognition of the environment may change as a result of new information from media, actual travel and social contacts, which may prompt him/her to adjust the aspiration level and actively explore new alternatives. Thus, choice set formation is conditional upon the context and dynamic in the sense that choice sets are updated each time an individual has executed an activity-travel schedule or when new information becomes available.

These considerations lead to the following three core parts of the proposed conceptual framework of modeling the dynamic process: (1) an aspiration level associated with the choice set that in combination with evaluation results determines whether the individual will start exploring or persist in habitual behavior, (2) an activation level of each location alternative that determines whether or not the alternative is included in the choice set in the next time step and, (3) an expected (utility) function that allows an individual to evaluate each alternative given current beliefs about the attributes of the alternative. Each of these elements is dynamic, which allows simulating habit formation and adaptation.

In the following, we will first identify key drivers that trigger changes in choice behaviors and describe how they are integrated in the decision making process. To depict mechanisms that influence such changes, we continue with describing their functions for cognitive updating based on principles of reinforcement and Bayesian learning. Then, we extend the system to incorporate social learning that involves social adaptation and information transfer. After an illustration case study of dynamics in shopping location choice sets, we complete with a conclusion and discussion for future research.

THE FRAMEWORK

The basic assumption is that an agent acts based on behavioral principles and mechanisms. (S)He holds beliefs (knowledge) about the environment during a certain life course, has preferences and basic needs, leading to plans, agendas and schedules. (S)He carries out those plans, agendas and schedules in time and space. When a deviation exists between his/her expectation and aspiration an agent may start exploring his/her environment for new alternatives. Thus, (s)he learns about the environment and the consequences of his/her actions, in this case the choice of activity locations, and is able to adapt to changing circumstances and improve less effective behavior. Based on experiences, an agent forms habits, reinforces memory traces, updates beliefs about attributes of alternatives, discovers the conditions under which certain states of the environment are more likely than others, and in so doing makes sense of the world around him/her. Moreover, through social contacts agents exchange information and adjust aspirations, which may trigger actions to explore new alternatives. Thus, for an agent, the composition of the choice set for a specific activity under certain contextual conditions is dynamic. The alternatives within the choice set will be expanded with newly discovered alternatives and reduced with old ones that are discarded or are no longer retrievable from memory. In this chapter, we will use location choice set as one of the activity-travel facets to illustrate the dynamics of our modeling approach.

Basic Drivers

An *aspiration level* is an agent's goal for the outcome of a decision (Payne et al., 1980; West and Broniarczyk, 1998). In theory, aspirations could be defined either at the level of choice alternatives (a bundle of attributes) or individual attributes. We assume that it is more plausible to define aspiration at the level of choice attributes as it is on that level that an agent may determine goals that give direction to exploration processes (e.g., find alternative stores with a lower price level rather than find stores that have higher utility for my purposes). Defined for an attribute, an aspiration serves as a subjective reference point, which determines what qualifies as a satisfactory outcome for that attribute. An aspiration level is agent-specific and, in case of a dynamic attribute, context-specific. The outcome of a comparison between aspiration and expected outcome given current knowledge provides a measure of an agent's satisfaction and willingness to explore new alternatives. A possible discrepancy between the expected outcomes derived from the alternatives within the current choice set and the agent's aspiration levels may trigger the agent to switch from habitual behavior to a conscious choice mode.

Generally, aspiration levels are context dependent. For example, satisfaction or tolerance about the crowdedness encountered at shopping locations may vary by day-of-the-week and shopping location's category type. Aspiration levels can relate to both (quasi)-static attributes and dynamic attributes (which may fluctuate as a function of the behavior of all agents in the system). Formally, we denote the set of current aspiration values as $A = \{A_k\}$, where $A_k = (c_k, e_k)$, $e_k = (e_{1k}, e_{2k} \ldots, e_{Jk})$, e_{1k} represents the aspiration value of the first attribute under the k-th condition, and $c_k = (c_{1k}, c_{2k}, \ldots, c_{Sk})$ defines the k-th condition as a set of states of S condition variables considered. Agents within similar social demographic classes or belonging to the same social network may have similar aspiration levels since they adapt their aspirations based on social comparison (as explained later).

Agents also judge what make up a satisfactory outcome, and have the ability to memorize situations and outcomes (i.e., events). In part, this is context dependent, that is, certain contextual conditions automatically activate particular memory traces that lead to particular levels of awareness. The *activation level* of a location alternative is the indicator of the strength of such a memory trace, and hence reflects the ease with which it can be retrieved from memory. As such, an activation level is associated with each alternative in the current choice set for each specific contextual condition, for example in case of location choice set, defined in terms of type of activity (i.e., purpose of the trip), the previous activity location (i.e., origin location of the trip), day-of-the-week and time-of-the-day.

By repeatedly performing certain behavior under same contextual conditions, agents develop habits. By forming and following habits, agents can reduce mental effort involved in constantly evaluating choice alternatives and making choices. By saving cognitive resources for the operation, habits help agents conserve mental resources and time, and free them for other tasks. Habits have been described as learned and scripted behaviors and are capable of being automatically activated by the contextual conditions that normally precede the behavior. As such, the activation level of a (location) alternative represents the degree of an agent's habit of choosing that (location) alternative under certain contextual conditions. In our framework, habitual behavior involves that agents consistently select from a choice set the alternative with the highest activation level under the given contextual condition at the moment a choice is to be made. In turn, we define the choice set in a given choice situation as the (location) alternatives that are retrievable from memory in that situation (i.e., condition). Formally, let i denote a particular (location) alternative, $W_i^t(z_m)$ be the activation level of the (location) alternative i under condition m, Q be the number of relevant condition variables, $z_m = (z_{1m}, z_{2m} \ldots, z_{Qm})$ represents the states of the Q condition variables under condition m, and ω be a minimum activation level for memory retrieval ability. Then, the (location) choice set is defined as:

$$\Phi^t(z_m) = \{i \,|\, W_i^t(z_m) \geq \omega\} \tag{1}$$

Note that, as implied by this equation, the definition of a (location) choice set may vary between situations. For example, the (location) choice set with the contextual condition of departure from home might be different from the choice set with the contextual condition of departure from work.

The attractiveness of a (location) alternative is in general influenced by values of its attributes. Depending on the targeted objective underlying the activity, the attributes that should be evaluated may be different. For example in case of shopping, the variety of stores is important for entertainment and purchase purposes, while a social need requires some familiarity with the location. Furthermore, the intention of resting attracts attention to spatial layout, while economic considerations emphasize quality and price. Thus, the impact of a (location) alternative may be diverse, that is, the combination of a (location) alternative and activity objectives determines the utility of a (location) alternative. Moreover, some of the attributes are (quasi)-static, X_j^s; while others are dynamic, X_j^d. The (quasi)-static attributes reflect characteristics of the (location) alternative that are in short term constant, for example in case of a shopping location, the size category, price level, parking space, and presence of stores for certain goods in a shopping centre. We assume that an agent will learn all the (quasi)-static attributes of a (location) alternative simply through observing them after implementing an activity with the chosen alternative. This knowledge will keep constant, and only change when the physical conditions are changed externally, for example, after a renovation of the shopping centre.

Dynamic attributes, such as crowdedness and travel time, are subjective and uncertain, and may be dependent on contextual conditions. We assume that for each dynamic attribute, X_j^d, the agent uses some classification, denoted as $X_j^d = \{x_{j1}, x_{j2}, \ldots, x_{jN}\}$, where $x_{j1} - x_{jN}$ represent possible states of X_j^d, and specifies his/her beliefs regarding (location) alternative i based on his/her current knowledge as a probability distribution across X_j^d denoted as $P_i^t(X_j^d)$, which sums up to 1. The degree of uncertainty is given by the degree of uniformity of $P_i^t(X_j^d)$. The more evenly the probabilities are spread across possible states, the larger the uncertainty is, and vice versa. For example, consider again the crowdedness of a shopping location. This is a dynamic attribute of a (location) alternative, and therefore, may involve uncertain knowledge. An agent could choose four states for crowdedness as {no, little, medium, very}, and specifies his/her beliefs regarding each (location) alternative i as a probability distribution across these four states.

In addition, the agent may discover that probabilities of states are conditional upon certain contextual variables. For example, the agent may discover that probabilities of crowdedness of a shopping location depend on day-of-the-week (e.g., weekday and weekend) and time-of-the-day (e.g., peak hours and non-peak hours). Learning that some variables have an impact on outcome-states means extending unconditional probabilities $P_i^t(X_j^d)$ to obtain conditional probabilities $P_i^t(X_j \,|\, C)$, where C stands for one or more condition variables.

A utility function allows the agent to evaluate each (location) alternative given his/her current beliefs about the attributes of the (location) alternative and his/her preferences. Using probabilities of the types $P_i^t(X_j|C)$ to describe the knowledge of the agent, the *expected utility* equation can be expressed as below:

$$EU_i^t(c_k) = EU_i^s + EU_i^d(c_k) \tag{2}$$

$$EU_i^s = \sum_{j^s} \beta_{j^s}^s X_{j^s}^s \tag{3}$$

$$EU_i^d(c_k) = \sum_{j^d} \sum_n \beta_{j^d n} x_{j^d n} P_{ij^d}^t(x_{j^d n}|c_k) \tag{4}$$

where EU_i^t is the expected utility of (location) alternative i at time t, $\beta_{j^s}^s X_{j^s}^s$ is the expected partial utility of (location) alternative i for static attributes and preferences, and $\beta_{j^d n} x_{j^d n} P_{ij^d}^t(x_{j^d n}|c_k)$ is the expected partial utility of (location) alternative i under possible states $x_{j^d n}$ with probabilities $P_{ij^d}^t(x_{j^d n}|c_k)$ and preference $\beta_{j^d n}$ regarding dynamic attribute j^d with state n. $c_k = (c_{1k}, c_{2k}, \ldots, c_{Sk})$ represents the values of relevant condition variables under the *k*-th condition. Thus, expected utility takes into account current beliefs regarding state probabilities as well as an agent's preferences. Of course, static attributes could also be dealt with as a special case of dynamic attributes where the believed state has a probability of 1. We use a different symbol here than in case of defining activation levels to indicate that condition states used for defining attribute belief (and aspiration) may not be the same as condition states used for defining activation levels.

Making a Choice

In the assumed choice making process, agents go through a mental process to arrive at a choice. They start with implementing their habitual behavior that requires least mental effort, and carry on with conscious choice that asks for more effort only if the habitual choice is not satisfactory, until they find a choice that is satisfactory. The decision making process is illustrated schematically in Figure 1. It is explained in details in the following paragraphs.

As the aspiration levels are the standards for determining whether an outcome is acceptable, they will try to find the alternative that meets the requirements within a tolerance range. The dissatisfaction tolerance is a predefined and agent specific parameter that reflects a characteristic of the agent. A large dissatisfaction tolerance indicates the agent strongly dislikes the mental effort involved to make better actions and is sooner happy with the current situation. Vice versa, a small dissatisfaction tolerance implies that on the one hand the agent is stricter in what is found acceptable, and on the other hand the agent may have a higher propensity to explore. In general, the larger an agent's dissatisfaction tolerance is, the higher the probability will be that the agent is satisfied with the expected performance of the current choice set. Being satisfied with the current situation means less desire to take a risk, invest effort, and change behavior. Thus, this is accompanied with a higher possibility of following habit. And consequently, also that it is less likely to explore and possibly make better choices in the future.

As implied by the definition of activation level, the alternative that has the highest activation level in the choice set is the one that is most easily retrieved from memory and requires the smallest amount of mental effort from an agent. In order to determine the level of satisfaction with the habitual choice, the attributes values of the (location) alternative with the highest activation level is compared to as-

Figure 1. The model scheme

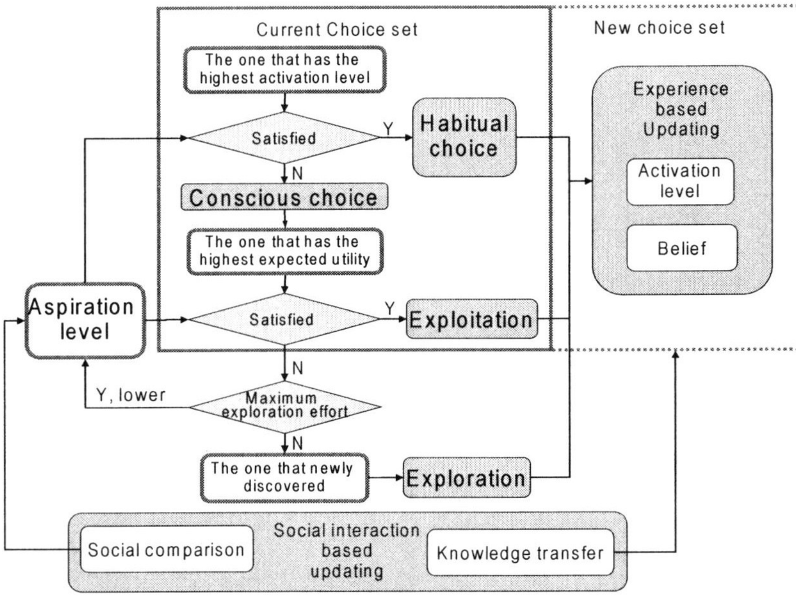

piration levels. We assume that if dissatisfaction (i.e., the difference between aspiration and expected level) regarding at least one attribute exceeds the tolerance range, an agent will switch to another mode of behavior and start searching consciously for better alternatives. On the other hand, if this range is not exceeded, we assume that no active search will take place and that the agent will exhibit habitual behavior executing the alternative that has the highest activation level.

We make a distinction between exploitation and exploration as alternative non-habitual modes of choice making. We assume that when acting in a conscious mode, an agent will first be engaged in exploitation and search within the current choice set (i.e., retrieve alternatives from the memory that have a lower awareness) for a better alternative under current conditions. With exploitation, the agent calculates the expected utilities (using equation 2-4) of all the alternatives within the choice set given current knowledge of the environment and under the given conditions, and compares the attributes of the one that has the highest expected utility with aspiration levels. When for none of the attributes dissatisfaction exceeds the tolerance range, we assume that no active exploration of new alternatives will happen and the agent will choose the (location) alternative that has the highest expected utility. If for at least one attribute there is a mismatch that exceeds the dissatisfactory tolerance, the agent will start to explore new alternatives that might solve the mismatch. We call this exploration. Thus, search for new alternative is not random, but rather directed. The attributes causing dissatisfaction will guide the agent in what to search for.

Exploration is a process by which new alternatives can enter the choice set. The probability of a (location) alternative to be discovered is modeled as a function of attractiveness of the (location) alternative regarding the attributes that are not satisfied by the alternative within the current choice set. Because agents are uncertain in such exploring situation due to limited information, we propose to use the Gibbs distribution/Boltzmann model (Sutton and Barto, 1998) to calculate discover probabilities across the universal choice set of (location) alternatives and simulate outcomes of search processes:

$$P(i|c_k) = \frac{\exp(V_i^{\prime}(c_k)/\tau)}{\sum_{i'} \exp(V_{i'}^{\prime}(c_k)/\tau)} \qquad (5)$$

where $V_i^{\prime}$ is a utility measure of (location) alternative i and τ is a parameter reflecting the availability of information in the selection of new (location) alternatives. The larger the value of the τ parameter is the lower the available information is and, hence, the more evenly discover probabilities are distributed across (location) alternatives, and vice versa. The parameter can be interpreted as the general (lack of) quality of information sources available to the agent, such as social network, public and local media and own observations during travel. $V_i^{\prime}$ is a utility calculated based on true values of attributes of (location) alternatives. Note that the utility depends on the objective of the search: by including only those attributes that are dissatisfactory with the current best (location) alternative, $V_i^{\prime}$ reflects the focus of the search. Moreover, a disutility of travel distance is included in the function for $V_i^{\prime}$ for two reasons: (1) the longer the travel distance is, the less likely information about the (location) alternative is available and, (2) the longer the travel distance is, the less likely the (location) alternative will be considered by the agent because of the higher generalized travel costs.

Having defined the discover probability distribution across (location) alternatives across the universal choice set, Monte Carlo simulation will be used to select a new (location) alternative that will be tried and may be added to the choice set. Once tried, the new (location) alternative receives an activation level reflecting memory trace strength and is subject to the same updating and learning process as other alternatives in the choice set as will be explained later.

In addition, an exploration effort counter is included to prevent an agent from getting trapped in continuous and endless exploration. We assume that an agent will keep a record of how many consecutive times it already tried exploring a new (location) alternative under the same contextual conditions. Every time a choice is made through exploration, it will add 1 unit of exploration effort. A habitual choice or an exploitation choice will break the chain of incrementing the score and restore it back to 0. We assume that when the exploration effort involved in search for a better alternative is built up and exceeds a predefined maximum, instead of continuing exploring, the agent will avoid further frustration by lowering the aspiration level (realizing that the current aspiration level is not realistic). The maximum exploration effort is another predefined and agent specific parameter that reflects a characteristic of the agent. A large maximum exploration effort indicates a higher willingness to spend effort to explore new alternative under the same contextual condition. It is accompanied with the possibility of a choice set including more alternatives.

Therefore, in the choice process, before engaging in exploration, the agent will check whether the accumulated exploration effort exceeds this maximum. If this maximum is not exceeded, the agent will continue exploring provided that no satisfactory alternative is found. When it is exceeded, the agent will replace the current aspiration levels with the attributes levels of the alternative that currently has the highest expected utility, to assure a relatively optimal outcome and maintain relatively high aspiration levels for future choices. As a consequence of choosing it, the activation level of this alternative will be increased.

As a consequence of the above mechanisms, an agent arrives at a selection of a single (location) alternative each time an activity is to be carried out. Depending on aspiration levels and evaluation result, this alternative could be the one that has the highest activation level (habitual choice), the one that has the highest expected utility (conscious exploitation choice), or the one that was newly discovered (conscious exploration choice).

Experience-Based Learning

Central to our dynamic process is the notion that choices are contingent upon the outcome of previous choices. By repeatedly making decisions, an agent acquires knowledge (and learns) about the environment and thereby forms expectations about attributes of the environment. It should be noted that adaptation and learning processes involve two operations. One concerns updating an agent's perception of the environment. Through repeated experience, agents will update their expectation of attributes of (location) alternatives (and routes), which are considered relevant for making choices, and discover conditions having an influence on outcomes. The other operation concerns the formation of habits to avoid the needless repetition of effortful memory retrieval and evaluation tasks. In this section, we will consider these two processes in turn, starting with habit formation.

A mechanism similar to reinforcement learning will be used for updating activation levels to simulate memory operation process. In line with evidence in cognitive psychology (Anderson, 1983), the basic assumptions are that an alternative that has higher utility stays longer in memory, and that memory is reinforced when an alternative is chosen and memory decays if it is not chosen. Every time a (location) alternative is chosen, the activation level of that (location) alternative will be incremented to simulate the strengthening of a memory trace. The reinforcement rate is an increasing function of the experienced utility of the chosen (location) alternative, which in turn is a function of its attributes. Limited memory retention capacity is simulated in the system by a parameter that determines rate of decay over time. If one alternative has not been chosen for some time, its activation level will decrease. When its activation level drops below some predefined minimum, it will be removed from the current choice set to reflect the limited human ability of memory retrieval. The minimum activation level indicates an agent's memory space. With a higher minimum activation level, an agent needs a strong memory trace to remember the alternative and is more easily forgetting things or discarding its memory, which may leads to more exploring. When the minimum activation level is extremely high, the choice set may not contain any alternatives, since none of the alternative meets the requirement. In such case, the agent tends to explore new alternatives every time a choice has to be made.

Formally, the strength of a memory trace of a particular activity (location) alternative *i* in the choice set is modeled as follows:

$$W_i^{t+1}(z_m) = \begin{cases} W_i^t(z_m) + \gamma U_i^t(z_m) & \text{if } I_i^t = 1 \\ \lambda W_i^t(z_m) & \text{otherwise} \end{cases} \tag{6}$$

where $W_i^t(z_m)$ is the strength of the memory trace (awareness) of location *i* at time *t* under a configuration of conditions z_m and $I_i^t = 1$, if the (location) alternative was chosen at time *t*, and $I_i^t = 0$, otherwise, $0 \le \gamma \le 1$ is a parameter representing a recency weight, which is relevant only when the location is chosen; and $0 \le \lambda \le 1$ is a parameter representing the retention rate. $U_i^t(z_m)$ is the experienced utility attributed to (location) alternative *i* that is calculated based on experienced states of the attributes of (location) alternative *i*, including both (quasi)-static and dynamic variables. The calculation (based on a utility function similar to the one represented by equation (2-4)) uses observed states of the dynamic attributes, such as crowdedness and travel time. Thus, at each time step the memory strength is reinforced or decays depending on whether the location alternative has been chosen in the last time step. The coefficients γ and λ determine the size of reinforcement and memory retention respectively and are parameters of the system. Based on the current value of memory strength, the system determines whether or not the location alternative is included in the choice set in the next time step based on the simple rule as described in equation (1), stating that it is included if it exceeds the minimum activation level and is not included, otherwise.

We assume that agents make personal observations and update their beliefs of their environment based on these observations in order to be able to make better predictions about what can be expected in the next time step. Each time a (location) alternative is chosen when an activity is implemented, the agent updates beliefs $P_i^t(X_j|C)$, where C is the condition or, if multiple condition variables are involved, the condition configuration experienced. Learning implies two processes: conditional learning and condition learning. The first process involves incrementally updating the conditional belief distributions across the possible states for each observed attribute of the (location) alternative after experiencing the actual states. The second process is aimed at discovering the conditions that have an influence on the likelihood of states of the system. Thus, the second process determines the form of the conditional probabilities that are kept up to date through the first process. This is done by periodically reconsidering splitting or merging condition states based on condition variables to update a tree structure that better predicts states based on observed outcomes. In the field of Bayesian perception updating, the two processes are generally known as parameter and structural learning respectively.

We will adopt the approach proposed in Arentze and Timmermans (2004b). In their approach, a method of parameter learning is used that is derived from Bayesian principles. Moreover, for structural learning, the proposed approach assumes a process of incrementally splitting and merging conditions based on events experienced in the past and stored in memory using some split criterion (Arentze and Timmermans, 2003b). In specific, the problem can be defined as a well-known problem considered by decision tree induction methods, namely as the problem of finding the most efficient way of splitting a set of known observations on predictor variables into partitions c_k that are as homogeneous as possible in terms of a response variable. For example in case of estimating the crowdedness of a location, the state of crowdedness is the response variable and time-of-the-day and day-of-the week serve as predictor variables. Then, the problem is to split the sample of observations on the condition variables such that observations within partitions are as homogeneous as possible in terms of crowdedness. Different criteria for finding the best splits, such as Chi-square or expected information gain can be used for this problem. Condition variables that are not significant in the current time step may become so at some next moment in time when more observations have been stored. Therefore, splitting and merging operations are periodically reconsidered. The result of a structural learning step, generally, is that subsequent parameter learning is based on a new belief structure. The new conditional probabilities can be derived from the event base in a straightforward way.

Social Learning

Agents are not isolated from each other, but participate in social networks. Participation in social networks may lead to adaptation of aspirations and diffusion of knowledge, which in turn may trigger changes in activity-travel choice behavior. Modeling the dynamic formation of social relationship between agents is beyond the scope of this chapter. (For a possible model of these processes, see Arentze and Timmermans, 2006.) In this section, we consider social links in agents' social network as given and focus on the impacts of social interactions on agents' aspiration levels and knowledge about activity (location) alternatives, and consequent dynamics in activity-travel patterns.

According to social comparison theory, people often obtain information about their performance by comparing themselves to others (Festinger, 1954). Social comparison theory posits that people are generally motivated to evaluate their opinions and abilities and that one way to satisfy this need for self-evaluation is to compare themselves to others. Information gathered from these social comparisons can then be used to provide insights into one's capacities and limitations, which may motivate them to achieve higher goals since people are motivated to maintain or increase positive self-evaluation.

Following this theory, we assume that when two agents P_1 and P_2 meet, agent P_1 will evaluate and update its aspiration levels based on the best performances of agent P_2, if P_2 belongs to the reference group of P_1. More specifically, for each contextual condition of which agent P_1 has defined aspiration levels, P_1 will ask P_2's best performance. Agent P_2 will provide as feedback the attribute information of the alternative that has the highest expected utility within its choice set under the corresponding conditions, since this alternative reflects its highest possible achievement given its current knowledge.

After receiving the information from agent P_2, agent P_1 first makes a decision on whether or not it will change its aspiration levels. For this, P_1 compares the expected utility that is calculated using attributes values from agent P_2's answer and its own preferences with the expected utility that is derived from its current aspiration levels. We assume that only if a positive discrepancy between the two expected utilities exist which exceeds the social deviation tolerance, ξ^{P_1}, of P_1 (i.e., $U(P_2) - U(P_1) > \xi^{P_1}$), then P_1 is willing to update its aspiration levels, and we say the agent is in an updating mode. If the discrepancy is not positive or the social deviation tolerance is not exceeded, we assume that no adjustment will take place implying that P_1 will leave its aspiration levels unchanged. Since the social deviation tolerance of comparing aspiration is used in turning the switch on upgrading aspiration levels, a higher social deviation tolerance indicates that an agent is more easily satisfied with its own current situation, despite the relative lower performance and the consequent position in the social network. As such, it may lead to a higher possibility of following habit, not socially adapting to higher references and investing effort to find better choices. A lower social deviation tolerance implies that an agent sets higher standards in what is found acceptable in social comparison, and probably has higher propensity in keeping in phase with the social network. It may lead to more adjustment, not only upgrading aspiration levels, but also more exploration because alternatives within the current choice set may not satisfy adapted aspiration levels.

We assume that when in an updating mode, P_1 will upgrade the aspiration levels on those attributes on which the alternative conveyed by P_2 has the better value. Note that, updating aspiration levels may lead to a switch from a habitual to a conscious choice mode, which in turn may lead to exploration of new alternatives and, hence, adaptation of the agent's choice set.

Besides social comparison, when two agents P_1 and P_2 meet, P_1 will also update its knowledge by integrating the new information provided by P_2. In the system, P_2 presents a list of all the (location) alternatives it knows to P_1. After receiving the list from P_2, P_1 checks the list with its knowledge to find out if the list of P_2 includes alternatives that are new to him. Each (location) alternative that is unknown to P_1 activates P_2 to provide further information about the attributes of the (location) alternative. Then, P_1 checks whether there are constraints (e.g., opening times, travel time) that limit the use of the new alternative, and add the new known alternative to context dependent choice sets, if any, for which the new alternative is appropriate. When added to a choice set, the new alternative is specified according to the attribute information conveyed by P_2 and an activation level is initialized based on the information acceptance of P_1 regarding P_2's information, $w_{P1 \to P2}$, and P_2's knowledge. A higher acceptance indicates that an agent is more inclined to view other's information as valid and assigns a higher initial activation level when added to its knowledge pool. This is accompanied with a higher possibility of not following habit. Once added, the new location is subject to the same selecting, updating and learning processes as other alternatives within the choice set. Consequently, a higher initial activation level of the newly added one implies longer time it needed to discard from the choice set even if it is not selected for some time and performs poorly in current situations.

In sum, social contacts provoke social learning that not only provides stimuli for adjusting aspirations to form partially common aspiration that may trigger changes in terms of exploring, but also for exchanging information in terms of adding new (location) alternatives to existing choice sets to form mutual choice sets. The properties of dyad relationships within social networks will influence dynamics of aspiration adaptation and knowledge diffusion. The agents that interact with each other within same network tend to have similar aspiration and maybe similar choice sets as the emerging results of the above mechanism.

ILLUSTRATION

To examine the behavior of the model, a series of numerical simulation were conducted. To reveal the separate impact of the various components contributing to the dynamics of the activity-travel patterns, a series of scenarios were set up starting with basic conditions and incrementally adding complexity. Due to space limitations and given the focus of the present chapter, this section presents only one simulation case study – a location choice of a shopping activity, which reveals some general dynamic properties of the proposed system.

Simulation Settings

The simulation considers an study area of 100 by 100 cells of 100 meter by 100 meter in size. There are 12 shopping locations including 6 small, 4 medium and 2 big shopping centers. The locations of these shopping centers are predefined across the study area. There are 6 agents with their residential location and work location pre-defined respectively. These locations are also possible origins of the agent for a shopping trip. The input schedules for the 6 agents are arbitrary generated with only one shopping activity a day for 72 days in total. A 2^3 full factorial design was used to generate 8 context condition profiles. The factors were: (1) day of the week (weekday or weekend), (2) time of the day (rush hour or non-rush hour), and (3) the origin of the trip (from home or from work). Schedules are constructed so that each profile occurred only once in every 8 days. Six static attributes of the shopping centre are included: (1) the size of the shopping centre (big, medium, or small), (2) store for the daily goods present (yes or no), (3) store for semi-durable goods present (yes or no), (4) store for durable goods present (yes or no), (5) price level (high, middle or low), and (6) parking space (yes or no). These attributes define the characteristics of each shopping centre. Only one dynamic attribute – crowdedness is included with four states as {No, Little, Medium, Very}. Travel time is calculated by physical distance at this simulation. The initial knowledge of each agent is based on a pre-period outcome using the same model starting with not knowing any of the locations and the highest aspiration level for each agent for every attribute.

The results reported here are the average results across 100 simulation runs. A simulation run considers a time period of 72 days. On each day, each agent considers choosing a location for its shopping activity. Dependent on its schedule, the agent checks out the alternatives in its context dependent choice set. Note that the choice set with the contextual condition of departure from home might be different from the choice set with the contextual condition of departure from work. The same applies to the rest of contextual conditions used to define the activation level. Based on its aspiration level of the day, the agent goes through a decision process as described in the framework section to arrive at a choice. Before going to the next day, the agent updates its knowledge, in particular, activation level and beliefs about the state of the environment. In reported results, the structure learning part is left out of consideration, and only parameter learning is considered. For every agent, the basic setting is: (1) the minimum activation level $\omega = 0.03$, the parameter for updating activation levels $\gamma = 0.99$ and $\lambda = 0.2$ (2) the maximum exploration effort is 3 units, (3) the aspiration dissatisfaction tolerance $\varepsilon = 1$ and the parameter of availability of information $\tau = 1$. Before the new day starts, the agent checks whether there is a scheduled social contact. When a social contact is scheduled, an agent randomly picks an encounter from the remaining 5 agents (of its network) and a one-way directed contact occurs, meaning that a situation may happen where agent P_2 has influence on agent P_1 since P_1 picks out agent P_2, while agent P_1 has no influence on agent P_2 if P_2 does not select P_1 according to its schedule. In reported results, a social contact is scheduled on an 8-day interval, since this is the minimum number of days to have a complete updated experiences given one activity a day and 8 predefined context condition profiles. In order to reveal clear influences of different types of social contact, comparing aspiration and exchanging information are taken place in turn. The social deviation tolerance is $\xi = 0.06$, while the acceptance of others information is $w = 0.16$.

Results

Figure 2 shows the general results of the basic case regarding each agent on 3 indicators: 1) average expected utility of the choice-set, 2) choice-set size and, 3) renewal rate (for the specific contextual condition that a decision is made). As expected, the expected utility of each agent's the choice set slightly increases across 72 days as a result of learning. The size of the choice set is not fixed, but shows a tendency to first decreases a bit and then to increase across agents on a 16-day interval. The range in the size is reasonable with an average around 2 for each agent. The waving curve showing the renewal rate explores the dynamics of the choice sets as the newly discovered alternative enters the choice set, and the ones not choosing for a long time are discarded. As it turns out, the size of the set as well as the renewal rate is bigger right after social contacts take place especially after exchanging information. It is in line with what we would expect that social learning brings about more dynamics to the choice set.

Even under the very basic conditions considered here, the emerging patterns in the behavior of the multi-agent system (in this case 6 agents) are already quite complex. On average among 72 choice occasions there are 47.08 habitual choices, 5.73 exploitation choices, and 19.19 exploration choices. As it turns out, the expected utility of habitual choices is not the highest among all the choice modes with an average value of 0.182; the expected utility of exploitation choices is the highest with an average value of 0.195. The expected utility of exploration choices is the lowest on average with a value of 0.096, because of the limited information in the search for new location alternatives. As it shows, the model is capable of incorporating social learning in addition to distinguishing habitual choice, exploitation choice and exploration choice. The frequency of social interaction will influence the social learning, as the differences may depend on the speed with which new experiences build up or old experiences decay, especially in the case of information exchange where differences are important. Thus, it provides a modeling approach for simulating habit formation and social adaptation under uncertainty. After a series simulation

Figure 2. The simulation results

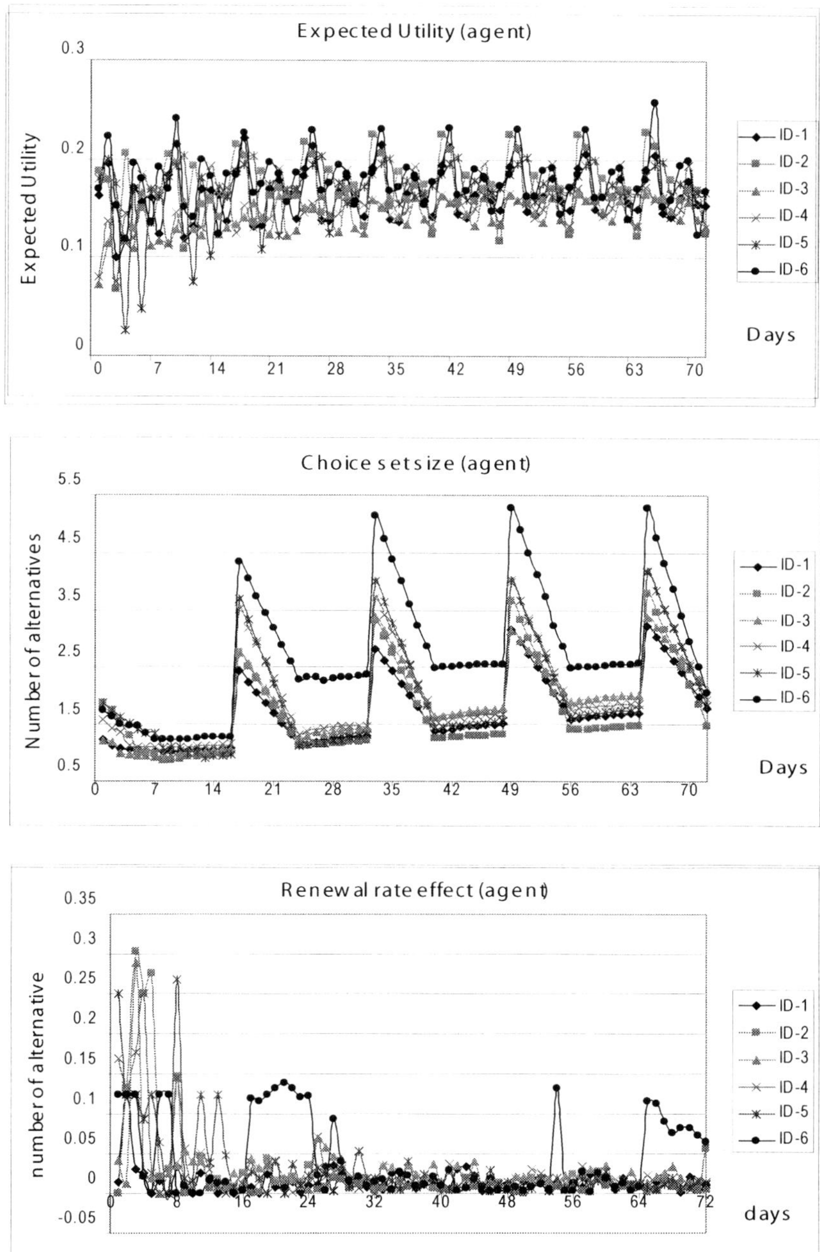

of various scenarios, the patterns of choice mode frequency, expected utility of different choice modes, size of the choice sets and renewal rate of the choice sets, average expected utility of choices and choice sets appear to respond in relatively unique ways to proposed parameters of the model.

CONCLUSION

This chapter has outlined the conceptual framework that will be used as a building block to model the dynamic process of agents' activity (location) choice in a large scale micro-simulation system. The framework considered the dynamic formation of the choice sets with the focus of location choice. It integrates cognitive learning and social learning. In the proposed approach, cognitive learning focuses on updating beliefs about a non-stationary environment that will impact the expected utility of alternatives and habit formation, while social learning emphasizes on deriving and updating aspirations that may trigger re-evaluating currently known alternatives (exploitation) or searching for new alternatives (exploration). As such, it provides a multi-agent modeling approach for predicting habitual choice, exploitation choice and exploration choice in activity-travel behavior as a function of discrepancies between dynamic, context-dependent aspirations and context-dependent expected utilities. A case study of shopping location choice is illustrated in this chapter. A similar framework can also be used for modeling other choice facets and learning behavior in activity-travel choices.

Our approach is scalable in the sense that it is applicable to study areas of large size (e.g., region wide). As expected, knowing the awareness set from which a choice is made may provide a parsimonious way in large scale micro-simulation in the areas of activity-based travel-demand modeling and integrated land-use – transportation systems. Some applications are straightforward. For example, conditions can be simulated under which learning leads to habitual behavior as well as what happens when moving to a new city. Likewise, the optimal location of a new shopping centre can be simulated. Also, spatial effects of the new shopping centre opening can be observed.

FUTURE RESEARCH DIRECTIONS

Attention should be paid to define the appropriate contextual conditions. Both aspiration and activation are context-dependent, but the condition states used for defining them may be different. As the departure location is an important contextual condition for defining activation and activity location choice sets, it may not be the same in case of defining aspiration. When it is used for other choice facets, the contextual conditions that define a choice set or aspiration needed additional attention to specify.

An activity-travel pattern is a complex product of decision making, including which activities are conducted, when, where, for how long, with whom, and the transport mode involved. If we take multiple facets of activity-travel behavior as a sequential choice making process, the contextual condition that defines the choice set (and aspiration) for the next choice facet could include previous decisions of the facets that already considered. By allowing iterative loop, the interdependences between these choice facets could be modeled. As such, the human reasoning process reflects well in the dynamic cognitions – aspiration, activation and beliefs that are conditional upon context conditions and subject to cognitive and social learning.

We assume that dynamics in behavior come about when a discrepancy between aspiration and expected utilities given current knowledge beyond some tolerance. It will trigger agents to explore new alternatives, which in turn may lead to changes in other facets of the activity-travel behavior. Many factors may cause such discrepancies: the travel environment may change; agent needs may change, etc. Waerden, et al., (2003) identified two important factors. One is critical incidents, unexpected events such as accidents or unexpected long delays which may cause agents to reconsider their habitual behavior. In addition, lifecycle or life trajectory events, such as the birth of a child, change of job, etc.

The proposed approach is well capable of dealing with changes of uncertain environment. In the later case, the discrepancy is increased because the set of conditions influencing satisfactory activity-travel choice has (dramatically) changed. We assume that an agent is likely to reconsider its current choices after the occurrence of a lifecycle event. Exploration for new alternative for one facet of the activity-travel pattern in long-term could be caused by dissatisfaction about current possible alternatives for the other facet. For example, because the resident and work location constraints the alternative travel mode, the agent may decide to move house. This reconsideration process can be properly modeled with an extension of the current system.

The current system already has information about conditions that trigger exploration and lowering aspiration (realize it is not realistic). The accumulated stress or incidents of lowering aspirations may increase one's need to make drastic changes, including changes in resources such as car availability (obtaining driving license or buying a new car), availability of public transport pass, or household income that may relax the constrains and increase the prospect of exploring new alternatives to add to current choice sets. As such changes may not solve the problem, more dramatic changes might be considered, such as changes in residential location, changes in work or study location that used as the contextual condition in defining choice sets and aspiration. Change in household composition could be modeled as external impact that changes constrains of the agent and its need, which will have the consequence of defining contextual conditions both for activation and aspiration.

A series of numerical simulation has been performed to assess the face validity of the system. It shows that the emerging patterns of choice behavior appear to response in relative unique ways for proposed parameters. It provides us not only avenues for improvement but also direction for the next steps in data collection and parameter calibration. Ideally, one would need continuous panel data covering a long period of time. However, for many reasons, such large scale panel data seems unrealistic. Alternatively, we argue that interactive computer experiments can be used successfully to capture some mechanisms underlying the dynamics. Moreover, when we consider single-day observations are nothing but one day realizations or manifestations of underlying dynamic processes, empirical cross section data or longitudinal data will be useful. We could adapt Bayesian principle or likelihood optimization method for parameter calibration, in an attempt to optimize the prediction to be closer to observation indicators in terms of the choice mode frequency, expected utility of different choice modes, size of the choice sets and renewal rate of the choice sets.

Some refinements of the system are also worth mentioning. Among many channels of social influence, we integrated only the influence from the social network. The information acceptance of others knowledge and the social deviation tolerance are pre-defined parameters in the current system. They can be extended into an expression of social relations and similarities between the two contacting agents. There are of course other mediates that could provide information and may have an impact on behavior, such as internet, TV, radio, etc.

In the current system, actions that produced positive rewards are reinforced and have a higher probability of being repeated in future choice occasions under similar conditions, while actions with negative outcome have a tendency to be ignored. In social contact, positive experiences are exchanged and have a higher probability of being adopted by others. It will also be interesting to see negative experience tend to be void, being exchanged and keeping out of choice sets.

Although the emerging result of a good alternative spreading among agents through network reflects the properties similar to the regeneration effect of genetic algorithm in evolution, it still needs theoretic prove. As one of the properties of social network the convergence of aspiration and knowledge and its impact on the activity-travel pattern as a whole need further investigation, especially whether the knowledge diffusion follows an S-shape as various literature reviewed.

The proposed system simulates the long-term dynamic aspect of activity-travel patterns, primarily habit formation and adaptation. The result of these behavior mechanisms are the evolution of choice-sets and choice patterns, reflecting emergent behavior in relation with non-stationary environment. It could be integrated with other activity generation & (re)scheduling approaches, such as need based theory or S-shape utility function, to comprehensively describe activity-travel patterns and uncertainties.

Modeling dynamics offers better understanding that behavior is context and situation dependent, but increases complexity, not only conceptually but also in model estimation, interpretation, application and data collection and hence puts forward major challenges. We hope to stimulate further research along this line to provide more insights into the direct and indirect effects of particular policy scenarios on future activity-travel patterns and their diverging on various segments of the population.

REFERENCES

Anderson, J. R. (1983). *The Architecture of Cognition*. Cambridge, Harvard University Press.

Arentze, T. A., & Timmermans, H. J. P. (2001). *Albatross: A Learning Based Transportation Oriented Simulation System*. EIRASS, Eindhoven, The Netherlands.

Arentze, T. A., & Timmermans, H. J. P. (2003a). Measuring impacts of condition variables in rule-based models of space-time choice behavior: method and empirical illustration. *Geography Analysis, 35*, 24-45.

Arentze, T. A., & Timmermans, H. J. P. (2003b). Modeling learning and adaptation processes in activity-travel choice. *Transportation, 30*, 37-62.

Arentze, T. A., & Timmermans, H. J. P. (2004a). A learning-based transportation oriented simulation system. *Transportation Research B, 38*, 613-633.

Arentze, T. A., & Timmermans, H. J. P. (2004b). *A theoretical framework for modelling activity-travel scheduling decisions in non-stationary environments under conditions of uncertainty and learning.* Paper presented at the Conference on Progress in Activity-Based Analysis, May 28 – 31, Maastricht, The Netherlands.

Arentze, T. A., Timmermans, H. J. P., Janssens, D., & Wets, G. (2006). Modeling short-term dynamics in activity-travel patterns: From Aurora to Feathers. Paper presented at the Innovations in Travel Modeling Conference, May 21-23, Austin, Texas.

Arentze, T. A., and Timmermans, H. J. P. (2006). Social networks, social interactions and activity-travel behavior: a framework for micro-simulation. *In Proceedings of the 85th Annual Meeting of the Transportation Research Board*, Washington, D.C. (CD-ROM: 18 pp.). To appear in Environment and planning B.

Balmer, M., Nagel, K., & Raney, B. (2004). Large scale multi-agent simulations for transportation applications. *In Proceeding of Behavior Responses to ITC*, Eindhoven (CD-Rom).

Ben-Akiva, M. E., & Boccara, B. (1995). Discrete choice models with latent choice-sets. *International Journal of Research in Marketing, 12*, 9-24.

Cascetta, E., & Papola, A. (2001). Random utility models with implicit availability/perception of choice alternatives for simulation for travel demand. *Transportation Research, C, 9*, 249-263.

Charypar, D., & Nagel, K. (2003). Generating complete all-day activity plans with genetic algorithms. *In Proceeding of the 10th International Conference of Travel Behavior Research*, Lucerne, Switzerland (CD-Rom).

Festinger, L. (1954). A theory of social comparison processes. *Human Relations, 7,* 117-140.

Hertkort, G., & Wagner, P. (2005). Adaptation of time use patterns to simulated travel times in a travel demand model. In H. J. P. Timmermans (Ed.), *Progress in Activity-Based Analysis,* Amsterdam, 161-174.

Janssens, D., Wets, G., Timmermans, H. J. P., & Arentze, T. A. (2007). Modeling short-term dynamics in activity-travel patterns: the Feathers model. *In Proceedings of WCTR Conference*, Berkeley, (CD-ROM: 24 pp.).

Joh, C.-H., Arentze, T. A., & Timmermans, H. J. P. (2006). Measuring and predicting adaptation behavior in multi-dimensional activity-travel patterns. *Transportmetrica, 2*, 153-173.

Payne, J. W., Laughunn, D. J., & Crum, R. (1980). Translation of gambles and aspiration level effects on risky choice behavior. *Management Science, 26*(10), 1039-1060.

Pellegrini, P. A., Fotheringham, S., & Lin, G. (1997). An empirical evaluation of parameter sensitivity to choice set definition in shopping destination choice models. *Papers in Regional Science, 76*, 257-284.

Rindsfuser, G., & Klugl, F. (2005). The scheduling agent using Sesam to implement a generator of activity programs. In H. J. P. Timmermans (Ed.), *Progress in Activity-Based Analysis, Amsterdam,* (pp. 115-137).

Rosetti, R. J. F., & Liu, R. (2004). A dynamic network simulation model based on multi-agent systems. *In Proceeding of the 3rd Workshop of Agents in Traffic and Transportation*, (pp. 88-93), AAMAS, New York.

Shocker, A. D., Ben-Akiva, M. E., Boccara, B., & Nedugadi, P. (1991). Consideration set influences on consumer decision-making and choice: issues, models and suggestions. *Marketing Letters, 2,* 181-197.

Sutton, R. S., & Barto, A. G. (1998). *Reinforcement Learning: An Introduction*. London: MIT Press.

Swait, J. (2001). Choice set generation with the generalized extreme value family of discrete choice models. *Transportation Research, B, 35*, 643-666.

Swait, J., & Ben-Akiva, M. E. (1987). Incorporating random constraints in discrete models of choice set generation. *Transportation Research, B, 21*, 91-102.

Thill, J. C., & Horowitz, J. L. (1997). Modeling non-work destination choices with choice sets defined by travel-time constraints. In M. M. Fisher & A. Getis (Eds.), *Recent Developments in Spatial Analysis-, -Spatial Statistics, Behaviour Modelling and Neuro-computing* (pp. 186-208). Heidelberg: Springer.

Timmermans, H. J. P., Arentze, T. A., & Joh, C. H. (2002). Analyzing space-time behavior: new approaches to old problems. *Progress in Human Geography, 26*(2), 175-190.

Timmermans, H. J. P., & Golledge, R. G. (1990). Applications of behavioral research on spatial problems II: preference and choice. *Progress in Human Geography, 14,* 311-354.

West, P., & Broniarczyk, S. (1998). Integrating multiple opinions: the role of aspiration level on consumer response to critic consensus. *Journal of Consumer Research, 25*(1), 38-51.

Waerden, P. J .H. J., Borgers, A. W. J., & Timmermans, H. J. P. (2003). The influence of key events and critical incidents on transportation mode choice switching behavior: a descriptive analysis. *In Proceedings of the IATBR Conference,* Lucerne, (CD-Rom: 24pp.).

ADDITIONAL READING

Aoki, M. (1995). Economic fluctuations with interactive agents: dynamic and stochastic externalities. *Japanese Economic Review, 46*, 148-165.

Arentze, T. A., & Timmermans, H. J. P. (2005). *A new theory of activity generation.* Paper presented at TRB Annual Meeting.

Axhausen, K. (2004). *Social networks and travel: some hypothesizes.* Paper presented at Conference on Progress in Activity-Based Analysis, Maastricht, The Netherlands.

Axhausen, K. (2005). *Activity spaces, biographies, social networks and their welfare gains and externalities: Some hypotheses and empirical results.* Paper for the PROCESSUS Colloquium, Toronto.

Axhausen, K., Dimitrakopoulou, E., & Dimitripoulos, I. (1995). Adapting to change: some evidence from a simple learning model. *Proceedings PTRC P392*, 191-203.

Beckman, R. J., Baggerly, K. A., & McKay, M. D. (1996). Creating synthetic baseline populations. *Transportation Research A, 30*, 415-429.

Ben-Akiva, M. E., De Palma, A., & Kaysi, I. (1991). Dynamic network models and driver information systems. *Transportation Research 25A*, 251-266.

Brock, W., & Durlauf, S. (2001). Interactions-based models. In J. Heckman & E. Leamer (Eds.), *Handbook of Econometrics V.* North Holland, Amsterdam.

Brock, W., & Durlauf, S. (2002). A multinomial choice model of neighborhood effects. *American Economic Review, 92*, 298-303.

Brock, W., & Durlauf, S. (2003). A multinomial choice model with social interactions. In L. Blume & S. Durlauf (Eds.), *The Economy as an Evolving Complex System III.* Oxford University Press.

Blume, L., & Durlauf, S. (2002). *Equilibrium Concepts for Social Interaction Models*, Working papers 7, Wisconsin Madison - Social Systems.

Breiman, L., Friedman, J. H., Olshen, R. A., & Stone, C. J. (1984). *Classification and regression trees*. Belmont, CA: Wadsworth.

Carrasco, J. A., & Miller, A. J. (2005). *Socializing with people and not places: modeling social activities explicitly incorporating social networks*. Paper presented at CUPUM 05, London, UK.

Dugundji, E., & Gulyas, L. (2005). *Socio-dynamic discrete choice on networks in space: impacts of agent heterogeneity on emergent outcomes*. Paper presented at CUPUM 05, London, UK.

Dugundji, E., & Walker, J. (2005). Discrete choice with social and spatial network interdependencies, an empirical example using mixed GEV models with field and "panel" effects. *Proceedings TRB Meeting*, Washington, D.C.

Emmerink, R. H. M. (1996). *Information and Pricing in Road Transport*. PhD dissertation, Tinbergen Institute Research Series, Vrije Universiteit, Amsterdam.

Fujii, S., & Kitamura, R. (2000). *Anticipated travel time, information acquisition and actual experience: The case of Hanshin Expressway Route Closure*. Paper presented at the 79th Annual Meeting of the Transportation Research Board, Washington, DC, USA.

Hackney, J. (2005). *Coevolving social and transportation networks*. Presentation Workshop Frontiers in Transportation: Social and Spatial Interactions, Amsterdam, The Netherlands.

Horowitz, A.J. (1984). The stability of stochastic equilibrium in a two-link transportation network. *Transportation Research, 18B*, 13-28.

Iida, Y., Akiyama, T., & Uchida, T. (1992). Experimental analysis of dynamic route choice behavior. *Transportation Research, 26B*, 17-32.

Kass, G. V. (1980). An exploratory technique for investigating large quantities of categorical data. *Applied Statistics, 29*, 119-127.

Mahmassani, H. S., & Chang, G. (1986). Experiments with departure time dynamics of urban commuters. *Transportation Research, 20B*, 297-320.

Nakayama, S., Kitamura, R., & Fujii, S. (1999). Drivers' learning and network behavior: A dynamic analysis of the driver-network system as a complex system. *Transportation Research Record 1676*, 30-36.

Nakayama, S., & Kitamura, R. (2000). *A route choice model with inductive learning*. Paper presented at the 79th Annual Meeting of Transportation Research Board, Washington DC, USA.

Nakayama, S., Kitamura, R., & Fujii, S. (2000). *Drivers' route choice heuristics and network behavior: a simulation study using genetic algorithms*. Paper presented at the IATBR Meetings, Gold Coast, Australia.

Osbay, K., Dattu, A., & Kachroo, P. (2001). *Modelling route choice behavior using stochastic learning automata*. Paper presented at the 80th Annual Meeting of the Transportation Research Board, Washington DC, USA.

Paez, A., & Scott, D. M. (2005). *Social influence on travel behavior: A simulation example of the decision to telecommute.* To appear in Environment and Planning A.

Polak, J., & Hazelton, M. (1998). The influence of alternative traveller learning mechanisms on the dynamics of transport systems. *Transportation Planning Methods, 1*, 83-95.

Polak, J., & Oladeinde, F. (2000). *An empirical model of travellers' day-to-day learning in the presence of uncertain travel times.* Unpublished manuscript, Imperial College of Science Technology and Medicine.

Swanen, T. (2005). *Managing uncertain arrival times through sociomaterial networks.* Presentation Workshop Frontiers in Transportation: Social and Spatial Interactions, Amsterdam.

Timmermans, H. J. P., Arentze, T. A., & Joh, C-H. (2000). Modeling learning and evolutionary adaptation processes in activity settings: Theory and numerical simulations. *Transportation Research Record, 1718*, 27-33.

Tversky, A. (1972). Elimination by aspects: a theory of choice. *Psychological Review, 79,* 281-299.

Van Berkum, E., & Van der Mede, P. (1993). *The Impact of Traffic Information: Dynamics in Route and Departure Choice.* PhD dissertation, Delft University of Technology.

Chapter III
MATSim-T: Architecture and Simulation Times

Michael Balmer
IVT, ETH Zürich, Switzerland

Marcel Rieser
VSP, TU Berlin, Germany

Konrad Meister
IVT, ETH Zürich, Switzerland

David Charypar
IVT, ETH Zürich, Switzerland

Nicolas Lefebvre
IVT, ETH Zürich, Switzerland

Kai Nagel
VSP, TU Berlin, Germany

ABSTRACT

Micro-simulations for transport planning are becoming increasingly important in traffic simulation, traffic analysis, and traffic forecasting. In the last decades the shift from using typically aggregated data to more detailed, individual based, complex data (e.g. GPS tracking) and the continuously growing computer performance on fixed price level leads to the possibility of using microscopic models for large scale planning regions. This chapter presents such a micro-simulation. The work is part of the research project MATSim (Multi Agent Transport Simulation, http://matsim.org). In the chapter here the focus lies on design and implementation issues as well as on computational performance of different parts of the system. Based on a study of Swiss daily traffic – ca. 2.3 million individuals using motorized individual transport producing about 7.1 million trips, assigned to a Swiss network model with about 60,000 links, simulated and optimized completely time-dynamic for a complete workday – it is shown that the system is able to generate those traffic patterns in about 36 hours computation time.

1. INTRODUCTION

By tradition, transport planning simulation models tend to be macroscopic or mesoscopic (e.g. de Palma & Marchal, 2002; PTV, 2008). Reasons for this are access to aggregated data only (e.g. traffic counts, commuter matrices, etc.) and limitations in computational hardware to calculate and store detailed computations. These limitations have changed in the last few decades. Performance of computer hardware has been continually growing—and still grows—while the cost of machines stays fixed. For transport planning, the relevant developments in computer hardware are:

- The capacities of fast random access memory (RAM) have increased dramatically.
- Multi processor hardware allows one to perform parallel computation without using (maintenance intensive) computer clusters.
- Shared memory architectures allow fast on-demand access to the physical memory for an arbitrary amount of processes.

In the same manner, the available data used in transport planning are getting more detailed and complex. Good examples for that are person diary surveys (e.g. Hanson & Burnett, 1982; Axhausen et al, 2002; Schönfelder et al, 2002), and analysis of individual transport behavior based on GPS data (e.g. Wolf et al, 2004).

Therefore, the demands made to transport planning software are getting more complex, too. Micro-simulation is becoming increasingly important in traffic simulation, traffic analysis, and traffic forecasting. Some advantages over conventional models are:

- Computational savings when compared to the calculation and storage of large multidimensional probability arrays necessary in other methods.
- Larger range of output options, from overall statistics to information about each synthetic traveler in the simulation.
- Explicit modeling of the individuals' decision-making processes.

The last point is important since it is not a vehicle that produces traffic; it is the person that drives the vehicle. Persons do not just produce traffic; instead they try to manage their day (week, life) in a satisfying way. They go to work to earn money, they go hiking for their health and pleasure, they visit their relatives for pleasure or because they feel obliged to do so, they shop to cook a nice dinner at home, and so on. Since not all of this can be done at the same location, they travel, which produces traffic. To plan an efficient day, many decisions have to be made by each person. They decide where to perform activities, which mode to choose to get from one location to another, in which order and at which time activities should be performed, with whom to perform certain activities, and so on. Some decisions are made hours (days, months) in advance while others are made spontaneously as reactions to specific circumstances. Furthermore, many decisions induce other decisions. Therefore, it is important to model the complete time horizon of the decision makers.

Transport simulation models should be able to implement (at least part of) such an individual decision horizon and assign the outcome to a traffic model, since it is the complete daily schedules (and the decisions behind that) that produce traffic. This chapter presents such a micro-simulation, called MATSim-T (Multi Agent Transport Simulation Toolkit), implemented as a Java application, usable on any operating system. The work is part of the research project MATSim (http://matsim.org). In the chapter here the focus lies on design and implementation issues as well as on computational performance of different parts of the system. On the basis of integrated (daily) individual demand optimization in MATSim-T, the system is extended such that it provides flexible handling of a large variety of input data; extensibility of models and algorithms; a simple interface for new models and algorithms; (dis)aggregation for different spatial resolutions; robust interfaces to third party models, programs, and frameworks; unlimited number of individuals; and an easily usable interface to handle new input data elements.

This chapter lays the focus on the modules. It analyzes how specific modules affect the functionality of the toolkit as well as how they affect the overall computational speed of the complete system. The chapter starts with an overview of MATSim-T and related work (Sec. 2). This is followed by sections about the modules of the iterative part of MATSim-T: the traffic flow simulation (Sec. 3), the scoring and plans selection modules (Sec. 4), the re-planning (Sec. 5), and finally a comprehensive view at the whole iterative process (Sec. 6). Sec. 7 sketches the computational demand of the initial demand generation. Although that process runs at the beginning of the study, some aspects of it are easier to explain after the iterative part of MATSim is laid out. The chapter closes with an overview of the current development processes which will enhance the system in size, speed and functionality.

2. OVERVIEW

2.1 MATSim-T

The term *multi agent micro-simulation* is used with different meanings in transport research. Often, the word "microscopic" is used to describe a "car following model" (e.g. Wiedemann, 1974) that is also used in some commercial products (e.g. VISSIM: PTV, 2008). In MATSim, the term is used to describe that each modeled person contains its completely individual settings. Each person is modeled as an *agent*, and the sum of all agents should reflect the statistically representative demographics of the region. The demand is modeled and optimized individually for each agent—not only for some parts of the demand like departure-time and route choice, but as a complete *temporal dynamic* description of the daily demand of each agent.

The demand of an agent is called a *plan* in MATSim. Figure 1 shows an example of one agent's daily plan, written in XML (W3C, 2008). This structure stays the same during all modeling and simulation of the demand. In particular, the assignment of the traffic demand does not only take single trips into account, but the complete daily plans—including the activities—are executed. Thus the term *micro-simulation* relates to the microscopic (individual) demand of each person in the scenario.

To produce individual plans for each agent with MATSim as shown in Figure 1 it is necessary to provide the user interfaces such that he/she is able to generalize and fuse the data available for the region of interest, so that a general dataset of the infrastructure, the population, and the demand can be created. To structure the process of demand creation/optimization, MATSim-T can be split up into four parts as shown in Figure 2:

- Scenario creation process
- Initial individual demand modeling process
- Iterative demand optimization process (including demand execution, scoring, and replanning)
- Post-process analysis

Since MATSim-T is a modular approach, all parts shown in Figure 1 (FUSION, IIDM, EXEC, SCORING and REPLANNING) are given as interfaces such that users are able to plug in their own modules.

The first two processes rely on the available data of the region of interest. Since the quality, quantity, and resolution of data can vary a lot from one scenario to another, the scenario creation and the initial

Figure 1. Description of the demand of a synthetic person (including demographic data) for a complete day. The agent with ID 393241 plans to leave home—located on link 58—to travel to his work place. He uses a route leading along 4 nodes (5 links) with an expected travel time of 25 minutes. The agent stays at work for 9 hours, then travels back home with an expected travel time of 14 minutes. The demand does not only describe single parts of the day, but the complete sequence for agent 393241 continually in time (Source: Balmer, 2007)

```
<plans name="example plans file">
        ...
        <person id="393241" sex="f" age="27" license="yes" car_avail="always"
                        employed="yes">

         <travelcard type="regional-abo" />

         <plan>

          <act type="home" link="58" start_time="00:00" dur="07:00" end_time="07:00" />

          <leg mode="car" dept_time="07:00" trav_time="00:25" arr_time="07:25">

            <route>1932 1933 1934 1947</route>

          </leg>

          <act type="work" link="844" start_time="07:25" dur="09:00" end_time="16:25"/>

          <leg mode="car" dept_time="16:25" trav_time="00:14" arr_time="16:39">

            <route>1934 1933</route>

          </leg>

          <act type="home" link="58" start_time="16:39" dur="07:21" end_time="24:00" />

         </plan>
        </person>
        ...
</plans>
```

demand modeling process steps can vary as well. MATSim-T therefore provides in its core only the resulting data representation of the infrastructure (network and facilities) and the population including each person's individual demand, plus parsers and writers for the XML data representation.

To clarify the functionality of a FUSION or IIDM module, here is an example: Let us assume that land-use information about the region of interest is given based on the resolution of municipalities, and that the number of work places is given for each municipality. The user implements a MATSim-FUSION module that parses this information and creates one facility (including the number of workplaces) per municipality. This gives a rough approximation of the existing work facilities and work places in the region. Let us now assume that at a later stage of the project the user has access to detailed buildings data including work facilities. The system allows one to add another module that replaces the already created work facilities with the new information and distributes the number of work places to the more realistic work facilities of that region.

While the resulting facilities of *both* situations are suitable for MATSim-T to start the third step of the overall process (demand optimization; Figure 2), the second version of the facilities delivers more detailed results. Even though the two described modules are implemented for a specific scenario, they can be part of the MATSim toolkit and therefore, another user with the same needs is able to reuse the modules for his/her own scenario.

The post-process analysis part of MATSim-T (fourth part of Figure 2) works in the same way, with the difference that now the input data follows MATSim standards (MATSim XML formats of the network, facilities, population and demand) and therefore is useable for any given scenario.

Figure 2. Process structure of MATSim-T

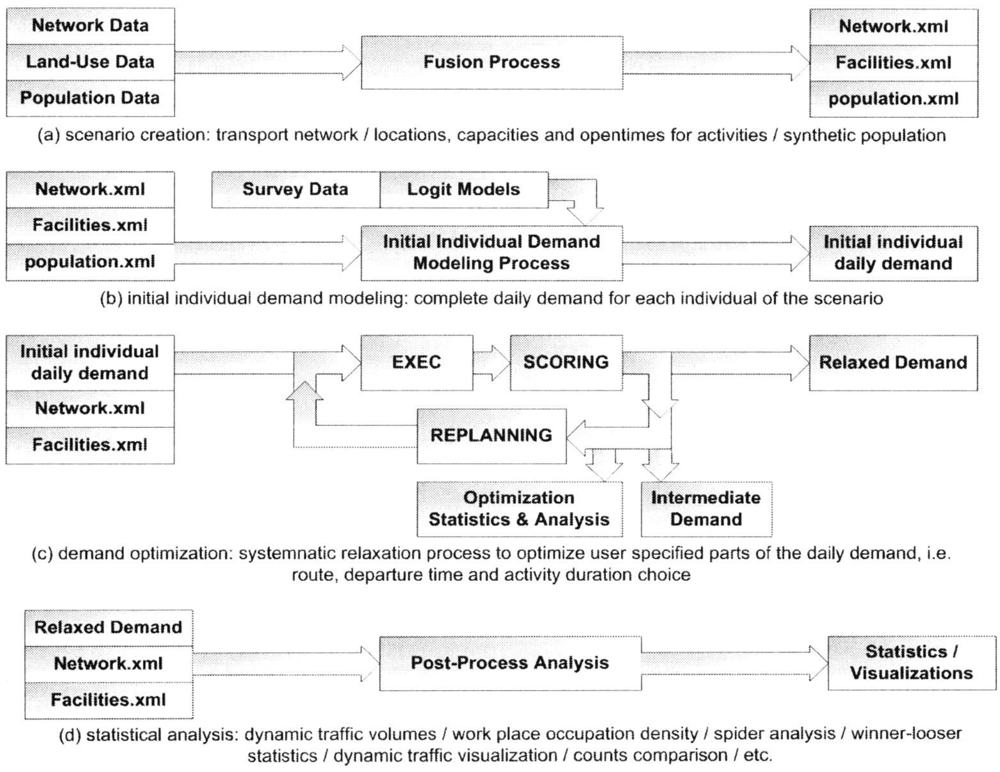

The *iterative demand optimization* process (third part of Figure 2) is, in a way, the core of MATSim-T. While all other steps are run once in a sequential order defined by the user, part three optimizes the demand for each individual synthetic traveler in the scenario such that they respect the constraints (network, facilities) of the scenario and the interaction with all the other actors of that region.

Usually, a method of relaxation is used to find an equilibrium state. For route choice the Wardrop equilibrium (Wardrop, 1952) describes such a relaxed state. But importantly, not only the routes are optimized in MATSim-T. Instead, the complete daily plan—including routes, times, locations, sequence of activities, activity types, and so on—of each agent is optimized. Each agent tries to execute its day with highest possible utility. The utility of a daily plan depends on infrastructural constraints (capacity of streets, opening times of shops, etc.) and on the daily plans of the other agents in the system. This implies that the effective utility of a daily plan can only be determined by the interaction of all agents. This is the place where *co-evolutionary algorithms* (Holland, 1992; Palmer et al, 1994) come into play. An evolutionary algorithm basically consists of the following steps:

1. `Initialize P(t=0)` — Create the population of individuals at time t=0
2. `Score P(t)` — Calculate the "fitness"
3. `Select P'(t) out of P(t)` — "survival of the fittest"
4. `Recombine and mutate P'(t)` — "crossover" and "mutation"
5. `P(t+1) = P'(t); t = t+1` — the next generation of individuals
6. `GOTO item 2`

Applied to the demand optimization (optimization of daily plans) in MATSim, this means:

1. Initialize / generate the daily plans for each agent in the system
2. Calculate the utility of the execution of the individual daily plans for each agent
3. Delete "bad" daily plans (the ones with a low utility)
4. Duplicate and modify daily plans
5. Make those plans the relevant plans for the next iteration; increase the iteration counter by one
6. Goto 2.

It is important to note that the "individuals" of the evolutionary algorithms are the plans, while the synthetic travelers are the entities that *co*-evolve.

Figure 2(c) shows this optimization loop. For each of the steps listed above, specific modules are available. The execution of the daily plans (EXEC) is handled by a corresponding *traffic flow simulation* module, in which the individuals interact with each other, i.e. individuals may generate congestion on streets of high usage. The SCORING module calculates the utility of all the executed daily plans. Plans with a high utility (high "fitness") survive, while plans with a low utility (e.g. caused by long travel times because of traffic jams) are eventually deleted.

The creation and variation of daily plans (REPLANNING) is distributed among different modules that are specialized on varying specific aspects of daily plans. The modifications in the plan of a single agent are completely independent on the re-planning of all the other agents' plans.

2.2 Related Work

Many models have implemented the concept of activity based demand generation (e.g. "VISEM": PTV, 2008; Vovsha et al, 2002; Bowman et al, 1999, Bhat et al, 2004; Pendyala, 2004; Arentze et al, 2000). But the results are typically delivered as (time-dependent) origin-destination matrices, which are used as input for static or dynamic traffic assignment models. Completely agent-based micro simulations (e.g. "mobiTopp": Schnittger & Zumkeller, 2004) are typically focused on telematics aspect or on effects of changes in infrastructure. Event driven simulations for transport planning (e.g. Axhausen, 1988; Balmer & Nagel, 2006; Axhausen & Herz, 1989) already presented the powers of micro-simulations, but they usually only work on small scenarios.

The work most related to the MATSim project is TRANSIMS (2008), which also generates individual activity schedules for large-scale scenarios. While the concepts are similar, there are some important differences. The most important differences are:

- MATSim is consistently constructed around the notion that travelers (and possibly other objects of the simulation, such as traffic lights) are "agents", which means that all information for the agent should always kept together in the simulation at one place. In this way, an agent in MATSim can access demographic characteristics or time pressure while he/she is moving around in the transport system. In TRANSIMS, such information is in principle available, but fragmented between many modules and many files.
- As a mirror of the coherent agent information, MATSim uses the hierarchical XML (W3C, 2006) format for the input or output of agent information. Because the file format is hierarchical, it can be filled out with different levels of detail. This means that in *all* places where agent information is exchanged between modules, the same file format is used. This has two important consequences: (i) *Arbitrary* modules can be combined to fill out the agent information. In TRANSIMS, the capabilities of the modules are given implicitly by the file formats. (ii) One DTD (Document Type Definition, see W3C, 2006) is sufficient to ensure correctness of all agent data files.
- As a consequence of the agent design, it is easy to maintain several plans per agent. This facilitates to interpret the iterative part of MATSim as a co-evolutionary algorithm, where every agent draws on a population of plans in order to find better solutions for him-/herself. Once more, this could be emulated in TRANSIMS, but it would be considerably more difficult to implement it, and in some sense the only option may be to add something similar to the MATSim agent database (Raney & Nagel, 2004) to TRANSIMS.
- The traffic flow simulations currently used in MATSim-T are simpler than that in TRANSIMS, and as a result run considerably faster, thus allowing meaningful runs in days instead of weeks. This is not really a conceptual difference, but it was an important design decision when starting MATSim: Iterations should essentially run over night.

Agent-based micro-simulation applications can also be found in related research fields to transport planning. Promising concepts in urban planning are land-use simulations, i.e. URBANSIM (Waddell et al, 2003), ILUTE (Salvini et al, 2005) or the models of Abraham/Hunt (Hunt et al, 2000).

2.3 Case Study ("all-of-Switzerland")

From a user point of view, it is of high interest how much time a simulation program needs to spend until results are produced. This chapter will present the performance measures of the toolkit on a typical large-scale transport planning study. Meister et al (2008) present the first results for the daily traffic for the whole of Switzerland created with MATSim-T. That case study will be used to present the computational performance of each part of the toolkit. The extents of the Swiss daily traffic demand study are:

- The national planning network ("Nationales Netzmodell": Vrtic et al, 2003) consists of ~24'000 nodes and ~60'000 links.
- Based on the enterprise census 2000 (SFSO, 2001) and the census 2000 (SFSO, 2000), ca. 1.7 million facilities are modeled. Up to five different activities ("home", "work", "education", "shop" and "leisure" activity) are assigned to each facility.
- With the census 2000 and the microcensus 2005 (SFSO, 2006), about 7 million synthetic persons (agents) are generated, incl. demographic attributes like age, gender, car license ownership, car availability, public transport ticket ownership and employed status.
- The generation of the initial, individual, time-dependent daily demand is described in detail in Ciari et al (2007) and Meister et al (2008). Overall, about 22 million trips are generated—about 7.1 million trips for motorized individual transport.
- The performance measures are produced on a machine with 8 dual-core processors with 2.2 GHz clock rate each.
- The case study needs about 22 GByte of RAM.

2.4 Computing Times

This chapter concentrates on the MATSim architecture and the resulting computing times. The above scenario is close to the largest that is currently feasible. Since it is possible to obtain plausible results with runs with 10% of the population, this means that scenarios up to 70 million people can currently be addressed. If hardware keeps improving in similar ways as in the past, simulating even large mega-cities or "all-of-Europe" seems within reach.

Computing times are given with respect to that specific scenario. Unfortunately, it has turned out consistently that finding simple predictive rules for the computational performance of MATSim is quite difficult (Nagel & Rickert, 2001; Cetin et al, 2003; Cetin, 2005). This has to do with the fact that interwoven aspects of hardware, implementation, scenario details, and scenario size play a role. For example, hardware, implementation and scenario size together determine how much of a scenario fits into cache or memory, and if the computation is I/O- or CPU-bound. Scenario details decide, say, during how much of the simulated time there is activity in all parts of the system (as opposed to activity on a small number of links). It might be possible to give worst-case complexities. These, however, in our experience are completely unrelated to the actual computing times. This chapter rather gives computing times for a specific scenario, plus information on how these times change when important aspects, such as the number of travelers or the number of network elements, change.

3. TRAFFIC FLOW SIMULATION

Despite considerable work over more than the last decade (e.g. Nagel & Schleicher, 1994; Nagel & Rickert, 2001; Gawron, 1998; Cetin et al, 2003; Charypar et al, 2007), the traffic flow simulation remains the module with the largest computing requirements for the problem at hand. The traffic flow simulation is responsible for executing the daily plans in a physical environment. In principle, arbitrary models could be used, e.g. the model by Wiedemann (1974) or a cellular automata model (e.g. Nagel & Schreckenberg, 1992), but both require still too large amounts of computing power. Transport planning is not so much interested in the detailed driving behavior, but in the dynamic amount of traffic, traffic that reflects traffic jams, tailbacks, the dissolving of traffic jams, etc. The queue model (Gawron, 1998) fulfills all these requirements. Every street is modeled as a queue in which vehicles have to wait for at least the free speed travel time on that street. In addition, both the flow and the storage capacity of each link are limited. The former causes congestion, the latter causes spillback since links can become full and then upstream links also become jammed.

The traffic flow simulations produce information about where each agent is at a specific time of the day and what it is doing at that time. Each agent generates for each of its actions (begin/end of an activity, entering or leaving a link, etc.) a temporally and spatially localized *event*.

3.1 Default Traffic Flow Simulation

The current default traffic flow simulation of MATSim-T is a single CPU Java re-implementation of the micro-simulation described by Cetin (2005). As an integral part of the toolkit it has the advantage that it can directly access all the data in the MATSim object database, saving time-consuming input and output of data. Because of the platform independence of Java, it runs on all major operation systems.

The default traffic flow simulation uses seconds as smallest entity of time. For each simulated second, all queues (all links of the network) synchronously get a new state assigned. As a result, the runtime is proportional to the number of links in the scenario:

$$T_{mobsim} \propto t_{sim} N_{links} / \Delta t$$

where t_{sim} is the real time window to be simulated (usually 1 day = 86,400 seconds), Δt the size of the time step (1 second), and N_{links} the number of links in the street network. There is, however, also some overhead to generate events, which depends on the number of agents in the system. Performing the Mobility Simulation on a 2.2 GHz processor, the computation time to simulate one day of the complete vehicular traffic of Switzerland (see above) takes about 70 minutes (ca. 20.5 times faster than real time).

The simulation performance in a naïve implementation of the queue model does not depend very much on the number of agents, respectively on their demand: Every link is processed once in every time step. This is acceptable for situations where all network elements are in use (e.g. morning rush hour), but the simulation will take just as long calculating low traffic during nightly hours. The current implementation in MATSim-T, however, switches off links that are completely empty, saving additional computing time but making it now even more dependent on the number of agents and their demand structure.

3.2 Deterministic, Event-Based Queue-Simulation (DEQSim)

DEQSim, an alternative traffic flow simulation, extends the queue-model. In addition to the FIFO (First-In, First-Out) behavior of the queue model, this traffic flow simulation imitates backwards-traveling gaps produced by vehicles that leave congestion. This leads to more realistic dynamics of congested links. Also the implementation differs. Rather than updating all links every second, it only operates whenever a link actually changes its state. Despite the improved dynamics, such state changes are rare. In a pure queue model, the state of a link only changes when a vehicle enters or a vehicle leaves, and since the earliest possible leaving time is known for every vehicle, the link can be processed at exactly those times. It was possible to add the improved dynamics in a similar way, by adding "holes" that travel backwards, and that have, in consequence, also pre-computed times of when they arrive at the upstream end of the link. Therefore, computing time is only used when agents produce *events* on links. As a side effect, the simulation does not have to stick to discrete time steps anymore. A detailed description of the DEQSim can be found in Charypar et al (2007). The performance is

$$T_{deqsim} \propto e(a, N_{links})$$

where the number of events e is proportional to the number of agents (a), respectively the number of executed plans, and depends on the street network (number of links N_{links}). On a high-resolution network of the same region, an agent's route contains more links than on a low-resolution network, thus generating more events.

For the case study described above, 162 million events are generated. The total computing time for the single-CPU implementation of the DEQSim takes about 50 minutes (real time ratio =~ 28). Additionally, the DEQSim also runs in parallel using multiple CPUs with distributed memory. The performance scales nearly linearly with the number of processors, implying that in 8 CPUs, the 50 minutes are reduced to less than 7 minutes.

In contrast to the default traffic flow simulation, DEQSim is written in the C++ programming language. This prevents the direct access to the data from the traffic simulation. Instead, the data needed to run the DEQSim is first written to disk and later read by DEQSim. Similarly, DEQSim writes its events to a file on disk, from where the MATSim toolkit reads them after DEQSim has finished. This file input and output (including the processing of the read in events) requires an additional 20 minutes in the given case study. The input is proportional to the number of agents, the output once more proportional to the number of the events. Maybe somewhat surprisingly, the main overhead does not stem from the physical disk I/O, but from the handling of the events while they are processed inside MATSim.

4. SCORING AND PLANS SELECTION

The events produced by the traffic flow simulation make it possible to calculate the effective utility of each daily plan, including the influences and effects of the interaction of other agents. The "success" of a daily plan is specified by an individual utility function. This function describes the goals of each agent, and with that its behavior. In principle, any arbitrary utility function could be used, for example one coming from *prospect theory* (Avineri & Prashker, 2003). MATSim currently uses a simple but effective utility function described in Charypar and Nagel (2005). It is related to the Vickrey bottleneck model (Vickrey, 1969; Arnott et al, 1993), but is modified in order to be consistent with the approach based on complete daily plans (Charypar & Nagel, 2005; Raney & Nagel, 2006).

Without going into detail, the elements of the utility function are:

- A positive contribution for the (usually) positive utility earned by performing an activity.
- A negative contribution (penalty) for traveling.
- A negative contribution for being late.

Intuitively, being early should also be punished, but it turns out that this is not necessary since "doing nothing" is already indirectly punished by the fact that something with a positive utility could be done instead in a better plan.

The utility function induces the behavior of the agent, because the agent searches in the solution space of the utility function for the best possible score, which implies the best possible daily plan. The agent cannot optimize outside of the solution space. This aspect is documented in more detail later.

Scores are computed in two ways, depending on the type of the traffic flow simulation:

- In the case of the integrated (default) traffic flow simulation, scores are computed when events from the traffic flow simulation reach the scoring module. The computational effort to compute the scores is smaller than the overhead caused by the events handling mechanism. Any effort to accelerate the computation at this end would need to accelerate the events handling mechanism first.
- In the case of the external DEQSim traffic flow simulation, scores are computed when Java events that are generated from the events file reach the scoring module. This ends up being the same problem: The main computational effort is caused by the events handling mechanism.

There is, thus, a computational cost of the events handling mechanism, that is either hidden in the default traffic flow simulation, or in the file I/O when the events are read from file. This may be an element of future improvements.

A small, but important step in the whole process is the deletion of a "bad" plan. As there are new plans generated in each iteration for a subset of all agents, the population of plans per agent increases up to a user-defined maximum (typically between 3 to 6 plans per agent). Before a new plan can be created for an agent that already has as many plans as the maximum defines, the worst plan (the one with the lowest score) is deleted from the population. As a consequence, only "good" plans survive. This step takes about 10 seconds for the all-of-Switzerland study.

5. PLANS VARIATION (RE-PLANNING)

The re-planning is responsible for making sure that every agent explores its solution space. This happens by duplicating an existing plan of an agent, varying (mutating) the copy, and executing and scoring it in the next iteration. Each re-planning module takes charge for a specific part in the optimization process. As an example, the *Router* module calculates the routes of a plan based on the amount of traffic from the last traffic flow simulation. The Time Allocation Mutator module modifies departure times and activity durations of a daily plan. This module varies the corresponding times randomly. Additional modules could change activities' locations, or change the sequence of activities. An important fact is that all these modules work *independently* from each other. This allows one to add an arbitrary number of re-planning modules to the optimization process.

A characterization of modules is whether they modify a plan randomly (*Random Mutation*) or whether they search for the best solution based on the results of the last traffic flow simulation (*Best Response*). The former has the advantage not to use any significant amount of computing power. Additionally, it searches—sooner or later—over the complete search domain. The disadvantage is that such modules require (too) many iterations until the optimization relaxes. Best Response modules on the other hand help to relax the system much faster, but they are usually more complex and computationally intensive.

5.1 Time Allocation Mutator

The Time Allocation Mutator is a typical example of a *Random Mutation* module. It varies randomly the departure times and durations of activities in a daily plan. The Time Allocation Mutator needs about 2 seconds to process the 10% re-planning agents (approx. 220 000 agents) per iteration.

5.2 Router Module

The Router Module calculates the best routes in a daily plan, given the departure times for each leg and the dynamic travel times of *all* streets (based on the last traffic flow simulation). The best route is defined as the one with the least negative utility. This *Best Response* module uses the complete and dynamic traffic load of the system for finding routes.

Currently, MATSim has three different implementations of the Router module. They are all based on a time-dynamic variant of Dijkstra's algorithm for finding shortest paths in networks, and they return all the identical results. Our newest implementation, the Landmarks-A* module (Lefebvre & Balmer, 2007), gives the best performance in average: For the given case study it needs in about 0.1 milliseconds to calculate one route in average. For the 7.1 million (motorized) routes of the "all-of-Switzerland" scenario and 10% route replanning rate this implies 71 seconds of computing time per iteration, which can, however, be shared between parallel CPUs.

Additional computational results for different routing algorithms and different networks sizes can be found in the paper by Lefebvre & Balmer (2007). Unfortunately, those results are not sufficient to make a prediction about the functional form of the average complexity of the Landmarks-A* implementation; the most probable fit may be $O(n^2)$ where n is the number of network nodes. It also plays a role that the Java implementation of the priority queue does not offer a fast "decrease-key" operation.

5.3 "planomat"

Another *Best Response* module available in MATSim is *planomat*, described in full detail in Meister (2004). This module not only optimizes one aspect of a daily plan, but all parts at the same time. It bases its assumptions heavily on the outcome (*events*) of the last execution of the traffic flow simulation (see above). Additionally, it is able to coordinate the daily plans of members of the same household (e.g. a common dinner at home). This module is written in C++, but can be called from the MATSim toolkit.

The C++-planomat is a genetic algorithm (GA) with a special encoding for activity sequences, activity locations, activity times, and activity participation. The encoding was constructed with the idea that a plan that is "good in the morning" and another plan that is "good in the afternoon" should be able to combine into a plan that is "good overall". This takes some input from the GA coding of the traveling salesman problem. One instance of the GA generates the plan(s) for one person or one household. As is common, the GA is not a particularly fast method to solve the problem, but it is extremely flexible with respect to the inclusion of additional constraints, for example facility opening times.

A simplified version of the planomat—written in JAVA—is an integral part of the MATSim toolkit and optimizes the time schedules. It is therefore a substitute of the *Time Allocation Mutator*. planomat uses an evolutionary algorithm for the optimization of departure times and activity durations. It is therefore far more computationally intensive than the Time Allocation Mutator module. In the above described case study, it uses about 5.7 milliseconds in average for the best response calculation of timing of a single daily plan. For the ca. 2.3 million (motorized) plans of the "all-of-Switzerland" scenario and 10% *planomat* replanning rate this implies 1331 seconds of computing time per iteration, which can, however, be shared between parallel CPUs.

5.4 Additional Modules

It is important to recall at this point that MATSim-T is not limited to the modules described above. Any user can add his or her own modules; additional modules are also added by the developers. The computational performance of such modules will be assessed in due time when such modules have proven their value with respect to the transport simulation problem.

6. SYSTEMATIC RELAXATION OF THE EVOLUTIONARY ALGORITHM

According to the user's needs it is now possible to combine all the previously mentioned modules. The optimization process, i.e., the iterative processing of single tasks, is done by the toolkit. However, with respect to the combination of modules one aspect has to be considered: Each additional re-planning module enlarges the solution space for the agent's day-plan. It is required that this solution space is completely covered by the utility function. Consider the following example:

If an agent is only allowed to optimize its route it would be feasible to reduce the above described utility function to

$$U_{total} = \sum_i U_{travel,i}$$

since this agent would not be capable to alter its time allocation.

However, if one adds a time allocation module, and therefore enlarges the solution space, this has to be considered by the utility function. On the other hand, it is legitimate to use the extended utility function for agents that consider only route choice, since it covers the complete solution space. On this account, the further functional development of the optimization process in MATSim-T (implementation of new re-planning modules) goes hand in hand with the extension of the agents' behavioral models.

In the following the relaxation behavior and the required computational time of the co-evolutionary algorithm will be analyzed.

6.1 Setup

For the suitable analysis of the relaxation process a typical and in the last years frequently used setup is used:

1. Each agent is capable of route-choice.
2. Each agent is capable of time allocation choice.
3. In each iteration, a randomly selected sample of 10% of all agents creates a new plan by altering the routes of an existing plan.
4. In each iteration, a randomly selected sample of 10% of all agents creates a new plan by altering the time allocation of an existing plan.
5. The remaining 80% of agents select an existing plan for repeated execution. The selection probability corresponds to the logit function $p_i = \exp(\beta S_i)/\sum_j \exp(\beta S_j)$, where S_i denotes the utility of plan i and β is an empirically estimated constant.
6. The utility function corresponds to the one given in the previous section.
7. The number of plans per agent is limited to a maximum of four.
8. The system will be considered relaxed once the trajectory of average utility per iteration represents a stationary process.

A detailed description of this setup with values for the parameters of the utility function can be found in Meister et al (2008).

6.2 Relaxation

The relaxed state of the co-evolutionary algorithm of MATSim-T is reached if the utility for each agent does not noticeably change through variation of the day plans. Since bad plans do not "survive", the utility of all remaining plans levels off eventually. Figure 3 depicts such a behavior. The light grey curve represents the utility of the plan that has been executed in the corresponding iteration averaged over all agents. The black and the medium grey curve, respectively, denote the average utility of the currently available best and worst plan, respectively. One can realize that in this example the utility converges to the "relaxed" state after iteration 70, and exhibits only a mean variance of approx. 2 units in iteration 100.

More noticeable is the behavior before iteration 70, especially in iteration 15. Figure 3 shows that the average scores for the executed plans (light grey curve) as well as for the worst plans (medium grey curve) are remarkably low in iteration 15, which can be ascribed to a so-called "network breakdown". Due to the optimization process and the given constraints (such as the time window for the starting a work-activity) it is possible that a lot of agents simultaneously try out similar plans, which in turn leads to high traffic densities on preferred roads and therefore to highly congested situations. Due to this temporal overload, this congestion cannot be absorbed by the surrounding road network due to the overall high traffic density. Spillbacks build up and spread over a large part of the network. The model requires a long time to resolve such congestion, resulting in high travel times, and therefore in large disutility for traveling. Since the last executed plan exhibits a low utility after such a network breakdown most of the agents switch their plans. Thus the last optimization step is discarded and the usage of more diverse plans will be reinforced. In the paper of Rieser & Nagel (2008) the "network breakdown" situations are analyzed in more detail.

Due to the diversification regarding departure times and route choice, average trips travel times decrease (black dotted curve in Figure 3), which in turn becomes noticeable in the resulting greater average utilities.

It appears that after iteration 70 a combination of plans arises which results in a stable traffic pattern that is robust towards variations of single agents. Good plans are duplicated during re-planning and the duplicates are kept if they also turn out to be good. Bad plans are discarded, so that finally only good plans will remain which can be observed in Figure 3 with the approximation of the medium grey curve to the other two curves.

6.3 Computational Time for Optimization Process

The total computational time of a single relaxation process consists of the sub-processes as described in Figure 2(c). Additional time is required for storing temporal results and analysis (intermediate demand, statistics and analysis shown in Figure 2(c)). This latter feature can be switched off by the user, so that only the final result will be saved. However, this feature helps to analyze the optimization process and

Figure 3. Average utility (score) and average trip travel time per iteration

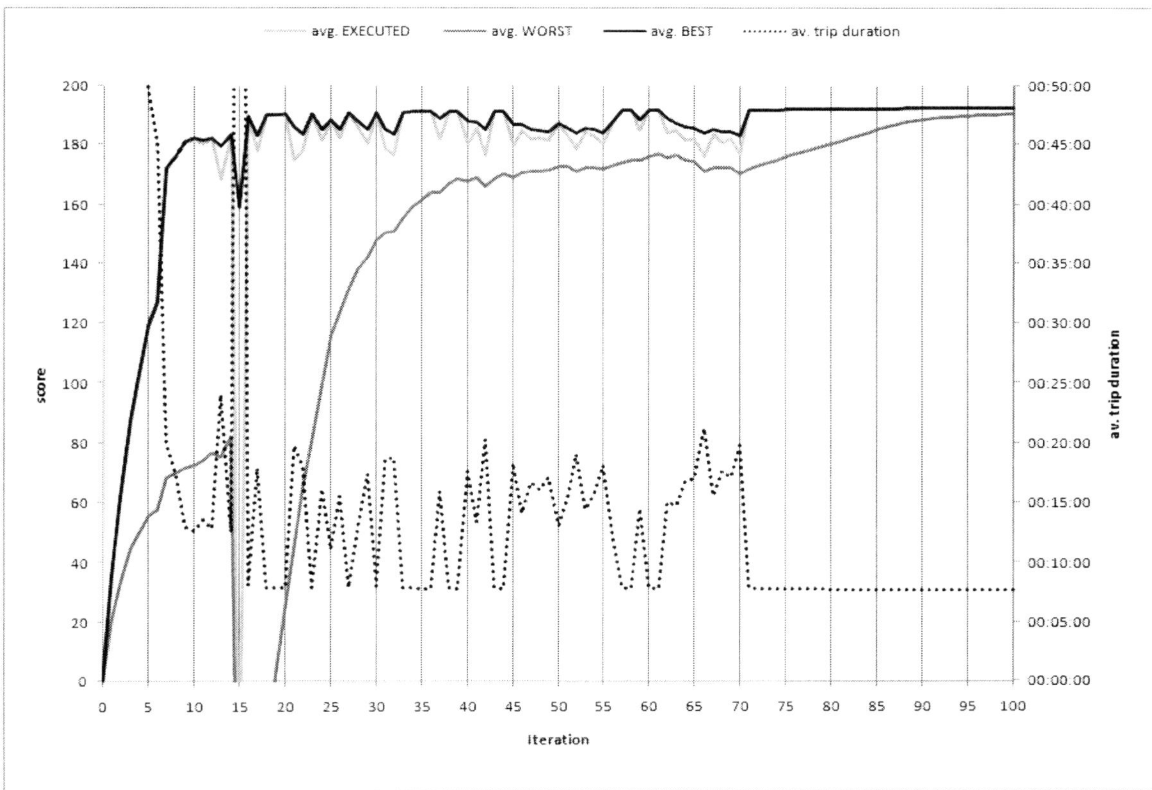

allows one to abort the process if needed. For that reason this part will not be excluded in the following discussion. In detail the process can be divided into the following chronological steps:

1. Initialization: Loading and managing of infrastructural data (network and facilities) and initial demand
2. Iteration 0: primary execution of the traffic flow simulation and calculating of utilities
3. Iteration 1 to *n*: re-planning, traffic flow simulation and scoring
4. Iteration 0, 1, 2 and every 10th iteration: saving of temporal results and analyses
5. Finalization: saving of final state (relaxed day-plans)

Additionally to these steps, certain modules require extra computational time for initialization and finalization. For instance, the initialization of the "Landmarks-A*" router module as described earlier takes some seconds for calculating the landmarks (see Lefebvre & Balmer, 2007). The DEQSim requires several minutes for loading the network and individual demand and for storing them in optimized data structures. In case of the parallel DEQSim, additional initialization time is required. Several java internal processes such as the garbage collector and hardware constraints (file I/O) induce additional delays.

Figure 4 shows the contributions of time to the total calculation time for the first 40 iterations of a relaxation process. In this setup the routing is done by eight parallel running "Landmarks-A*" router modules. Time allocation is done by another eight "Time Allocation Mutator" modules. For the execution of the demand the parallel version of the DEQSim with eight threads is used. This setup results in a relaxation behavior as shown in Figure 2(c).

It turns out that the DEQSim requires in average 8-10 minutes per iteration as shown in Figure 4. There is, however, an additional overhead of 20 minutes for data exchange with the other modules of MATSim-T. Re-planning (time allocation and routing) requires about 90 seconds computational time,

Figure 4. Share of the overall computation time by process steps

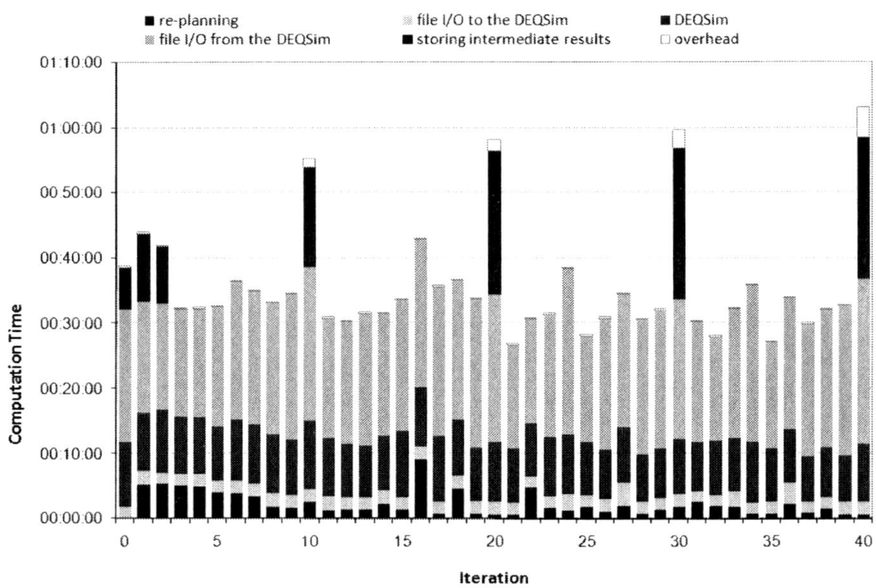

where the main fraction is consumed by the "Landmarks-A*" routing modules. The re-planning after a "breakdown" situation as shown in Figure 3 causes a significant increase in calculation time for the router (approx. nine minutes; iteration 16). The cause for this is that the performance of the Landmarks-A* router decreases if link travel times differ significantly between the uncongested and congested traffic state.

One iteration of the co-evolutionary algorithm requires about 32 minutes for the calculation of the individual time-variant daily demand consisting of 7.1 million trips on a 60,000 links network of Switzerland. In addition, every 10 iterations 22 minutes are used for saving temporal results and analysis. Taking into account that the system reaches a relaxed state after about 100 iterations, the total time for calculating the resulting demand and the corresponding traffic takes about 3.2 days.

6.4 Combinations

It is possible to run the relaxations faster when using other modules. Table 1 lists a set of possible combinations of modules and their required average and total runtime to reach a stable state. With the replacement of the *random mutation* module (Time Allocation Mutator) with a *best response* module

Table 1. Computing times of different combinations of modules

traffic flow simulation	**routes**	**times**	**# of iterations**	**run time on computer**
DEQSim (1 CPU)	Landmarks-A*	Time Allocation Mutator	100	~5.2 Days
default traffic flow simulation (1 CPU)	Landmarks-A*	Time Allocation Mutator	100	~5.5 Days
DEQSim (1 CPU)	Landmarks-A*	planomat	30	~1.9 Days
default traffic flow simulation (1 CPU)	Landmarks-A*	planomat	30	~2.1 Days
DEQSim (8 CPU)	Landmarks-A*	Time Allocation Mutator	100	~3.2 Days
DEQSim (8 CPU)	Landmarks-A*	planomat	30	~1.5 Days

Module	Average runtime	Remarks
DEQSim (1 CPU)	ca. 70 minutes	ca. 50 minutes DEQSim and ca. 20 minutes I/O Overhead
DEQSim (8 CPU)	ca. 28 minutes	ca. 8 minutes DEQSim and ca. 20 minutes I/O Overhead
default traffic flow simulation	ca. 70 minutes	
Landmarks-A*	ca. 1.5 minutes	significantly longer after „break-downs" (ca. 6–10 minutes)
Time Allocation Mutator	10 sec	
planomat	ca. 22 minutes	

(planomat) a significant reduction of the number of iterations can be achieved. On the other hand, the planomat requires in average 30 minutes computational time per iteration. However, finally this trade-off pays off (the total runtime halves).

If one includes the additional overhead for data exchange between MATSim-T and the DEQSim, the performance of the default traffic flow simulation (in Java) and the DEQSim are equivalent. With the parallel run of the DEQSim one can achieve a remarkable gain in performance, however, the overhead for file-I/O remains the same.

Finally it is worthwhile to mention that in terms of computational performance the results clearly show the applicability (large scenarios, time-dynamic and detailed) of micro simulations for transport planning.

7. INITIAL INDIVIDUAL DEMAND MODELING

In Fig. 2 the initial demand is stated to be a prerequisite for the optimization in MATSim-T. This section describes how the toolkit can be used to create the initial daily demand for each individual. The reason why this pre-process is introduced at the end of this article is that the solution space – as defined by the setup of the optimization – determines which aspects of the plan do not need to be modeled by the pre-process.

Thus the task of the *initial individual demand modeling* is to model aspects of the agents' plans that cannot be handled by the iterative optimization process. To get a best possible mapping of the real demand, this part is built upon knowledge, surveys and socio-demographic data of the investigation area. MATSim-T is built in such a way that it can operate on various types of input data. Depending on the scenario, existing input data can vary in quality, level of detail, and quantity. The modules for the initial demand modeling are adopted correspondingly, or replaced. For this reason the runtime for generating the initial demand varies. Basically, this pre-process is of sequential nature. All required modules need to be used only one time.

The modeling of the individual initial demand for the "all-of-Switzerland" application can be found in detail in Ciari et al (2007) and Meister et al (2008). The required runtime is about 14.4 hours.

MATSim-T operates on disaggregated information, i.e., the infrastructure is based on coordinates rather than aggregates, such as zones (districts, communes, etc.). Activities – and hence the facilities in which activities are preformed – are mapped to the links of the network. Since the network has a particular resolution, it defines the level of detail of the modeling. In other words, ultimately the investigation area has as many zones as the network has (directed) links, in the above case 60,000 zones. For high-resolution networks the number of zones can increase to more than one million. Since the raw data are typically of aggregated nature, they need to be disaggregated. MATSim-T provides several aggregation layers to store such data and to disaggregate them on to facilities, activities and persons if needed.

The modeling of the initial demand can be split up into several steps depending on the available raw data and the user's needs. Each of these processes is implemented in one module. These modules can be arbitrarily used, extended, replaced or skipped.

At each point of time during the modeling process it is possible to output intermediate results. This is important, since it is typically required to statistically validate the results of the model implemented in a module. The intermediate results can be used as input data for further modeling steps.

A further important aspect is the so-called streaming process for the generation of an initial demand. While infrastructural data (facilities, network and even aggregation layers) require relative little memory, the demand (= the initial plans) requires several gigabytes of memory. However, since the demand is generated individually for each synthetic person in the investigation area, it is possible to reduce the memory consumption: One loads the agent into memory, applies the demand-modeling module, writes the demand to file, and frees the memory. Then the next agent is loaded and so on. This allows one to model the individual demand for an unlimited amount of agents on standard consumer hardware. A detailed description of the features of the MATSim-T initial demand modeling can be found in the dissertation of Balmer (2007) and also in Balmer, Axhausen, & Nagel (2006).

8. DISCUSSION AND OUTLOOK

This work shows that the development of the last years considering hardware architecture, CPU performance, and optimization of programming implementations allows one to handle large-scale scenarios for transport planning with agent-based micro simulations in reasonable time. Furthermore it shows that the optimum of performance has not been reached yet. For instance, a re-implementation of the parallel DEQSim in JAVA as an integrated part of MATSim-T would avoid the overhead per iteration caused by the data exchange between DEQSim and MATSim-T (about 20 min for the discussed application), which in turn would decrease the total runtime by about 40%. A scenario of the magnitude of complete Switzerland could be handled in approximately one day.

The setup of the optimization process offers further possibilities of optimization. For instance, it is possible to reduce the number of iterations until the system becomes relaxed by introducing adaptive re-planning rates. Also the re-planning modules offer potential for optimization, in particular the routing module and the planomat. All these optimizations are to be aspired, since further functional extension, such as location and mode choice will certainly consume more computational time, be it because of the complexity of these modules or because more iterations will be required until the system reaches a relaxed state.

Finally it is worthwhile to mention that the results of MATSim-T are not only traffic patterns, but also rather a detailed description on the single agent level. In other words, it is possible to determine for each synthetic person at each point in time where she/he is and what she/he does. Still, the results should not be interpreted on the level of single agents, but rather at the level of aggregated sub-populations.

REFERENCES

Arentze, T., Hofman, F., van Mourik, H., & Timmermans, H. (2000). *ALBATROSS: A multi-agent rule-based model of activity pattern decisions*. Paper 00-0022, Transportation Research Board Annual Meeting, Washington, D.C.

Arnott, R., Palma, A. D., & Lindsey, R. (1993). A structural model of peak-period congestion: A traffic bottleneck with elastic demand. *The American Economic Review, 83*(1), 161–179.

Avineri, E., & Prashker, J. (2003). Sensitivity to uncertainty: Need for paradigm shift. *Transportation Research Record, 1854*, 90–98.

Axhausen, K. (1988). *Eine ereignisorientierte Simulation von Aktivitätenketten zur Parkstandswahl.* Ph.D. thesis, Universität Karlsruhe, Germany.

Axhausen, K., & Herz, R. (1989). Simulating activity chains: German approach. *Journal of Transportation Engineering, 115*(3), 316–325.

Axhausen, K., Zimmermann, A., Schönfelder, S., Rindsfüser, G., & Haupt, T. (2002). Observing the rhythms of daily life: A six-week travel diary. *Transportation, 29*(2), 95–124.

Balmer, M. (2007). *Travel demand modeling for multi-agent transport simulations: Algorithms and systems.* Ph.D. thesis, Swiss Federal Institute of Technology (ETH) Zürich, Switzerland.

Balmer, M., Axhausen, K., & Nagel, K. (2006). Agent-based demand modeling framework for large scale micro-simulations. *Transportation Research Record, 1985*, 125–134.

Balmer, M., & Nagel, K. (2006). Shape morphing of intersection layouts using curb side oriented driver simulation. In J. Van Leeuwen, & H. Timmermans (Eds.), *Innovations in Design & Decision Support Systems in Architecture and Urban Planning* (pp. 167–183).

Bhat, C., Guo, J., Srinivasan, S., & Sivakumar, A. (2004). A comprehensive econometric microsimulator for daily activity-travel patterns. *Transportation Research Record, 1894*, 57–66.

Bowman, J., Bradley, M., Shiftan, Y., Lawton, T., & Ben-Akiva, M. (1999). Demonstration of an activity-based model for Portland. In *World Transport Research: Selected Proceedings of the 8th World Conference on Transport Research 1998, 3*, 171–184. Elsevier, Oxford.

Cetin, N. (2005). *Large-Scale parallel graph-based simulations.* Ph.D. thesis, Swiss Federal Institute of Technology (ETH) Zürich, Switzerland.

Cetin, N., Burri, A., & Nagel, K. (2003). A large-scale agent-based traffic microsimulation based on queue model. In *Proceedings of Swiss Transport Research Conference (STRC).* Monte Verita, CH. See www.strc.ch. Earlier version, with inferior performance values: Transportation Research Board Annual Meeting 2003 paper number 03-4272.

Charypar, D., Axhausen, K., & Nagel, K. (2007). An event-driven parallel queue-based microsimulation for large scale traffic scenarios. In *Proceedings of the Word Conference on Transport Research.* Berkeley, CA.

Charypar, D., & Nagel, K. (2005). Generating complete all-day activity plans with genetic algorithms. *Transportation, 32*(4), 369–397.

Ciari, F., Balmer, M., & Axhausen, K. (2007). Mobility tool ownership and mode choice decision processes in multi-agent transportation simulation. In *Proceedings of Swiss Transport Research Conference (STRC).* Monte Verita, CH. See www.strc.ch.

de Palma, A., & Marchal, F. (2002). Real case applications of the fully dynamic METROPOLIS toolbox: An advocacy for large-scale mesoscopic transportation systems. *Networks and Spatial Economics, 2*(4), 347–369.

Gawron, C. (1998). An iterative algorithm to determine the dynamic user equilibrium in a traffic simulation model. *International Journal of Modern Physics C, 9*(3), 393–407.

Hanson, S., & Burnett, K. (1982). The analysis of travel as an example of complex human behaviour in spatially-constraint situation: Definition and measurement issues. *Transportation Research Part A: Policy and Practice*, *16*(2), 87–102.

Holland, J. (1992). *Adaptation in Natural and Artificial Systems*. Bradford Books. Reprint edition.

Hunt, J., Johnston, R., Abraham, J., Rodier, C., Garry, G., Putman, S., & de la Barra, T. (2000). Comparisons from Sacramento mode test bed. *Transportation Research Record*, *1780*, 53–63.

Lefebvre, N., & Balmer, M. (2007). Fast shortest path computation in time-dependent traffic networks. In *Proceedings of Swiss Transport Research Conference (STRC)*. Monte Verita, CH. See www.strc.ch.

Meister, K. (2004). *Erzeugung kompletter Aktivitätenpläne für Haushalte mit genetischen Algorithmen*. Master's thesis, IVT, ETH Zürich. See www.ivt.ethz.ch/docs/students/dip44.pdf.

Meister, K., Rieser, M., Ciari, F., Horni, A., Balmer, M., & Axhausen, K. (2008). Anwendung eines agentenbasierten Modells der Verkehrsnachfrage auf die Schweiz. In *Proceedings of Heureka '08*. Stuttgart, Germany.

Nagel, K., & Rickert, M. (2001). Parallel implementation of the TRANSIMS micro-simulation. *Parallel Computing*, *27*(12), 1611–1639.

Nagel, K., & Schleicher, A. (1994). Microscopic traffic modeling on parallel high performance computers. *Parallel Computing*, *20*, 125–146.

Nagel, K., & Schreckenberg, M. (1992). A cellular automaton model for freeway traffic. *Journal de Physique I France*, *2*, 2221–2229.

Palmer, R., Arthur, W. B., Holland, J. H., LeBaron, B., & Tayler, P. (1994). Artificial economic life: a simple model of a stockmarket. *Physica D*, *75*, 264–274.

Pendyala, R. (2004). *Phased Implementation of a Multimodal Activity-Based Travel Demand Modeling System in Florida. Volume II: FAMOS Users Guide*. Research report, Florida Department of Transportation, Tallahassee. See www.eng.usf.edu/~ pendyala/publications.

PTV (accessed 2008). Traffic Mobility Logistics. See www.ptv.de.

Raney, B., & Nagel, K. (2004). Iterative route planning for large-scale modular transportation simulations. *Future Generation Computer Systems*, *20*(7), 1101–1118.

Raney, B., & Nagel, K. (2006). An improved framework for large-scale multi-agent simulations of travel behaviour. In P. Rietveld, B. Jourquin, & K. Westin (Eds.), *Towards better performing European Transportation Systems* (p. 42). London: Routledge.

Rieser, M., & Nagel, K. (2008). Network breakdown "at the edge of chaos" in multi-agent traffic simulations. *European Journal of Physics*. Doi 10.1140/epjb/e2008-00153-6.

Salvini, P., & Miller, E. (2005). ILUTE: An operational prototype of a comprehensive microsimulation model of urban systems. *Network and Spatial Economics*, *5*(2), 217–234.

Schnittger, S., & Zumkeller, D. (2004). Longitudinal microsimulation as a tool to merge transport planning and traffic engineering models - the MobiTopp model. In *Proceedings of the European Transport Conference*. Strasbourg.

Schönfelder, S., Axhausen, K., & Antille, M., N.and Bierlaire (2002). Exploring the potentials of automatically collected GPS data for travel behaviour analysis – a Swedish data source. In J. Mölthen, & A. Wytzisk (Eds.), *GI-Technologien für Verkehr und Logistik – IfGI, 13*, 155–179. Münster, Germany: Institut für Geoinformatik.

SFSO (2000). Eidgenössische Volkszählung. Swiss Federal Statistical Office, Neuchatel.

SFSO (2001). Eidgenössische Betriebszählung 2001 - Sektoren 2 und 3. Swiss Federal Statistical Office, Neuchatel.

SFSO (2006). Ergebnisse des Mikrozensus 2005 zum Verkehrs. Swiss Federal Statistical Office, Neuchatel.

TRANSIMS (accessed 2008). TRansportation ANalysis and SIMulation System. See transims.tsasa.lanl.gov.

Vickrey, W. S. (1969). Congestion theory and transport investment. *The American Economic Review, 59*(2), 251–260.

Vovsha, P., Petersen, E., & Donnelly, R. (2002). Microsimulation in travel demand modeling: lessons learned from the New York best practice model. *Transportation Research Record, 1805*, 68–77.

Vrtic, M., Froehlich, P., & Axhausen, K. (2003). Schweizerische Netzmodelle für Strassen- und Schienenverkehr. In T. Bieger, C. Lässer, & R. Maggi (Eds.), *Jahrbuch 2002/2003 Schweizerische Verkehrswirtschaft* (pp. 119–140). St. Gallen: SVWG Schweizerische Verkehrswissenschaftliche Gesellschaft.

W3C (2006). *eXtensible Markup Language (XML)*. World Wide Web Consortium (W3C). See www.w3.org/XML.

Waddell, P., Borning, A., Noth, M., Freier, N., Becke, M., & Ulfarsson, G. (2003). Microsimulation of urban development and location choices: Design and implementation of UrbanSim. *Networks and Spatial Economics, 3*(1), 43–67.

Wardrop, J. (1952). Some theoretical aspects of road traffic research. *Proceedings of the Institute of Civil Engineers, 1*, 325–378.

Wiedemann, R. (1974). *Simulation des Straßenverkehrsflusses*. Schriftenreihe Heft 8, Institute for Transportation Science, University of Karlsruhe, Germany.

Wolf, J., Schönfelder, S., Samaga, U., Oliveira, M., & Axhausen, K. (2004). Eighty weeks of Global Positioning System traces. *Transportation Research Record, 1870*, 46–54.

Chapter IV
TRASS:
A Multi-Purpose Agent-Based Simulation Framework for Complex Traffic Simulation Applications

Ulf Lotzmann
University of Koblenz, Germany

ABSTRACT

In this chapter an agent-based traffic simulation approach is presented which sees agents as individual traffic participants moving in an artificial environment. There is no restriction on types of players, such as car drivers or pedestrians. A concept is introduced which is appropriate to model different kinds of traffic participants and to have them interact with each other in one single scenario. The scenario may not only include roads, but also stadiums, shopping malls and any other situations where pedestrians or vehicles of any kind move around. Core theme of the chapter is an agent model that is founded on a layered architecture. Experiences with implementation and usage of the agent model within the universal multi-agent simulation framework TRASS will be explained by means of several application examples which also support discussion about validation of concept and implementation.

INTRODUCTION

Since agent-based simulation began to be widely used to treat problems in the field of traffic and transportation, several approaches have been established based on different definitions of the term "agent". This has led to a large number of implementations which are more or less narrowly coupled to dedicated software tools. There are two main kinds of implementations, both of which depend upon the user community:

- In scientific research, the typical procedure is to build models for various components of a particular (type of) traffic system. Often general purpose programming languages are used, together with more or less universal development environments or frameworks, most of which are open source. These systems can be adapted to the requirements of the model or class of models under consideration. This approach requires a high level of computer science expertise.
- Traffic planners and other practice-oriented users prefer the application of specialized, mainly commercial tools. Any adaptation of such a tool is restricted to representing the real-world target system with the available features. Often an integration with analytical methods and real-world traffic control systems is desired.

In any case, two components of traffic simulation with characteristic features can be identified:

- A model of an environment as a network or a landscape, i.e. a topography with typical static entities such as streets, sidewalks, static obstacles,
- A representation of the traffic flow within the modeled environment where attributes are the geometric forms and sizes of moving entities, and motion obeys physical laws (either macroscopically as flows or microscopically as individual entities).

The development of the TRASS framework presented in this chapter was guided by the goal to have a universal platform to design multi-agent-based simulations with maximum flexibility, representing the properties of traffic described, and its environment. Special attention was devoted to the agent model representing traffic participants. These are seen as human beings who participate in traffic in many different roles. This makes the integration of social and behavioral science aspects relevant for the agent model.

The following presents TRASS as a framework that can be used for research and practice-oriented purposes.

BACKGROUND

The literature offers a very broad spectrum of contributions on topics related to traffic simulation. Authors from diverse scientific backgrounds – social scientists, physicists, mathematicians, computer scientists and many others – are engaged in that field and hold even more diverse views on that theme.

In the following literature review we will present a few articles on traffic simulation, which represent the context of microscopic agent-based models and show some evolutions in this specific field. We pass on a description of important and well-known contributions on macroscopic (Lebacque, 2003) or Cellular Automata-based (Nagel & Schreckenberg, 1992) approaches.

One direction of evolution is the increasing complexity of simulation models and scenarios. An early conversion of a traditional approach on traffic simulation into an agent-based model was done by Klügl et al. (2000). Based on their simulation platform SeSAM the original Cellular Automaton design car following model by Nagel and Schreckenberg was implemented. Klügl et al. successfully replicated the original model behavior and predictions. This project marked a starting point for further activities in the field of traffic simulation (e.g. Klügl & Bazzan, 2004).

Since agent-based simulation became more established in the traffic simulation domain, numerous specific aspects of traffic systems were studied, employing different kinds of software tools.

One example copes with traffic lights coordination using the SWARM platform (Oliveira & Bazzan, 2006). Another example is the simulation of self-organization effects in groups of pedestrians based on the "social-force model" by Helbing & Johansson (2007). In this model, the individuals are seen as particles like molecules in liquids or gases, which influence each other through some kind of force. This agent concept matches the established definitions of the term "agent" only to a certain extent: Agents in this simulation approach are components of a deliberately "simplified model of pedestrian interactions", as Helbing and Johansson (2007, p. 624) explicitly put it. They are interested in the behavior of large crowds where "uncertainties about the individual behaviors is averaged out" (p. 626).

A step towards more complex simulation scenarios and more versatile agent behavior is presented in the contribution of Banos & Charpentier (2007). The authors also deal with the simulation of pedestrians, this time in the complex environment of subway stations including gangways, halls, ticket offices and so on. The Netlogo-based implementation exhausts the possibilities of grid-based topography models.

Besides enhancing the complexity of simulation scenarios, the trend to larger simulation worlds and increasing numbers of agents is another observable development.

One approach in this context is SUMO (Krajzewicz et al., 2006). This platform is built upon a simple, efficient but precise space-continuous model for car traffic, and was used for simulation of traffic control systems in major cities, running on a stand-alone PC.

In contrast, Perumalla & Bhaduri (2006) place emphasis on more complex models and comparatively huge scenarios, running on massive distributed hardware.

For the field of systems related to practical applications of simulation, the commercial system VISSIM® by "PTV Planung Transport Verkehr AG" shall be mentioned.

Focus in this domain is the frictionless integration of different applications for traffic planning purposes like systems for analytical planning and forecast, logistics, traffic light controls and so on. Another requirement is the high-quality visual presentation of simulation runs and results. These aspects are implemented in VISSIM® including the ability to simulate all relevant types of traffic participants, based on a psycho-physical car-following model by Wiedemann (1974).

In this chapter we show the concept and implementation of a simulation framework which on the one hand offers better conditions for the realization of traffic participants and their environment than general-purpose tools, but which on the other hand is still sufficiently general to include numerous features of simulation models, including those mentioned above. One innovation is the application of a continuous environment with an efficient topography model as the basis for precise and realistic design, combined with a structured agent reference architecture.

As the literature review above already shows, the scientific community suffers from the lack of a standardized definition for the term "agent". Thus, it is all the more important to clearly state on which definition a particular contribution relies on. Our favorite definition of agent is the one presented in Gilbert & Troitzsch (2005), which encompasses four important properties: An agent

- is an autonomous entity that cannot be controlled by external interference;
- can interact with other agents by means of a "language";
- can perceive his environment and react on changes or events within this environment by some action on this environment;

- is able to take the initiative and perform action on the environment with the intention to reach one ore many goals.

THE TRASS CONCEPT

In the following sections, concepts will be presented that have been realized in terms of running simulation software and that have proven their practical applicability in several models. We are aware that there are always numerous possibilities to design a particular feature and that our choices are not in all cases optimal. But we attempted at creating reasonable compromises for solving the partial problems. What follows is an overview of the different aspects of the system and of the problems that play a role in implementing and using the system. Afterwards the chosen concepts and solutions are presented in detail.

The Impact of Modeling

The starting point of any simulation is a repertoire of appropriate models of the target system under consideration. In the context of traffic simulation, this means that appropriate models of the main components, i.e. of environment and traffic participants, have to be found.

Models are representations of real-world entities or systems whose attributes represent only part of the properties of the latter, whereas other properties of real-world entities are neglected. The selection of properties represented by attributes should be done pragmatically. If, for instance, types of vehicles are to be modeled for the purposes of analyzing and simulating traffic within a city, then sizes, weights, power, pollution class or the minimum turning radius might be important. It is unimportant, however, to know and represent color or whether a car stereo is installed.

An important instrument to manipulate complex systems using models with restricted parameter and attribute spaces is to split up complex model elements into smaller units that are easier to handle. This concept is important wherever modeling is done. Thus a large traffic environment can be split up into smaller districts in order to

- make models clearer,
- realize a third dimension, using several two-dimensional planes,
- introduce distributed simulation execution.

The inner structure of an agent can also be determined by splitting it into functional units on different layers. A possible design for such a partitioning will be discussed in greater detail in section "The TRASS Agent Model".

The consistent application of the agent-oriented paradigm implies at least one aspect of this partitioning, as all the dynamics is concentrated in the agent unit. This is also reflected in the implementation presented later on, as agents are seen as entirely autonomous units and executed in parallel program threads.

The process of modeling cannot reasonably be separated from the process of model validation. For some components this might be possible analytically, but for the system as a whole this is only possible by comparing simulation output to empirical data or by replication of already validated models.

The following subsection tries to offer solutions to all the problems discussed so far.

The TRASS Topography Model

The characteristic of agent-based traffic simulation is the involvement of agents for traffic participants, situated in an environment with heterogeneous topographic structures. These structures can be described by regions, where each region is suitable and thus accessible usually by a subset of all defined agent types.
In a simulation scenario of an urban district, for example, incorporating car traffic as well as "mobile obstacles" (like pedestrians), the topography model must provide a network of paths consisting of at least:

- roads with different lanes and intersections for cars, which might be entered by pedestrians;
- sidewalks only for pedestrians;
- "forbidden zones" (static obstacles, like buildings), which must not be entered by any agent.

A possible and frequently used approach to implement such topographical models relies on a regular grid structure. In this case, the simulated environment is divided into squares of equal size, and each of these cells is assigned a meaning. The choice of the size of a cell obviously has an immense influence on simulation results: With cells that are too large, only a very imprecise representation of the real-world topography is possible. In many cases this is a serious problem, as geometric details may have great influence on the behavior of pedestrians and vehicles. On the other hand, the choice of very small cells increases the computational expense of the simulation, as a large number of cells have to be evaluated and updated in every simulation step. A short example as an illustration: The perceived environment of an agent is defined by some neighborhood, and all cells in this — possibly large — neighborhood have to be evaluated by the agent in question. If the agent has a range of vision of, say, 20 meters, and this range is shaped like a cone, then some 40 cells have to be evaluated if a grid cell represents one square meter. If it represents 0.1 sq.m., then the number of cells to be included in the calculation increases to approximately 4000.

Thus it makes sense to use a continuous space model instead. In this model, positions and dimensions are described by floating point values. The maximum extension and the smallest measurable distance are determined only by the floating point data type used (in TRASS a double-precision data type with 15 significant digits). In principle, each object can hold any position and size within these limits.

With adequate techniques, such a model accomplishes the necessary calculations with maintainable effort that depends only on the number of objects involved in the simulation scenario.

The continuous topography model that is implemented in TRASS will now be used to make clear what a structure appropriate for traffic simulation might look like.

The elements of the topographic structure are labeled as regions. A region defines a limited polygon-shaped spatial area with homogeneous attributes. The complete topography is composed of an arbitrary number of regions. The polygons of two neighboring regions share the intersecting vertices and edges, so that all regions together form a polygon mesh. This mesh structure is alike an irregular Cellular Automaton, but with the difference of free mobility of the agents within the "region-cells".

Each region is parameterized by attributes for identifying its type, designated usage, default direction or any other property an application may require. The attribute values are valid for the entire region, hence labeled as "homogeneous attributes".

The agents designed for this topography model are equipped with sensor units able to perceive borders and attributes of regions.

The limitation to polygons and hence the abnegation of arcs as part of a topography implies a model abstraction introduced mainly due to performance reasons. Nevertheless it is feasible to achieve sufficient precision for any purpose by assigning a large number of segments to a polygon and thus approximate the shape of an arc.

Figure 1 shows the simple model of a road crossing as an example for a simulation topography. The software-supported generation and editing of topography models is a subject of the "Concept Validation" section where we will demonstrate how city maps, road maps and aerial views can be transformed (semi-) automatically in a simulation topography.

The TRASS Agent Model

After having designed a simulation environment with an appropriate topography, it's about time to shed some light on the "inhabitants".

Starting point for designing the agent model is the premise that an agent is a fully autonomous entity. Autonomy in this context means that the environment is perceivable by an agent, but the competence to decide when and how to interact with the environment is situated solely within the agent. A topographic region called "road" within an agent's environment can, for instance, be interpreted as a region where driving a car is allowed, but it is the agent's decision to leave the road and continue its way onto the neighboring meadow.

The dynamics according to the application domain is brought into a universal simulation framework by implementation of one or more agent types. A well-defined and functional structured framework for agent design is indispensable to cope with complexity of numerous real world applications.

A widely accepted and used scheme for structuring complex problems into simple entities is the layer concept. The starting point for such a segmentation is the definition of discrete levels of abstraction for the given topic. Each identified abstraction level is packed into a corresponding layer, and interfaces are specified between the layers.

Figure 1. Example for polygon-structured simulation topography (model of a crossing)

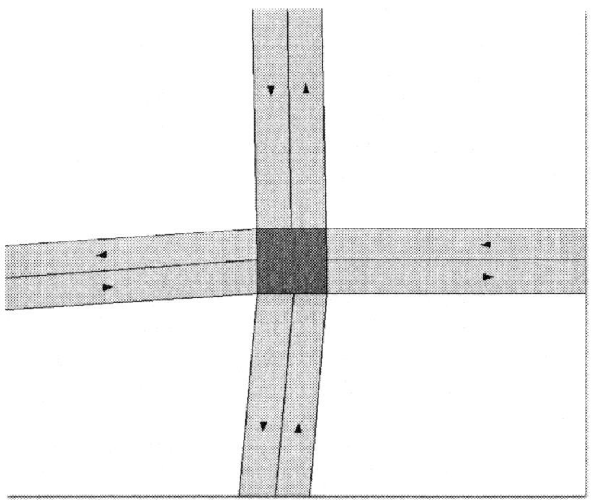

Layered architectures have often proven successful not only for reducing complexity, but also in increasing the flexibility and adaptability of the respective implementations. A very famous example is the TCP/IP protocol stack software, which is the basis of the internet; many examples can also be found in agent design history (Bonasso et al., 1997; Kendall et al., 1997; more references in the additional reading section). Here we will present an agent model with a fine-granular, hierarchically-structured layer architecture that was developed in conjunction with TRASS (Figure 2).

Three layers can be distinguished: the "Physical layer", the "Robotics layer" and the "AI layer".

The Physical layer subsumes properties like agent shape, style of the sensor unit and a number of physical attributes. The specification of these properties must, of course, be compatible with the TRASS topography model and satisfy requirements for adequate computing performance. The Physical layer determines the perceptive capabilities of the agent and projects the effects of actions into the topographic environment.

The properties provided by the Physical layer are the basis for describing the agent behavior.

Behavior is revealed mainly by variation of the physical attributes during the lapse of time. A hierarchy of two different layers is considered for behavior modeling: The Robotics layer represents actions and sequences of actions (activities) together with the perceptive processes triggering these actions. The evaluation of perception input and the controlling of actions are done by the AI layer.

Figure 2. Layered architecture of the TRASS agent model

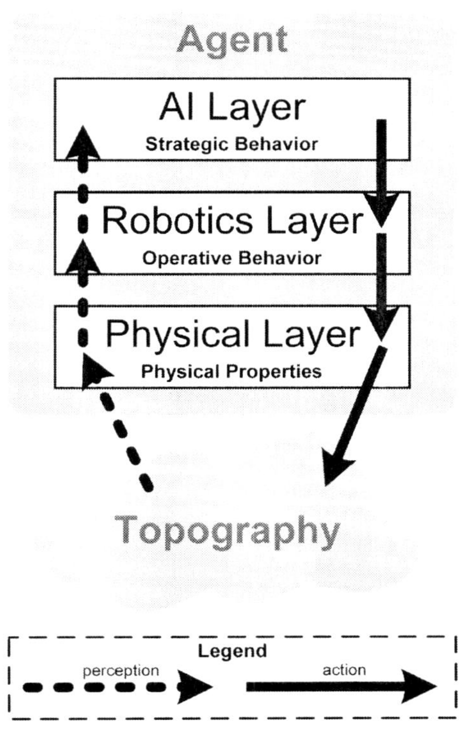

Background for this segmentation of behavioral aspects is a distinction in

- more or less automatic, reflex actions and heuristically controlled activities (which can be learnt habitually, unthinkingly) in the Robotics layer, and
- thoughtful, planned actions on the other hand in the AI layer.

Among others, economists use models with a similar structure. Max-Neef (1992), for instance, distinguishes human needs according to the evolution of the brain into

- survival needs, controlled by the brain stem and cerebellum (reptile), resulting in automated decisions,
- social needs, controlled by the limbic system (mammal) and involving social heuristics, and
- identity needs, controlled by the neo-cortex (primate) via individual heuristics.

Norris & Jager (2004) show the usability of this concept for the simulation of markets.

Within the Robotics layer (and, depending on the design, also the AI layer), a further splitting into sub-layers might be reasonable (see section "Robotics Layer").

In the following, the introduced layers of the TRASS agent model are presented in detail.

Physical Layer

As already mentioned, one has to consider several prerequisites for model abstractions when designing a simulation framework. This concerns the precise modeling of real-world details, but also sufficiently large scenarios that can be modeled and computationally simulated in a reasonable length of time. The abstractions which we have used will be described in the following, not neglecting the peculiarities of a simulation in continuous space and the necessary geometric operations within the software.

The agents considered will usually represent real-world physical objects existing and moving in an environment. Every agent has an attribute position within this environment (reference point in Cartesian coordinates) and parameters such as direction as well as a geometric shape. As we restrict ourselves to a two-dimensional world, both position and direction can be expressed by vectors with two components (x, y) and (dx, dy) respectively. Regarding the agent shape, only the outline is of interest - the height does not play a role. This outline could just be a polygon, as in the case of regions. However, during the simulation a large number of distance calculations (including the test whether two agents might collide) and a frequent update of the outlines of the agents are necessary, both of which need a large amount of computing resources. We therefore chose another, additional abstraction which allows for a very efficient calculation: The outline of an agent is approximated by circles with an arbitrary level of detailing. This means that it is up to the user to appoint the number of circles used for the approximation. The shape can even be changed during a simulation run, as the number and size of the circles which form the outline can be changed.

Thus the shape is described by a set of circles with different radii whose centers are defined by polar coordinates with respect to the Cartesian reference point of the agent. This makes it easy to perform a collision test by simply calculating the distance between two circles.

The second important component is the sensor, which allows the agent to perceive its environment, i.e. other agents and the topography. Whereas in the case of grid-based topographies neighboring cells

Figure 3. Agent shape (dark gray) consisting of three circles with reference point (x, y) and sensor (light gray) with direction φ, central angle α and radius r

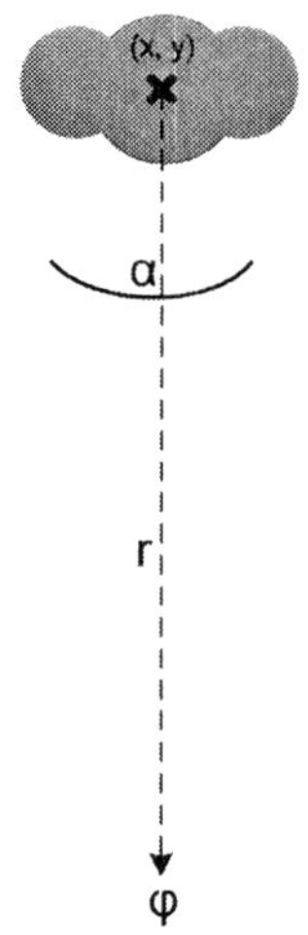

are defined by von Neumann or Moore neighborhoods, a continuous world makes it necessary to define a geometric structure for the perceivable part of the environment.

Much like in the case of the shape, here, too, the sector of a circle is an abstraction that complies with real-world details (e.g. the range of vision of a human) and additionally allows for a simple and computationally fast modeling. Exactly as in the case of the shape-determining circles, an agent can be equipped with an arbitrary number of circle sectors by defining the centre of the sector, but additionally by defining the direction φ and the angle of vision α. Which sector is used in a particular situation is determined adaptively. A car driver might have different perception sectors: wide angle forward, narrow angle to the left and right, backward mirrors etc. Figure 3 displays the geometric details of shapes and perception sectors of the agent model.

The function of the sensor unit is to identify whether another agent or a border of a topographic region is within the vision range of an agent. The perception process consists of several steps. The first step is the calculation of the distance between the centre of the sensor sector and the circle of the shape of the other agent. If this yields the distance d smaller than the length of the sector, then a second step calculates the angle θ between the direction of the perceiving agent and the sight line towards the perceived agent. If the value is within the interval of the perception sector, then the shape circle of the other agent is within this agent's perception area (Figure 4).

Detecting the boundary of a topographic region is done with a similar procedure. Only two steps are needed in order to find out whether there is an overlap between region and sensor. If in the first step an overlap between the circular hull of the polygon and the sensor is found, only then will all edges of the polygon be tested for intersection or inclusion with the sectors of the sensor.

The results of all calculations are stored in collections separately for agents and regions. Further investigation of details about the perceived objects is done via inter-agent communication when such information is needed within the layers of the behavior model.

The communication area is an additional feature similar to perception sectors which is important for attribute retrieval. During processing of a request from agent B, agent A checks whether agent B is

Figure 4. Perception process: The agent on top checks whether the agent at bottom right is situated within the sensor area by calculating distance d and angle θ

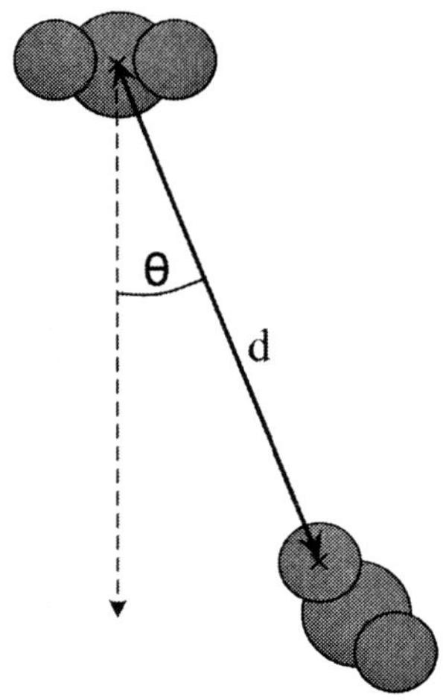

situated within one of agent A's communication areas. Depending on the result of this check, agent A possibly sends different information or does not respond.

The purpose of this concept can be illustrated by an example. A stop sign at a road junction (as a physical object) can be perceived by all agents heading towards the sign. But only for agents arriving from one distinct road will the sign hold the special message to stop in front of the junction. For all other agents the sign is just an obstacle which should not be knocked down.

All components and attributes of the physical agent model are summarized in a class diagram (Figure 5). While the specification of values for reference point, orientation and velocity is obligatory, an agent can utilize any subset of the other features (shape, sensor and communication area) according to the requirements.

Robotics Layer

In accordance with the structure presented in Figure 2, the robotics layer fills the gap between the "low-level" attributes of the physical layer and the abstract strategic decisions treated by the AI layer. Thus, the design of the robotics layer is extremely important for the entire system.

In many cases, activities performed by humans are not initiated explicitly by reasoning processes, but rather by a repertory of heuristics of action learnt by experience. Therefore it seams reasonable to transfer such heuristics into artificial agents, which is the purpose of the robotics layer.

An approach often applied in the domain of autonomous robots is the Finite State Machine (FSM), originated by Hopcroft & Ullman (1979). A commonly used graphical representation of an FSM is Harel's statechart (Harel, 1987).

An enhanced type of Finite State Machine – inspired by the extensions proposed by Harel – was chosen as the methodological basis for the robotics layer. A notation similar to the UML state diagram is used here for graphical representation (Figure 6).

The FSM applied in TRASS is formed by hybrid automata (hybrid because of the active character of the state transition, as described below) and consists of a finite set of states of which one can be exclu-

Figure 5. Physical agent properties

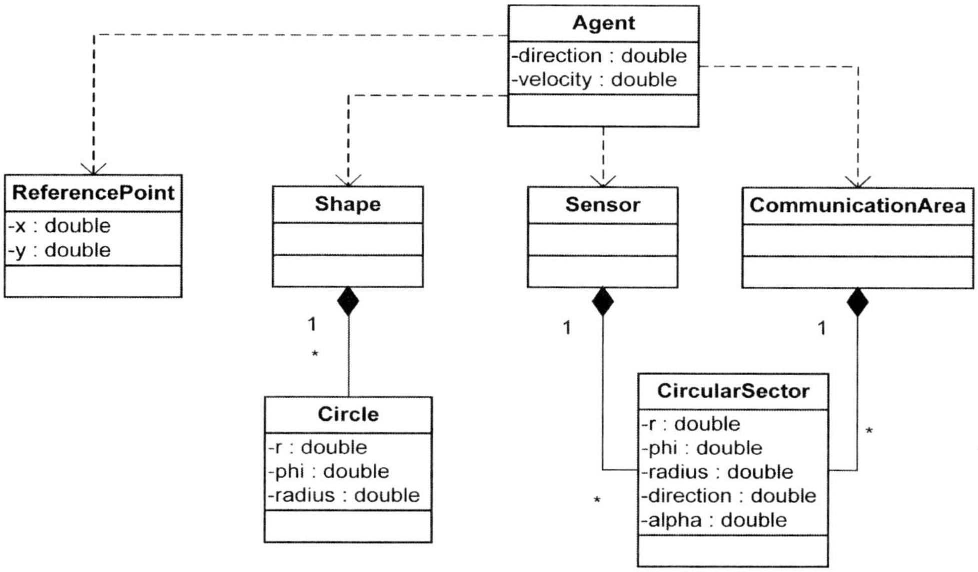

Figure 6. Symbol for State with explicit declaration of the Transition Function

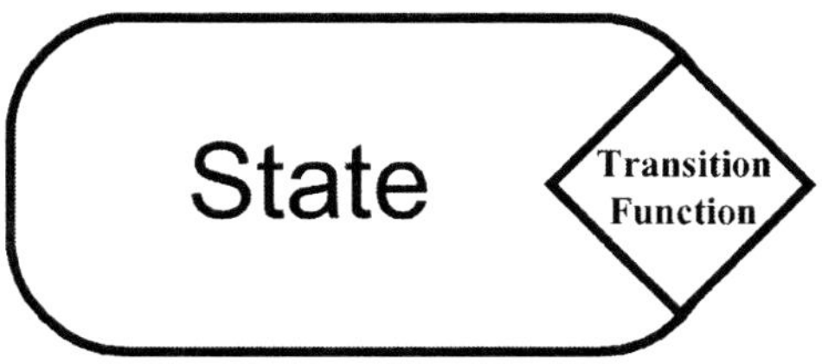

sively activated at any time. The transition from one activated state to another is described by functions for each state and initiated by events.

For each state, an entry-action, an exit-action and an activity, as well as associated parameter sets, can be defined. The entry-action is executed when the state is activated as successor of a different state. Accordingly, the exit-action is executed for a state that is deactivated with a different state as successor. In contrast, the activity is executed every time a state is activated or re-activated, i.e., even if the result of the transition is to retain the actual active state.

The state transition process is divided into three steps:

- Within the first step, information as source for an estimation function is collected and pre-processed. The pre-processing aims to categorize the information and filter out irrelevant data.
- The second step carries out the estimation function with the pre-processed information as input and the reference to the successive state as output.
- The third step finally configures and activates the successor of the actual state.

A distinctive feature of the FSM concept realized in connection with TRASS is the nesting of several automata in a hierarchy of levels. Each level contains a complete automaton, whose state-actions and state-activities affect the automaton on the level below by supplying the state transition functions and by assigning values to the state parameters.

During execution of a state transition function, a trigger signal for state change on the automaton at the level above might arise. This happens when the transition function cannot handle the input information (because of missing or unknown data).

Special attention needs to be paid to the automata on top and at the bottom of the hierarchy. While the former is controlled by an external entity (in TRASS by the AI layer), the latter operates on another entity outside the robotics layer (in TRASS the physical layer).

The actual implementation of the robotics layer in TRASS makes use of a three-stage hierarchy of automata levels (Figure 7). The specific functions of the levels are defined as follows:

1. Level one represents basic actions which are usually conducted without thinking by humans (e.g. turning the steering wheel). The state activities on this level directly modify the physical agent attributes during the lapse of time. Due to the time-discrete simulation model, the activities are

Figure 7. Nested FSM structure for the Robotics layer; $T_\emptyset$ is the symbol for "transition template"; {T} and {A} are sets of transitions and actions, respectively

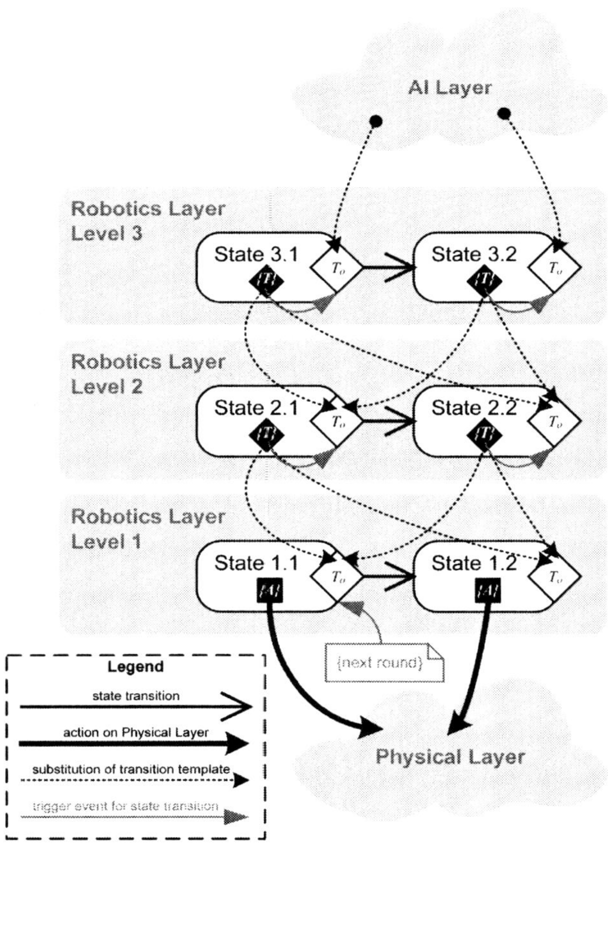

adjusted for the duration of a discrete time step. Furthermore, state activity and transition function are executed at every time step.

2. Level two deals with activities composed of basic actions (e.g. hold the centre of a lane). The state-activities at this level configure the transitions for the level 1 automaton. A possible state transition at level 2 is triggered by the transition functions of level 1.
3. Level three subsumes all required plans for complex activities a human is aware of when executing (e.g. lane change operation). The impact of activities and the mechanism for triggering a transition is analogue to the level 2 automaton. The transition functions for level 3 are provided by the AI layer.

Each layer is equipped with a perception filter for processing the data from the perceived environment. The filter is fed with perception data from the sensor unit or the underlying level respectively. The output data is reduced to significant information and presented in appropriate structures.

In order to demonstrate the utilization of this concept in the context of behavior modeling for agents, a slightly more comprehensive example will conclude this subsection. The technical capabilities of this sample agent type are restricted to driving in lanes and cross roads, and conducting lane changes.

First, the specific automata on the three levels of the robotics layer will be discussed.

On level 1 the automaton state activities manipulate the physical agent attributes velocity v and direction φ during lapse of time. The states corresponding to the basic actions and the relation between them are shown in Figure 8. A basic action conducted automatically by a skilful driver in this context is e.g. to apply adequate pressure on the brake or accelerator pedal to achieve a desired velocity.

Meaning and effect of the respective states are:

- **Idle:** The vehicle remains in the current status. v and φ remain constant.
- **Accelerate:** The driver accelerates or decelerates the vehicle. Parameter "a" determines the actual intensity of acceleration and "v" the desired velocity. v is modified, φ remains constant.

Figure 8. Sample FSM for Robotics Action Level

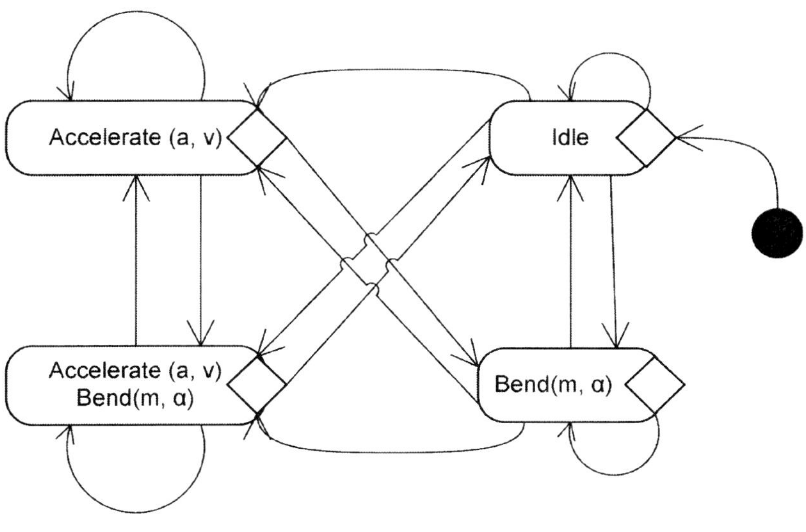

- **Bend:** The driver turns the steering wheel with the move determined by parameter "m" in order to gain a new direction specified by "α". v remains constant, φ is modified.
- Accelerate/Bend is a combination of the states Accelerate and Bend. Both v and φ are modified.

The states at level 2 describe several basic driving maneuvers (Figure 9). The following activities (parameter "v" indicates the desired velocity) are considered:

- **OffLane:** The driver steers the vehicle onto a topographic region not marked as a road (e.g. crossing, parking area), heading towards a target "tg".
- **LaneCenter:** The driver steers the vehicle in the centre of a lane.
- **LaneCenter/LaneEndAhead:** The driver steers the vehicle in the centre of a lane and reaches the end of the lane (because of a bend or a crossing ahead).
- **Bend:** The driver steers the vehicle into a bend.
- **LaneBorder:** The driver steers the vehicle in the direction of the (left or right) lane border with angle "α". This maneuver might be part of a lane change for example.

The level 3 automaton covers complex driving maneuvers (Figure 10), such as:

- **GoAhead:** The driver follows the course of the road with intended velocity "v" taking traffic rules and road situation into account.
- **Cross:** The driver passes an intersection with intended velocity "v", heading towards outbound lane "out".
- **ChangeLane:** The driver performs a lane change to the left or right side according to parameter "dir".

Figure 11 shows a snapshot of a typical situation for car traffic. Agent a1 is approaching from north at a crossing and has switched on a widespread perception sector to examine the crossing area. The actual state of a1 is {L1:Idle; L2:LaneCenter/LaneEndAhead; L3:GoAhead}.

Figure 9. Sample FSM for Robotics Basic Operations

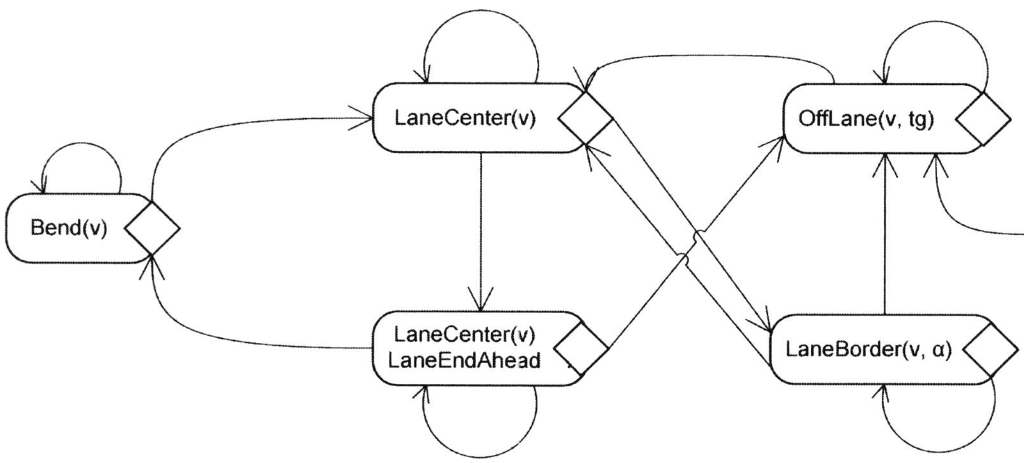

Figure 10. Sample FSM for Robotics Complex Operations

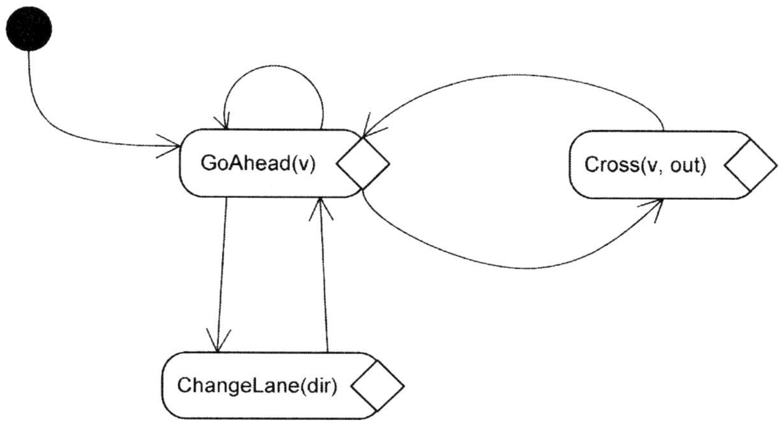

Figure 11. Sample simulation snapshot

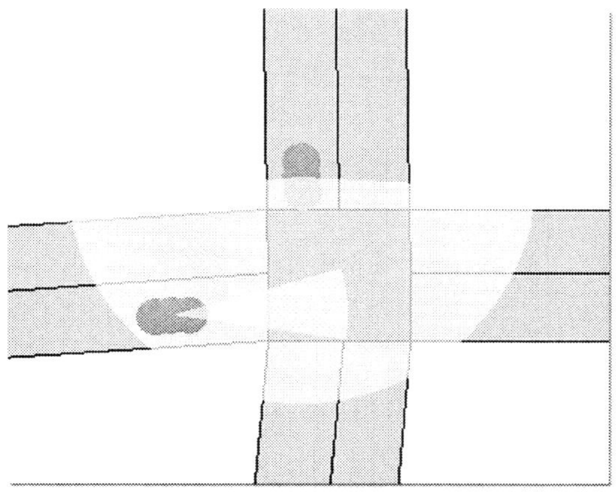

From west another agent a2 is driving towards the crossing with presumably high velocity (because a2 will be perceived by a1 in the actual time step for the first time).

The program flow within the robotics layer of agent a1 for this distinct time step is explained in Figure 12.

AI Layer

Beyond the physical properties and technical abilities of the agents described so far, another "level of mind" exists in human beings that make this species unique from most other creatures: the ability to reason about problems and make strategies and plans for solving them. These capabilities are subsumed under the keyword of (human) intelligence.

Figure 12. UML sequence diagram showing an incident that causes state changes at all automata levels (Abbreviations of object names: Sensor=sensor unit of the agent; RL_levelx=level-x automata of robotic layer; SensorProc_Lx=perception filter for level x)

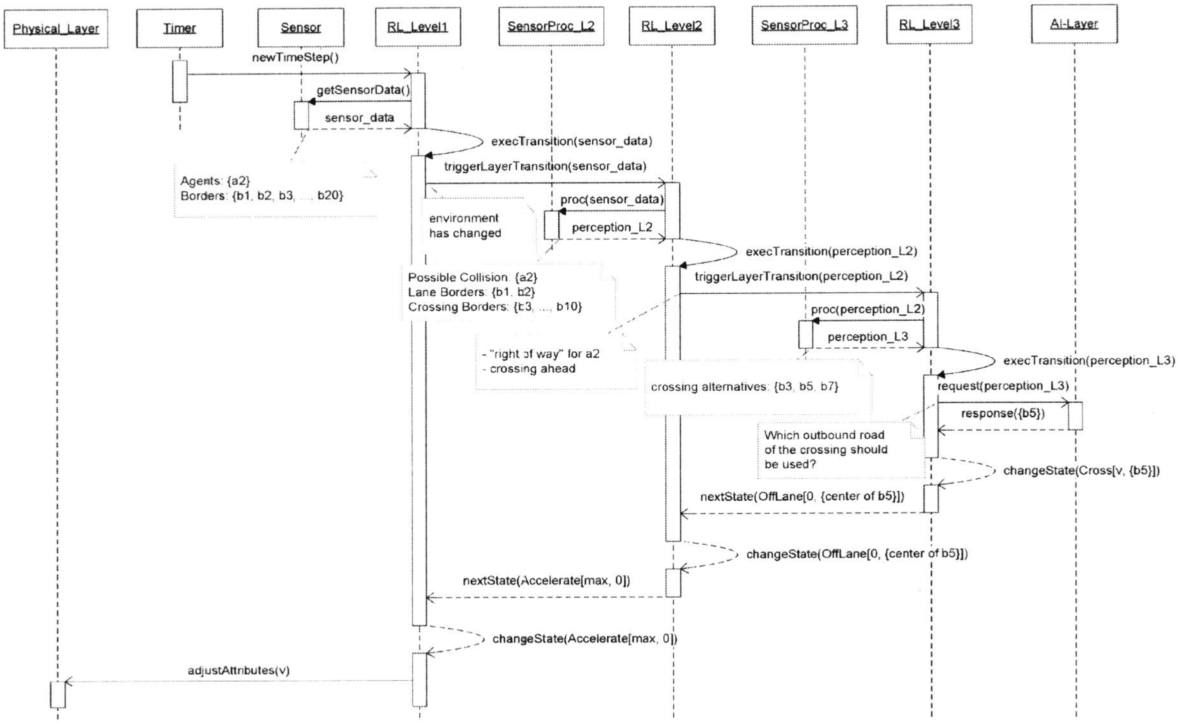

Scientific research on this topic during the past decades has originated in numerous systems where intelligent behavior is achievable for restricted domains.

The agent-based concept represents an approach on implementing such systems. A keyword often mentioned in this context is Distributed Artificial Intelligence (DAI).

Characteristic for this concept is not to create single entities with very high complexity, but to build systems with a number of communicating entities of simple structure, which are able to let intelligent structures emerge as a result of the interaction process.

The concept presented below aims at defining a model of mental activity for agents on top of a given set of technical abilities. While the implementation of the technical issues is based on finite state machines within the robotics layer, a variety of approaches and methods is feasible for a reasoning unit. In TRASS, the enclosure for these methods is the AI layer.

As shown in the previous subsections, the interface of the AI layer is given by the top-level automaton of the robotics layer. Communication on this interface is limited to the following:

- The robotics layer sends a notification about an event in the perceived environment that cannot be handled on a technical level. This could be a route decision at a crossing, the decision whether to get ahead of or to stop in front of an obstacle or similar incidents.
- Configuration and controlling of the robotics layers' top-level automaton.

Hence, the AI layer has to cope with the following tasks:

- observation and analysis of events reported by the robotics layer;
- inter-agent communication in order to gather information from other agents or to respond on requests;
- reaction as a result of observation and communication;
- action on the basis of layer-internal processes.

The operation of the AI layer will be illustrated by means of a few examples with respect to the functionality of the robotics layer configuration from the previous section. This simulation model comprises traffic flow scenarios in a simplified urban environment, consisting of crossroads and road junctions, linked via road segments.

An agent is driving on a road with the GoAhead state active. As soon as the agent arrives at a crossing the robotics layer signals this event to the AI layer, which has to decide which outbound road to choose. The decision could be the result of a stochastic process and the road alternatives might be annotated with certain probabilities (like 0.3 for left bend, 0.2 for right bend and 0.5 for straight on). Just as well a rule-based (follow the preceding car) or a knowledge-based decision is conceivable. In the latter case, the agent must feature a memory which for instance contains a map of the topography (represented by a graph data structure). The graph might be incomplete or partly defective, but could be completed or corrected while driving through the road network.

In a road network paths to desired targets can be calculated applying well-known graph algorithms. Once a path description (e. g. on intersection x go right, on junction y go left, etc.) is available, the actual decisions can be extracted directly when needed. By calculating different path to a given target, game theory-based approaches as suggested by Chmura et al. (2005) which already work for two-way-selection could be used to study more complex route choice behavior.

The examples above only describe reactive elements of agent behavior.

To achieve a more realistic agent habitus, proactive behavior must be added to the model. For instance different driver characteristics (e. g. aggressive drivers conducting frequent lane changes) or driving mistakes (driving against the traffic) could be simulated.

Examples of realizations of the AI-layer are presented in the subsection "Research Prototypes".

Using TRASS

All concepts presented so far are incorporated in the TRASS software system and various associated simulation models. The purpose of this section is to point out particular properties of TRASS from the user's perspective.

While the TRASS framework is basically suited for a notably wide range of applications, the model properties involved foster certain types of simulation scenarios. Especially these scenarios will be part of simulation projects, where

- a medium quantity (up to several hundreds) of agents of different types and rather high complexity are to interact within a fine-structured topographic environment;
- the topography can be projected into one (or more) two-dimensional planes with polygon-shaped structures (this is usually true for any kind of geographic maps);

- a free mobility of agents within the topography is important (e. g. to simulate authentic pedestrian appearance, or incidents resulting from drivers bad or illegal conduct);
- a simple and efficient but at the same time graphic visualization of the simulation runs is needed;
- after a rapid construction period, first results have to be achieved and experimentation is to start.

In contrast, TRASS will not be the best solution for simulation scenarios involving a very high amount of (simple and uniform) agents or when the relationship between the agents cannot be represented by the predetermined topography model.

The TRASS software system is composed of several components, which are shown in Figure 13.

The TRASS core is built around the simulation kernel, associated with the topography model, the physical agent model and a powerful message-oriented communication system.

The communication system is used primarily for inter-agent communication including unicast (two parties), multicast (group receiver) and broadcast (all components) message delivery, but also for administration and controlling of simulation runs (messages from and to the "observer").

Thus, the interface to the TRASS core includes the following aspects:

- An API-like interface to the physical agent model which can be utilized directly by Java programs.
- Configuration of actual simulation scenarios is described by XML files.
- While running a simulation, the access to time step-related attribute assignments is also available via XML-shaped data.
- Simulation runs are administered by messages which must be entered into the communication system.

Figure 13. UML component diagram of TRASS

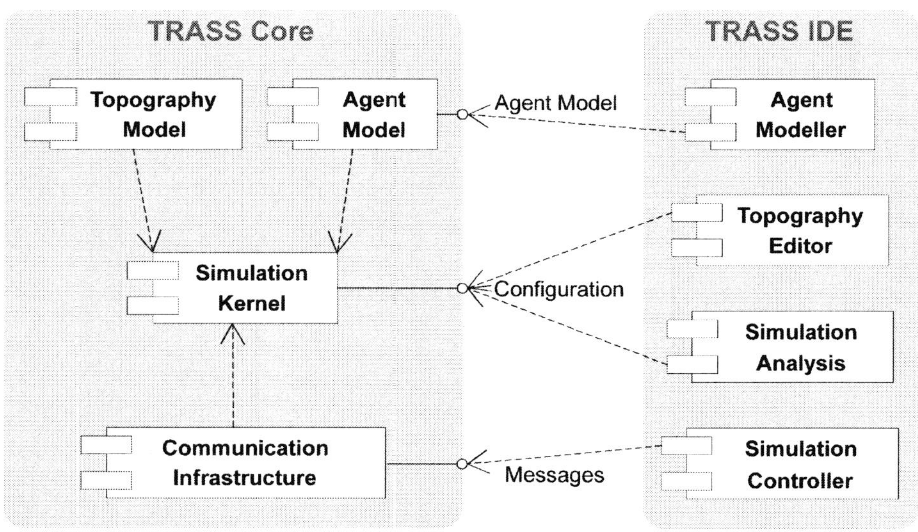

An Integrated Development Environment (IDE) is provided in addition to the TRASS core, offering the user convenient access to all parts of the core functionality. The IDE features a Graphical User Interface (GUI) for:

- interactive construction and editing of topography and agents with many powerful drawing tools available
- the simulation timer,
- animated visualization of simulation runs within the editor (with special functions for debugging purposes); all geometric elements used for calculation can be displayed,
- (statistical) analysis.

All these components of the TRASS software, a lot of sample models and the accompanying documentation can be downloaded and used by interested researches and practitioners. A common release of the source code will be considered in the future. Please contact the author for further information.

Concept Validation

This section presents three examples of simulation models that were implemented using the TRASS framework. These models are part of the validation process for the concepts described in the previous sections.

The first example is a real-life application with real topography and parameters for configuration derived from empirical data. Thus, this scenario is of foremost significance for validation.

The other two examples show realizations of simulation models from the social sciences, ported in the context of traffic simulation.

Real-Life Application

The subject of the practical example is the simulation of traffic flows within an urban area in maximum load situations. Different types of traffic participants had to be considered. This simulation was part of a project in cooperation with the local city administration.

In the first project phase, the analysis and simulation of the current state was conducted for model calibration and validation. In the following phase, the effects of various traffic planning concepts for the projected reconstruction of this area were examined and visualized.

For each of the scenarios, detailed topographic information is available as well as empirical and analytical load data.

Since it is crucial for this example that the reproduction of the topography is to scale and as detailed as possible, raw data from different air photographs and maps was used.

A road network graph was extracted from the raw data using the editing facility of the TRASS IDE. All nodes (representing crossroads and intersections) and edges (representing road sections) were parameterized according to the given geometry, lane count etc. With this graph as an input, the IDE generated an appropriate simulation topography automatically. Some fine-tuning was performed on this topography to achieve the maximum similarity to the real shape (Figure 14).

After completion of the topography design, agents were added to the simulation environment. The agent types incorporated in this example are

Figure 14. Creating the simulation topography: Based on an air photograph a road network graph was drawn (left), automatically converted into a polygon structure (middle) and afterwards manually fine-tuned (right)

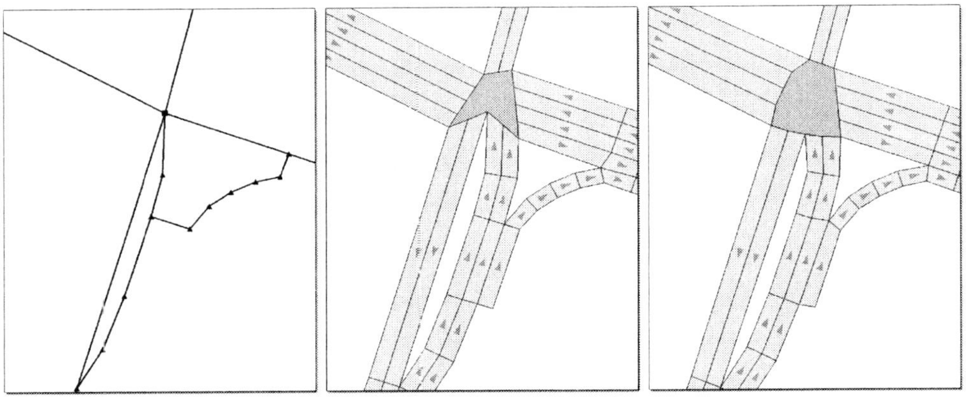

- "sources", producing different types of traffic participants (like cars, buses, pedestrians etc.) with an adjustable arrival rate;
- "sinks", removing traffic participants from the simulation world;
- agents serving as signposts (mainly traffic lights and guideposts).

At the moment, the simulation scenario contains a detailed topography, several calibrated agent types (cars, busses, pedestrians) and a traffic management system with interconnected traffic lights. In this context, TRASS was used and evaluated by numerous students within the scope of seminar papers and bachelors and masters theses. Further development in this field is expected to be funded within the MEET ("Mobility Environment Event Traffic") project currently being applied for in an EU programme (as of June 2008).

Research Prototypes

In line with an EU FP6 project called EMIL ("Emergence in the Loop: Simulating the two-way dynamics of norm innovation"; Andrighetto et al., 2007), two other simulation examples were designed.

The EMIL project especially focuses on the modeling of norm innovation processes in societies by introducing intelligent agents on different levels, which allows both modeling and observing the emergence of properties at a macro-level and their immergence into the micro-level (Castelfranchi, 1998).

One scenario pursues the constitution of a uniform driving mode – either left-hand or right-hand traffic – and is built upon stylized facts. Each car driver has the goal to move within a road network preferably without colliding with other cars. This is ensured by conducting evasion maneuvers in case of a conflict (reactive behavior) on the one hand; on the other hand conflicts can be avoided by (proactive) adaptation of the driving mode that the majority of all drivers hold.

The implementation of the car drivers' AI-layer is based on the simple opinion formation model by Weidlich & Haag (1983).

The original version of this model includes two levels. At the macro level the current ratio of left-hand to right-hand drivers is calculated embracing all existing agents, and stored within a scaled real-valued variable x:

$$x = \frac{leftHand_{total} - rightHand_{total}}{leftHand_{total} + rightHand_{total}}$$

The range of x is within [-1, 1], where the minimum stands for "all agents go right-hand", the maximum for "all agents go left-hand", respectively.

At the micro level each agent "decides" (relying on the macro variable x) in every time step whether to change or keep its actual mode. This is done by calculating the probability for an opinion change. The probability function depends on the actual mode and reads for right-hand traffic as

$$\mu = \nu \exp(\pi + \kappa x)$$

and analogously for left-hand traffic:

$$\mu = \nu \exp(-(\pi + \kappa x))$$

The meaning of the parameters involved is as follows:

- ν – flexibility (higher values increase the probability for opinion changes);
- π – preference (for left-hand traffic when > 0, for right-hand traffic when < 0);
- κ – coupling (higher values increase the influence of variable x).

Basis for the mode change decision is the comparison of the calculated probability with a random number m (in the range [0; 1]).

Simulation runs with this model show expected results for typical parameter assignments. Within a short period of time (which depends on the number of participating agents), a convergent state stabilizes with equilibrium of emerging left-hand versus right-hand systems among the simulation runs.

An interesting variation of the Weidlich-Haag model transfers the variable x into each agent, thus not modeling the macro level explicitly. The value of the internal x-variable of agent A is no longer determined on the base of all agents in the world, but on the agents which are perceived by A in the respective time step t:

$$x_t = \frac{leftHand - rightHand}{leftHand + rightHand}$$

The value calculated is smoothed by the present experience:

$$x = \frac{999x + x_t}{1000}$$

Simulation runs with this model show a slightly different picture in comparison to the original model. At first a convergent state emerges and stabilizes for a spell of time. At certain situations (many small car clusters with large gaps in between) a single agent, driving on the "wrong" side by chance, is able to overturn the current state and establish a new stable state with the opposite sign.

Another, more sophisticated, scenario in the context of the EMIL project examined the formation of norms in a traffic scenario incorporating two different types of traffic participants (Lotzmann & Möhring, 2008). One of the key deliverables of EMIL is an architecture template for normative agents in combination with a message-based communication concept (Andrighetto, Campennì & Conte, 2007). This message concept involves messages which can be presented in different modals, amongst others:

- assertion, to utter a fact (e. g. information about an environmental state);
- behavior, to inform about an action carried out;
- deontic, to communicate obligations, forbearances or permissions;
- sanction, to issue a moral evaluation on an action or state.

This architecture was already employed in other applications (e. g. norm emergence in a Wikipedia community; Troitzsch, 2008). The purpose of the re-implementation within the AI layer of TRASS-agents was to advance the development of a general method for describing the dynamics of normative agents and the process of norm formation.

The traffic scenario consists of a simple topography: a straight one-way road and two meadows to the left and right of the road. A small segment of the road has a special mark (much like a crosswalk). Situated within both meadows, a number of pedestrian agents (which is constant during a simulation run) move around. From time to time each pedestrian gets an impulse to reach a target point on the opposite meadow within a given period of time. For this activity, the agent can choose between the direct way to the target or a detour via the crosswalk. The road is populated by car agents who attempt to reach the end of the road at a given point of time.

For both types of agents, the deviation from the allowed duration leads to a valuation of the recent agent activity: a penalty when more time was required and accordingly a gratification when the target was reached early.

Due to the interaction between agents, occasional (near-) collisions are likely to happen. Such an incident, when occurring between a car and pedestrian, is classified as undesirable. Observations of a collision provoke other agents to issue sanctions against the blamable agents. The strength of the sanction is determined by various factors reflecting the environmental situation (e.g. the road section in which the collision occurred) and the normative beliefs of the valuating agent (e.g. a collision on a crosswalk might result in a harder sanction than on the rest of the road). Sanctions lead to a temporary stop of motion for the involved agents. Hence, to avoid sanctions is a competing goal to the aforementioned aims (reaching the target point or end of the road, respectively, in due time).

Several steps were conducted to map this informal concept into the frame given by the theoretical foundation developed in the EMIL project.

A classification of the expected information transfer to the message types defined in EMIL is necessary in a first step. Each agent is able to perceive its environment and to conduct actions in order to adjust its own parameters of motion or perception. These agent-internal processes can be mapped to EMIL messages with modals "assertion" and "behavior". In addition, agents can send and receive messages to and from other agents. The content of the messages are different kinds of notifications such as positive or negative valuations and sanctions, presented in modals "sanction", "deontic" or "valuation" in the EMIL frame. While these message exchanges are either intra-agent matters or speech acts between exactly two agents, another agent property is important for a norm formation process. This is the ability to listen to the communication of other agents in order to gain information about the experience of those agents, and to learn from this information.

With regard to this message classification, rules for the specification of agent behavior are defined in a second step. The structure of a rule follows an event-action scheme where a set of events triggers with certain probabilities actions from an action set. In this model, all events are coupled with message receiving and most actions are expressed by message sending activities. Furthermore, additional actions for learning are defined.

All rules are constructed with the help of event-action trees, a special type of decision tree. For each event an arbitrary number of action groups is defined. Each action group represents a number of mutually exclusive actions. All edges of the tree are attached to selection probabilities for the respective action groups or actions. While the structure of the scenario-specific (initial) event-action trees is static, the selection probabilities may change during the simulation in a learning process. Figure 15 shows an example for this kind of learning process involving event-action trees suitable for car driver agents (with events E10, E20 and E30, action groups G1 and G3, and actions A10, A11, A12 and A30).

During a simulation, relations between specific events will appear. The recognition of such relations leads to the linking of several event-action trees. This can be considered as a higher stage of the learning process, resulting in the formation of more complex rules.

With these behavioral rules, norm candidates emerge as soon as the majority of the population start to use the same rules with similar probability distributions. Agents usually start to defend the norms

Figure 15. Learning process for a car driver

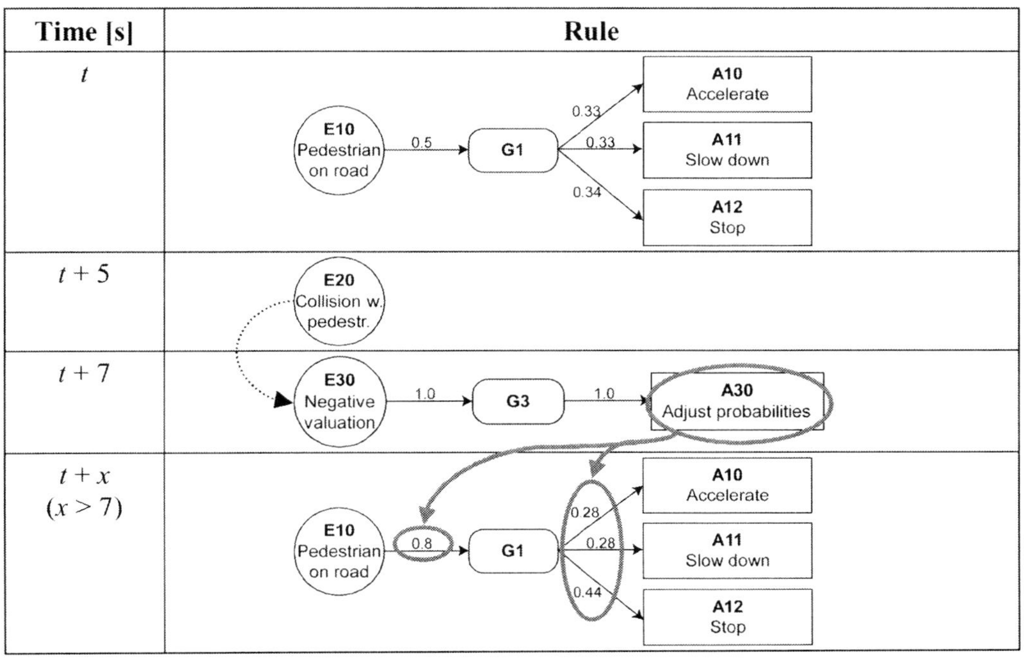

they are aware of (via sanctioning other agent's abnormal behavior), leading to a further spreading of the respective rules. This distributed norm candidate will then be transformed into a norm as soon as the number of norm defenders exceeds another threshold.

CONCLUSION

In this chapter, an approach for traffic simulation is presented that incorporates an agent model built upon mental and cognitive processes of humans. These properties in combination with the overall agent autonomy allow for interaction between agents representing human traffic participants in multiple roles within an artificial environment.

The distinct aspects of the agent model are represented by different layers:

- Physical layer for agent appearance within the topography model,
- Robotics layer covering technical aspects of agent behavior and
- AI layer for representing mental properties of human traffic participants.

A realization of this agent model was implemented on the base of the universal multi-agent framework TRASS. While the physical layer is an integral part of the TRASS framework, the other layers constitute a specialization of the core framework which in turn can serve as a behavior reference model for a broad range of traffic application: from research prototypes for exploring of social phenomena to large-scale scenarios for traffic planning analysis and forecast.

Hence, the TRASS multi-agent core framework, specialized with the behavior model and in combination with a convenient user interface, can be considered as an optimal platform to efficiently design various kinds of simulation scenarios in an interactive manner and to test, visualize and analyze simulation runs.

The aim of the chapter was to give the reader an introduction to an approach for handling more complex traffic simulation tasks, where individual behavior – which cannot be specified by the explanatory power of a simple mathematical car following model – of autonomous traffic participants is the key to reproducing manifold emergent phenomena observable in real traffic systems. Besides, whoever might want to use simulation software will find suggestions to aid the decision-making on which tool can be used for their own project.

FUTURE RESEARCH DIRECTIONS

Two directions for future work are of special interest:

- continued development of the framework towards distributed execution and towards a visual programming of agents,
- trying out additional simulation applications within our simulation framework in order to find out what has to be improved and extended, particularly with respect to the agent behavior models.

Within all future work on and with the tool, a simulation specific aspect of parallel and distributed execution will have to be dealt with (beside the known challenges of, for instance, synchronization): the replicability and validation of round-robin based simulation models. Many simulation models contain stochastic processes which are simulated with the help of pseudo-random number generators. Initializing these generators with appropriate seeds leads to identical initial conditions which result in identical simulation results, but only in the case of sequential execution. Only in this case, model parameters can be calibrated, and validation and sensitivity analysis can be carried out reasonably.

Any true parallel execution makes appropriate measures necessary to replicate simulation runs. It is the process scheduler of the operating system that decides which part of a parallel program is executed by which processor. As a rule, the application has no influence on the process scheduler. This is why distributed parallel execution of stochastic models, which use pseudo random number generators, introduces an additional stochastic effect. This additional stochastic effect, which is not under control of the modeler, is undesirable in the context of model replication, verification and validation, but might be useful in practical scenarios with sufficiently tested, verified and validated simulation models.

Another direction for further development is the trend to visual programming and model driven architectures (MDA). Interactively "drawing" structures — much like the one already implemented for modeling topographies — should also be possible for the automata of the robotics layer and particular approaches of the AI layer. This further development might be inspired by recent developments of tools such as Repast Simphony (North et al., 2005).

On the other hand, the development of new and the enhancements of existing agent models on the AI layer level is important. One approach is a widespread introduction of normative agents in diverse simulation applications. These agents should be able to learn traffic rules by observation and experience and save them as explicitly internalized norms. This approach would start from the examples presented in the concept validation section of this chapter.

Moreover, many other challenges, such as evacuation of buildings and city quarters, and the optimization of public transport, can be found for agent-based simulation which can be (or even have been) implemented within the TRASS framework.

REFERENCES

Andrighetto, G., Conte, R., Turrini, P., & Paolucci, M. (2007). Emergence in the loop: Simulating the two way dynamics of norm innovation. In *Proceedings of the Dagstuhl Seminar on Normative Multi-agent Systems, March 2007*. Dagstuhl.

Andrighetto, G., Campennì, M., & Conte, R. (2007). *EMIL-M: Models of norms emergence, norms immergence and the 2-way dynamic* (Tech. Rep. No. 00507). Rome: The National Research Council (CNR), Laboratory of Agent Based Social Simulation (LABSS) at The Institute of Cognitive Science and Technology (ISTC).

Banos, A., & Charpentier, A. (2007). Simulating pedestrian behavior in subway stations with agents. In F. Amblard (Ed.), *Fourth Conference of The European Social Simulation Association (ESSA)* (pp. 611-621). Toulouse: Institut de Recherche en Informatique de Toulouse (IRIT).

Bonasso, R. P., Firby, R. J., Gat, E., Kortenkamp, D., Miller, D. P., & Slack, M. G. (1997). Experiences with an architecture for intelligent, reactive agents. *Journal of Experimental and Theoretical Artificial Intelligence,* 9(2-3), 237-256.

Castelfranchi, C. (1998). Simulating with cognitive agents: The importance of cognitive emergence. In J. Sichman, R. Conte, & N. Gilbert (Eds.), *Multi-Agent Systems and Agent-Based Simulation* (pp. 26-44). Berlin: Springer.

Chmura, T., Pitz, T., Möhring, M., & Troitzsch, K. G. (2005). Netsim. A software environment to study route choice behavior in laboratory experiments. In *Representing Social Reality* (pp. 339-344). Fölbach: European Social Simulation Association.

Gilbert, N., & Troitzsch, K. G. (2005). *Simulation for the social scientist* (2nd ed.). McGraw-Hill, Maidenhead: Open University Press.

Harel, D. (1987). Statecharts: A visual formalism for complex systems. In *Science of Computer Programming* (pp. 231–274). North-Holland: Elsevier Science Publishers.

Helbing, D., & Johansson, A. (2007). Quantitative agent-based modeling of human interactions in space and time. In F. Amblard (Ed.), *Fourth Conference of The European Social Simulation Association* (ESSA) (pp. 623–637). Toulouse: Institut de Recherche en Informatique de Toulouse (IRIT).

Hopcroft, J., & Ullman, J. (1979). *Introduction to automata theory, languages and computation* (1st ed.). Boston, MA: Addison-Wesley.

Kendall E. A., Malkoun, M. T., & Jiang, C. H. (1997). The layered agent pattern language. In *Proceedings of the Conference on Pattern Languages of Programs (PLoP'97).* Monticello, IL.

Klügl, F., Wahle, J., Bazzan, A. L. C., & Schreckenberg, M. (2000). Towards anticipatory traffic forecast – modelling of route choice behaviour. In *Proceeding of the workshop "Agents in Traffic Modelling" at the Autonomous Agents 2000.* Barcelona.

Klügl, F., & Bazzan, A.L.C. (2004). Route decision behaviour in a commuting scenario: Simple heuristics adaptation and effect of traffic forecast. *Journal of Artificial Societies and Social Simulation (JASSS), 7*(1).

Krajzewicz, D., Bonert, M., & Wagner, P. (2006). The open source traffic simulation package sumo. In *Proceedings of RoboCup 2006 Infrastructure Simulation Competition.* Bremen.

Lebacque, J.-P. (2003). Intersection modeling, application to macroscopic network traffic flow models and traffic management. In S. P. Hoogendoorn, S. Luding, P. V. L. Bovy, M. Schreckenberg, & D. E. Wolf (Eds.), *Traffic and Granular Flow 2003* (pp. 261–278). Berlin: Springer.

Lotzmann, U., Möhring, M., & Troitzsch, K. G. (2008). Simulating norm formation in a traffic scenario. In *Proceedings of the Fifth Annual Conference of the European Social Simulation Association (ESSA).* Brescia.

Max-Neef, M. (1992). Development and human needs. In P. Ekins, & M. Max-Neef (Eds.), *Real-life economics: Understanding wealth creation.* London, New York: Routledge.

Nagel, K., & Schreckenberg, M. (1992). A cellular automaton model for freeway traffic. *Journal de Physique I, 2*, 2221–2229.

Norris, G. A., & Jager, W. (2004). Household-level modeling for sustainable consumption. In *Proceedings of the Third International Workshop on Sustainable Consumption*. Tokyo.

North, M. J., Howe, T. R., Collier, N. T., & Vos, R. J. (2005). The repast simphony runtime system. In *Proceedings of Agent 2005 Conference on Generative Social Processes, Models and Mechanisms*. Argonne, IL: Argonne National Laboratory.

Oliveira, D., & Bazzan, A. L. C. (2006). Traffic lights control with adaptive group formation based on swarm intelligence. In M. Dorigo, L. M. Gambardella, M. Birattari, A. Martinoli, R. Poli, & T. Stützle (Eds.), *ANTS Workshop* (pp. 520-521). Berlin: Springer.

Perumalla, K. S., & Bhaduri, B. L. (2006). On accounting for the interplay of kinetic and non-kinetic aspects in population mobility models. In A. G. Bruzzone, A. Guasch, M. A. Piera, & J. Rozenblit (Eds.), *International Mediterranean Modeling Multiconference (I3M)* (pp. 201-206). Barcelona.

Troitzsch, K. G. (2008). Simulating Collaborative Writing: Software Agents Produce a Wikipedia. In *Proceedings of the Fifth Annual Conference of the European Social Simulation Association (ESSA)*. Brescia.

Weidlich, W., & Haag, G. (1983). *Concepts and models of a quantitative sociology: The dynamics of interacting populations* (Series in Synergetics, 14). Berlin: Springer.

Wiedemann, R. (1974). *Simulation des Verkehrsflusses* (Tech Rep. No. 8). Karlsruhe: Universität (TH) Karlsruhe, Instituts für Verkehrswesen.

ADDITIONAL READING

Traffic Simulation

Balmer, M., Cetin, N., Nagel, K., & Raney, B. (2004). Towards truly agent-based traffic and mobility simulations. In *Proceedings of the Third International Joint Conference on Autonomous Agents and Multiagent Systems (AAMAS '04)* (pp. 60-67). Washington, DC: IEEE Computer Society.

Bazzan, A. L. C., Wahle, J., & Klügl, F. (1999). Agents in traffic modelling - from reactive to social behaviour. In *Proceedings of the 23rd Annual German Conference on Artificial Intelligence (KI '99)* (pp. 303-306). Berlin: Springer.

Chmura, T., Kaiser, J., Pitz, T., Blumberg, M., & Brück, M. (2007). Effects of advanced traveller information systems on agents' behaviour in a traffic scenario. In T. Pöschel, A. Schadschneider, R. Kühne, M. Schreckenberg, & D. E. Wolf (Eds.), *Traffic and Granular Flow ,05* (pp. 631–640). Berlin: Springer.

Helbing, D. (2001). Traffic and related self-driven many-particle systems. *Reviews of Modern Physics*, 73(4), 1067–1141.

Helbing, D. (2007). Self-organization and optimization of pedestrian and vehicle traffic in urban environments. In S. Albeverio, D. Andrey, P. Giordano, & A. Vancheri (Eds.), *The dynamics of complex urban systems: An interdisciplinary approach* (pp. 287–309). Berlin: Springer.

Schreckenberg, M., & Selten, R. (Eds.). (2004). *Human behaviour and traffic networks*. Berlin: Springer.

Agent-Based Simulation with Connection to Social Sciences

Axelrod, R. (1997). Advancing the art of simulation in the social sciences. In R. Conte, R. Hegselmann, & P. Terna (Eds.), *Simulating social phenomena* (pp. 21–40). Berlin: Springer.

Epstein, J., & Axtell, R. (1995). *Growing artificial societies: Social science from the bottom up*. Cambridge, MA: MIT Press.

Epstein, J. M. (1999). Agent-based computational models and generative social science. *Complexity*, 4(5), 41-60.

Gilbert, N., & Abbott, A. (Eds.). (2005). *Special issue: Social science computation*, 110(4). Chicago, IL: University of Chicago Press.

Gilbert, N., den Besten, M., Bontovics, A., Craenen, B. G., Divina, F., Eiben, A., Griffioen, R., Hévízi, G., Lorincz, A., Paechter, B., Schuster, S., Schut, M. C., Tzolov, C., Vogt, P., & Yang, L. (2006). Emerging artificial societies through learning. *Journal of Artificial Societies and Social Simulation (JASSS)*, 9(2).

Kampis, G. and Gulyás, L. (2006). Emergence out of interaction: Developing evolutionary technology for design innovation. *Advanced Engineering Informatics*, 20(3), 313-320.

Uhrmacher, A. M., & Weyns, D. (Eds.). (2009). *Multi-agent systems: Simulation and applications*. London: Taylor and Francis.

Normative Agents

López, F., Luck, M., & d'Inverno, M. (2006). A normative framework for agent-based systems. *Computational & Mathematical Organization Theory*, 12(2-3), 227-250.

Silva, V. T. (2008). From the specification to the implementation of norms: An automatic approach to generate rules from norms to govern the behavior of agents. *Autonomous Agents and Multi-Agent Systems*, 17(1), 113-155.

Layered Agent Models

Fischer, K., Müller, J. P., & Pischel, M. (1995). Unifying control in a layered agent architecture. In V. R. Lesser, & L. Gasser (Eds.), *Proceedings of the First International Conference on Multiagent Systems* (pp. 446–474). San Francisco, CA.

Müller J. P., & Pischel, M. (1994). An architecture for dynamically interacting agents. *International Journal of Cooperative Information Systems (IJCIS)*, 3(1), 25-46.

Sarjoughian, H. S., Zeigler, B. P., & Hall, S. B. (2001). A layered modeling and simulation architecture for agent-based system development. *IEEE Proceedings*, 89(2), 201-213.

Software Development

Bloch, J. (2008). *Effective Java (The Java Series)* (2nd ed.). Upper Saddle River, NJ: Prentice Hall PTR.

Booch, G., Jacobson, I., & Rumbaugh, J. (1999). *The Unified Modeling Language User Guide*. Boston, MA: Addison-Wesley.

Horstmann, C., & Cornell, G. (2007). *Core Java(TM), Volume I–Fundamentals (8th Edition) (Sun Core Series)*. Upper Saddle River, NJ: Prentice Hall PTR.

Miles, R., & Hamilton, K. (2006). *Learning UML 2.0*. Sebastopol, CA: O'Reilly.

Chapter V
Applying Situated Agents to Microscopic Traffic Modelling

Paulo A. F. Ferreira
University of Porto, Portugal

Edgar F. Esteves
University of Porto, Portugal

Rosaldo J. F. Rossetti
University of Porto, Portugal

Eugénio C. Oliveira
University of Porto, Portugal

ABSTRACT

Trading off between realism and too much abstraction is an important issue to address in microscopic traffic simulation. In this chapter the authors bring this discussion forward and propose a multi-agent model of the traffic domain where integration is ascribed to the way the environment is represented and agents interoperate. While most approaches still deal with drivers and vehicles indistinguishably, in the proposed framework vehicles are merely moveable objects whereas the driving role is played by agents fully endowed with cognitive abilities and situated in the environment. The authors discuss on the role of the environment dynamics in supporting a truly emergent behaviour of the system and present an extension to the traditional car-following and lane-change models based on the concept of situated agents. A physical communication model is proposed to base different interactions and some performance issues are also identified, which allows for more realistic representation of drivers' behaviour in microscopic models.

INTRODUCTION

Using simulation is imperative for planning and realising the correct relation among parameters of any application domain. In the traffic engineering domain, however, most analyses are carried out on an individual basis as an attempt to reduce the number of variables observed and to simplify the process of finding out their correlations. This brings about the issue of how different standpoints from which the domain is viewed could be coupled in the same model and simulation environment in order to allow for wider analysis perspectives (Barceló, 1991; Grazziotin et al., 2004; Rossetti & Bampi, 1999). This is not a recent concern, though. The basic general framework for a fully transportation theory identifies two different concepts, borrowed from Economics, which encompass all aspects related to demand formulation and supply dynamics within the framework, including multi-modal selection and activities planning (McNally, 2000).

Arguably, realistic models are the first instrument to allow the integration of different analysis perspectives in virtually any application. However, modelling is not an easy task and abstraction is often necessary in order to make thinks feasible. The autonomous agent metaphor has been increasingly used in this way and offers a great deal of abstraction while important cognitive and behavioural characteristics of the system entities are preserved. Also, advances in engineering environments for multi-agent systems (MAS) have fostered the idea of overall system behaviour that emerges from the interaction of microscopically modelled entities.

Some examples of MAS applied to the field of traffic and transportation engineering can be found in the literature (e.g. Bouchefra, Reynaud, & Maurin, 1995; Roozemond, 1999; Davidsson *et al.*, 2005; Oliveira & Duarte, 2005, to mention some) and are further discussed elsewhere in this book. However, most of the applications are concerned with the control system, even though it is possible to recognise an increasing interest in the representation of the driver elements and the way they interact and communicate (e.g. Burmeister, Haddadi, & Matylis, 1997; Rossetti *et al.*, 2000; Rossetti *et al.*, 2002; Dia 2002). To increase complexity, transportation systems have recently evolved so quickly as Intelligent Transportation Systems (ITS) start to make part of everyone's daily life. According to Chatterjee & McDonald (1999), the underlying concept of ITS is to ensure productivity and efficiency by making better use of existing transportation infrastructures. Now, a wide range of novel technologies is presented to the user and start to directly affect the way individuals perceive their surrounding environment and ultimately make decisions, which must also be taken into account.

In very basic terms, the moving element in this whole picture is the vehicle that moves from one point to another throughout the network. Disregarding the importance of pedestrians in the first stage of this work, we consider bicycles, motorcycles, automobiles, trucks and buses as examples of moving elements. However, they are actually moving objects steered by their drivers and sometimes occupied by many other passengers that are people with a trip purpose. Also, their decision concerning how the trip will be carried out in most cases seeks to minimize some individual sense of cost. Therefore, we make a clear distinction between travellers and vehicles, as we shall see later on in this chapter.

Indeed, from a transport planning perspective, the inhabitants of urban areas are potential travellers with specific trip needs. Prior to each journey, travellers must make some options basically regarding mode of transport (whether to drive their own cars or to take a public transport service, for instance), the itinerary and a departure time. To the contrary, in the traffic system perspective, flow is actually formed of each single vehicle. As a simplification then, drivers and vehicles are dealt with indistinguishably in virtually the totality of microscopic models (Gipps, 1981; Gipps, 1986; Hidas, 2005). In the microscopic point of view, however, it is the driver behaviour that influences traffic flow. Actually, drivers manifest an interesting yet implicit social interaction – they compete for the limited resources of the network infrastructure. These different interactions may emerge on an aggregate perspective as these properties will become available in terms of natural stimuli to the inhabitants, who will behave accordingly as they have different perception capabilities and different goals. A proper modelling of the environment then is imperative to allow such specificities of human behaviour to be represented in microscopic traffic simulations.

In this chapter we bring this discussion forward and ascribe to the environment the responsibility for coping with the complexity inherent in the transportation domain, more specifically in the field of traffic modelling and simulation, in order to provide engineers and practitioners with an adequate framework for integrated analyses. Complexity is expected to emerge from the interaction of simpler self-cantered autonomous entities in pursuit of maximizing some individual or collective utility measure. We start by presenting the environment as an interaction means, in next section, where a detailed explanation on the interaction mechanism used to support the implementation of situated agents is presented. In the following section we conceive the architecture of a system to support practical simulation of traffic scenarios on the basis of the concepts discussed, which is followed by some interaction examples to illustrate the approach proposed. In last section, some conclusions are drawn and presented with important considerations for future developments.

THE ENVIRONMENT AS AN INTERACTION MEANS

Agents and their Environment

The perspective over **environments** for MAS has been changing in the direction of an increasing importance of this entity. Weyns *at al.* (2005a) stress out the importance of considering the environment as a first-order abstraction in the engineering process of developing MAS. The authors recall a classical definition of **autonomous agent**: "a system situated within and a part of an environment that senses that environment and acts on it, over time, in pursuit of its own agenda so as to effect what it senses in the future" (Franklin & Graesser, 1996). From this definition, elsewhere Weyns *at al.* (2005b) state "the importance of the environment as the medium for an agent to live, or the first entity the agent interacts with". He also recalls the notion of embodiment as "the fact that an autonomous agent has a "body" that delineates it from the environment in which the agent is situated".

Let us take a better look at the definitions above (of autonomous agent and of embodiment). The first states that the agent is not only situated in the environment: it is a component part of that environment. While not contradicting, this is diverse from the second definition which presents the agent and the environment as separate (and separable) entities. We could redefine an agent's body as a subset of the environment. This allows us to clearly define the agent (the agent still has a body) while providing a wider and more complex notion of environment. We will refer to the agent's body as the agent's internal environment. The environment without the agent's body is the agent's external environment.

Odell, Parunak, & Fleischer (2003) differentiate between *physical environment* and *communication environment*. The physical environment models the physical existence of objects and agents, whereas the communication environment includes the structures that support exchange of information (knowledge). These include roles, groups and communication protocols. They further define *social environment* as a restriction to the set of communication environments. A social environment is "*a communication environment in which the agents interact in a coordinated manner*". Note how the definition somehow restricts the forms of communication that may occur in a MAS. Tummolini *et al.* (2004) introduce *Behavioural Implicit Communication*, in which case communication clearly occurs at the physical level (via perception), diverging from the definition presented by Odell, Parunak, & Fleischer (2003).

Both views can be unified by extending communication to the *physical environment*. We then classify communication into two main modes: *implicit communication*, occurring in the physical environment and *explicit communication*, occurring in the communication environment and regulated by high-level protocols (out of the scope for this chapter). We further classify implicit communication into two distinct forms: *physical communication* (related to the observability of objects and agents) and *behavioural communication* (related to the observability of agent's actions). For the rest of this chapter, we focus on the implicit forms (physical and behavioural).

Physical communication occurs when an agent produces influences over its external environment, these influences produce a state change in that environment, that state change is perceived and interpreted by another agent (could be more than one), and this agent possibly changes its own behaviour in face of the interpretation of the perceived facts. As an example of physical communication, consider the following scenario. When a driver wishes to communicate a lane change to its neighbouring agents, it switches the appropriate car light on. This implies **producing an influence** that will most likely result in a **state change** of the vehicle object controlled by the driver. This change will be detected by the agents that "*pursuing their own agenda*", are scanning the environment for visual perception. Some of the agents will **interpret** the state change as an intention of the peer driver and possibly change their own behaviour in face of the peer's intentions.

Behavioral communication works the same way around, with the difference that it occurs when an agent produces influences over its own internal environment. Examples would be a semaphore controller agent switching the signals, or a flagman waving his arms. The action consists of a list of influences over the agent's own body (the *internal environment*), although success or failure may still depend on the *external environment* (i.e., a power failure would prevent the semaphore controller from switching the lights). This is a very important feature of the model. An agent does not fully control its *internal environment*, since it is also a part of the coexisting agent's *external environment*, and so the agent may be "forced" by these external actions, at least up to some extent. Finally, an action may influence both the internal and external environments at the same time. This is transparent in our model, since both forms of communication are leveled by the way agents send their influences and receive the "messages" (via perception).

The Environment Model

With these notions of environment and communication in mind, we will elaborate a definition of *physical environment* that stresses on the fact that an agent (and all other agents and objects) is *part of* the environment, instead of being merely inserted in it. We define it as collection of entities and laws. Entities may be objects (inanimate, yet possibly reactive or interactive) and agents (animate and partially *autonomous*). These entities and their interactions are ruled by a set of laws about their own properties and about the environment. All these collections are dynamic (objects may be created, consumed, transformed into other objects; agents may enter or leave the environment, "die" or be "born"). As a draft of a more formal approach we may say that:

Env(t) = {*Objs(t)*, *Ags(t)*, *Laws(t)*}

An object is characterized by:

- A set of *perceptible features* (*PF*), representing all possible features that may be perceived by agents. A feature may or may not be active. We could identify the set of active features in a given time *t* as *PF(t)*. It should also be possible to provide the features with operational (run-time) parameters. As an example, consider the lights of a car. They are always perceptible but the current state of the light may change in each time step (it makes sense that a light is a feature that is always active but it may be "on" or "off" e.g., it is a run-time parameter of the feature).
- A set of *interactive features* (*IF*), representing interfaces that provide the environment access to modify the object state. Agents will not have direct access to the *IFs*. The set of active features in a given time *t* is *IF(t)*.
- A *set of properties* (*SP*), representing part of the internal state of the object. Note that we do not restrict the internal state to *SP*. Instead, we consider *SP* is part of the entities' internal state (which also includes the *PF* and *IF* sets).

To limit the agents' autonomy, reflecting the fact that agents are conditioned by the environment which they are part of, and allow for *influences* of the environment over themselves, we define *agent* as a subclass of *object*. The agent may at best have partial control over these influences. This is fundamental to our approach. We long for a highly complete model to accommodate complex environments, allowing agents and agent's actions to be perceived by other agents (agents' actions also have perceptible features) and forced influences from the environment to be performed on the agents. Besides the inherited sets, an agent has:

- A set of *perception abilities* (*PA*), that the agent uses to send messages to the environment expressing the current *foci* of the preceptors and receive messages from the environment with perceptual representations (we will elaborate on this). For performance reasons, only one message is sent/received at each time step, possibly containing several foci/representations.
- A set of *action abilities* (*AA*), that the agent uses to send messages to the environment expressing influences over the agents internal and/or external environments (again, we will restrict agents to send only one message in each time step, though possibly expressing several influences).

Figure 1 illustrates how these sets provide the interface with the external environment of both agents and objects.

The Interaction Mechanism

To connect the basic concepts of our model, we illustrate the relations among the fundamental entities and roles in a class diagram where the roles are specified as "interfaces" (see Figure 2).

The basic design of the suggested architecture is to consider that every entity in the traffic system is an object that influences the PF's of agents (which are also objects). To achieve a desired perception of a part of the environment an agent becomes a listener by sending its *foci* to a mediator. All objects within the listener foci become its casters (becoming a caster of a listener means that the listener must perceive

Figure 1. Primary interfaces of objects and agents with the external environment

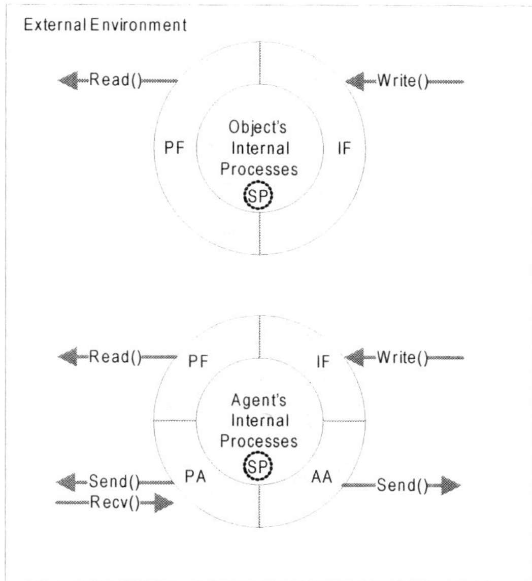

Figure 2. Class diagram with the fundamental entities and their roles

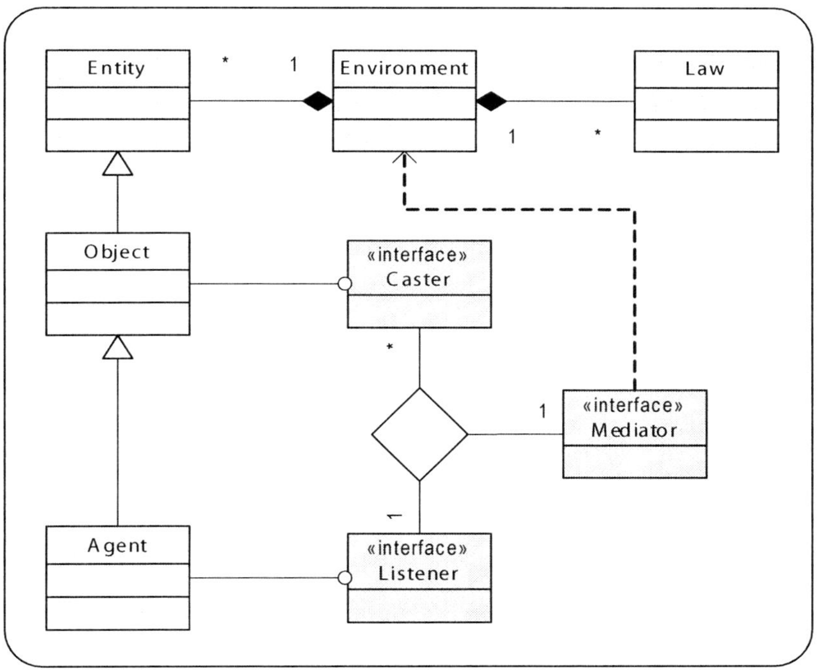

the casters' PF). The mediator is responsible for translating the casters' PF according to the current state and laws that rules the world and sending the set of perceptions to the listener accordingly. The interpretation of the set of PF's received by each listener are stored or updated in the knowledge base of the respective agent. We name this type of knowledge as the agent cognitive memory, or just memory. In some other approaches, this information could integrate the beliefs set of an agent (Wooldridge, 1999). Anticipating performance issues, and relating model elements to real world counterparts, we consider that the traffic environment can be divided into zones, each of which will be assigned a mediator. More on this topic will be discussed later on in this chapter.

Thus, each mediator contains a representation of all entities inside its zone. So a listener sends its influence (for example a car that accelerates influences the external environment) and its foci to the mediator that updates its zone representation. The mediator contains a representation of every agent structure, allowing the correct interpretation of the agent's set of PF and all its internal states, and updates the necessary information into that structure based on the influence sent by the listener. The influence created by an agent affects the entire surrounding environment and consequently the perception of it. The mediator is also responsible for finding the correct casters for each listener based on the foci sent by each agent in every time step. All objects inside a given foci become casters to that listener. These casters are basically the entities that exist in the mediator, representing agents or objects in its environment zone that are inside the agent *foci*.

The mediator is able to read and write the state of any object including access to the objects' PF's (for example, if an agent is a car and it decides to turn on the lights it will change the characteristics of the front vehicles because they become more illuminated, so a perception feature (illumination) was changed in those vehicles because of an influence made by the listener. Therefore the mediator needs to access those objects internal perception feature set, search for the illumination perception feature in it and change it to a new value accordingly). If necessary, then the mediator alters the perception features of the casters based on the influence provoked by the listener and after it reads all of the casters perceptions features. With this information it builds the perception of the listener having into account the laws of the environment (for example, if a listener is looking at its front and has a truck and a car as its casters, if the caster car is in front of the truck and the listener car is very near to the truck the listener car cannot receive perception of the caster car, unless the law of the environment rules that trucks are transparent).

With these perceptions received the listener must update its cognitive memory. Such a memory can be understood as a human driver mental perception of objects surrounding its vehicle (e.g. other cars, traffic signs, traffic lights, people, buildings, and so on). An agent's memory is dynamically updated according to the perceptual representations received by its PA and by the execution of AA. After finishing updating all agents' memory, a time step cycle is concluded. When a new one begins each agent has to decide on the action (influence) it must take, and where to focus its attention on. The decisions are made based on the information of the agent's memory updated on the last time step, and on its own desires (desired destiny, desired speed, desired sight, and so forth). Sometimes the information contained in its memory is not enough for an agent to transform a desire into an intention causing the agent to engage in a course of actions (e.g. it desires a left lane change but the left back vehicle perception is too old to risk it without updating it first). In these cases, an agent can continue its movement and focus its attention to the desired scene in order to obtain the necessary information to fulfil its desires.

The model of the interaction mechanism explained is depicted in Figure 3, in form of a sequence diagram, and the concepts herein presented are illustrated in a more concrete way through example scenarios later on.

OVERVIEW OF THE SYSTEM

According to what has been discussed so far a distributed system is defined to support the implementation of a microscopic simulation engine (MSE). The MSE contains all the world states and objects, and the laws of the world. It also contains the mediators that will translate and send the updated perceptions of objects to the agents that need them. The necessity of a distributed system is a must to guarantee system efficiency and also as a natural way to implement the entities of our application domain.

System Architecture

Basically the world is represented by a set of zones each containing a mediator. Each zone runs independently, having a centralised process that is responsible for the coordination of the world time steps (it guarantees that every zone processes the correct time step, meaning that it is not possible to have different zones processing simultaneously at different time steps). This synchronous process is also

Figure 3. Sequence diagram, detailing the interactions

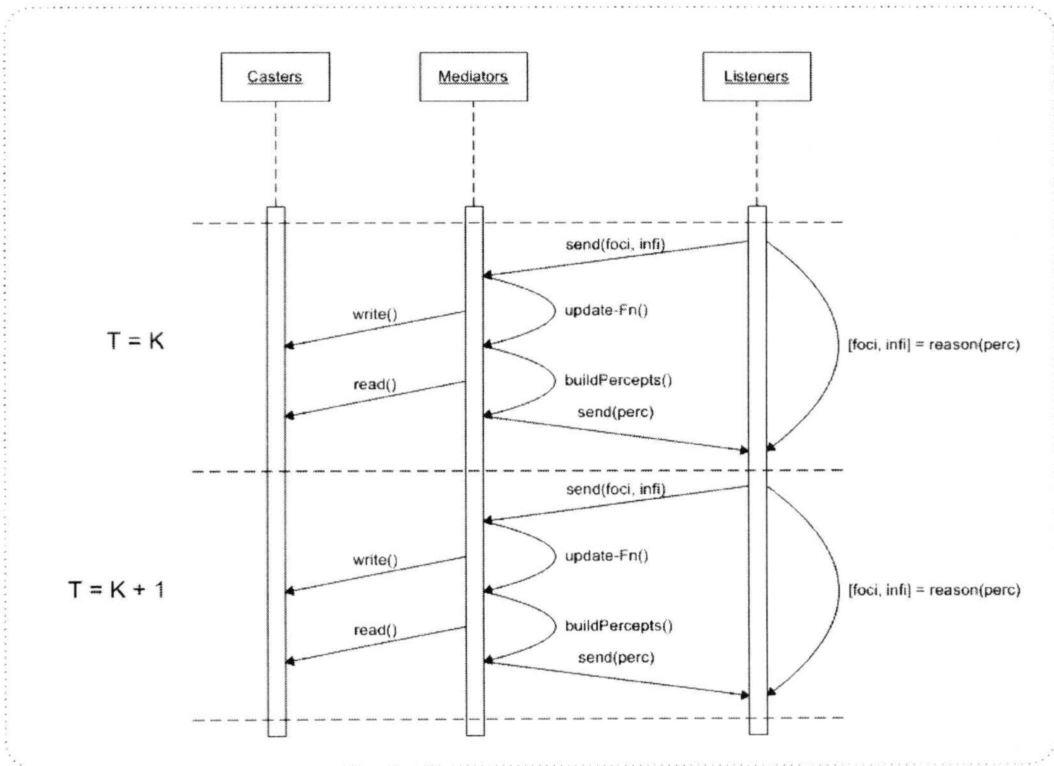

responsible for reading the topology of the networks, dividing them into zones, receiving the registration of mediators, and assigning them an appropriate zone. A simulation cannot be started until each zone has been assigned a mediator. It also allows registration of graphical interface modules providing them, in the registration, with the address of each mediator. Such a structure also allows for the simulation to run with no graphical support, which can contribute to speed up simulation studies.

Each mediator provides a communication interface responsible for sending the updated zone states to the different graphical interface modules so they can create real-time graphical representations of the simulation in runtime. There is also a centralised process that provides a communication interface for these graphical modules in order to allow them to stop or to start simulation runs, change environment characteristics such as the set of laws, load different networks, save simulation states, and so on. Such a centralised process is managed by the simulation engine controller (SEC).

In Figure 4, it is possible to identify the domain entities, as well as their relations. A connection between two nodes represents a road. A road is a set of road segments. The division of a road into road segments depends on the different number of lanes or the different geometry a road can have. For example, if in the beginning of a road there are two lanes, but in the middle of the road it passes to have only one lane, it means that the road has two road segments – a road segment with two lanes and another road segment with just one lane. The world objects are decomposed into two different entities, namely the entities that have the capacity to move (vehicles and people, for instance) and the ones that are static (traffic controllers, road signs – both horizontal and vertical, road obstacles, and so on). In every given time an entity is situated in a lane.

Figure 4. Class Diagram of the environment domain

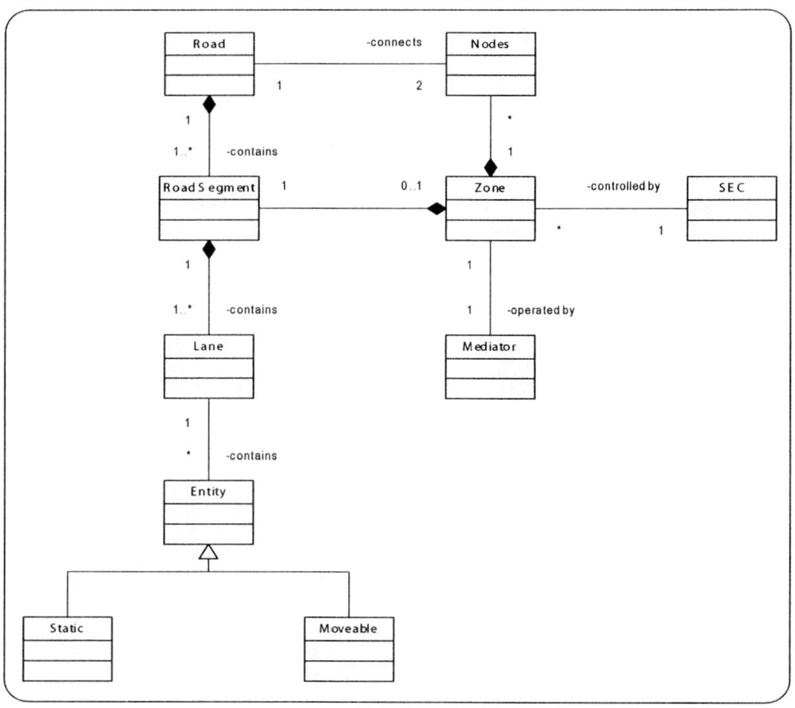

A system physical overview is represented in the diagram of Figure 5. In that structure a mediator has always the necessary information to construct perceptions for the correct behaviours of the world agents. Their interaction will follow the mechanism proposed in this research.

Environment Zones

Since the perception treatment and communication can be a heavy load for overcrowded networks the distribution of the environment perceptions becomes critical in order to improve the global efficiency of the simulation.

In order to assure that each agent receives the world perception efficiently in every time step, we assume that the process that delivers it has a limited capacity of the number of agents it has to inform. So a distributed division of the environment is a question of defining the correct capacity limit and number of perceptions a mediator will be dealing with. Such an organisation easily resembles the concept of traffic zones, used in control and management systems in most urban areas.

As defined before, the entities responsible for the delivery of the perceptions are the *Mediators*. Analyzing the scenario presented in Figure 6 and assuming that M1 has a limit capacity of 3000 vehicles, M2 of 2000 vehicles and M3 of 1500 (the limit capacity of Mediators is calculated based on the processing capacity of the machine in which they are instantiated). The division into different Mediator zones is easy to obtain. Each link (Road Segment) has a physical capacity, limiting the quantity of vehicles it can contain. This means that in the worst scenario each road segment will only ensure that number of vehicles. So a mediator zone is defined as a set of road segments, whose sum of their capacities is equal or lower to the vehicle capacity of its mediator.

Figure 5. System physical overview

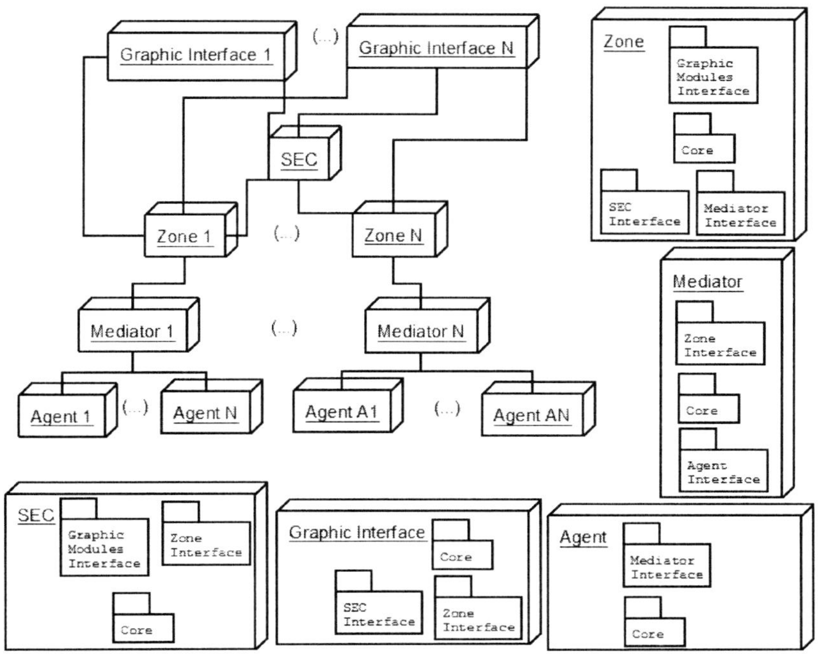

Figure 6. Example of a possible network

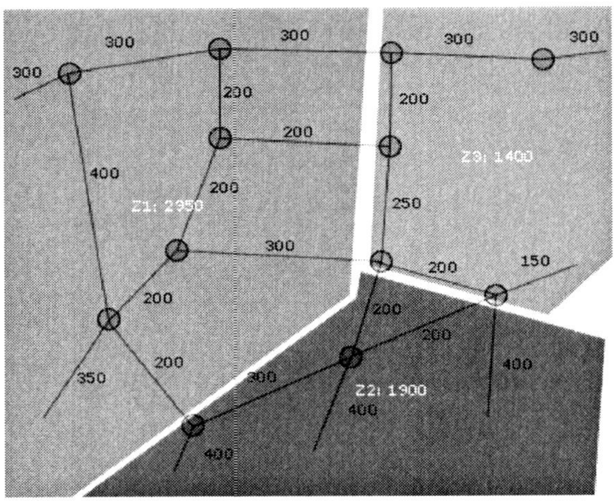

This way it is possible to guarantee that the mediator will process, in the worst scenario, the world perceptions of a number of drivers equal or lower to its own capacity. Translating it to the scenario of Figure 6, M1 will be assigned zone 1 (Z1 in the figure), M2 will be assigned Z2 and M3 will be assigned Z3. This means that each of the Mediators will be responsible for translating the zone objects' perceptible features to all agents inside its assigned zone that ask for it.

ILLUSTRATIVE SCENARIOS

Consider the following scene as depicted in Figure 7, representing the current state of the environment and already populated with all the casters and listeners that will interact throughout the examples. The visual focus (for the current time step) of the agents controlling vehicles A, B, C and D is represented by the highlighted circle slices. In fact we opted to represent all of these vehicles to explain different situations that occur in traffic simulations and also to explain the concepts related to the "car follow-

Figure 7. An example of a time-step of the simulation

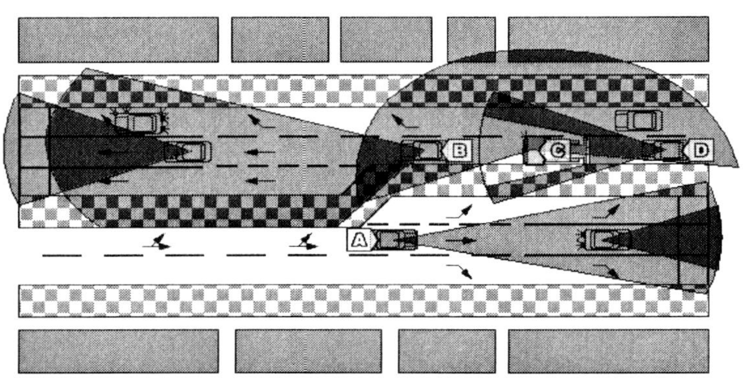

ing" (CF) and "lane changing" (LC) behaviours found in most microscopic traffic models. Let us call the agents by the letters on the vehicles they drive. Along with their foci, they have also expressed the influences over the environment for this time step.

To ease the understanding of the concept of CF let us centre on agent A. Since it wants to go in front, its foci are naturally the front area. Take into account that as it becomes a listener the front vehicle becomes its caster. In the meantime the memory of the agent (see Figure 8) is updated according to the interpretation of the received PF's (given by the Mediator). This information is given with regard to the object which is being observed by the subject driver, so perceptions are enclosed into balloons attached to the object being observed.

For this specific example agent A will take the particular action of "BRAKING". That's because it does not have any previous deduction (previous time step) of the other vehicle's velocity ("REALLY FASTER"; "FASTER"; "SLOWER" and "REALLY SLOWER"). In the next time step it will send that action to its mediator.

More complex situations can occur, like demonstrated by agents B and C. Agent B was having the same attitude as the one demonstrated before but new variables will make it to change (see Figure 9). Assuming that it wants to go in front, a new lane appears in that direction and the front vehicle was evaluated as going "SLOWER". It will take the action "CHANGE_TO_LEFT_LANE" then. This kind

Figure 8. Memory entries resulting from perception of agent A

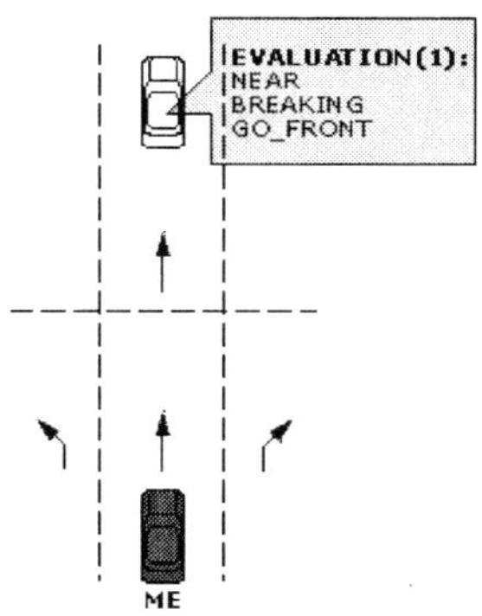

Figure 9. Memory entries resulting from perception of agent B

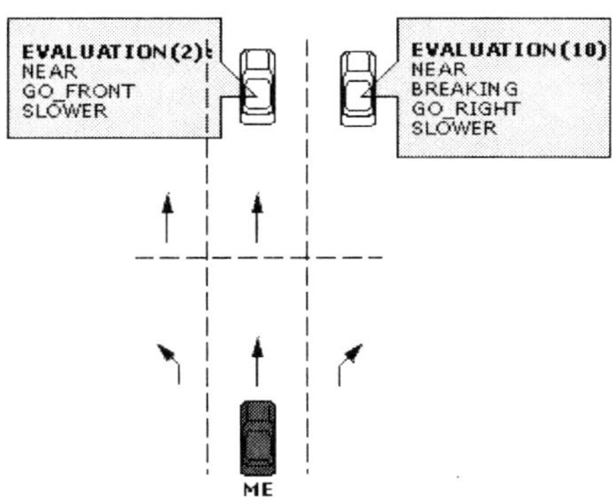

of actions transpires when an agent wants to maintain or achieve its desired velocity and is inherited from the lane changing concept.

The previous figure also illustrates a representation of a "front right vehicle" that is having the intension of turning right. If in the next time step it transforms its intention into an influence, it will be deleted from agent B's memory.

At the same time agent C is in a delicate situation. It needs to go to the right lane to accomplish its path direction previously defined (supposedly). Like in real situations, in which we need to look into the mirrors and take care with the front vehicle, it sets its foci to the front, back and right sides. The Mediator informs it about all the casters positions, velocities, acceleration and intentions (PF's) and the evaluation of its memory will be like the one represented in Figure 10. The fact that the back right vehicle is being faster than itself will not permit the lane changing in the current time step (according to its own AA's) forcing it to wait for the next time step. If in future steps the back right vehicle does not pass him or new similar situations occur it will be impossible to make that action and agent C will be forced to stop.

Taking into account that in human behaviours there is also a factor of cordial attitudes, it is agreeable to think that agent C can try to change to another lane to let pass the back vehicle (since its velocity evaluation is "FASTER"). This kind of actions is also inherited from the lane changing concept (Gipps, 1986).

The representation of agent D is intended to illustrate two different kinds of laws in the present scenario (transparent and opaque objects). The PA's are affected by these laws since the Mediator interprets the PF's according to them and to the agent' foci. As a consequence the casters are not the same in the two different configurations. In the case of agent D there are two vehicles directly in front of it and inside its foci (C and B). If the laws of the environment are configured to opaque objects then the mediator will not give D the PF's of object B (vehicle C is a truck and blocks the visibility of agent D). Otherwise if the laws are configure to allow transparent objects then PF's of both C and B will be included in the information that the mediator will send to agent D. This example shows the influence that the laws of the environment can have over perceptions captured by each agent.

In Figures 8, 9, and 10, notice the numbers that appear inside the parentheses and after the evaluation word. Those numbers represent the last time that the evaluation of that caster was done. That means

Figure 10. Memory entries resulting from perception of agent C

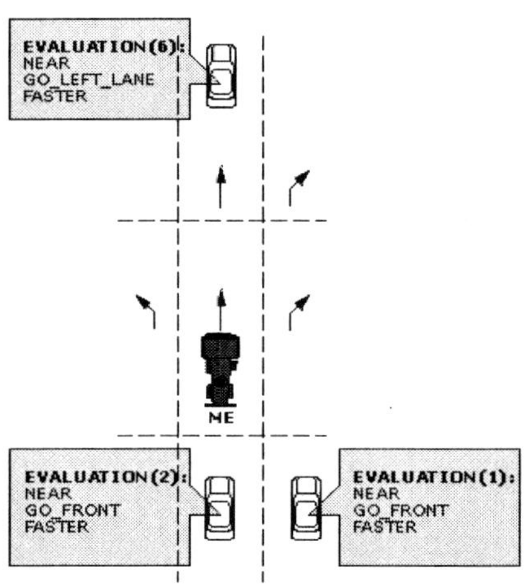

those agents have more or less trust on their evaluations according to their PA's (for instance, if a back car is FAR and SLOWER the agent does not need to verify whether it is near every time step). It is possible to have a factor in each agent that dictates how each agent will trust on predicting future positions of its surrounding objects. For example, consider that an agent looks back in time step n and gets the perception of an agent called X. If in the time step n+20 the agent needs the information about agent X to perform an influence, it must decide whether to have to look back to update the perception of X in its memory or if it trusts its future prediction just on information perceived 30 time steps ago.

A prototype of the proposed system is being developed and some basic features of the communication mechanism were implemented, demonstrating the potential of this approach in extending traditional car-following and lane-changing behaviours. The environment is a first-order abstraction that plays an imperative role in this framework being developed. An example of its interface is depicted in Figure 11.

CONCLUSION

In this chapter we present a multi-agent model to cope with the complexity inherent in microscopic traffic simulation modelling in order to provide engineers and practitioners with an adequate framework for integrated analyses. The physical conceptualization of the environment using the interaction mechanisms presented as the basis for every interaction among agents and the environment itself allows for different perception abilities of individuals to be implemented and assessed, which is expected to have a direct influence in the emergence of the system overall performance in different circumstances. Therefore, a truly agent-based microscopic simulation approach must necessarily be build on the basis of the concept of situated agents and consider the environment as a first-order abstraction, playing as relevant roles as other entities in the system. In this way, as drivers are integrant parts of the environment and interact directly with it, more realistic behaviours can now be considered. With such a concept

Figure 11. Prototype of the simulation environment

of environment, traditional car-following and lane-changing models can be extended to feature more contemporary performance measures, which can include influence of road-side parking, collisions, interaction with traveller information systems, en-route decision-making, and many others. This is just possible as different perception abilities of drivers can now be considered in the way they interact with their environment. An initial prototype with very simple features of the presented model has been implemented, to demonstrate car-following and lane-changing behaviours. The very next steps in this research include the improvement of this prototype to fully demonstrate all the potential of the concept of situated agents and the role of the environment in implementing more realistic microscopic traffic simulations. Also in the agenda, we expect to devise an appropriate methodology for validating and calibrating such agent-based microscopic traffic models. Following this, some simulations and analyses of performance measures will be carried out as well.

ACKNOWLEDGMENT

We gratefully acknowledge the financial support from the Department of Electrical and Computer Engineering at Faculty of Engineering, University of Porto, and from the GRICES-CAPES bilateral cooperation programme.

REFERENCES

Barceló, J. (1991). Software environment for integrated RTI simulation systems. In Proceedings of the DRIVE Conference, *Advanced Telematics in Road Transport, 2,* 1095-1115. Amsterdam: Elsevier.

Bouchefra, K., Reynaud, R., & Maurin, T. (1995). IVHS viewed from a multiagent approach point of view. *In Proceedings of the IEEE Intelligent Vehicles Symposium.* Piscataway, NJ: IEEE. 113-117

Burmeister, B., Haddadi, A., & Matylis, G. (1997). Application of multi-agent systems in traffic and transportation. *IEE Proceedings on Software Engineering, 144*(1), 51-60.

Chatterjee, K., & McDonald, M (1999). Modelling the impacts of transport telematics: current limitations and future developments. *Transport Reviews, 19*(1), 57-80.

Davidsson, P., Henesey, L., Ramstedt, L., Törnquist, J., & Wernstedt, F. (2005). An analysis of agent-based approaches to transport logistics. *Transportation Research, 13C*(4), 255-271.

Dia, H. (2002). An agent-based approach to modelling driver route choice behaviour under the influence of real-time information. *Transportation Research, 10C*, 331-349.

Franklin, S., & Graesser, A. (1996). Is it an agent or just a program? A taxonomy for autonomous agents. In Intelligent Agents III, *Agent Theories, Architectures, and Languages* (LNAI, No.1193). 21-35.

Gipps, P. G. (1981). A behavioural car-following model for computer simulation. *Transportation Research*, 15B, 105-111.

Gipps, P. G. (1986). A model for the structure of lane-changing decisions. *Transportation Research, 20B*, 403-414.

Grazziotin, P. C., Turkienicz, B., Sclovsky, L., & Freitas, C. M. D. S. (2004). CityZoom – A Tool for the Visualization of the Impact of Urban Regulations. *In Proceedings of the 8th Iberoamerican Congress of Digital Graphics*. (pp. 216-220).

Hidas, P. (2005). Modelling Vehicle Interactions in Microscopic Simulation of Merging and Weaving. *Transportation Research, 10C*, 37-62.

McNally, M. G. (2000). The four step model. In Handbook of Transport Modelling (pp. 35-52). Oxford: Pergamon Press.

Odell, J., Parunak, H. V. D., & Fleischer, M. (2003). Modeling Agents and their Environment: the communication environment. *Journal of Object Technology, 2*(3), 39-52.

Oliveira, E., & Duarte, N. (2005). Making way for emergency vehicles. In the European Simulation and Modelling Conference (pp. 128-135). Ghent: EUROSIS-ETI.

Roozemond, D. A. (1999). Using autonomous intelligent agents for urban traffic control systems. *In Proceedings of the 6th World Congress on Intelligent Transport Systems.*

Rossetti, R. J. F., & Bampi, S. (1999). A Software Environment to Integrate Urban Traffic Simulation Tasks. *Journal of Geographic Information and Decision Analysis, 3*(1), 56-63.

Rossetti, R. J. F., Bampi, S., Liu, R., Van Vliet, D., & Cybis, H. B. B. (2000). An agent-based framework for the assessment of drivers' decision-making. *In Proceedings of the IEEE Conference on Intelligent Transportation Systems* (pp. 387-392). Piscataway, NJ: IEEE.

Rossetti, R. J. F., Bordini, R. H., Bazzan, A. L. C., Bampi, S., Liu, R., & Van Vliet, D. (2002). Using BDI agents to improve driver modelling in a commuter scenario. *Transportation Research, 10C*, 373-398.

Tummolini, L., Castelfranchi, C., Omicini, A., Ricci, A., & Viroli, M. (2004). Exhibitionists and Voyeurs do it better: a Shared Environment for Flexible Coordination with Tacit Messages. *In 1st International Workshop on Environments for Multiagent Systems (LNAI, No.3374).* (pp. 215-231).

Weyns, D., Parunak, H., Michel, F., Holvoet, T., & Ferber, J. (2005a). Environments for multiagent systems, State-of-the-art and research challenges. *In 1st International Workshop on Environments for Multiagent Systems (LNAI, No.3374).* (pp. 1-47).

Weyns, D., Schumacher, M., Ricci, A., Viroli, M., & Holvoet, T. (2005b). Environments in multiagent systems. *The Knowledge Engineering Review, 20*(2), 127-141.

Wooldridge, M. (1999). Intelligent Agents. In Weiss, G. (Ed.), *Multiagent Systems: A modern approach to distributed artificial intelligence.* Cambridge, MA: The MIT Press.

Chapter VI
Fundamentals of Pedestrian and Evacuation Dynamics

Andreas Schadschneider
Universität zu Köln, Germany

Hubert Klüpfel
TraffGo HT GmbH, Germany

Tobias Kretz
PTV AG, Germany

Christian Rogsch
University of Wuppertal, Germany

Armin Seyfried
Forschungszentrum Jülich GmbH, Germany

ABSTRACT

Multi-Agent Simulation is a general and powerful framework for understanding and predicting the behaviour of social systems. Here the authors investigate the behaviour of pedestrians and human crowds, especially their physical movement. Their aim is to build a bridge between the multi-agent and pedestrian dynamics communities that facilitates the validation and calibration of modelling approaches which is essential for any application in sensitive areas like safety analysis. Understanding the dynamical properties of large crowds is of obvious practical importance. Emergency situations require efficient evacuation strategies to avoid casualties and reduce the number of injured persons. In many cases legal requirements have to be fulfilled, for example, for aircraft or cruise ships. For tests already in the planning stage reliable simulation models are required to avoid additional costs for changes in the construction. First, the empirically observed phenomena are described, emphasizing the challenges they pose for any modelling approach and their relevance for the validation and calibration. Then the authors review the basic modelling approaches used for the simulation of pedestrian dynamics in normal and emergency situations, focussing on cellular automata models. Their achievements as well as their limitations are discussed in view of the empirical results. Finally, two applications to safety analysis are briefly described.

INTRODUCTION

Understanding and predicting the dynamical properties of large human crowds is of obvious practical importance (Schadschneider et al., 2009). Especially emergency situations and disasters require efficient evacuation strategies to avoid casualties and reduce the number of injured persons. In many cases legal regulations have to be fulfilled, e.g. for aircraft or cruise ships. For tests already in the planning stage reliable simulation models are required to avoid additional costs for changes in the construction. But even, if changes in the construction are not possible, simulations can be very helpful for organizational issues like the design of evacuation routes, where full-scale tests are either too expensive or too dangerous.

Multi-Agent Simulation provides a general and powerful framework for understanding and predicting the behaviour of social systems. In this contribution, we describe its application to the dynamics of human crowds, especially their physical movement. The latter restriction allows us to focus on the operational and tactical levels of the agents' decisions. 'Operational' in this context means the proper body motion, i.e. avoiding collisions and the movement within a short time-span (e.g. one second). 'Tactical' means that putting this in the well-established BDI-framework (beliefs, desires, intentions), only the intentions (like getting to the closest exit in the case of an evacuation) are explicitly modelled. Desires and beliefs are either neglected or modelled implicitly, e.g. by assuming that everyone wants to get out as fast as possible and representing orientation as following the gradient of a static floor field (for details, please refer to the following sections). Furthermore, the multi-agent paradigm is flexible enough to cover the model extensions that belong to the tactical and strategic realm.

Having said that, the fact that there are such distinct models as cellular automata and molecular dynamics-like simulations used in the field, gives strong hint to the need for a thorough understanding of basic model characteristics, their scope and limitations. This part can be addressed by investigating the models themselves without making reference to empirical data. This is useful but of course not sufficient. Therefore, we will cover the latter in this contribution, too.

In recent years several models of different sophistication have been developed. Macroscopic approaches use a coarse-grained description in terms of densities. In contrast in microscopic models, which are the focus of this review, different agents[1] are distinguished. This allows to equip them with different properties reflecting demographics.

In this contribution we will try to give a compact introduction to the most important empirical results and theoretical approaches. All of these are relevant for most agent-based simulations of pedestrian dynamics. We will emphasize the importance of a close interplay of empirical observations and data with theoretical modelling approaches. We demonstrate how the realism of the model dynamics can be improved by taking into account qualitative and quantitative empirical observations. Such validation and finally calibration is extremely important, e.g. for the applications in safety analysis mentioned above.

In Sec. "Empirical Results" we will give an overview of the experimental observations. A variety of interesting dynamical properties and collective effects (fundamental diagram, behaviour at bottlenecks, lane formation in counterflow, flow oscillations etc.) have been found that provide information about the basic interactions and can be used as a kind of benchmark test for any modelling approach. Quantitative results are used to obtain the parameters specifying the interactions between the agents.

We then review in Sec. "Modelling of Pedestrian Dynamics" microscopic approaches to model pedestrian dynamics in normal and emergency situations. Our focus will be cellular automata based models, especially those related to the floor field model. Although the dynamics is often stochastic the cellular automata approach allows an intuitive specification of the motion of pedestrians. It can be easily extended to include not only interactions between different agents but also with the infrastructure, e.g. doors, stairs or walls. In more complex situations an extension to a multi-agent model is possible, e.g. by specifying origin-destination matrices etc.

Although the currently available modelling approaches give a quite accurate representation of pedestrian motion and crowd dynamics, certain aspects need to be improved. For cellular automata models these are often connected with the discreteness in space and time. This will be discussed in Sec. "Validation and Extension of CA Models". In Sec. "Application of Models" two concrete applications related to safety analysis are discussed. Finally, we will discuss future challenges and research directions by identifying the most important open problems.

EMPIRICAL RESULTS

An important part of empirical results are qualitative observations of collective effects which are often know from everyday experience. Quantitative results, on the other hand, are much more difficult to obtain. Controlled experiments are rare and sometimes it is questionable whether results obtained under laboratory conditions can be transferred to realistic scenarios.

Collective Phenomena

One of the reasons why physicists are interested in pedestrian dynamics is the large variety of interesting collective effects and self-organization phenomena that can be observed. These macroscopic effects reflect the individuals' microscopic interactions and thus give also important information for any modelling approach. Any model that does not reproduce these effects is missing some essential part of the dynamics.

- **Jamming:** Jamming and clogging typically occur for high densities at locations where the inflow exceeds the capacity. Such locations with reduced capacity are called *bottlenecks*. Typical examples are exits (Fig. 1) or narrowings. Clogging is not related to the microscopic dynamics of the agents, but rather a consequence of an *exclusion principle*: space occupied by one particle is not available for others. Identification of possible jamming locations is very important for practical applications, especially evacuation simulations. A different kind of jamming occurs in counterflow when two groups of pedestrians moving in opposite directions mutually block each other. This happens typically at high densities and when it is not possible to turn around and move back, e.g. when the flow of people is large.

- **Density waves:** Density waves in pedestrian crowds can be generally characterised as quasi-periodic density variations in space and time. A typical example is the movement in a densely crowded corridor (close to the density that causes a complete halt of the motion) where phenomena similar to stop-and-go vehicular traffic can be observed, e.g. density fluctuations in longitudinal direction that move backwards (opposite to the movement direction of the crowd) through the corridor. Recently, the occurrence of stop-and-go waves has been reported for the Hajj pilgrimage in Makkah (Helbing et al., 2007a).
- **Lane formation:** In counterflow, (dynamically varying) lanes are formed where people move in just one direction (Oeding, 1963; Navin et al., 1969; Yamori, 1998) (see Fig. 1). In this way, motion becomes more comfortable and allows higher walking speeds, since strong interactions with oncoming pedestrians are reduced. The occurrence of lane formation does not require a preference for moving on one side. It also occurs in situations without left- or right-preference. However, such a preference, e.g. due to cultural aspects, can have an influence on the structure of lanes.There are only a few quantitative empirical studies of lane formation (Yamori, 1998; Kretz et al., 2006a). Most results are based on qualitative observations which e.g. show that the number of lanes can vary considerably with the total width of the flow. It often fluctuates in time, even for small changes in the total density. Furthermore the number of lanes in opposite directions is not necessarily identical. A surprising results of the quantitative experiments is the occurrence of surprisingly large flows: the sum of (specific) flow and counterflow can even exceed the specific flow for one-directional motion.
- **Oscillations:** In counterflow at bottlenecks (e.g. doors) often oscillatory changes of the direction of motion are observed. Once a pedestrian has passed the bottleneck it is easier for others to follow in the same direction. This changes when somebody is able to pass (e.g. through a fluctuation) the bottleneck in the opposite direction.
- **Patterns at intersections:** When several streams of pedestrians moving in different directions intersect, various collective patterns of motion can be formed. Typical examples are short-lived roundabouts at four-way crossings which make the motion more efficient. Even if they are connected with small detours the formation of these patterns can be favourable since they allow for a "smoother" motion.
- **Emergency situations, "panic":** In emergency situations various collective phenomena have been reported that have sometimes misleadingly been attributed to *panic behaviour*. However, there is no precise accepted definition of *panic* although in the media usually aspects like selfish, asocial

Figure 1. Schematic representation of collective phenomena observed in pedestrian dynamics. (left) clogging at a bottleneck (exit); (right) lane formation in counterflow.

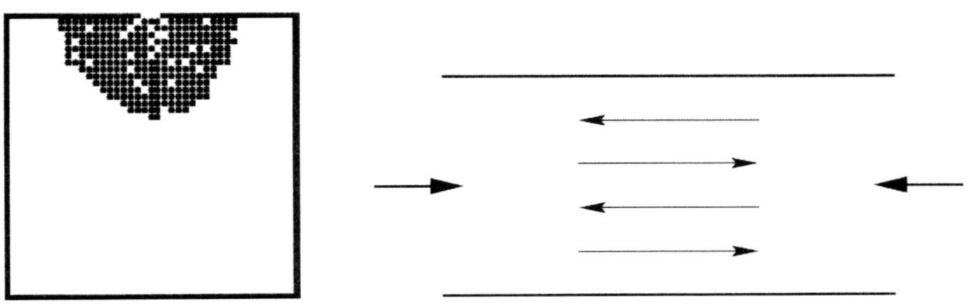

or even completely irrational behaviour and contagion that affects large groups are associated with this concept (Keating, 1982). However, in many emergency situations it has been found that these characteristics have played almost no role (see e.g. Johnson, 1987) and that the reasons for these accidents are much simpler. Therefore the term "panic" should be avoided, *crowd disaster* being a more appropriate characterisation. Related concepts like "herding" and "stampede" seem to indicate a certain similarity of the behaviour of human crowds with animal behaviour. This terminology is also quite often used in the public media. *Herding* has been described in animal experiments (Saloma, 2006) and is difficult to measure in human crowds. However, it seems to be natural that herding exists in certain situations, e.g. limited visibility due to failing lights or strong smoke when exits are hard to find. Although empirical data on crowd disasters exist, e.g. in the form of reports from survivors or even video footage, it is almost impossible to derive quantitative results from them. Models that aim at describing such scenarios make predictions for certain counter-intuitive phenomena that should occur. In the *faster-is-slower effect* (Helbing et al., 2000b) a higher desired velocity leads to a slower movement of a large crowd. In the *freezing-by-heating effect* (Helbing et al., 2000a) increasing the fluctuations can lead to a more ordered state. For a thorough discussion we refer to (Helbing et al., 2000b; Helbing et al., 2002) and references therein. However, from a statistical point of view there is no sufficient data to decide the relevance of these effects in real emergency situations, also because it is almost impossible to perform "realistic" experiments.

Quantitative Results

We now introduce the fundamental observables which are the foundation of any quantitative description of pedestrian dynamics. A more detailed critical discussion and list of references can be found e.g. in (Schadschneider et al., 2009).

Flow and Density

The flow J of a pedestrian stream gives the number of pedestrians crossing a fixed location of a facility per unit of time. There are various methods to measure the flow, e.g. via the average time gap $\langle \Delta t \rangle$ between two consecutive pedestrians which is directly related to the flow

$$J = \frac{1}{\langle \Delta t \rangle}. \tag{1}$$

Alternatively the flow of a pedestrian stream through a facility of width b can be determined in analogy to fluid dynamics using the average density ρ and the average speed v of a pedestrian stream as

$$J = \rho v b = J_s b, \tag{2}$$

where the specific flow[2]

$$J_s = \rho v \tag{3}$$

gives the flow per unit-width. This relation is also known as *hydrodynamic relation*.

Measuring densities is usually more difficult. One possible way is by counting the number of pedestrians N within the selected area A. One can then associate a local density $\rho = \frac{N}{A}$ with the center of the area.

Another definition of density considers the ratio of the sum of the projection area f_j of the bodies and the total area of the pedestrian stream A, defining the (dimensionless) density $\tilde{\rho}$ as

$$\tilde{\rho} = \frac{1}{A}\sum_j f_j \tag{4}$$

which is known as *occupancy* in vehicular traffic.

Other ways to quantify the pedestrian density have been proposed, e.g. the "pedestrian area module" (Fruin, 1971) or the "inter-person distance" (Thompson et al., 1994).

The use of various density definitions in the literature make quantitative (and sometimes even qualitative) comparisons of different results difficult. This has also consequences for the calibration of modelling approaches. In the modelling section (Sec. "Modelling of Pedestrian Dynamics") we will focus on CA models where a natural definition of density is given by the fraction of occupied cells.

Fundamental Diagram

The fundamental diagram describes the relation between density ρ and flow J. Due to the hydrodynamic relation (3) there are three equivalent forms: $J_s(\rho)$, $v(\rho)$ and $v(J_s)$. In applications the relation is a basic input for engineering methods developed for the design and dimensioning of pedestrian facilities (Fruin, 1971; Nelson et al., 2002; Predtechenskii et al., 1978) and it serves as quantitative benchmark for models of pedestrian dynamics.

The empirically obtained fundamental diagrams vary considerably. All three characteristic values of the fundamental diagram, namely its maximum (the *capacity*) $J_{s,\max}$, the density ρ_c where the capacity is reached, and the density ρ_0 where the velocity approaches zero due to overcrowding, differ strongly in various studies (Fig. 2).

The problems with the measurements of density and flow described in Sec. "Flow and Density" are only one possible reason for these discrepancies. But also cultural and population differences, differences between uni- and multidirectional flow or the type of traffic (commuters, shoppers) could be of relevance. However, in all diagrams velocity decreases with increasing density. For the movement of pedestrians along a line a linear relation between speed and the inverse of the density was measured in (Seyfried et al., 2005). The speed of walking pedestrians depends linearly on the step size (Weidmann, 1993). Since the inverse of the density can be regarded as the required length for a pedestrian to move, smaller step sizes caused by a reduction of the available space with increasing density is, at least for a certain density region, one origin of the observed decrease of speed.

Bottleneck Flow

The flow of pedestrians through bottlenecks shows a rich variety of phenomena, e.g. the formation of lanes at the entrance to the bottleneck (Hoogendoorn et al., 2003b; Hoogendoorn et al., 2005; Kretz et al., 2006b; Seyfried et al., 2007), clogging and blockages at narrow bottlenecks (Predtechenskii et al., 1978; Muir et al., 1996; Helbing et al., 2005; Kretz et al., 2006b) or some special features of bidirectional

Figure 2. Fundamental diagrams for pedestrian movement in planar facilities. The lines refer to specifications according to planning guidelines Adapted from: (SFPE Handbook (Nelson et al., 2002)), Predtechenskii and Milinskii (PM) (Predtechenskii et al., 1978), Weidmann (WM) (Weidmann, 1993)). Data points give the range of experimental measurements (Older (Older, 1968) and Helbing (Helbing et. al, 2007b)).

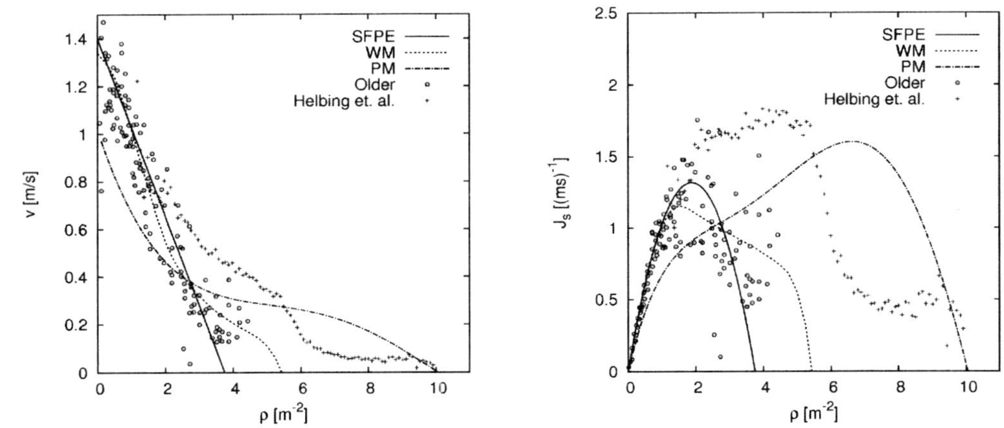

bottleneck flow (Helbing et al., 2005). Moreover, the estimation of bottleneck capacities by the maxima of fundamental diagrams is an important tool for the design and dimensioning of pedestrian facilities.

One of the most important practical questions is how the capacity of the bottleneck increases with increasing width. Although this has been investigated for a long time now, it is still discussed controversially. Two different scenarios are possible: the capacity can either increase stepwise or continuously.

At first sight, a stepwise increase of capacity with width appears to be natural in case of lane formation. Then capacity will increase only when an additional lane appears. A recent empirical study (Hoogendoorn et al., 2003b; Hoogendoorn et al., 2005) has found indications for lane formation and due to the *zipper effect*, a self-organization phenomenon leading to an optimization of the available space and velocity, these lanes are not independent and thus do not allow passing (Fig. 3), implying a stepwise increase of capacity.

In contrast, the study (Seyfried et al., 2007) found a lane distance which increases continuously as illustrated in Fig. 3. This leads to a very weak dependence of the density and velocity inside the bottleneck on its width. Thus in reference to Eq. (2) the flow does not necessarily depend on the number of lanes and increases continuously.

Blockages in Competitive Situations

By definition a bottleneck is a limited resource and it is possible that under competitive situation pedestrian flow through bottlenecks is different from the flow in normal situations. A qualitative difference to normal situations is the occurrence of blockages, e.g. in the form of stable wedges. These obstructions occur due to the formation of arches in front of the bottleneck under high pressure which is very similar to the well-known phenomenon of *arching* occurring when granular materials flow through narrow openings (Wolf et al., 1996).

Figure 3. A sketch of the zipper effect with continuously increasing lane distances in x: The distance in the walking direction decreases with increasing lateral distance. Density and velocities are the same in all cases, but the flow increases continuously with the width of the section.

Figure 4. Influence of the width of a bottleneck on the flow. Experimental data Adapted from: (Kretz et al., 2006b; Müller, 1981; Muir et al., 1996; Nagai et al., 2006a; Seyfried et al., 2007) of different types of bottlenecks and initial conditions. All data are taken under laboratory conditions where the test persons are advised to move normally.

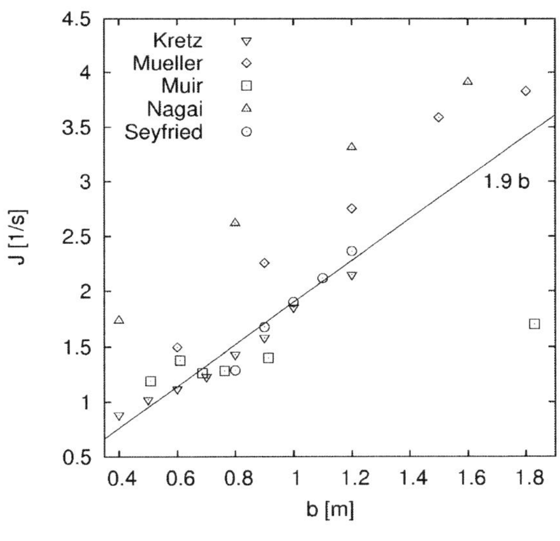

Systematic studies including the influence of the shape and width of the bottleneck (Fig. 4) and the comparison with flow values under normal situations have shown that funnel-like geometries support the formation of arches and thus blockages (Müller, 1981; Muir et al., 1996). Especially in emergency situations the flow through bottlenecks shows strongly intermittent behaviour since the formation of blockages might lead to zero flow temporarily.

To reduce the occurrence of blockages and thus evacuation times, it has been suggested to put a column (asymmetrically) in front of a bottleneck (Helbing et al., 2000b). It should be emphasized that this theoretical prediction (see also Sec. "Conflicts and Friction") was made under the assumption that the system parameters, i.e. the basic behaviour of the pedestrians, does not change in the presence of the column. This is highly questionable in real situations where the columns can be perceived as an additional obstacle or even make it difficult to find the exit.

Evacuations: Empirical Results

So far we have focussed on empirical results for pedestrian motion in rather simple scenarios like corridors or single bottlenecks. As we have seen there are many open questions where no consensus has been reached, sometimes even about the qualitative aspects. This becomes even more relevant for full-scale descriptions of evacuations from large buildings or cruise ships. These are typically a combination of many of the simpler elements. Therefore a lack of reliable information is not surprising.

Evacuation Experiments

In the case of an emergency, the movement of a crowd usually is more straightforward than in the general case. Commuters in a railway station, for example, or visitors of a building might have complex itineraries which are usually represented by origin-destination matrices. In the case of an evacuation, however, the aims and routes are known and usually the same, i.e. the exits and the egress routes. This is the reason why an evacuation process is rather strictly limited in space and time, i.e. its beginning and end are well-defined (sound of the alarm, initial position of all persons, safe areas (final position of all persons), and the time, the last person reaches the safe area.

In the past some evacuation trials for complete buildings have been performed. These trials show that the *response time* (or *pre-movement time*) is, especially in cases with very low densities, a very important factor for evacuation times. This influence is very hard to forecast in evacuation models, because it can not be determined in a mathematical way, thus a "correct" forecast of evacuation times and comparison to real evacuation trials is very hard. To eliminate the influence of response time, evacuation experiments had been performed under laboratory conditions, thus these experiments are a good basis for calibrating models to predict movement of people.

Another important aspect in this regard is the fact that although empirical data on crowd disasters exist, e.g. in the form of reports from survivors or even video footage, it is almost impossible to derive quantitative results from them. Therefore evacuation trials and simulations are very important for our understanding of crowd behaviour in emergency situations.

Legal Regulations

For evacuation scenarios various legal regulations exist, which differ for different facilities and countries. For aircraft an evacuation test is mandatory and there is a time limit of 90 seconds that has to be complied to in an evacuation trial (FAA, 1990). In many countries there is no strict criterion for the maximum evacuation time of buildings. The requirements are often based on minimum exit widths and maximum escape path lengths.

To avoid expensive reconstruction or changes of the layout, computer simulation of evacuation processes become more and more important in the early stages of planning. In order to produce reliable results, especially quantitatively, the underlying models for pedestrian motion have to be able to reproduce the empirical observations described in Sec. "Collective Phenomena" and "Quantitative Results".

However, often evacuation exercises are just too expensive, time consuming, and dangerous to be a standard measure for evacuation analysis. Therefore evacuation simulations based on properly validated and calibrated models will become more and more important in the future.

MODELLING OF PEDESTRIAN DYNAMICS

Modelling of pedestrian dynamics has a long history in various fields ranging from engineering to physics and applied mathematics. Many different model classes have been proposed which can roughly be classified as follows:

- **Microscopic vs. macroscopic:** In *microscopic models* each pedestrian is represented separately whereas in *macroscopic models* the state of the system is described in terms of densities.
- **Discrete vs. continuous:** Each of the three basic variables for a description of a system of pedestrians, namely space, time and state variable (e.g. velocities), can be either discrete (i.e. an integer number) or continuous (i.e. a real number). Cellular automata are fully discrete whereas fluid-dynamic models are continuous in all variables. But also combinations are possible, e.g. models which are continuous in space and state variable, but discrete in time.
- **Rule-based vs. force-based:** Interactions between the pedestrians can be implemented in at least two different ways: In a *rule-based approach* pedestrians make "decisions" based on their current situation and that in their neighbourhood as well as their goals etc. These rules are therefore often motivated by arguments from psychology. In contrast, *force-based models* specify interactions directly on the level of equations of motion, similar to classical mechanics although the forces are not necessarily "physical" forces.
- **Deterministic vs. stochastic:** The dynamics of pedestrians can either be deterministic or stochastic. In the first case the behaviour at future times is completely determined by the past. In stochastic models, the behaviour is controlled by certain probabilities such that the agents can react differently in the same situation. This "intrinsic" stochasticity should be distinguished from "noise" which is sometimes added to the *macroscopic* observables, like position or velocity. Often the main effect of these noise terms is to avoid certain special configurations which are considered to be unrealistic. Otherwise the behaviour is very similar to the deterministic case. For true stochasticity, on the other hand, the deterministic limit usually has very different properties from the generic case.

It should be mentioned that a clear classification according to the characteristics outlined here is not always possible.

Fluid-dynamic models (Henderson, 1974; Helbing, 1992; Hughes, 2000; Hughes, 2002) belong to the macroscopic approaches and are characterized by continuous variables and deterministic, force-based dynamics. The equations of motion are similar to the Navier-Stokes equations although modifications due to the special properties of the "pedestrian fluid" are essential. A more general approach are so-called *gaskinetic models* (Helbing, 1992) which also allow for a more fundamental justification of fluid-dynamic approaches.

In recent years, especially due to the interest of physicists, modern approaches adopted from statistical physics have become quite popular (Chowdhury et al., 2000; Chopard et al., 1998). In statistical physics many powerful methods to deal with interacting many-particle systems have been developed (Chowdhury et. al, 2008). Among those, descriptions based on *cellular automata (CA)* have become most fruitful owing to the relative simplicity and flexibility of this approach, e.g. since it easily possible to simulate even large crowds efficiently. CA are microscopic models that are discrete in all variables. Usually the dynamics is rule-based and stochastic. Especially the fact that the dynamics can be implemented in the form of intuitive "rules" has allowed to include rather complex aspects (e.g. psychology) in a rather simple way.

Cellular Automata Models

As mentioned above, cellular automata are discrete in space, time and state variable which in the case of traffic and transport models usually corresponds to the velocity. The discreteness in time means that the positions of the agents are updated *synchronously (in parallel)* in well defined timesteps. The timestep corresponds to a natural timescale Δt which could e.g. be identified with some reaction time. This can be used for the calibration of the model which is essential for making quantitative predictions. A natural space discretization follows from the maximal densities observed in dense crowds which gives the minimal space requirement of one person. In CA each cell can only be occupied by one agent (exclusion principle) and thus a maximal density of 6.25 P/m (Weidmann, 1993) leads to a cell size of 40×40 cm. The exclusion principle and the modelling of humans as non-compressible particles mimicks short-range repulsive interactions, i.e. the "private-sphere" or "personal space".

The dynamics is usually defined by stochastic rules which specify the transition probabilities for the motion to one of the neighbouring cells (Fig. 5). The models differ in the specification of these probabilities as well as in that of the "neighbourhood". For deterministic models all except of one probability are zero.

The first cellular automata models (Fukui et al., 1999b; Muramatsu et al., 1999; Klüpfel et al., 2000; Blue et al., 2000) for pedestrian dynamics can be considered two-dimensional variants of the *asymmetric simple exclusion process (ASEP)* (for reviews, see (Derrida, 1998; Schütz, 2001) or models for city or highway traffic (Chowdhury et al., 2000; Chowdhury et al., 2008) based on it. Most of these

Figure 5. A particle, its possible directions of motion and the corresponding transition probabilities p_{ij} for the case of a von Neumann neighbourhood.

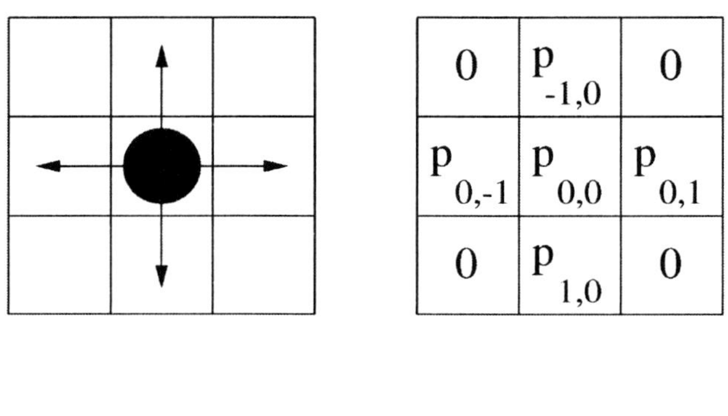

models represent pedestrians by particles without any internal degrees of freedom. They can move to one of the neighbouring cells based on certain transition probabilities which are determined by three factors: (1) the desired direction of motion, e.g. to find the shortest connection, (2) interactions with other pedestrians, and (3) interactions with the infrastructure (walls, doors, etc.).

Fukui-Ishibashi Model

The CA model proposed by Fukui and Ishibashi (Fukui et al., 1999a; Fukui et al., 1999b) is based on a two-dimensional variant of the ASEP. They have studied bidirectional motion in a long corridor where particles moving in opposite directions are updated alternatingly. Particles move deterministically in their desired direction, only if the desired cell is occupied by an oppositely moving particle they make a random sidestep.

Various extensions and variations of the model have been proposed, e.g. an asymmetric variant (Muramatsu et al., 1999) where walkers prefer lane changes to the right, or the possibility of backstepping (Maniccam, 2005). The influence of the shape of the particles has been investigated in (Nagai et al., 2006b). Also other geometries (Muramatsu et. al, 2000b; Tajima et al., 2002) and extensions to full 2-dimensional motion have been studied in various modifications (Muramatsu et al., 2000a; Maniccam, 2003; Maniccam, 2005)

Blue-Adler Model

The model of Blue and Adler (Blue et al., 2000; Blue et al., 2002) is based on a multi-lane variant of the Nagel-Schreckenberg model (Nagel et al., 1992) of highway traffic. The update is performed in four steps which are applied to all pedestrians in parallel. First, each pedestrian chooses a preferred lane. Then lane changes are performed. In the third step the velocities are determined based on the available gap in the new lanes. Finally, pedestrians move forward according to the velocities determined in the previous steps. Motion is not restricted to nearest-neighbour sites. Instead pedestrians can have different velocities v_{max} which correspond to the maximal number of cells they are allowed to move forward.

In counterflow head-on-conflicts occur which are resolved stochastically by allowing (with some probability) opposing pedestrians to exchange positions within one timestep. Note that the motion of a single pedestrian (not interacting with others) is deterministic otherwise.

Gipps-Marksjös Model

In the model suggested by Gipps and Marksjös (Gipps et al., 1985) interactions between pedestrians are assumed to be repulsive anticipating the idea of social forces. The pedestrians move deterministically on a grid of rectangular cells. To each cell a score is assigned based on its proximity to other pedestrians. This score represents the repulsive interactions and the actual motion is then determined by the competition between this repulsion and the gain of approaching the destination. Applying this procedure to all pedestrians, to each cell a potential value is assigned which is the sum of the individual contributions. A pedestrian then selects the cell of its nine neighbours (Moore neighbourhood, including current position) which leads to the maximum benefit. This benefit is defined as the difference between the gain of moving closer to the destination and the cost of moving closer to other pedestrians as represented by the potential.

The updating is done sequentially to avoid conflicts of several pedestrians trying to move to the same position. In order to model different velocities, faster pedestrians are updated more frequently.

Floor Field CA

The floor field CA (Burstedde et al., 2001; Schadschneider, 2002; Burstedde et al., 2002; Kirchner et. al, 2002) can be considered as an extension of the ASEP where transition probabilities to neighbouring cells are no longer fixed, but vary dynamically. This is motivated by the process of chemotaxis (see (Ben-Jacob, 1997) for a review) used by some insects (e.g. ants) for communication. They create a chemical trace to guide other individuals to food sources. In this way a complex trail system is formed that has many similarites with human transport networks.

In the approach of Burstedde et al. (2001) the pedestrians also create a trace. In contrast to chemotaxis, however, this trace is only virtual and mainly a technical trick which reduces interactions to local ones that allow efficient simulations in arbitrary geometries, although one could assume that it corresponds to some abstract representation of the path in the mind of the pedestrians. The locality becomes important in complex geometries as no algorithm is required to check whether the interaction between particles is screened by walls etc. The number N of interaction terms always grows linearly with the number of particles, whereas in force-based models they are typically of order N^2.

Mainly the transition probabilities are determined by the preferred walking direction of the pedestrians which depends on his/her origin and destination. This information is encoded in the so-called *matrix of preference*. Its matrix elements M_{ij} are directly related to observable quantities, namely the average velocity and its fluctuations (Burstedde et al., 2001).

The translation into local interactions is achieved by the introduction of so-called *floor fields*. The transition probabilities for all pedestrians depend on the strength of the floor fields in their neighbourhood in such a way that transitions in the direction of larger fields are preferred. The *dynamic floor field* D_{ij} corresponds to a virtual trace which is created by the motion of the pedestrians and in turn influences the motion of other individuals. Furthermore it has its own dynamics, namely through diffusion and decay, which leads to a dilution and finally the vanishing of the trace after some time. The *static floor*

Figure 6. Left: Static floor field for the simulation of an evacuation from a large room with a single door. The door is located in the middle of the upper boundary and the field strength is increasing with increasing intensity. Right: Snapshot of the dynamical floor field created by agents leaving the room.

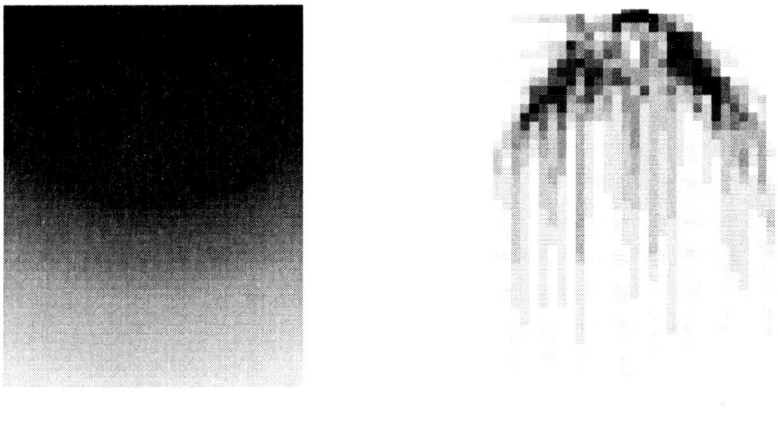

field S_{ij} does not change with time since it only takes into account the effects of the surroundings. It allows to model e.g. preferred areas, walls and other obstacles. Fig. 6 shows the static floor field used for the simulation of evacuations from a room with a single door. Its strength decreases with increasing distance from the door. Since the pedestrians prefer motion into the direction of larger fields, this is already sufficient to find the door.

Coupling constants control the relative influence of both fields. For a strong coupling k_S to the static field pedestrians will choose the shortest path to the exit. This corresponds to a 'normal' situation. A strong coupling k_D to the dynamic field implies a strong herding behaviour where pedestrians try to follow the lead of others. This often happens in emergency situations.

The model uses an entirely parallel update. Therefore conflicts can occur where different particles choose the same destination cell. Their role will be discussed in more detail in Sec. "Conflicts and Friction".

The update rules of the full model, including the interaction with the two floor fields, consist of the following five steps (Burstedde et al., 2001):

1. The dynamic floor field D is modified according to its diffusion and decay rules (Burstedde et al., 2001), controlled by parameters α and δ. In each timestep of the simulation each single boson of the whole dynamic field D decays with the probability δ and diffuses with the probability α to one of its neighbouring cells.
2. For each pedestrian, the transition probabilities p_{ij} for a move to a neighbour cell (i,j) (Fig. 5) are determined by the local dynamics and the two floor fields. The relative influence of the fields D and S is controlled by sensitivity parameters $k_S \in [0,\infty[$ and $k_D \in [0,\infty[$. This yields

$$p_{ij} = NM_{ij}e^{k_S S_{ij}}e^{k_D D_{ij}}(1-n_{ij})\xi_{ij}. \qquad (5)$$

   The occupation number n_{ij} is 0 for an empty and 1 for an occupied cell[3], which reflects the exclusion principle. The obstacle number $\xi_{ij} = 0$ for forbidden cells, e.g. walls, and $\xi_{ij} = 1$ otherwise, and the normalization N ensures the normalization $\sum_{ij} p_{ij} = 1$ of the probabilities.
3. Each pedestrian chooses randomly a target cell based on the transition probabilities p_{ij} determined in Step 2.
4. Conflicts arising if two or more pedestrians attempt to move to the same target cell are resolved (see Sec. "Conflicts and Friction").
5. D at the origin cell (i,j) of each moving particle is increased by one: $D_{ij} \to D_{ij} + 1$.

The above rules are applied to all pedestrians at the same time (parallel update). The sensitivity parameters can be interpreted in a simple way. The coupling k_D to the dynamic floor field controls the tendency to follow in the footsteps of others, e.g. to reduce interactions with oncoming pedestrians. In the absence of a matrix of preference, k_S determines the effective velocity of a single agent in the direction of its destination.

Other Models

Fluid-Dynamic and Gaskinetic Models

Pedestrian motion has some obvious similarities with the dynamics of fluids. E.g. the motion around obstacles appears to follow "streamlines". Since the middle of the 1950's pedestrian movement has been calculated in the field of engineering in analogy to the behaviour of liquids. In this approach, known as *handcalculation method* (Togawa, 1955; Predtechenskii et al., 1978; Nelson et al., 2002), pedestrian motion is described as a "flow" of particles in a building, similar to liquids flowing through a pipeline network with links of limited capacities. With equations analogous to those of fluid dynamics (e.g. the continuity equation expressing conservation of mass) it is possible to forecast congestions and even evacuation times for buildings (or parts of it).

More complex fluid-dynamic models have been developed later (Henderson, 1974; Helbing, 1992, Hughes, 2000; Hughes, 2002), e.g. by taking inspiration from gas-kinetic theory. Typically these macroscopic models are deterministic and force-based. In contrast to real fluid the assumption of conservation of energy and momentum is not true for interactions between pedestrians which in general do not even satisfy Newton's Third Law ("actio = reactio"). Several other differences to normal fluids are relevant, e.g. the anisotropy of interactions or the fact that pedestrians usually have an individual preferred direction of motion.

Lattice-Gas Models

In (Marconi et al., 2002) a mesoscopic approach inspired by *lattice-gas models* for hydrodynamics (Rothman et al., 1994; Rothman et al., 1997) has been suggested as a model for pedestrian dynamics. These models are similar to cellular automata, but the exclusion principle is relaxed: Particles with different velocities are allowed to occupy the same site. In analogy with the description of transport phenomena in fluids (e.g. the Boltzmann equation) the dynamics is then based on a succession of "collision" and "propagation".

In the lattice-gas model developed in (Marconi et al., 2002) pedestrians are modelled as particles moving on a triangular lattice which have a preferred direction of motion $\mathbf{c}_F$. However, the particles do not follow strictly this direction by have also a tendency to move with the flow. In the *propagation step* each pedestrian moves to the neighbour site in the direction of its velocity vector. In the *collision step* the particles interact and new velocities (directions) are determined. In contrast to physical systems, momentum etc. does not need to be conserved during the collision step. These considerations lead to a collision step that takes into account the favourite direction $\mathbf{c}_F$, the local density (the number of pedestrians at the collision site), and a quantity called mobility at all neighbour sites which is a normalized measure of the local flow after the collision.

Social-Force Models

The *social-force model* (Helbing et al., 1995) is a deterministic microscopic continuum model in which the interactions between pedestrians are implemented by using the concept of a *social force* or *social field* (Lewin, 1951). It is based on the idea that changes in behaviour can be understood in terms of fields or forces. Applied to pedestrian dynamics the social force $\mathbf{F}_j^{(\text{soc})}$ represents the influence of the environment (other pedestrians, infrastructure) and changes the velocity $\mathbf{v}_j$ of pedestrian *j*. Thus it is responsible for acceleration which justifies the interpretation as a force. The basic equation of motion for a pedestrian of mass m_j is then of the general form

$$\frac{d\mathbf{v}_j}{dt} = \mathbf{f}_j^{(\text{pers})} + \mathbf{f}_j^{(\text{soc})} + \mathbf{f}_j^{(\text{phys})} \tag{6}$$

where

$$\mathbf{f}_j^{(\text{soc})} = \frac{1}{m_j}\mathbf{F}_j^{(\text{soc})} = \sum_{l \neq j} \mathbf{f}_{jl}^{(\text{soc})}$$

is the total (specific) force due to the other pedestrians. $\mathbf{f}_j^{(\text{pers})}$ denotes a "personal" force which makes the pedestrians attempt to move with their own preferred velocity $\mathbf{v}_j^{(0)}$ and thus acts as a driving term. In high density situations also physical forces $\mathbf{f}_{jl}^{(\text{phys})}$ become important, e.g. friction and compression when pedestrians make contact.

The most important contribution to the social force $\mathbf{f}_j^{(\text{soc})}$ comes from the territorial effect, i.e. the private sphere. Pedestrians feel uncomfortable if they get too close to others, which effectively leads to a repulsive force between them. Similar effects are observed for the environment, e.g. people prefer not to walk too close to walls.

The two-dimensional optimal-velocity model (Nakayama et al., 2005) is also of a form similar to (6). Here physical forces are usually neglected and the acceleration is determined by the deviation $\sum_l \mathbf{V}(x_l(t) - x_j(t))$ of the actual velocity $\mathbf{v}_j$ from an optimal-velocity $\mathbf{V}(\mathbf{x}_l - \mathbf{x}_j)$ that depends on difference to the positions $\mathbf{x}_l$ of the other pedestrians.

The appeal of the social-force model is given mainly by the analogy to Newtonian dynamics. For the solution of the equations of motion of Newtonian many-particle systems the well-founded molecular dynamics technique exists. However, a straightforward implementation of the equations of motion can lead to unrealistic movement of single pedestrians, e.g. negative velocities in the main moving direction. Therefore not only different specifications of the forces have been used (Helbing et al., 1995; Helbing et al., 2000b; Werner et al., 2003), but also other modifications were proposed (Lakoba et al., 2005; Seyfried et al., 2006; Yu et al., 2005) To prevent this effect additional restrictions for the degrees of freedom have to been introduced, see for example (Helbing et al., 1995).

Surprisingly the qualitative behaviour of the social force model and the floor field model (Sec. "Floor Field CA") is very similar despite the fact that the interactions are very different. Apart from the ad hoc introduction of interactions the structure of the social-force model can also be derived from an extremal principle (Hoogendoorn et al., 2003a). It follows under the assumption that pedestrian behaviour is determined by the desire to minimize a certain cost function which takes into account not only kinematic aspects and walking comfort, but also deviations from a planned route.

Relation with Multi-Agent Systems

The models described above focus on the interactions between pedestrians which influence the collective dynamic of pedestrian movement. Examples for qualitative and quantitative phenomena related to these interactions are the formation of lanes in bidirectional streams or the decrease of the velocity with the density. Often these phenomena are assigned to an operational level. But the tactical level, for example which goal a pedestrian wants to reach or how the environment influences the decision of pedestrians, is only treated marginally by these models. In other words the models consider pedestrians as a less individual particle of a pedestrian crowd. Examples in applications are the evacuation of buildings in case of an emergency. There the dynamics is determined by the fundamental diagram and the capacity of bottlenecks. But the goal to leave the building and thus the direction of movement are more or less the same for all pedestrians. In such situations there is only little leeway for individual decisions.

However there are a lot of other applications where the individual character of a pedestrian becomes more important. Examples are the movement of pedestrians through a shopping area or a museum. Under uncrowded conditions the movement of the agents is less influenced by other pedestrians but more by the environment. Here *Multi-Agent models* (Ferber, 1999) focusing on the individual properties and decisions are needed.

All microscopic model approaches described above can be extended to a full multi-agent system. Additional properties are assigned to the particles that describe their mutual interactions, interactions with the environment, their goals etc. (see e.g. (Klügl et al., 2007)). Thus particles are gradually transformed into a collection of heterogeneous agents. A very promising framework for multi-agent modelling is based on *situated cellular agents*, see e.g. (Bandini et al., 2004; Bandini et al., 2007).

VALIDATION AND EXTENSION OF CA MODELS

We now discuss some issues that are relevant for realistic CA models and thus applications. A comparison with empirical results show that not all of them are reproduced in a way that is necessary for sophisticated applications, e.g. in safety analysis. Therefore extensions of the basic models are necessary to reproduce empirical observations more accurately.

Conflicts and Friction

As has been emphasized earlier, the use of synchronous (parallel) dynamics is essential for the calibration of the models since it introduces a timescale. However, synchronous motion often leads to *conflicts* where two or more particles choose the same destination cell. These conflicts have to be resolved to respect the exclusion principle, e.g. by choosing one particle randomly which is allowed to move whereas the others stay at their positions.

Conflicts might appear to be undesirable effects which reduce the efficiency of simulations and should therefore be avoided by choosing a different update scheme, e.g. by updating pedestrians sequentially instead of synchronously. However, this leads to other problems, e.g. the identification of the relevant timescale. Therefore it has been suggested (Kirchner et al., 2003b) to take these conflicts seriously as an important part of the dynamics. Although conflicts are local phenomena they can have a strong influence on global quantities like evacuation times. They become most important in clogging situations encountered in large crowds and at high densities, especially near intersections and bottlenecks. In real life this often leads to dangerous situations and injuries during evacuations.

For a more realistic description of clogging effects the floor field model described in Sec. "Floor Field CA" needs to be modified only in Step 4, the resolution of conflicts (Kirchner2003c). In real life,

Figure 7. Refused movement due to the friction parameter μ *for a conflict involving four particles*

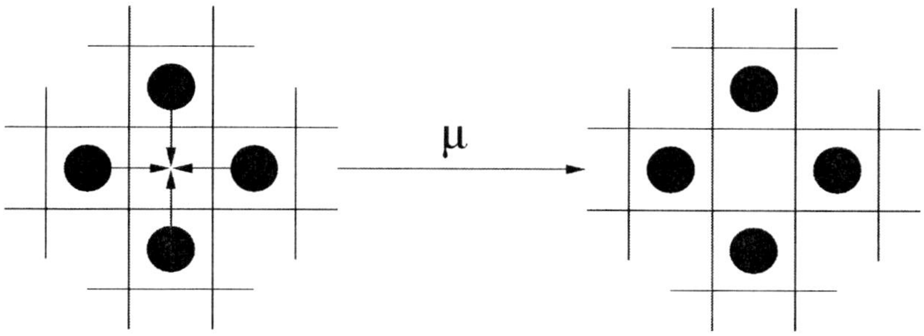

conflict situations often lead to a moment of hesitation where the involved agents hesitate before trying to resolve the conflict. This reduces on average the effective velocities of all involved particles. Therefore Step 4 is modified such that with some probability μ the movement of all involved particles is denied, i.e. all pedestrians remain at their site (see Fig. 7). This means that with probability $1-\mu$ one of the individuals moves to the desired cell. Which particle actually moves is then determined by the rules for the resolution of conflicts as described in Step 4 in Sec. "Conflicts and Friction". This effect has been called *friction* and μ *friction parameter* since its consequences are similar to contact friction, e.g. in granular materials. The velocity of a freely moving particle is not reduced and effects only show up in local interactions.

Obviously the use of a parallel update is essential. Any random or ordered sequential update will disguise the real number of arising conflicts between the pedestrians in the system. In any model with continuous time these effects have to be implemented in a different way, e.g. through contact friction.

The influence of friction effects has been investigated for a simple evacuation scenario in (Kirchner et al., 2003c). In general evacuation times increase with increasing friction parameter μ. Conflicts close to the exit are most important since they have a direct influence on the evacuation time by reducing the outflow. In large density situations and for large μ the pressure between the pedestrians becomes so strong that any motion is almost impossible (Helbing et al., 2000b).

Emergency situations can be characterized by large k_S and large μ. Then an ordered outflow is inhibited due to local conflicts near bottlenecks or doors, resulting in strongly increased evacuation times.

However, also counterintuitive phenomena occur. As mentioned earlier, the coupling strength k_S to the static floor field determines the effective velocity in the direction of the exit for a single pedestrian. Therefore one would expect that the evacuation time decreases with increasing k_S. However, this is not the case for large friction constants μ (see Fig. 8) where it leads to stronger clogging and thus more conflicts at the exit. This non-monotonic dependence of the evacuation time on the free velocity is known as *faster-is-slower effect* (Helbing et al., 2002; Helbing et al., 2000b).

This behaviour is also reflected in the time evolution of the evacuation, i.e. the number of people $N(t)$ who left the room up to time t. In the absence of friction, $N(t)$ is linearly increasing, whereas for large μ, $N(t)$ is not only smaller than for $\mu=0$ but also shows intermittent behaviour. In Fig. 9 small plateaus can be observed which are formed stochastically and where over short time periods no persons leave the room. This irregular behaviour is well-known from granular flow and is typical for clogging situations (Wolf et al., 1996).

Another important aspect for applications is the strong increase of the variance of evacuation times. This can be seen in Fig. 9 which shows besides curves averaged over different samples also the two

Figure 8. Evacuation time as function of S for different values of μ and $k_D = 0$. For μ = 0.9 the faster-is-slower effect occurs (Adapted from (Kirchner et. al, 2003c)).

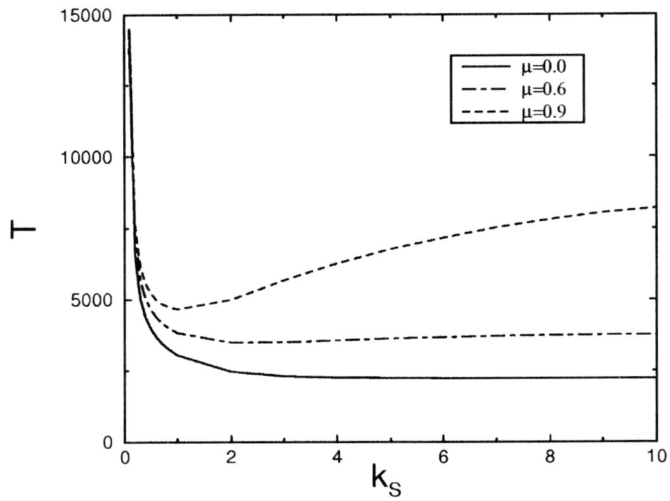

Figure 9. Typical time-dependence of the number N of evacuated persons in the absence of friction (graph 1: μ = 0) and for strong friction (graph 2: μ = 0.9). Shown are results for the longest, shortest and averaged process which are almost indistinguishable for small μ (Adapted from (Kirchner et al., 2003c)).

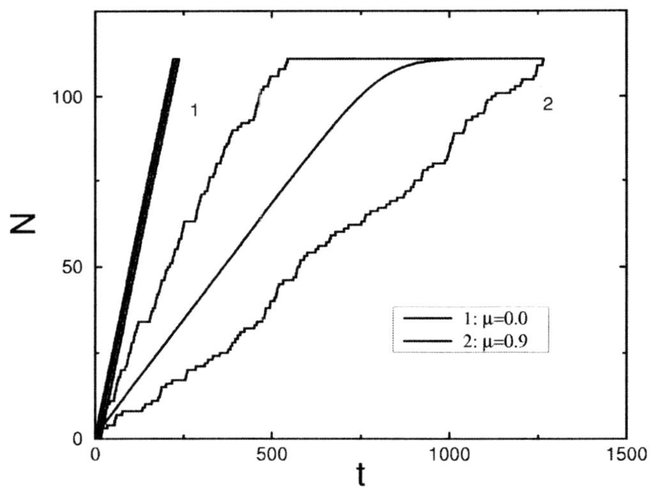

extremal curves with minimal and maximal evacuation times. For μ=0 the evacuation process is almost deterministic and fluctuations (due to the random initial conditions and the dynamics) are very small. With increasing μ the number of conflicts increases and the enveloping curves differ clearly from the averaged curves. This indicates that the evacuation time is no longer a meaningful quantity for safety estimates.

Figure 10. Fundamental diagrams of the floor field model for $v_{max} = 1,\ldots,5$ *(left) and* $v_{max} = 1$ *with two different space discretizations (right) (Adapted from (Kirchner et al., 2004))*

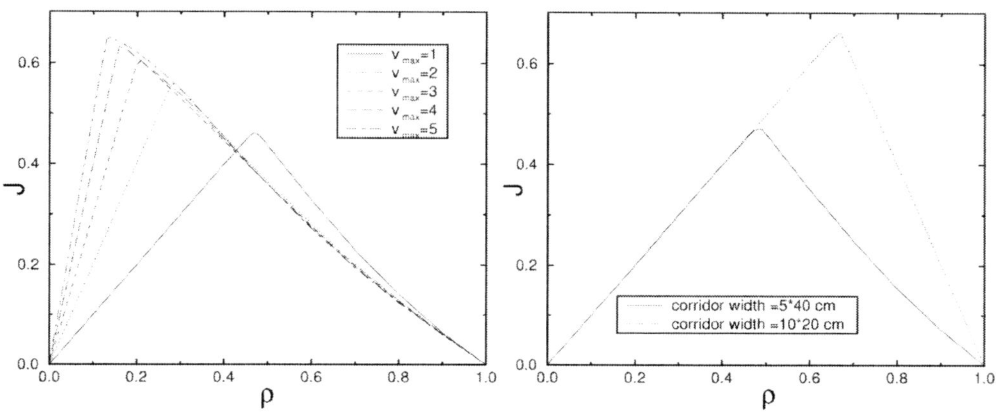

In (Helbing et al., 2000b) it has been proposed to place an additional column in front of the exit. Surprisingly, this can lead to a reduction of evacuation times (Helbing et al., 2002; Helbing et al., 2000b). This is confirmed by simulations of the floor field model (Kirchner et al., 2003c). However, it is questionable whether this effect is relevant for real situations (see Sec. "Blockages in Competitive Situations").

Higher Velocities

The comparison of empirical results obtained in various experiments with theoretical predictions of most models have shown that even some qualitative features of the fundamental diagram (see Sec. "Fundamental Diagram") are not reproduced correctly. The origin of this discrepancy is the restriction to models with nearest-neighbour interactions which do not capture essential features like the dynamic space requirement of the agents that depends on their velocity (and thus density).

Modifications of the floor field model (Kirchner et al., 2004; Kretz et al., 2006c) take this effect into account. Here motion is not restricted to nearest-neighbour cells, but also to farther cells. This is equivalent to a motion at different instantaneous velocities $v = 0, 1,\ldots, v_{max}$ where v is the number of cells an agent moves. Then $v_{max} = 1$ corresponds to the case where motion is allowed only to nearest neighbours. Note that different extensions of this type are possible, depending on how one treats crossing trajectories of different agents (Kirchner et al., 2004). But in all cases, the fundamental diagrams become more realistic since the maximum of the flow is shifted towards smaller densities with increasing v_{max} (Fig. 10, left), in accordance with the empirical observations.

Finer Space Discretization

For applications the space discretizations poses sometimes practical problems. Using a cell size of 40×40 cm, should a corridor of 60 cm width be represented by one or two cells? Therefore, in order to fit the geometry of the environment better the use of smaller cell sizes seems to be necessary.

Apart from this practical problem, also the dynamics is not reproduced faithfully in all details if the size of a cell corresponds to the space requirement of a single agent. For instance the zipper effect described in Sec. "Bottleneck Flow" can not be reproduced since overlapping lanes are not possible with this discretization.

If smaller cell sizes are used, agents occupy more than one cell. e.g. 2×2 cells of size 20×20 cm. This has effects on the dynamical quantities (Kirchner et al., 2004). As shown in Fig. 10 the maximum of the fundamental diagram is shifted to higher densities because the space needed for unimpeded movement in one timestep corresponds to only half the length of the particle size. Therefore, for a realistic description, a finer space discretization has to be combined with higher velocities (Sec. "Higher Velocities"). Furthermore non-local conflicts can occur that are not restricted to single cells and involve many agents. In fact this leads to a even more realistic representation of clogging effects in evacuation scenarios (Kirchner et al., 2004).

APPLICATION OF MODELS

We will now discuss some applications of the models to pedestrian crowds in normal and emergency situations, especially egress and evacuation scenarios. Evacuation scenarios are simple in one respect: One can assume that there is one main desire, namely to get out as soon as possible. The strategical level of goal creation therefore only has the task to decide for one of the exits for each of the agents. The simulation of normal situations has the advantage that one does not have to worry about extreme types of behaviour (commonly called "panic", see Sec. "Collective Phenomena"). To get realistic results one therefore has to create a far more elaborate model for desires and intentions.

In the following two practical applications of models for crowd dynamics are discussed briefly. The first application refers to surprising results found in evacuation trials with aircraft. The second example shows how the models can be applied to analyse crowd motion in the rather complex infrastructure of a football stadium.

Figure 11. Left: Egress time as function of the door width for competitive and non-competitive behaviour (Adapted from (Muir et al., 1996)). Right: Results of computer simulations using the floor field model with friction (Adapted from (Kirchner et al., 2003a)).

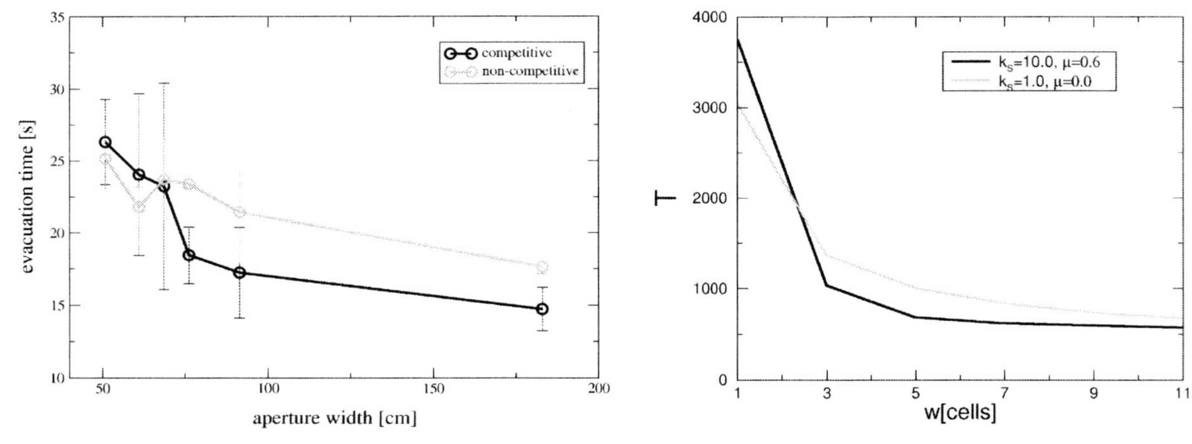

Egress from Aircraft

Friction effects are responsible for an interesting experimental result (Muir et al., 1996) which shows that the motivation level (competitive vs. non-competitive or cooperative) of passengers has a significant influence on the egress time from an aircraft. The experiment was carried out with groups of 50 to 70 persons, where in one case (competitive) a bonus was paid for the first 30 persons. The time of the 30th person reaching the exit was measured for variable exit widths w. The main result found is that $t_{\text{comp}} > t_{\text{non-comp}}$ for $w < w_c$, whereas $t_{\text{comp}} < t_{\text{non-comp}}$ for $w > w_c$. The critical width was determined experimentally as $w_c \approx 70$ cm (Fig. 11). Thus competition is beneficial only for wide exits, but harmful for narrow ones.

Within the framework of the floor field model, competition can be described as an increased assertiveness (large k_S) and at the same time strong hindrance in conflict situations, i.e. large friction μ. Cooperation is represented by small k_S and μ=0. In (Kirchner et al., 2003a) the experimental results have been reproduced within a simplified scenario using the floor field model. Fig. 11 shows typical average evacuation times for the non-competitive and the competitive regime. Clearly the simulations are able to reproduce the observed crossing of the two curves at a small door width qualitatively. Without friction (μ=0), increasing k_S alone always decreases T. The effect is therefore only obtained by increasing both, k_S and μ.

Thus there are two factors that determine the egress of persons and the overall evacuation time in this scenario: On one hand, walking speed (controlled by the parameter k_S) and, on the other hand, friction (controlled by μ). These parameters depend in a different way on the door width: the influence of the friction dominates for very narrow doors which leads the crossing shown in Fig. 11. It should again be emphasized that conflicts close to the exit are most important since they have a direct influence on the evacuation time. Therefore, in case of competitive behaviour and narrow doors it is important to find other means in order to reduce the number of conflicts occurring at the exits.

Figure 12. The Westfalenstadion Dortmund: Outside view and general arrangement plan (Borussia Dortmund KGaA, www.borussia-dortmund.de)

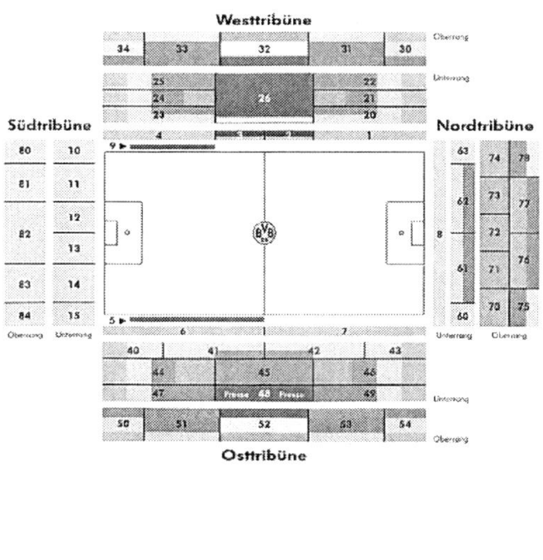

Egress from Football Stadium

As another example for the application of pedestrian flow simulation and analysis we briefly discuss the non-emergency egress from a football stadium.

The Westfalenstadion is a football stadium for national and international games and was a venue for the Worldcup 2006. The aim of the evacuation analysis was to check whether the available safe evacuation time ASET (determined by fire and smoke simulations and the ventilation available, i.e. the existence of a sufficient smoke free layer) is larger than the required safe evacuation time RSET (determined by the evacuation simulation), i.e. ASET > RSET. The analysis showed that this was (and is) the case. The analysis focused on the extension phase 3 (towers in the corners) which increased the spectator capacity from 60,000 to 80,000. In order to validate the simulation results, the egress from the stadium after a match between Germany and Scotland was videotaped and analysed (results are shown below and compared to simulation results).

The model used in the simulations is a cellular automaton similar to the floor field model, but has $v_{max} > 1$ and no dynamic floor field. Table 1 contains the parameter values for the standard population used in the Westfalenstadion simulation. The reaction time distribution was deliberately chosen to be very low in order to get a worst case scenario. It is well known from empirical observations that immediate detection of and reaction to an alarm leads to the highest rates of congestion.

Concerning quantitative verification, movement patterns provide a valuable tool to investigate the reliability of simulation results. This can be done by comparing video footage to simulations, especially concerning overall egress time (non-emergency). The video footage was taken at an international match between Germany and Scotland. For the details of the underlying model and its application to the problem (and further results) we refer to (Klüpfel et al., 2003a; Klüpfel et al., 2003b; Klüpfel, 2006; Klüpfel, 2007).

In Fig. 13 the first six minutes of the video footage and the first three minutes of the simulation are shown. The reason for the different time spans is that the real persons react slower. However, due to their effectiveness and group formation which is not represented in the simulation, the motion is more synchronized than in the simulation. Therefore, the snapshots were chosen such that the situations are comparable even though the times might be different.

For the second half of the egress shown in Fig. 14 this difference vanishes and after 13 minutes, the situation is very much alike for reality and simulation. It is remarkable that after less than 15 minutes, the normal egress is nearly complete. One important pattern that can be identified is the sequence of egress from the rows. The lower rows are emptied first. This pattern is represented nicely by the simulation. An important aspect in the egress from football stadiums is the V-like shapes that are formed because the egress from the lower seating rows is faster.

Table 1. Parameters of the standard population

Parameter	Minimum	Maximum	Mean	Std. Dev.	Unit
Free Walking Speed	0.8	2.0	1.2	0.4	m/s
Dawdling Probability	0	0.3	0.15	0.05	-
Reaction Time	0	10	5	2	s

Figure 13. Comparison of the results for the video analysis (left column) and the simulation (right column) at the beginning of the egress. The video snapshots are taken at (from top to bottom) t = 2 and t = 6 minutes for the videos and t = 20 seconds and t = 3 minutes for the simulation.

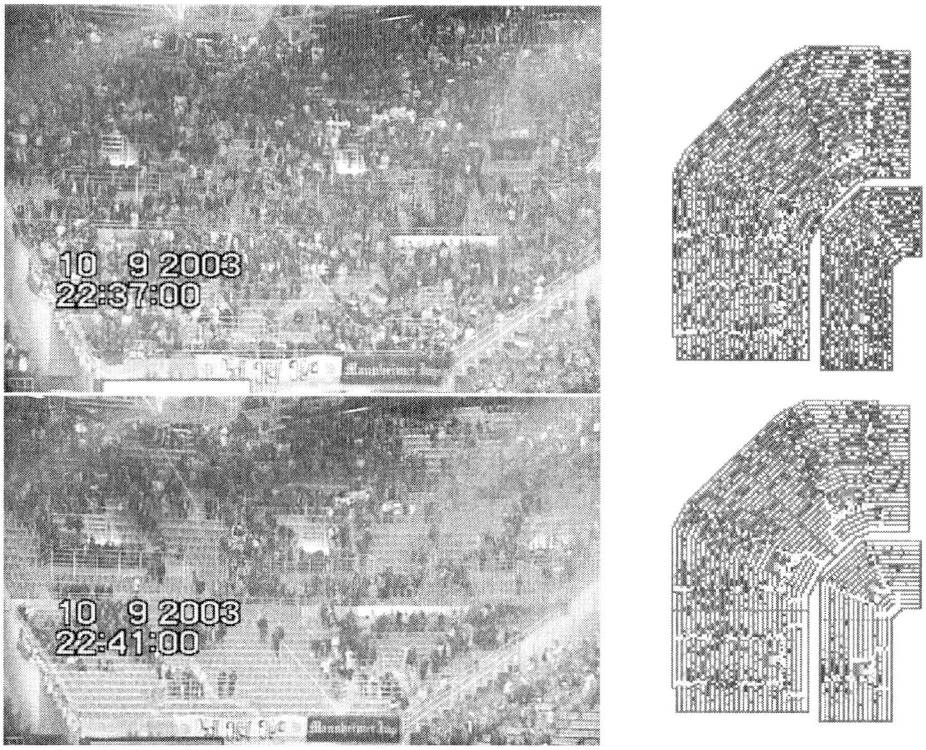

CONCLUSION AND PERSPECTIVES

In this Chapter we have discussed the theoretical and practical aspects of multi-agent simulations of crowd movement. We have tried to make a connection between empirical observations of pedestrian dynamics and modelling approaches on the operational level.

As discussed in Sec. "Empirical Results", empirically for several of the observed phenomena no consensus about the essential properties exists, sometimes not even qualitatively. This is partially related to a lack of controlled experiments. In contrast to vehicular traffic, so far the possibilities of getting empirical data in an automated way are rather limited.

On the modelling side we have focussed on cellular automata approaches which, due to their rule-based approach, form an ideal basis for extensions to multi-agent systems. The latter is necessary for investigations of more complex scenarios as in applications like safety analysis.

A careful validation is obviously absolutely essential if models should provide reliable qualitative or even quantitative predictions, e.g. for evacuation times. Our experiences so far show that a model has to be tested on different length and time scales, i.e. for simple geometries (single rooms, stairs, or hallways) *and* for complex scenarios like the evacuation of a football stadium. Neglecting one of those scales might lead to a false sense of trust into the model predictions.

Figure 14. Same as Figure 13 but video (left column) and simulation snapshots (right column) are taken (from top to bottom) t = 10 minutes and t = 13 minutes

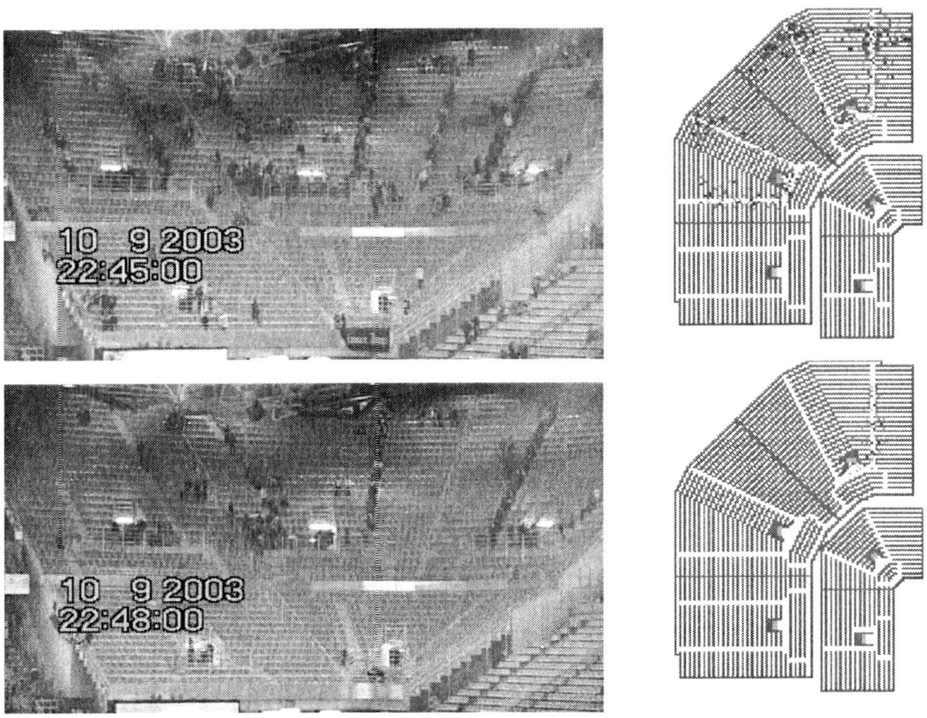

Thus, currently the situation is not very satisfying, both on the empirical and the theoretical side. Due to the lack of consensus about empirical observations a proper validation or calibration of models is almost impossible. Indeed different models which are frequently used might make rather different predictions. This has been investigated in detail for some commercially available software tools in (Rogsch et al., 2007).

Future Research Directions

What are the future tasks and challenges in modelling crowd dynamics? First of all the lack of detailed empirical data is serious limitation which makes validation and calibration of models difficult or even impossible. However, in the near future it will become possible to extract routinely motion data by an automated analysis of video data. Together with experiments under well-controlled conditions this will provide a much better data basis. An important aspect is that in such a way even trajectories of individual pedestrians can be determined automatically which will provide much more detailed information than aggregated data like densities.

The currently existing modelling approaches appear to be flexible enough to allow validation and even calibration, if proper empirical data exist. However, all model classes suffer from certain problems. In the case of cellular automata models these often are related to the discreteness of space. But also details of the basic interactions need to be understood better, e.g. for the description of bottleneck flows.

On the other hand one needs to be careful with models that are too simplistic. These can often produce misleading results which is especially dangerous if these models are used as basis for software tools for applications in safety analysis (Rogsch et al., 2007).

Currently much effort is put into the formulation of models and performing simulations. Neither calibration nor validation are sufficiently addressed from our point of view. This should become mandatory for any model, at least on the operational level, which is to be used commercially especially in safety analysis.

Beyond the operational level, the general multi-agent framework discussed in the introduction provides sort of a roadmap for further developments. Most psychological aspects have not been taken into account explicitly here. This is justified by the basic nature of the investigations and the fact that in the applications presented, they can be modelled implicitly. What's got to be done next, one can either call it desires, beliefs, and intentions or the strategic and tactical decisions of the agents. This is done already in road traffic simulations, where origin-destination matrices and itineraries are generated from census and other statistical data. Such itineraries will be necessary to simulate longer times and more complex scenarios, e.g. an airport for a complete day.

REFERENCES

Bandini S., Federici, M. L., & Vizzari, G. (2007). Situated cellular agents approach to crowd modeling and simulation. *Cybernetics and Systems*, *38*, 729.

Bandini, S., Manzoni, S., & Vizzari, G. (2004). Situated cellular agents: A model to simulate crowding dynamics. *IEICE Trans. Inf. & Syst.*, *E87-D*, 726.

Ben-Jacob, E. (1997). From snowflake formation to growth of bacterial colonies. Part II. Cooperative formation of complex colonial patterns. *Contemp. Phys.*, *38*, 205.

Blue, V. J., & Adler, J. L. (2000). Cellular automata microsimulation of bi-directional pedestrian flows. *J. Trans. Research Board*, *1678*, 135–141.

Blue, V. J., & Adler, J. L. (2002). Flow capacities from cellular automata modeling of proportional splits of pedestrians by direction. In M. Schreckenberg & S. D. Sharma (Eds.), *Pedestrian and Evacuation Dynamics.* Berlin Heidelberg: Springer.

Burstedde, C., Kirchner, A., Klauck, K., Schadschneider, A., & Zittartz, J. (2002). Cellular automaton approach to pedestrian dynamics – applications. In M. Schreckenberg & S. D. Sharma (Eds.), *Pedestrian and Evacuation Dynamics* (pp. 87–98). Berlin Heidelberg: Springer.

Burstedde, C., Klauck, K., Schadschneider, A., & Zittartz, J. (2001). Simulation of pedestrian dynamics using a two-dimensional cellular automaton. *Physica A*, *295*, 507–525.

Cellular Automata - 7th International Conference on Cellular Automata for Research and Industry, acri 2006. (2006).

Chopard, B., & Droz, M. (1998). *Cellular Automata Modeling of Physical Systems.* Cambridge University Press.

Chowdhury, D., Nishinari, K., Santen, L., & Schadschneider, A. (2008). *Stochastic Transport in Complex Systems: From Molecules to Vehicles*. Elsevier.

Chowdhury, D., Santen, L., & Schadschneider, A. (2000). Statistical physics of vehicular traffic and some related systems. *Physics Reports*, *329*(4–6), 199–329.

Derrida, B. (1998). An exactly soluble non-equilibrium system: The asymmetric simple exclusion process. *Phys. Rep.*, *301*, 65.

FAA, F. A. A. (1990). *Emergency evacuation - cfr sec. 25.803* (Regulation No. CFR Sec. 25.803). : Federal Aviation Administration.

Ferber, J. (1999). *Multi-agent systems*. Addison-Wesley.

Fruin, J. J. (1971). *Pedestrian Planning and Design*. New York: Metropolitan Association of Urban Designers and Environmental Planners.

Fukui, M., & Ishibashi, Y. (1999a). Jamming transition in cellular automaton models for pedestrians on passageway. *J. Phys. Soc. Jpn.*, *68*, 3738.

Fukui, M., & Ishibashi, Y. (1999b). Self-organized phase transitions in cellular automaton models for pedestrians. *J. Phys. Soc. Jpn.*, *68*, 2861.

Galea, E. R. (Ed.). (2003). *Pedestrian and Evacuation Dynamics 2003*. London: CMS Press.

Gipps, P. G., & Marksjö, B. (1985). A micro-simulation model for pedestrian flows. *Mathematics and Computers in Simulation*, *27*, 95–105.

Helbing, D. (1992). A fluid-dynamic model for the movement of pedestrians. *Complex Systems*, *6*, 391–415.

Helbing, D., Buzna, L., Johansson, A., & Werner, T. (2005). Self-organized pedestrian crowd dynamics: Experiments, simulations, and design solutions. *Transportation Science*, *39*, 1–24.

Helbing, D., Farkas, I., Molnár, P., & Vicsek, T. (2002). Simulation of pedestrian crowds in normal and evacuation situations. In M. Schreckenberg & S. D. Sharma (Eds.), *Pedestrian and Evacuation Dynamics* (pp. 21–58). Berlin Heidelberg: Springer.

Helbing, D., Farkas, I., & Vicsek, T. (2000a). Freezing by heating in a driven mesoscopic system. *Phys. Rev. Let.*, *84*, 1240–1243.

Helbing, D., Farkas, I., & Vicsek, T. (2000b). Simulating dynamical features of escape panic. *Nature*, *407*, 487–490.

Helbing, D., Johannson, A., & Al-Abideen, H. (2007a). Crowd turbulence: the physics of crowd disasters. In *The Fifth International Conference on Nonlinear Mechanics (ICMN-V)* (pp. 967-969). Shanghai.

Helbing, D., Johansson, A., & Al-Abideen, H. Z. (2007b). The dynamics of crowd disasters: An empirical study. *Phys. Rev. E*, *75*, 046109.

Helbing, D., & Molnár, P. (1995). Social force model for pedestrian dynamics. *Phys. Rev. E*, *51*, 4282-4286.

Henderson, L. F. (1974). On the fluid mechanics of human crowd motion. *Transpn. Res.*, *8*, 509–515.

Hoogendoorn, S. P., & Bovy, P. (2003a). Simulation of pedestrian flows by optimal control and differential games. *Optim. Control Appl. Meth.*, *24*, 153.

Hoogendoorn, S. P., & Daamen, W. (2005). Pedestrian behavior at bottlenecks. *Transportation Science*, *39 2*, 0147-0159.

Hoogendoorn, S. P., Daamen, W., & Bovy, P. H. L. (2003b). Microscopic pedestrian traffic data collection and analysis by walking experiments: Behaviour at bottlenecks. In E. R. Galea (Ed.), *Pedestrian and Evacuation Dynamics '03* (pp. 89–100). CMS Press, London.

Hughes, R. L. (2000). The flow of large crowds of pedestrians. *Mathematics and Computers in Simulation*, *53*, 367–370.

Hughes, R. L. (2002). A continuum theory for the flow of pedestrians. *Transportation Research Part B*, *36*, 507–535.

Johnson, N. R. (1987, Oct). Panic at "The Who Concert Stampede": An Empirical Assessment. *Social Problems*, *34*(4), 362–373.

Keating, J. P. (1982). The myth of panic. *Fire Journal*, *May*, 57-62.

Kirchner, A., Klüpfel, H., Nishinari, K., Schadschneider, A., & Schreckenberg, M. (2003a). Simulation of competitive egress behavior: Comparison with aircraft evacuation data. *Physica A*, *324*, 689.

Kirchner, A., Klüpfel, H., Nishinari, K., Schadschneider, A., & Schreckenberg, M. (2004). Discretization effects and the influence of walking speed in cellular automata models for pedestrian dynamics. *J. Stat. Mech.*, *10*, P10011.

Kirchner, A., Namazi, A., Nishinari, K., & Schadschneider, A. (2003b). Role of conflicts in the floor field cellular automaton model for pedestrian dynamics. In E. R. Galea (Ed.), (p. 51). London: CMS Press.

Kirchner, A., Nishinari, K., & Schadschneider, A. (2003c). Friction effects and clogging in a cellular automaton model for pedestrian dynamics. *Phys. Rev. E*, *67*, 056122.

Kirchner, A., & Schadschneider, A. (2002). Simulation of evacuation processes using a bionics-inspired cellular automaton model for pedestrian dynamics. *Physica A*, *312*, 260.

Klügl, F., & Rindsfüser, G. (2007). Large-scale agent-based pedestrian simulation. *Lect. Notes Comp. Sc.*, *4687*, 145.

Klüpfel, H. (2006). The simulation of crowds at very large events. In A. Schadschneider, T. Pöschel, R. Kühne, M. Schreckenberg, & D. Wolf (Eds.), *Traffic and Granular Flow '05* (p. 341). Berlin: Springer.

Klüpfel, H. (2007). The simulation of crowd dynamics at very large events — calibration, empirical data, and validation. In N. Waldau, P. Gattermann, H. Knoflacher, & M. Schreckenberg (Eds.), (p. 285). Berlin: Springer.

Klüpfel, H., & Meyer-König, T. (2003a). Models for crowd movement and egress simulation. In S. F. et al. (Ed.), *Traffic and Granular Flow '03* (pp. 357–372). Berlin: Springer.

Klüpfel, H., & Meyer-König, T. (2003b). Simulation of the evacuation of a football stadium. In S. F. et al. (Ed.), *Traffic and Granular Flow '03* (pp. 423–430). Berlin: Springer.

Klüpfel, H., Meyer-König, T., Wahle, J., & Schreckenberg, M. (2000). Microscopic simulation of evacuation processes on passenger ships. In S. Bandini & T. Worsch (Eds.), *Theory and Practical Issues on Cellular Automata.* Berlin Heidelberg: Springer.

Kretz, T., Grünebohm, A., Kaufman, M., Mazur, F., & Schreckenberg, M. (2006a). Experimental study of pedestrian counterflow in a corridor. *J. Stat. Mech.*, P10001.

Kretz, T., Grünebohm, A., & Schreckenberg, M. (2006b). Experimental study of pedestrian flow through a bottleneck. *J. Stat. Mech.*, P10014.

Kretz, T., & Schreckenberg, M. (2006c). Moore and more and symmetry. In N. Waldau, P. Gattermann, H. Knoflacher, & M. Schreckenberg (Eds.), (pp. 317–328). Berlin: Springer.

Lakoba, T. I., Kaup, D. J., & Finkelstein, N. M. (2005). Modifications of the Helbing-Molnár-Farkas-Vicsek social force model for pedestrian evolution. *Simulation, 81 5*, 339–352.

Lewin, K. (1951). *Field Theory in Social Science.* Harper.

Maniccam, S. (2003). Traffic jamming on hexagonal lattice. *Physica, A321*, 653.

Maniccam, S. (2005). Effects of back step and update rule on congestion of mobile objects. *Physica A, 346*, 631.

Marconi, S., & Chopard, B. (2002). A multiparticle lattice gas automata model for a crowd. *Lecture Notes in Computer Science, 2493,* 231.

Muir, H. C., Bottomley, D. M., & Marrison, C. (1996). Effects of motivation and cabin configuration on emergency aircraft evacuation behavior and rates of egress. *Intern. J. Aviat. Psych.*, *6*(1), 57–77.

Müller, K. (1981). *Zur Gestaltung und Bemessung von Fluchtwegen für die Evakuierung von Personen aus Bauwerken auf der Grundlage von Modellversuchen.* Dissertation, Technische Hochschule Magdeburg, in German.

Muramatsu, M., Irie, T., & Nagatani, T. (1999). Jamming transition in pedestrian counter flow. *Physica A, 267*, 487–498.

Muramatsu, M., & Nagatani, T. (2000a). Jamming transition in two-dimensional pedestrian traffic. *Physica A, 275*, 281–291.

Muramatsu, M., & Nagatani, T. (2000b). Jamming transition of pedestrian traffic at crossing with open boundary conditions. *Physica A, 286*, 377–390.

Nagai, R., Fukamachi, M., & Nagatani, T. (2006a). Evacuation of crawlers and walkers from corridor through an exit. *Physica A, 367*, 449–460.

Nagai, R., & Nagatani, T. (2006b). Jamming transition in counter flow of slender particles on square lattice. *Physica A, 366*, 503.

Nagel, K., & Schreckenberg, M. (1992). A cellular automaton model for freeway traffic. *Jrl. Physique I, 2*, 2221.

Nakayama, A., Hasebe, K., & Sugiyama, Y. (2005). Instability of pedestrian flow and phase structure in a two-dimensional optimal velocity model. *Phys. Rev. E, 71*, 036121.

Navin, P. D., & Wheeler, R. J. (1969). Pedestrian flow characteristics. *Traffic Engineering, 39*, 31–36.

Nelson, H. E., & Mowrer, F. W. (2002). Emergency movement. In P. J. DiNenno (Ed.), *SFPE Handbook of Fire Protection Engineering* (Third ed., p. 367). Quincy MA: National Fire Protection Association.

Oeding, D. (1963). *Verkehrsbelastung und Dimensionierung von Gehwegen und anderen Anlagen des Fußgängerverkehrs* (Forschungsbericht No. 22). : Technische Hochschule Braunschweig, in German.

Older, S. J. (1968). Movement of pedestrians on footways in shopping streets. *Traffic Engineering and Control, 10*, 160-163.

Predtechenskii, V. M., & Milinskii, A. I. (1978). *Planing for Foot Traffic Flow in Buildings*. Amerind Publishing, New Dehli. (Translation of: Proekttirovanie Zhdanii s Uchetom Organizatsii Dvizheniya Lyuddskikh Potokov, Stroiizdat Publishers, Moscow, 1969)

Rogsch, C., Klingsch, W., Seyfried, A., & Weigel, H. (2007). How reliable are commercial software-tools for evacuation calculation? In *Interflam 2007 - Conference Proceedings* (pp. 235–245).

Rothman, D., & Zaleski, S. (1997). *Lattice-gas Cellular Automata*. Cambridge University Press.

Rothman, D. H., & Zaleski, S. (1994). Lattice-gas models of phase separation: Interfaces, phase transitions, and multiphase flow. *Rev. Mod. Phys., 66*, 1417.

Saloma, C. (2006). Herding in real escape panic. In N. Waldau, P. Gattermann, H. Knoflacher, & M. Schreckenberg (Eds.), *Pedestrian and Evacuation Dynamics 2006*. Berlin: Springer.

Schadschneider, A. (2002). Cellular automaton approach to pedestrian dynamics – theory. In M. Schreckenberg & S. D. Sharma (Eds.), *Pedestrian and Evacuation Dynamics* (pp. 75–86). Berlin Heidelberg: Springer.

Schadschneider, A., Klingsch, W., Klüpfel, H., Kretz, T., Rogsch, C., & Seyfried, A. (2009). Evacuation dynamics: Empirical results, modeling and applications. In B. Meyers (Ed.), *Encyclopedia of Complexity and System Science*. Springer.

Schadschneider, A., Pöschel, T., Kühne, R., Schreckenberg, M., & Wolf, D. (Eds.). (2006). *Traffic and Granular Flow '05*. Berlin: Springer.

Schreckenberg, M., & Sharma, S. D. (Eds.). (2002). *Pedestrian and Evacuation Dynamics*. Berlin Heidelberg: Springer.

Schütz, G. M. (2001). Exactly solvable models for many-body systems. In C. Domb & J. L. Lebowitz (Eds.), *Phase Transitions and Critical Phenomena, Vol. 19*. Academic Press.

Seyfried, A., Rupprecht, T., Passon, O., Steffen, B., Klingsch, W., & Boltes, M. (2007). Capacity estimation for emergency exits and bootlenecks. In *Interflam 2007 - Conference Proceedings*.

Seyfried, A., Steffen, B., Klingsch, W., & Boltes, M. (2005). The fundamental diagram of pedestrian movement revisited. *J. Stat. Mech.*, P10002.

Seyfried, A., Steffen, B., & Lippert, T. (2006). Basics of modelling the pedestrian flow. *Physica A, 368*, 232-238.

Tajima, Y., & Nagatani, T. (2002). Clogging transition of pedestrian flow in T-shaped channel. *Physica A, 303*, 239–250.

Thompson, P. A., & Marchant, E. W. (1994). Simulex; developing new computer modelling techniques for evaluation. In *Fire Safety Science – Proceedings of the Fourth International Symposium* (pp. 613–624).

Togawa, K. (1955). *Study on Fire Escapes Basing on the Observation of Multitude Currents* (Report of the Building Research Institute).: Ministry of Construction, Japan.

Waldau, N., Gattermann, P., Knoflacher, H., & Schreckenberg, M. (Eds.). (2007). *Pedestrian and Evacuation Dynamics 2005.* Berlin: Springer.

Weidmann, U. (1993). *Transporttechnik der Fußgänger - Transporttechnische Eigenschaften des Fußgängerverkehrs (Literaturauswertung)* (Schriftenreihe des IVT No. 90). : ETH Zürich. (Second Edition, in German)

Werner, T., & Helbing, D. (2003). The social force pedestrian model applied to real life scenarios. In E. R. Galea (Ed.), (p. 17). London: CMS Press.

Wolf, D., & Grassberger, P. (Eds.). (1996). *Friction, Arching, Contact Dynamics.* Singapore: World Scientific.

Yamori, K. (1998). Going with the flow: Micro-macro dynamics in the macrobehavioral patterns of pedestrian crowds. *Psychological Review, 105*(3), 530–557.

Yu, W. J., Chen, R., Dong, L. Y., & Dai, S. Q. (2005). Centrifugal force model for pedestrian dynamics. *Phys. Rev. E, 72*, 026112.

Chapter VII
"Social Potential" Models for Modeling Traffic and Transportation

Rex Oleson
University of Central Florida, USA

D. J. Kaup
University of Central Florida, USA

Thomas L. Clarke
University of Central Florida, USA

Linda C. Malone
University of Central Florida, USA

Ladislau Boloni
University of Central Florida, USA

ABSTRACT

The "Social Potential", which the authors will refer to as the SP, is the name given to a technique of implementing multi-agent movement in simulations by representing behaviors, goals, and motivations as artificial social forces. These forces then determine the movement of the individual agents. Several SP models, including the Flocking, Helbing-Molnar–Farkas-Visek (HMFV), and Lakoba-Kaup-Finkelstein (LKF) models, are commonly used to describe pedestrian movement. A systematic procedure is described here, whereby one can construct and use these and other SP models. The theories behind these models are discussed along with the application of the procedure. Through the use of these techniques, it has been possible to represent schools of fish swimming, flocks of birds flying, crowds exiting rooms, crowds walking through hallways, and individuals wandering in open fields. Once one has an understanding of these models, more complex and specific scenarios could be constructed by applying additional constraints and parameters. The models along with the procedure give a guideline for understanding and implementing simulations using SP techniques.

INTRODUCTION

Modeling traffic and transportation requires consideration of how individuals move in a given environment. There are three general aspects to consider when looking at movement: reactive behaviors, cognitive behaviors and constraints due to environmental factors. Individual drivers and pedestrians have a general way of dealing with certain situations, some of which comes from experience and some from personality. In this situation, there is generally only one specific response for any given agent. In other situations, one needs to allow an individual to choose from a set of various possible decisions based on how they affect movement and path planning. A final consideration is how the environment will constrain the general movement of the individual.

Much of an individual's movement, especially when driving a vehicle, is reactive. This is due to the fact that most actions are reactions to the conditions of the road and events which are occurring nearby. This is similar to pedestrian movement since walking becomes routine for people. Individuals do not think about every step that they are going to make and every possible outcome, they simply step forward and know the general outcomes they expect. When things deviate from the expected, then their movements are adjusted. Individuals transporting cargo, have a defined origin and destination which requires some decision making such as route planning. There is a goal they are trying to reach, and decisions are made along the way to achieve this goal. We will refer to these as cognitive behaviors, due to the fact that they take some conscious thought to achieve the goal. Techniques of path planning, seeking or organization can be used to represent these choices. The final aspect of movement is the definition of the environment. The individuals need to know where obstacles are and how they interact with them in order to avoid collisions and other unwanted contact.

In multi-agent systems there are numerous techniques which can be used to describe how each agent makes decisions and moves, such as Genetic Programming, Reinforced Learning, Case Based Reasoning, Rules Based Reasoning, Game Theory, Neural Network, Context Based Reasoning, Cellular Automata, and SP. The two primary techniques which are used to represent the decisions of individuals in pedestrian simulations are Cellular Automata and SP.

This chapter will focus on SP techniques for modeling and how to use it to represent individuals' desires and movements during a simulation. A description of the technique is given along with a detailed example of constructing a model from scratch. This will give some insight into the elements of the technique and the process which must be taken to use it effectively. There are a few commonly used models which represent pedestrian movement: Flocking (Reynolds, 1987), HMFV (Helbing, 2002), and LKF (Lakoba, 2005). A brief description of these models will be given along with the forces which are used in the model. Then cognitive behaviors will be discussed which can be added to any of the existing models to create specific desired movements in the individuals. Next, a description of different techniques used to interact with the environment is given. We then conclude by looking at how to apply this technique to more than individuals' movements.

BACKGROUND

Individuals tend to move in predictable manners due to the fact that walking in an environment becomes an automatic process where decisions are made instinctively (Helbing, 2005). People are familiar with walking and the paths they tend to follow. This fact allows for the construction of models which should represent the movement of individuals in reasonably simple terms. The same could be said for traffic and transportation movements, except that the possible movements for these are constrained more than for individuals. Nonetheless, the same techniques can be used for both systems.

One manner of looking at how an object moves is to relate it to the physical forces acting on the object, referred to as Newtonian Mechanics. The SP technique represents the movement of sentient beings by artificial forces between an individual and the environment in the same way Newtonian Mechanics represents movements via physical forces. The SP technique was originally developed as a way of modeling individuals' decisions. One of the earliest uses was in modeling flocks of birds and schools of fish (Reynolds, 1993). Then the techniques were applied to robotic movement and path planning (Herbert, 1998; Lee, 2003; Reif, 1999). The technique is set up to allow the behavior of an individual to be defined through a collection of simple force-like rules. These artificial forces sometimes relate the social interaction between individuals and therefore the name "Social Potential" was given to describe the modeling technique (Reif, 1999). The SP technique originally used potentials to calculate forces and then used these forces to determine the movement of the individuals. Research in the field has shown that forces other then potential based forces might be required (Helbing, 2002) to simulate the movement of some individuals. Therefore we will refer to any model which uses forces to determine the direction of movement as an SP model.

In order to use this technique, the causes of the movements must be identified and then artificial forces representing their effects must be designed. The appropriately designed forces will then define how the individual reacts to each of these causes. The final movement of an individual is then taken to be a superposition of all influencing forces. This separation into an individual force for each cause allows for a simple definition of the individual forces whose sum creates the specific movements in the individuals.

The SP technique treats each individual like a particle; these particles are attracted or repelled from points, obstacles, other individuals, and areas of interest. This technique creates interactions on a microscopic level by simulating the movement of each individual. Treating each individual as a particle allows the creator of the model to focus on what influences a given individual and how to define the reaction of the individual to these influences. This allows for simple definitions and relates these influences to a commonly used technique, Newtonian Mechanics. Groups of individuals can then be simulated by placing numerous individuals into a common area and allowing these individuals to interact. The sum total of all the individual movements then gives rise to the emergent behaviors which is referred to as macroscopic "Crowd Dynamics."

Simple Example

Consider searching for a place to eat when visiting a new location. This would have to be a place where you have never been before therefore you have no previous knowledge of the location of possible places to eat. Now assume that you intend to find a place by wandering around; in this way you will also get to know the area. What factors are going to be important to you?

1. Desire to stay close to the hotel, or where you are staying.
2. Attraction to visible restaurants.
3. Slight repulsion from other individuals.
4. Repulsion from crowded restaurants.

These four factors are identified as the causes for the movements of the individual. Assuming that there are no constraints on where you can walk (no walls or buildings) then there is a simple set of rules governing the movement. These rules are built as a set of forces representing the previously defined factors.

Since you want to stay near the hotel, the further you get away from the hotel, the larger the attraction to the hotel should be. The force keeping you near the hotel should have a form which increases as you get further away, like $f = -a \cdot r_{hotel}$ or $f = -a \cdot e^{b \cdot r_{hotel}}$ where r_{hotel} is the distance from the individual to the hotel and *a and b* are parameters. As you approach an eating establishment your attraction to the establishment should grow in the opposite manner, so the force should have something like the form $f = \frac{-c}{r_{\text{restaurant}}}$ or $f = -c \cdot e^{-d \cdot r_{\text{restaurant}}}$ where *c and d* are parameters with $r_{\text{restaurant}}$ being the distance from the individual to the restaurant. Everyone has a certain amount of personal space they attempt to maintain, so they are generally repelled from nearby individuals by something like $f = \frac{g}{r_{individual}}$ or $f = g \cdot e^{-h \cdot r_{\text{restaurant}}}$. If you are currently hungry, you know that any crowd at a restaurant generally means a long wait time, so you should be repelled from crowded restaurants. You could represent this by using a force of the form $f = j \cdot (\#of\_individuals_{\text{restaurant}})$.

In general it is easiest to start out with the simpler polynomial type forces then try the exponential forces second. However these different forms could produce distinctly different behaviors. Choosing the simple polynomial functions to represent the forces will give a general idea of the movement, so we would take

$$f_1 = -a \cdot r_{hotel}$$

$$f_2 = \frac{-c}{r_{\text{restaurant}}}$$

$$f_3 = \frac{g}{r_{individual}}$$

$$f_4 = j \cdot (\#of\_individuals_{\text{restaurant}}).$$

The above illustrates the four general forces which one would use to begin simulating the above scenario. There are other forces which should be present, such as a small random force to start the individual looking for a restaurant, and to keep them from moving in perfectly straight lines. There could also be other interactions between individuals as well as interactions to prevent the individual from walking into buildings or other obstacles.

Generally, the approach taken in constructing an SP model involves the three broad steps discussed above. These are:

1. Define the important aspects which need to be modeled.
2. Decide on the types of forces and their functional form which would represent their causes.

3. Determine the appropriate values for the free parameters in the forces which would best represent the system you are trying to model.

If there is no driving reason for choosing a certain functional form for the forces then start as simple as you can. Begin with a simple polynomial and test the application to see if the individuals move in the general manner that you require. Get close approximations of the parameters then see if you need to adjust the types of forces, or possibly even add new forces. These three steps will be iterated numerous times before completing the construction of a model. Since this process can be very time consuming it can be helpful to start from an existing model.

CURRENT SP MODELS

There are only a few standard SP models being used to describe pedestrian movement. There are also models which have been developed for robotic movement (Khatib, 1985; Reif, 1999) which can also be used, but since each model for robotic movement is constructed for a specific goal, we will focus on the general models which are currently used for pedestrian movement. Each model has particular strengths as well as disadvantages, but they can be used as a starting point on which to build your model. These models already have the forces defined for basic movement and certain parameters have been set or bounded. This allows for a simple starting point and reduces the number of free parameter values which one would need to set (or determine) to represent a specific simulation.

Flocking/Herding

Flocking was one of the first recognized models using the SP technique. Craig Reynolds in the 1980s was trying to find a new way of defining movement of computer simulated individuals (Reynolds, 1987). Up to that time the movement of each individual was constructed by hand; this made simulating large numbers of individuals difficult and labor intensive. Reynolds found that he could represent these movements by four simple forces: cohesion, avoidance, flock centering, and a small random force. This simulation was called Boids and did an amazing job of representing both bird flocking and fish schooling

Of the social forces used in this model, cohesion is the force which causes the individuals to stick together; it is a mild attractive force toward other individuals within a local neighborhood of the individual. Avoidance is a repulsive force which balances the cohesive force so as to keep the individuals from running into one another. Flock centering is a force used to bring the individuals into a unified entity. The force representing the flock centering causes each individual to try to get into the center of the individuals it can see. This would give the individual the most protection from the surrounding elements and enemies. A small random force is necessary to prevent an individual from walking in a straight line. This randomness makes the simulation more accurate in portraying the life-like pattern of humans walking.

Flock centering is very noticeable in schools of fish. Since the fish on the edge of the school are most likely to be eaten, these fish constantly push themselves toward the center, thereby pushing the other fish out to the edges (Seghers, 1974). The constant pushing toward the center creates the shape of the school and causes the location of any individual in it to be constantly changing, not only in regard to its surroundings but also with regard to the school itself. Flock centering behavior is not as recognizable in flocks of birds, so in this case, this force is less important and can be given less influence. However, an exception to this is found in penguins. The emperor penguins guard their eggs over the long cold winter; the birds on the edge constantly move in towards the center causing the same cycle-type motion as mentioned above in fish. In this way the penguins keep the entire collection of birds at a reasonable temperature instead of leaving the edge to freeze (Gilbert, 2006).

Current implementations for pedestrian movement generally contain various forms of the above three types of social forces, excluding the random force. Since people do not generally have the need for protection whereby they would struggle to get toward the center, a centering concept as in flocking is not needed. In place of flocking a "consistency force" is added, keeping each individual moving in the direction he/she was generally moving.

A distinction in this model is that velocities are fixed and the forces are only used to determine the *direction* which the individuals will move. The collection of these forces is sometimes called a herding model, since the individuals loosely clump together and thereby act as a single collection, or herd. These forces would typically be of the form:

$$\vec{f}_{consistency} = c \cdot \vec{v}$$

$$\vec{f}_{avoidance} = \frac{\vec{r} \cdot a}{r^3}$$

$$\vec{f}_{cohesion} = -s \cdot \vec{r}$$

$$\vec{F} = \vec{f}_{consistency} + \vec{f}_{avoidance} + \vec{f}_{cohesion}$$

$$\Delta\vec{X} = \vec{F} \cdot \frac{v \cdot \Delta t}{\|F\|}$$

HMFV Model

Helbing, Molnar, Farkas and Vicsek realized that the representation of an individual's movements in a physical environment must consider standard physical forces because contact can occur with other pedestrians or objects (Helbing, 2002). In this model, an individual generally has both types of forces acting on him/her: the physical forces and the social forces. The physical forces are actual forces, like frictional and pushing forces, which occur when two individuals run into or otherwise contact each other, or when an individual collides with an obstacle. The social forces are those which represent how a self-determined individual would want to move. Both classes of forces are necessary in order to obtain realistic movement of individuals and realistic interaction between an individual and obstacles in the environment.

The HMFV model uses three primary forces: social, frictional, and pushing. The social force represents the personal space an individual wishes to keep open around them; it is modeled using exponential decay. The force of friction occurs when the individual contacts another individual or an obstacle. The frictional force on a pedestrian is tangential and opposite to the relative motion between them and the object or other individual. The pushing force occurs due to the fact that in a crowd, packed individuals are slightly compressible and therefore spring, or push back, when pressing on another individual

Table 1. Variables for Flocking model

$\vec{r}$	The distance between two individuals, directed to the individual on which the force acts
r	The magnitude of the distance between two individuals
$\vec{v}$	The velocity of the individual of interest
v	The magnitude of the velocity
$\Delta\vec{X}$	The change in position of an individual
Δt	The time step used for the simulation
c, a, s	Free parameters to adjust strength of the individual forces

or obstacle. The normal, pushing force is modeled by Hooke's law. The forms for these forces in the HMFV model are given below:

$$\vec{f}_{social} = \vec{r} \cdot a \cdot e^{-r/b}$$

$$\vec{f}_{friction} = \kappa \cdot \vec{N} \times \left[\vec{N} \times \vec{v}\right].$$

$$\vec{f}_{pushing} = \vec{N} \cdot c \cdot \delta r$$

LKF Model

Lakoba, Kaup, and Finkelstein modified the HMFV model by including more physically realistic parameter values in the physical forces (Lakoba, 2005). However when this was done, new issues arose, especially when dealing with different densities of individuals in the simulations. New social forces had to be included in order to create more physically realistic simulations for all densities. The new social forces dealt with the directionality of interactions between individuals as well as the "excitation level" of an individual. The physical forces kept the same basic form as in the original HMFV model (Lakoba, 2005), which are the first two equations listed below.

Table 2. Variables for HMFV model

$\vec{N}$	The outward normal vector from the object or other individual, located at the point of contact
δr	The magnitude of overlap between an individual on the object of interest
a, b, c, k	Free parameters to adjust the strengths or ranges of the various forces

A description of the other forces introduced by LKF is in order. There are two different forms for the social forces: one for the social force acting on an individual (ind) due to any obstacle (obs) and another for the social force acting on an individual due to the presence of another individual. For the former, the force is a repulsive force along the line from the individual to the center of the obstacle. Its magnitude is given by $wF1(obs) \cdot \max faceToBack \cdot e^{\frac{-r}{B}}$, with the coefficient $wF1$ as an orientation factor. If the obstacle can be seen then $wF1$ is unity, otherwise as the angle increases from $\pi/2$, the value of the $wF1$ decreases to the value of b, which is defined below. The value of b is reached when the angle is π. The quantity $\max faceToBack$ represents the maximum value of this force when the individual is facing the back of an obstacle. For the social force between individuals, the form is the same except that the additional factor $wF3$ is included and has the effect of replacing $\max faceToBack$ with $\max faceToFace$ when the individual can see the other. The velocity of an individual is defined by the excitation of the individual, their current speed, and the average speed of nearby individuals. The excitation of the individual is also allowed to change over time, and this is based on the current excitation and the ability of the individual to move at their initial velocity. The function $wF3$ is defined as $\max faceToBack$, or $\max faceToFace$ depending on if the entity in question can be seen. This notation given here is a change from the notation in Lakoba (2005), but without changing the value of any force they used. This only simplifies the definition of the social force acting on an individual, and causes $wF3$ to be nothing more than a switch from $\max faceToFace$ to $\max faceToBack$.

$$\vec{f}_{friction} = \kappa \cdot \vec{N} \times \left[\vec{N} \times \vec{v}\right]$$

$$\vec{f}_{pushing} = \vec{N} \cdot c \cdot \delta r$$

$$\vec{f}_{social\_obs}(obs) = \vec{N} \cdot wF1(obs) \cdot \max faceToBack \cdot e^{\frac{-r}{B}}$$

$$\vec{f}_{social\_ind}(ind) = \vec{N} \cdot wF1(ind) \cdot wF3(ind) \cdot \left[\frac{e^{\frac{-r}{B}}}{m}\right]$$

where

$$wF1(entity) = \begin{cases} 1 & canSee(entity) \\ 1 - \frac{2}{\pi}\left[\theta - \frac{\pi}{2}\right][1-b] & else \end{cases}$$

$$wF3(entity) = \begin{cases} \max faceToFace & canSee(entity) \\ \max faceToBack & else \end{cases}$$

$$\max faceToFace = F\left[1 + \frac{1}{e_{\max}}\right][[1-\tilde{\rho}] + k2 \cdot \tilde{\rho}]$$

$$\max faceToBack = w_0 \cdot m \cdot [1 + e_{\max}] \cdot e^{\frac{-1}{B}\left[D - \frac{1}{\rho_{\max}}\right]}[k_0[1-\tilde{\rho}] + k1 \cdot \tilde{\rho}]\frac{1}{\tau[1-b]}$$

$$\vec{v}_0(t) = \vec{\theta}_g [1 + E(t)] v_{preferred} [1-p] + p \cdot \vec{v}_{local}$$

$$E(t) = E(t - \Delta t) - \Delta t \cdot \left[\frac{E(t-\Delta t)}{T} + \frac{e_{\max}}{T}\frac{1 - \|\vec{v}\|}{v_{preferred}}\right]$$

Some of these symbols have already been defined above. The new symbols introduced are given in Table 3 just below along with the values of the parameters used in the original LKF model.

COGNITIVE BEHAVIORS

Cognitive behavior forces are forces that can be added to the individual to create specific directional choices. These are things like wandering, seeking, following a path, or following a wall. They are considered cognitive behavioral forces due to the fact that the individual is making a decision using these forces; they are not purely reactive style forces.

Wander

Wander is sometimes referred to as a random walk. This type of force is generally needed in order to keep an individual from walking in a perfectly straight line. Basically it creates small deviations from the path the individual would otherwise take (Reynolds, 1999).

One method of applying this technique is to choose a small maximum angle (θ max) of deviation inside of which one would place an artificial attraction point and then add the force from the attraction point to the other forces acting on the individual (Figure 1). The strength of the force can be adjusted by choosing the distance (d) the artificial attraction point is placed from the center of the individual. For example:

$$\theta = (random * 2\theta \max) - \theta \max$$
$$f = (d * \cos\theta, d * \sin\theta)$$

where *random* is a randomly selected number between 0 and 1.

This force should never be so large that the individual will not follow the path at all; this is supposed to be small deviations in the movement of the individual as they follow the main path. The main path should still be selected by other forces.

Table 3. Variables for LKF model

$b = 0.3$	Back-to-front ratio of perception
$B = 0.5m$	Approximate fall off length (personal space) for the social forces
$D \approx 0.7m$	The diameter of the individual
E	The excitation state of the individual
$e_{\max} = \frac{v_i}{w_0} \approx 1$	The maximum value allowed for the excitement parameter (E)
$m \approx 80kg$	The average mass of an individual
$p \in (0,1)$	The parameter representing the independence of an individual (does not change through the simulation)
ρ	The number of individuals inside a circle of radius B around the individual of interest, divided by the area πB^2
$\tilde{\rho} = \rho \cdot \frac{\pi D^2}{4}$	The non-dimensionalized density of individuals
$\rho_{\max} = 5.4 people / m^2$	Maximum allowable density of people per square meter (Weidmann, 1992)
$T = 2s$	The lag time for excitement to return to initial state when unaffected
$\tau = 0.2s$	The average reaction time of a person
θ	The angle between θ_g and r
$\vec{\theta}_g$	The vector representing the direction the individual is looking
$\vec{v}$	The velocity of the individual
$v_{preferred}$	The individual's preferred speed. The values used in LKF were 1.5, 3.0, and 4.5 m/s.
$\vec{v}_0$	The preferred velocity of the individual
$\vec{v}_{local}$	The average velocity of individuals in the local neighborhood
$w_0 = 1.34 m/s$	Average walking speed of a non-panicked individual
$k_0 = 0.3$ $k_1 \in [1.2, 2.4]$ $k_2 = 1.5$	Parameters to adjust high density corrections for face-to-back orientation
c, F, k	Free parameters to adjust strength of the individual forces

Figure 1. Wander example

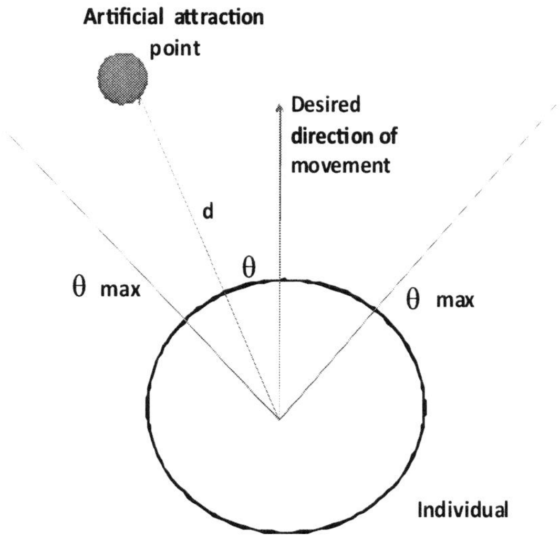

The previous example is capable of creating a jittery movement in the individual. For a smoother movement, one could pick θ such that it would have a pattern instead of being purely random (Hebert, 1998). For example: $\theta(t) = \theta \max * \cos(t)$, would create a smooth, wave-like motion around the path instead of the jitter due to a random selection (Ueyama, 1993). The trigonometric function can be adjusted to modify the frequency of the wander.

Seek (Flee)/Pursue (Evade)

Seek (Flee)/Pursue (Evade) occurs when an individual either tries to head toward an individual of interest or away from an individual of interest. This is different from the standard attraction and repulsion between individuals in that it is a *selected* attraction or repulsion. If a man saw someone selling fruits when he was looking for an apple, then he would be attracted to that particular vendor, hence seeking or pursuing the vendor. If someone was being followed and was trying to not be caught, then they would be evading or fleeing. This is a technique used in predator/prey style simulations (Isaacs, 1999). The key feature to these behaviors is to predict where the individual (either following or being followed) will be at some point of time in the future. The point of attraction will actually be to the projected position and not the current position. If the pursuer goes to the point where the evader is currently at, then no matter how fast he is traveling he will never reach the evader. This is because the evader will have moved a little bit, and therefore will be just outside of the reach of the pursuer. This is why the pursuer must move to the projected location of the evader. For the evader, the force would be structured like

Figure 2. Seek Flee example

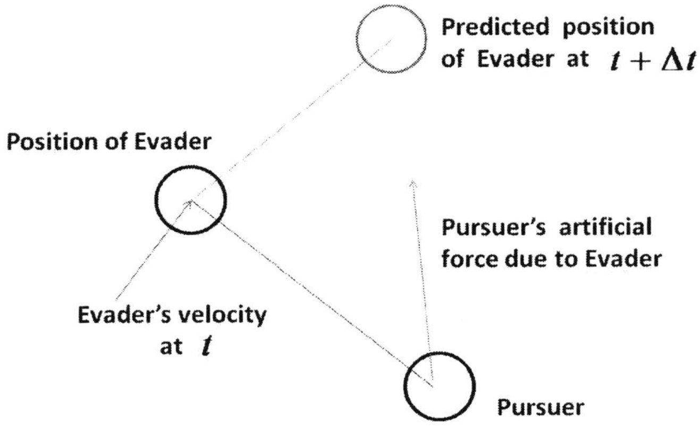

$$\vec{f} = \frac{a \cdot \vec{r}}{\|\vec{r}\|^b}$$

where

$$\vec{r} = \vec{X}_{evader(t)} - \vec{X}_{pursuer(t+\Delta t)}.$$

Similarly, for the pursuer the force would be

$$\vec{f} = \frac{-a \cdot \vec{r}}{\|\vec{r}\|^b}$$

where

$$\vec{r} = \vec{X}_{evader(t)} - \vec{X}_{pursuer(t+\Delta t)}.$$

Path Following

Path following is also sometimes referred to as "way-point based path planning". This is the ability to set up distinct way points to define a path that an individual will follow as he/she progresses to a destination point. In some ways, this goes against the idea of SP technique movement models in that the path is *not* determined by the forces. This technique can be very useful in planning out available routes that an individual can choose or to give an individual an idea of where movement should occur in an environment. The individual following the path must know the waypoints and the order in which to follow them. At the start, the individual gets an attraction force to the first waypoint. Once the individual gets close to the waypoint, the first attraction force is turned off and the attraction to the next waypoint in the list is turned on. This progression continues until the individual has passed all of the waypoints in the path. This is a way to create queues or lines in a simulation.

Figure 3. Path following example

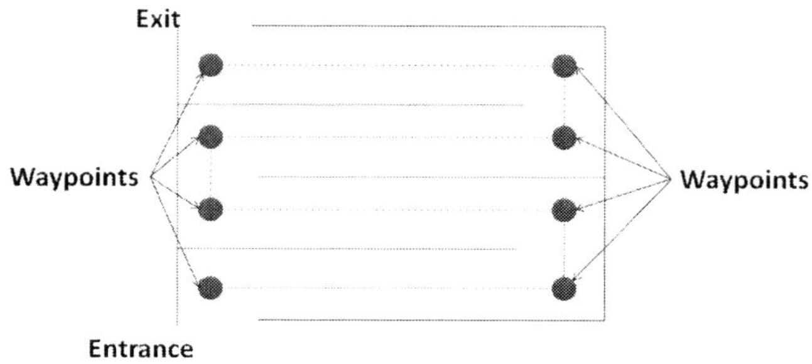

Wall Following

Wall following is a method which has been used for years to get out of mazes. Upon entering a maze, place a hand on one of the walls that touches the entry way, and continue to follow that wall. If you had started from the beginning of the maze then you are guaranteed to find the exit. On the other hand, if you were dropped into the middle of the maze, you could still use this principle. First, you would have to place a hand on a wall and mark where you are. Then follow that wall and if you found that you returned to the exact same spot, then you would move to the other wall and repeat the scenario. If you found that you returned once again to the exact same spot then both walls are interior walls and the technique fails because you are basically stuck inside a room with no doors. Otherwise, you will eventually find your way out.

In simulations, wall following becomes useful because when one is using social forces to represent the movement, an individual can become stuck in closed areas and at corners. The individuals have to get out of these areas before they can reach their goal (e.g. there could be an obstacle between the individual and their goal that they would have to go around before they could reach their goal). Wall following can create the necessary break-out condition to move the individual out of these trapped situations and allow them to continue toward the intended goal.

One way to do this in a simulation is to set up an artificial attraction point which is parallel to the obstacle and in the direction of the individual's movement (Figure 4). This new point is there to pull the individual along the wall. This force should become active only when the individual is within a given distance of a wall and then should flip to repulsive once he/she gets too close to the wall. This will allow the individual to keep a given distance from the obstacle that the individual is walking along.

The second option (Figure 5) for applying wall following does not use an artificial point of attraction, but rather just modifies the calculated forces to cause the desired movement. First, the movement is calculated as originally defined to get a direction and magnitude; this is the calculated force vector. Next, the line is found which goes through the center of the individual of interest and is parallel to the obstacle of interest. Finally, the calculated force vector is projected onto the parallel line. This forces all movements to be parallel to the obstacle of interest. Using this approach, the individual will not follow the wall at all times, but will only follow a wall when the wall is impeding the individual's movements toward a given goal.

Both of these techniques can be very useful when trying to manuever around obstacles and explore environments. Some decisional logic must sometimes be included when two obstacles touch each other so that the individual will interact with the correct obstacle.

Figure 4. Wall following option 1

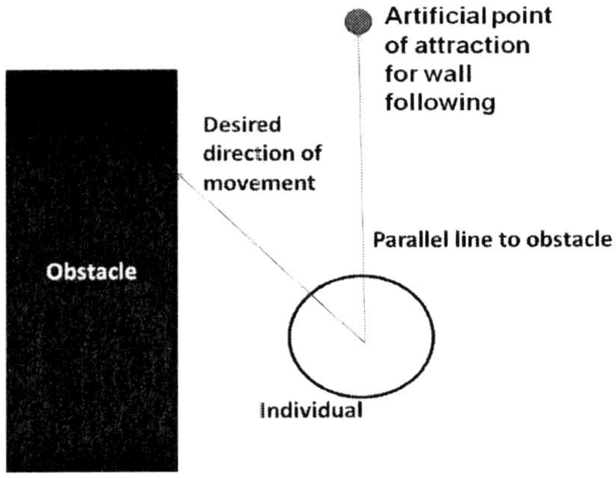

Figure 5. Wall following option 2

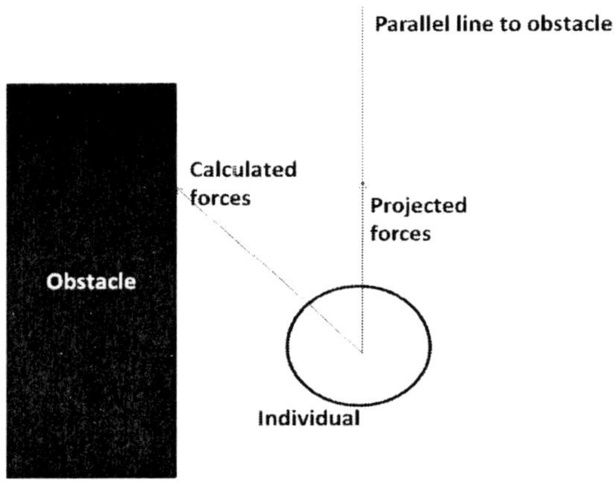

ENVIRONMENTAL FEATURES

The environment is a collection of geometric objects the individual must interact with, usually by avoiding them. The following are obstacles found in the simulation that define the environment in which individuals must maneuver.

Obstacles

An obstacle should have an external shape described in some manner such that the distance to points on it can be found. Also, obstacles should have a center. It is best to keep the definition of the obstacles to simple structures like rectangles and circles. Using pixilation principles defined for computer graphics, it is reasonably easy to represent all possible shapes by these two primitive structures (Pineda, 1988).

Walls

Walls are simply rectangular obstacles placed where a wall should occur in the simulation. There are some key points to consider though, primarily, what happens at the intersection of two walls. You do not want individuals walking between two connected walls, so make sure that there is no gap whatsoever between the two walls. Even a gap of a few centimeters could possibly be recognized and the individuals could attempt to squeeze between the two walls. This scenario can cause many problems in the simulation, and is sometimes very difficult to recognize. A simple solution to this is to always have the walls overlap slightly. This removes all possibility of an individual squeezing in between the walls.

Paths

Paths were described previously as a collection of waypoints the individual follows. Paths can be constructed as part of the environment and then handed to individuals when they need to use them. Consider an amusement park with five different rides. Each ride has a waiting line, and therefore each ride would have a path associated with it. These paths could reside as part of the environment. Once an individual decides to go on a given ride, a copy of the ride's associated path gets assigned to the individual. In this manner, the paths are part of the environment and the individuals only use these paths when they become of interest or are needed by the individuals being simulated (Lee, 2003).

Moving Obstacles

There is nothing that restricts an obstacle from moving. It is possible to define a simulation where the obstacles move regularly, like a train at a train station, or with a more complicated description, like vehicles at an intersection. As long as the descriptions for the movement are defined on the same time step (Δt) as the SP models, the two different entities can interact simultaneously. Also, if any obstacles need to move, they could be defined as a different type of entity having a given movement pattern with all other obstacles being stationary. Either approach is valid; it depends on what is being modeled and which approach fits the scenario the best.

Regions

Sometimes there are areas, or regions, in an environment where certain events should happen or where certain effects occur on an individual. These can be constructed in a manner similar to the technique used in video games where a region of effect is created and all individuals within that region are affected. To do this, define a region as an obstacle in the environment which has no attraction or repulsion. Associate a given effect with this obstacle. This effect could be a speed reducer to represent tough terrain, or it could be a more mild repulsive force to represent an area where an individual would not like to enter. These regions could be associated with given individuals or all individuals to allow for a large variation in simulation scenarios.

Interactions

How an individual recognizes other entities and obstacles in the environment is a very important aspect to the simulation. There are a few different techniques used: centroidal, subdivision, force field, axial, and centroid with axial.

Centroidal

Traditionally the obstacles are treated as point masses (Reynolds, 1987) and are usually located at the center of the obstacle. This is similar to the way an introductory course in physics simplifies the features of Newtonian Mechanics. In progressing through the levels of physics, one learns that dealing with everything as only point masses is a drastic over-simplification to the system. This simplification can cause erroneous results or leave out important dynamics of the system.

Subdivision

Here, the environment is subdivided into small cells. Once the grid is developed, the obstacles are intersected with the grid and any cell of the grid intersecting with an obstacle is considered to be an obstacle. The grid divisions need to be chosen according to the size of the obstacles and their general shapes. SP models are computationally dependent on the number of entities in the simulation since forces are calculated for each obstacle and for each individual. Because of these two factors, this grid-type division of the environment makes the calculations of movement for an individual in the simulation much more time consuming.

Force Field

If the environment is static, a force field can be generated from the environment definition. This technique can combine the information on strength of attraction/repulsion and overall shape of all obstacles in the environment (Gazi, 2005; Khatib, 1985). A map of the obstacles and their forces can then be used to determine the social forces on an individual due to the static environment. Once the map is generated, it can be referenced by the location of the individual, and the values for the social forces would be retrieved. The disadvantage here is that it takes a lot of work to define the environment and then to pre-calculate the necessary force fields to represent that environment.

Axial

This technique is based upon ray tracing concepts in computer graphics, but only a discrete number of rays are shot. This technique was used by Craig Reynolds who would shoot a single ray in the direction that the individual was moving and then check to see if it intersected any obstacles in the environment (Reynolds, 1993). The point of contact with the ray and the obstacle is the point used to calculate the interactions. Checking in the four axial directions gives a better idea of what was happening around an individual, instead of just what was occurring in front of the individual. The ray in the direction of movement could still be included but was not found to be that useful. This technique works reasonably well, but can miss a large number of obstacles which should be "seen" by an individual.

Centroid with Axial

Since all objects in a given vicinity of the individual are important, only checking in the axial directions is insufficient. The centroid with axial technique starts with gathering a collection of all obstacles in the known vicinity of the individual. The centroidal distance to the first obstacle is calculated. Next, one checks the four axial directions and calculates the distance to that obstacle. Then, one takes the minimum of the centroidal and axial distances and uses the point associated with that distance to calculate the social forces. Repeat this process for all obstacles in the vicinity of the individual. In this way, one is guaranteed to locate at least one point of interest for any obstacle near the individual.

Interacting with Various Models

A given SP model can be used simultaneously with a different SP model or even a different type of model altogether. SP models are continuous models discretized in time. The key point in ensuring that two continuous models work reasonably well together is they must have the same time step (Δt). In contrast, if the other model is a discrete model, like a Cellular Automata, the time step for the discrete model should be a multiple of the time step being used for the continuous models. If that is not possible, have the discrete model execute the first time step occurring after its execution should have occurred. Take care to make sure that the speeds and sizes of the individuals are in agreement between the models and then they can work reasonably well together.

CONCLUSION

SP techniques are very useful in describing the movements of individuals. The procedures described have been used to implement various models and to look at how individuals might be expected to react to given environments. However, it can easily be expanded and applied to individuals driving a car, riding a bicycle, etc. Recently Majid Ali Khan, Damla Turgut and Ladislau Bölöni (Khan 2008) have demonstrated the use of the SP technique for simulating trucks driving in highway convoys.

The mathematics of the models presented has been condensed, where needed, to allow for simpler implementations and easier understanding of the process of the SP techniques. These simplifications allowed relationships between interactions with obstacles and with other individuals to be apparent and quickly defined. Anytime individuals are in control of their movement and need to make decisions while simultaneously being constrained by the environment, SP models can be constructed to represent how individuals would tend to move.

Environments representing exiting rooms, walking in hallways, exiting gated areas, and wandering in a room have been visualized and simulated using this technique. By adding new parameters to existing models, ages and certain social characteristics were represented (Jaganthan, 2007; Kaup, 2006; Kaup, 2007). This has allowed the exploration of how environmental changes can affect different types of individuals. Differing exit strategies have been studied to see if environmental factors can be used to increase the efficiency of an exit. All of these results demonstrate the usefulness and applicability of the procedures described for the SP technique.

It also provides the possibility of eventually testing and validating social interaction theories. Given any theory, one could directly model that theory by programming a simulation so that the agents would respond per that theory. Then by running the simulation, one could observe what social structure(s) would arise.

FUTURE RESEARCH

Plans exist to continue to study additional parameters which can be included in current models to allow the design of better simulations for describing cultural and social differences. To do this correctly, one needs to have some reasonable measure by which one could determine whether or not two different simulations were sufficiently similar, as well as how close any one given simulation would compare to a real world event. Such methods need to be designed as quantitatively as possible.

As a first approach in this direction, videos have been created and gathered of various pedestrian movements in various venues, with the intention of gathering data from these videos which could be used for comparing simulations of these venues to real world videos of the same venue. A technique for doing this has been developed which is still in the testing phase. Preliminary results are encouraging.

REFERENCES

Gazi, V. (2005). Swarm aggregations using artificial potentials and sliding-mode control. *IEEE Transactions on Robotics* , (pp. 1208–1214).

Gilbert, C., Robertson, G., Le Maho, Y., Naito, Y., & Ancel, A. (2006). Huddling behavior in emperor penguins: Dynamics of huddling. *Physiology & Behavior , 88* (4-5), 479-488.

Hebert, T., & Valavanis, K. (1998). Navigation of an autonomous vehicle using an electrostatic potential field. *In Proceedings of the 1998 IEEE International Conference on Control Applications , 2*, 1328–1332.

Helbing, D., Buzna, L., Johansson, A., & Werner, T. (2005). Self-Organized Pedestrian Crowd Dynamics: Experiments, Simulations, and Design Solutions. *Transportation Science , 39*, 1-24.

Helbing, D., Farkás, I. J., Molnár, P., & Vicsek, T. (2002). Simulation of pedestrian crowds in normal and evacuation situations. In M. Schreckenberg, & S. D. Sharma (Eds.), *Pedestrian and Evacuation Dynamic* (pp. 21–58). Berlin, Germany: Springer.

Isaacs, R. (1999). *Differential Games: A Mathematical Theory with Applications to Warfare and Pursuit, Control and Optimization.* New York: Dover Publications.

Jaganthan, S., Clarke, T. L., Kaup, D. J., Koshti, J., Malone, L., & Oleson, R. (2007). Intelligent Agents: Incorporating Personality into Crowd Simulation. *I/ITSEC Interservice/Industry Training, Simulation & Education Conference.* Orlando.

Khan, M. A., Turgut, D., & Bölöni, L. (2008). A study of collaborative influence mechanisms for highway convoy driving. 5th Workshop on AGENTS IN TRAFFIC AND TRANSPORTATION. Estoril, Portugal.

Khatib, O. (1985). Real-time obstacle avoidance for manipulators and mobile robots. *In IEEE International Conference on Robotics and Automation , 2*, 500–505.

Kaup, D. J., Clarke, T. L., Malone, L., & Oleson, R. (2006). Crowd Dynamics Simulation Research. *Summer Simulation Multiconference.* Calgary, Canada.

Kaup, D. J., Clarke, T. L., Malone, L., Jentsch, F., & Oleson, R. (2007). Introducing Age-Based Parameters into Simulations of Crowd Dynamics. *American Sociological Association's 102nd Annual Meeting* . New York.

Lakoba, T., Kaup, D., & Finkelstein, N. (2005). Modifications of the Helbing-Molnr-Farkas-Vicsek Social Force Model for Pedestrian Evolution. *Simulation , 81*, 339–352.

Lee, J., Huang, R., Vaughn, A., Xiao, X., Hedrick, K., Zennaro, M., et al. (2003). Strategies of path-Planning for a UAV to track a ground vehicle. *AINS Conference* .

Pineda, J. (1988). A parallel algorithm for polygon rasterization. *SIGGRAPH '88: Proceedings of the 15th annual conference on Computer graphics and interactive techniques*, (pp. 17-20).

Reif, J., & Wang, H. (1999). Social potential fields: A distributed behavioral control for autonomous robots. *Robotics and Autonomous Systems, 27*(3), 171-194).

Reynolds, C. W. (1993). An Evolved, Vision-Based Behavioral Model of Coordinated Group Motion. *From Animals to Animats, Proc. 2nd International Conf. on Simulation of Adaptive Behavior.* Cambridge, MA: MIT Press.

Reynolds, C. W. (1987). Flocks, herds and schools: A distributed behavioral model. *SIGGRAPH Computer Graphics , 21*, 25-34.

Reynolds, C. W. (1999). Steering behaviors for autonomous characters. In AYu (Ed.), *Proceedings of the 1999 Game Developer's Conference* (pp. 763 – 782). San Francisco, CA: Miller Freeman.

Seghers, B. H. (1974). Schooling Behavior in the Guppy (Poecilia reticulata): An Evolutionary Response to Predation. *Evolution , 28*, 486-489.

Ueyama, T., & Fukuda, T. (1993). Self-organization of cellular robots using random walk with simple rules. *In Proceedings of 1993 IEEE International Conference on Robotics and Automation , 3*, 595-600.

Weidmann, U. (1992). Transporttechnik der Fussgänger. *Zürich: Institutfür Verkehrsplanung.*

ADDITIONAL READING

Bachmayer, R., & Leonard, N. (2002). Vehicle networks for gradient descent in a sampled environment. *In Proceedings of the 41st IEEE Conference on Decision and Control, 1*, 112–117.

Barraquand, J., Langlois, B., & Latombe, J. (1991). Numerical potential field techniques for robot path planning. *In Fifth International Conference on Advanced Robotics, 2*, 1012–1017.

Breder JR., C. M. (1954). Equations Descriptive of Fish Schools and Other Animal Aggregations. *Ecology*, *35*(2), 361-370.

Flacher, F., & Sigaud, O. (2002). Spatial Coordination through Social Potential Fields and Genetic Algorithms. *In In From Animals to Animats 7: Proceedings of the Seventh International Conference on Simulation of Adaptive Behavior* (pp. 389–390). Cambridge, MA: MIT Press.

Ge, S., & Cui, Y. (2000). New potential functions for mobile robot path planning. *IEEE Transactions on Robotics and Automation*, *16*(5), 615–620.

Hamacher, H., & Tjandra, S. (2002). Mathematical Modelling of Evacuation Problems: A State of the Art. In M. Schreckenberg, & S. D. Sharma (Eds.), *Pedestrian and Evacuation Dynamcis* (pp. 228-266). Berlin: Springer.

Heigeas, L., Luciani, A., Thallot, J., & Castagne, N. (2003). A physically-based particle model of emergent crowd behaviors. *Graphikon*.

Helbing, D. (1991). A mathematical model for the behavior of pedestrians. *Behavioral Science* (36), 298–310.

Hoogendoorn, S., Bovy, P., & Daamen, W. (2002). Microscopic Pedestrian Wayfinding and Dynamics Modelling. In M. Schreckenberg, & S. Sharma, *Pedestrian and Evacuation Dynamics* (pp. 123-154). Berlin: Springer.

Kachroo, P., Al-nasur, S. J., Wadoo, S. A. & Shende, A. (2008). Pedestrian Dynamics: Feedback Control of Crowd Evacuation. Berlin: Springer.

Kirkwood, R., & Robertson, G. (1999). The Occurrence and Purpose of Huddling by Emperor Penguins During Foraging Trips. *Emu*, 40–45.

Kitamura, Y., T. Tanaka, F. K., & Yachida, M. (1995). 3-D path planning in a dynamic environment using an octree and an artificial potential field. *In International Conference on Intelligent Robots and Systems*, *2*, 474–481.

Leonard, N. E., & Fiorelli, E. (2001). Virtual Leaders, Artificial Potentials and Coordinated Control of Groups. *Proceedings of the 40th IEEE Conference on Decision and Control*, (pp. 2968–2973).

Luce, R. D., & Howard, R. (1989). *Games and Decisions.* New York: Dover.

Millonig, A., & Schechtner, K. (2005). Decision Loads and Route Qualities for Pedestrians - Key Requirements for the Design of Pedestrian Navigation Services. In N. Waldau, P. Gattermann, H. Knoflacher, & M. Schreckenberg, *Pedestrian and Evacuation Dynamics 2005* (pp. 109-118). Berlin: Springer.

Rimon, E., & Koditschek, D. (1982). Exact robot navigation using artificial potential functions. *IEEE Transactions on Robotics and Automation*, *8*, 501–518.

Russo, F., & Vietta, A. (2002). Models and Algorithms for Evacuation Analysis in Urban Road Transportation Systems. In M. Schreckenberg (Ed.), & S. D. Sharma (Ed.), *Pedestrian and Evacuation Dynamcis* (pp. 315-322). Berlin: Springer.

Schreckenberg, M. (Ed.), & Sharma, S. D. (Ed.) (2001). Pedestrian and Evacuation Dynamcis. Berlin: Springer.

Thalmann, D., Musse, S. R., & Kallmann, a. M. (1999). Virtual humans behaviour: Individuals, groups, and crowds. In Proceedings of Digital MediaFutures. Bradford In.

Waldau, N. (Ed.), Gattermann, P. (Ed.), Knoflacher, H. (Ed.), & Schreckenberg, M. (Ed.) (2007). Pedestrian and Evacuation Dynamcis 2005. Berlin: Springer.

Vasudevan, C., & Ganesan, K.. (1996). Case-based path planning for autonomous underwater vehicles. *Autonomous Robots*, *3*(2-3), 79-89.

Chapter VIII
Towards Simulating Cognitive Agents in Public Transport Systems

Sabine Timpf
University of Augsburg, Germany

ABSTRACT

In this chapter, the authors present a methodology for simulating human navigation within the context of public, multi-modal transport. They show that cognitive agents, that is, agents that can reason about the navigation process and learn from and navigate through the (simulated physical) environment, require the provision of a rich spatial environment. From a cognitive standpoint, human navigation and wayfinding rely on a combination of spatial models ("knowledge in the head"), (default) reasoning processes, and knowledge in the world. Spatial models have been studied extensively, whereas the reasoning processes and especially the role of the "knowledge in the world" have been neglected. The authors first present an overview of research in wayfinding and then envision a model that integrates existing concepts and models for multi-modal public transport illustrated by a case study.

1. INTRODUCTION

In transport planning, simulation is an established tool and traditionally comprises four sequential steps: trip generation, trip distribution, modal choice, and traffic assignment (Ortuzar, 2001). These macro-models were critiqued, mainly for the strict sequence of the steps (re-planning is not possible) and the strong focus on individual motor car traffic (Meier, 1997). In contrast, simulations of public transport systems require an integration of different modes of transport, where each mode has specific properties and peculiarities.

The current trend in traffic simulation is towards activity oriented micro simulations and the inclusion of other modes of transport besides private cars (Widmer, 2000; Nagel, 2001; Raney et al., 2002a). Raney et al. (2002b) simulate the navigation of many travelers at once, which is needed to forecast the load on the transportation networks (Nagel, 2002). The focus in these simulations is on properties of the whole system (traffic loads, traffic flows), not on the individual traveler. However, when more than one mode of transport is involved, the cognitive processes of the individual traveler clearly matter. Hence, there is a need for models that handle the details of transfers between different modes of transport and provide agents with minimal cognitive processing capabilities.

We simulate the navigation process from the perspective of the user of the public transport system. The focus is on the user as a *cognitive agent*, i.e., an agent who can reason about the navigation process and navigate through and learn from the environment. From a cognitive standpoint, human navigation and wayfinding rely on a combination of spatial models ("knowledge in the head"), (default) reasoning processes, and knowledge in the world. There is a consensus in the spatial cognition community that many different models exist, but that there is a need to integrate them before the whole navigation process can be adequately described. Up to date, no integration effort has been undertaken because of the diversity of the existing models and theories.

Our goal is to design a modular system/framework in which different theories (and their resulting models) can be tested. The idea is that the system will allow the exchange of modules (implemented models) according to which theory currently prevails and for the purpose of comparing different approaches. Currently, many theories of spatio-temporal knowledge processing are known from research in psychology, geography and robotics, but they exist in separate models or even separate research communities. There are computational models for navigation or aspects thereof, which were built for the purposes of either proving psychological theories or for the purpose of robot navigation. As we will discuss in section 3, none of these models can be used for our specific case, although we build on insights from the TOUR (Kuipers, 1979) and NAVIGATOR models (Gopal et al. 1989). A multi-agent simulation system where each agent has cognitive and spatial processing capabilities seems to be an ideal basis for integrating models of navigation and test their effectiveness.

We will thus use a multi-agent methodology to simulate the complete navigation process, i.e. the wayfinding and the locomotion processes for a public transport system. This is different from other approaches where either the wayfinding aspect alone is modeled for a single agent (Raubal, 2001; Frank, 2001; Pontikakis, 2006) or the research is focused on pure locomotion described as physical models (e.g., Helbing & Molnar, 1995; Raney et al., 2002a). The integration of existing models will be a challenge in itself – thus we will work with a case study of navigation in a public transport system and especially with a study of the transfer process. The need to transfer from one means of transport to another at a specific time and place is mentioned as stressful by 70% of public transport users in our case study; 50% admitted to being annoyed by the need to transfer and 63% would rather travel a longer route than transferring (Heye, 2002).

The spatial and spatiotemporal reasoning processes during the transfer situation are rather complex (Raubal 2001; Heye & Timpf 2003). Formally modeling the basic process is possible and revealing (Rüetschi 2007). Such a model forms the basis for a multi-agent model of public transport. Navigation is an integrative process and requires many different sub-processes, which are well researched. An integrated model dealing with all aspects of navigation at the personal level has not yet been established. In this research, we are working towards such an integrated model in which different theories and different combinations of models of sub-processes can be experimented with in a single framework and a consistent fashion.

In the remainder of the chapter we will first describe the process of navigation as we understand it now while reviewing pertinent literature on the perception and representation of spatial information. We will pay special attention to the role of the environment, the processing of spatial information and the building of cognitive models. Section 3 presents and discusses computational models of navigation and wayfinding. Section 4 studies a case of navigation in a public transport system. Finally, in section 5 we provide our conclusions and present future work.

2. THE PROCESS OF NAVIGATION

Navigation is a process that includes (1) the thought processes going on while planning a trip and carrying it out; and (2) the physical locomotion along the route. The information processing is also known as wayfinding. As a scientific discipline, wayfinding is a relatively young field that originated in architecture, more *precisely*, with Kevin Lynch's (Lynch, 1960) book "The Image of the City". Lynch carried out empirical research on how individuals perceive and navigate the urban landscape. He described wayfinding as the "consistent use and organization of definite sensory cues from the external environment".

Wayfinding takes place in large-scale space (Kuipers, 1978), that is, space that cannot be seen and apprehended from a single vantage point. Research on wayfinding deals with the investigation of spatial abilities, wayfinding tasks, and means to solve the tasks (Allen, 1999, Golledge book 1996). Wayfinding is defined as spatial problem-solving with the purpose of reaching a destination (Arthur and Passini, 1992). It is also problem-solving under uncertainty (McDermott and Davis, 1986), because the traveler does not know and cannot plan for all the details of the trip. These unknown details are (hopefully) provided by the (physical, social and institutional) environment in which the wayfinding takes place. Thus, the problem-solving process is a continuous matching of perceptual input from the environment with existing knowledge, and in the lack thereof, with default (Barkowsky, 2002) or common sense knowledge (Kuipers, 1979). Therefore, wayfinding requires an interplay between "knowledge in the head" and "knowledge in the world" (Norman, 1988). In fact, the knowledge in the world could be seen as a type of "external storage" of spatial knowledge.

Locomotion is the physical movement in space with a body, describing the "action" part of the navigation process. While humans move about in space they avoid static and moving obstacles, automatically find paths around larger obstacles, determine trajectories and react with their own body movements to all these processes. Locomotion could also be described as the subconscious processes in navigation. Within the context of modeling navigation of humans, research on locomotion is scarce. In contrast to computer graphics, where the actual movements of the body parts are of interest (see e.g., Thalman computer graphics book), we are looking for algorithms describing human behavior from a bird's eye view. Another source of locomotion algorithms can be found in robotics research. However, due to the different sensory equipment, robot locomotion algorithms cannot easily be applied to human movement. In evacuation research, moving humans are treated as particles following physical laws. The social force approach (Helbing and Molnar, 1995) for example requires careful calibration of the model in order to yield natural (i.e., human-like) behavior.

2.1 The Role of the Environment

The physical environment plays the role of information storage on one hand; On the other hand, the traveler is bombarded by sensory information from the physical environment that needs to be sorted out. The traveler needs to find a balance between information input and focused information retrieval. Finding this balance can be influenced greatly by the structure and dynamics of the environment.

Bovy and Stern (1990) describe three objective factors that determine individual travel behavior: the physical environment, the socio-demographic environment (people around us) and the normative environment (knowledge about rules of behavior). In addition, a subjective factor influences the perception of these objective factors. In route choice and planning, the physical environment has the largest influence. The same is true for route descriptions or route instructions: the physical environment in the form of landmarks plays the most important role in producing good route instructions (Denis 1997).

Gärling (1986) proposed a system for classifying environments to predict the extent of wayfinding problems. Weismann (1981) recommended similar classes of environmental variables that influence wayfinding performance (meant for buildings). According to Gärling, the following facets of the environment are important for successful wayfinding:

- degree of architectural differentiation,
- degree of visual access, and
- complexity of spatial layout.

The degree of architectural differentiation is less relevant for the public transportation environment than it is for the building literature, except for underground environments. Travelers need to differentiate between transfer points, but this is usually made easy with signs stating the name of the station. By design those names are unambiguous within a specific transportation system. For our specific case study each station differs from others by the urban environment in which they are set.

Visual access is important for the traveler. The start and goal of a route within a city are usually not visually accessible from a single vantage point, because the space we are dealing with is at a geographic or environmental scale (Montello 1993). Visual access is important anywhere along the route, however it is especially important within transfer points. Travelers need to be able to see the stop where they are supposed to board the transportation means. In our case study most stops within a transfer point are visible or almost visible (i.e., walking a few paces will make the stops visible).

The complexity of spatial layout refers to the environmental size and the number of possible destinations and routes. "A simple layout should facilitate both the formation and execution of travel plans by making it easier to choose destinations and routes, to maintain orientation, and to learn about the environment" (Gärling 1986). In our case study the transfer station Regensbergbrücke (cf. Fig.2) retains a simple layout: it is a prototypical street crossing.

The complexity of spatial layout and visual access are linked: a complex layout may mean a visually cluttered environment; conversely a visually legible environment may not mean a simple layout. Lynch (1960) has emphasized the importance of the legibility of the environment, of which visual access is one part and maybe the complexity of spatial layout another. Good legibility of the environment improves perception and orientation and thus wayfinding.

Lynch (1960) found that humans persistently organize space for orientation purposes using five elements: nodes, paths, landmarks, edges, and districts. Nodes are places where several paths come together

- they represent important decision points along a route. Paths represent streets, pathways or waterways along which a traveler can proceed. Landmarks provide recognizable environmental objects, often over larger distances, that help with orientation. Edges have the function to demarcate one area from another - for example the edge along the river, but also barriers such as overpasses can be considered edges. Finally, districts define recognizable areas within the urban landscape. It needs to be pointed out that objects within an urban environment can belong to several of these elements. Their interpretation depends on the traveler's viewpoint, e.g., an overpass can be classified as a path by the car traveling over it, but perceived as a barrier by the pedestrian crossing underneath. Passini (1992) expanded Lynch's concepts to include signage and other graphic communication - spatial clues inherent in the environment.

2.2 Perceiving the Environment

Travelers perceive the environment within the context of navigation, i.e., their attention is focussed on finding cues and retrieving information from the environment. Two main theories exist about the perception of the environment: the first one is the theory on affordances by Gibson (1986). Affordances are what objects can offer the perceiving person. For example a chair affords sitting by its form, a path affords following, a barrier affords stopping (negative affordance).

Image schemata (Johnson 1087) are a second way to perceive and interpret the environment. Image schemata are mental patterns, which provide a structured understanding of our experiences. For navigation, the image schemata "path", "part-whole", "link" and "gateway" are the most important ones.

In Raubal & Worboys (1999), image schemata are augmented with action and information affordances to describe the physical environment as perceived by a human. This results in a wayfinding graph. The nodes of the wayfinding graph represent states of knowledge and the current location, whereas links represent transitions between those. In Rüetschi (2007) image schemata are used to produce a different wayfinding graph consisting solely of image schemata (see section 3) and their interconnections.

2.3 Representations: Mental Models and Cognitive Collages

Mental models store the knowledge gained from experiences in geographic space (Lynch, 1960; Golledge, 1999). They are spatial mental models in the sense of Johnson-Laird: "A mental model is an internal representation of a state of affairs in the external world" (1992). When traveling, the perceived environment is compared with the existing mental map. Differences between perception and representation are noted and updated in the mental map. This update process is crucial for learning about the environment.

Lynch (1960) reported that urban inhabitants understand their surroundings in a predictable way, forming mental maps with usually five elements: paths (streets, sidewalks), edges (perceived boundaries such as walls, shorelines), districts (relative large urban regions with distinct properties and resulting identity), nodes (intersections or focal points), and landmarks (readily identifiable objects which serve as reference points for orientation). The most important elements for wayfinding are paths, nodes, and landmarks, the minimal route information relies on paths and nodes.

There is a consensus in the literature that a mental map can only be metaphorically compared to a real map, i.e., a scaled and coherent representation of the geographic world. Mental maps can rather be seen as cognitive collages (Tversky, 1993): information along routes is added to the mental representation but it is not integrated into a coherent whole. Imagine that you are walking along a street one day from one direction, the other day from the opposite direction. You two paths did not coincide and thus you did not make the connection that the two streets could be the same. So, two separate information pieces were added to your mental map.

Mental maps are distorted in that the angles of street corners or the overall orientation of a street within a city are usually falsely remembered. Humans tend to remember approximate angles and distances and an abstraction process is carried out that fixes angles to multiples of 30 or 45 degrees and represents distances in time units. Other mental processes such as hierarchization (Hirtle & Heidorn, 1993) and orientation further distort the spaces perceived while traveling. In addition, attention plays an important role in what and how well we perceive. When traveling we pay attention to our route and some additional information along the route. This type of knowledge has been termed route knowledge (Siegel & White, 1975). Mental maps are, metaphorically speaking, containers in which many different route information pieces are stored. Sometimes we are able to fit several pieces together into a coherent whole and reason about this space correctly. At other times, the information gap is such that we cannot come to a spatial image that fits our current perception. In these cases we need to rely solely on our perception and interpretation of the environment.

2.4 Reasoning while Navigating

Goal-directed wayfinding consists of three main actions or tasks: planning, tracking, and assessing (Infopolis2 1999). Planning answers the question 'where do I need to go?', tracking deals with the question 'where am I compared to the plan?', and assessing argues 'how good has my travel plan and subsequent execution been?'. Planning mostly takes place before the trip, tracking goes on during the trip, and assessing is the main task for the last part of the trip or after the trip. In addition orientation plays a major role. The difference between orientation and tracking is that orientation answers the question as to where I am, whereas tracking compares the current location to the planned location.

The goal-directed navigation process has two principal reasoning processes: a planning process and a traveling process. The planning process can be described as pure information processing, whereas the traveling prcoess comprises both, the locomotion aspect (i.e., the movement across space itself) and the information processing aspect. Actions within the information processing and the locomotion are further subdivided into operations, but are themselves part of a specific activity (Kaptelinin, 1999; Nardi, 1996). This leads to a hierarchical organization of activities or tasks (Freksa, 1991) and their corresponding spatial models (Timpf et al, 1992; Timpf, 2002, Timpf & Kuhn, 2003).

In spatial information processing the existence of multiple representations (models) provides a crucial element when dealing with humans: each additional solution to a problem resulting from a different representation adds to the human's confidence in the correctness of the solution.

Table 1. Activity Model of Wayfinding, derived from Infopolis2 (1999)

Activity	Wayfinding: get from place A to place B		
Tasks	Planning	Tracking	Assessing
Operations	Information gathering, find routes, determine constraints, determine complexity, produce instructions	Orienting, track location, compare to plan, orient yourself	Compare needed to planned time, assess instructions, determine complexity of route

The representation of space when dealing with the activity wayfinding is called a spatial mental model (Tversky, 1993). Spatial mental models are used to store information about and experiences with geographical space. Subsequently to this learning phase information about a specific geographical space and about inferred properties are retrieved to solve wayfinding and locomotion tasks. It is not exactly clear how the process of applying information to a new situation works. Currently, there is much debate about the role of analogical reasoning in psychology, which might be a key to this knowledge application.

3. COMPUTATIONAL MODELS OF NAVIGATION AND WAYFINDING

Computational models provide the researcher with a means to test theories and models. Many different models of navigation and wayfinding exist. TOUR (Kuipers, 1978) was the first computational model of wayfinding. It illustrates the wayfinding process with incomplete knowledge about the environment using a view, action, view structure. Other models include TRAVELLER (Leiser, 1989), NAVIGATOR (Gopal, 1989), and ELMER (McCalla, 1982). All these models have a strong focus on individual learning about the environment, that is, they simulate how the cognitive map is built, but they do not intend to simulate the complete navigation process.

This is also true of more recent models. PLAN (Chown, 1995) pursues a "head-up" approach, starting from the individual that is looking around and perceives what is called a scene. It also tries to embed the wayfinding process into human cognition in general. The Spatial Semantic Hierarchy SSH (Kuipers, 2000), a sequel to TOUR, identifies different layers of reasoning more clearly than the original model. Both PLAN and SSH are being applied to research in robotics.

The MOSES software agent (Maass, 1995) is a computational model for the generation of incremental route descriptions. The agent travels through a 3D environment, selects visuospatial information and generates appropriate route descriptions. The model is based on two cognitive abilities, which are visual perception and natural language, whereby the agent adapts his linguistic behavior to spatial and temporal constraints. The focus however, is on the production of language, not on the navigation process.

Timpf et al. (1992) developed a model for navigation in interstate networks based on hierarchical spatial reasoning in task graphs. The reasoning of this model was implemented to show which minimal information is needed in a data structure for the reasoning to work at all levels of detail (Timpf & Kuhn 2003).

Raubal (2001) simulated a wayfinder navigating in an airport, represented as a graph and annotated with signage information. He stresses the importance of an investigation of the information needs of travelers, following a research direction suggested by Gluck (1991). The model can be used to determine where and why people face wayfinding difficulties in buildings, which was illustrated for signage problems in airports.

The ODEON software is a computational model of wayfinding in complex spaces using image schemata (Rüetschi 2007). In contrast to all other approaches, this research dealt with scene spaces, i.e. spaces where there is no network provided by the environment. The program requires an image schemata network as input, on which the reasoning is performed.

Research by Pontikakis (2006) simulates a cognitive agent who moves in a network space where different states occur for the agent. This thesis integrates wayfinding processes and business processes (such as buying a ticket). The model uses affordances to determine potential actions.

Caduff (2007) implemented a framework for navigating in a urban environment using (automatically derived) landmarks. The system starts with the perception of scenes, extracts salient objects according to perceptual, cognitive and task-related criteria and produces a ranking of potential landmarks for that scene. Using many agents with different paths, the individual ranking of landmarks can be extended to a global ranking.

Common to these models is the aspect of navigation in the real world, as opposed to navigation in virtual space. However, all the models (except Pontikakis) assume one single modality throughout the whole process, which seldom represents reality in urban spaces, where travelers have the option to choose between several transportation means (i.e. walking, driving, taking the train, etc.). With the change in transportation means the change in environment and the human ability to switch from one interpretation to another comes to the fore. Hence, further research is required to understand the cognitive process involved with multimodal traveling and to simulate this process accordingly.

In summary, one can say, that a wealth of models of wayfinding and navigation exist. However, each model deals with one or two special aspects of navigation and tests a single hypothesis. In order to model the whole process of navigation and to compare different modalities, a single framework needs to be provided that allows for integrating existing models. The next section presents such a framework using a special case study - that of public transport - where most of the heretofore discussed models need to be integrated.

4. CASE STUDY ZÜRICH REGENSBERGBRÜCKE

In contrast to transportation simulations where the focus is on the whole system, in our navigator model, we will focus on the single agent and his capabilities to orient in space, plan a route, determine instructions, perceive affordances, make use of image schemata, increment the mental model and reason using knowledge from the head and the environment. A situation in which these capabilities all play a role is the transfer process in public transportation.

In our case study a public transport user travels from Oberwiesenstrasse to Bad Allenmoos, changing means of transportation at the transfer node Regensbergbrücke. A process model of this short route should encompass the agent's perception of the physical environment (pertaining to the navigation), storage of spatial information in a mental map, spatial information processing concerning the transfer process and the (simulated) locomotion from start to goal.

4.1 Modeling the Physical Environment

The physical setup consists of four nodes in a (real) transportation network (see Fig. 1): Oberwiesenstrasse, Regensbergbrücke, Bahnhof Oerlikon and Bad Allenmoos. Three of these nodes (Oberwiesenstrasse <-> Regensbergbrücke <-> Bahnhof Oerlikon) are part of bus line 62, and again three nodes (Bahnhof Oerlikon< -> Regensbergbrücke <-> Bad Allenmoos) are part of tramway line 11. Each node contains at least two stops (one per direction) with stops being different for tram and bus even though they may be (and often are) in the same spatial area/region. There also exists a network for walking - the pedestrian network. Here we just show the network present at each node in order to enable walking from one stop to the next, additional edges linking the stops over larger distances and paralleling the tram or bus lines are not shown, although they are part of the model.

The last network is one that enables entering and leaving a transport vehicle, i.e. it facilitates changing the means of transport. This network consists of a set of links, namely links from a stop of type1, e.g., of type bus, to a stop of type2, e.g., of type pedestrian. Only links from pedestrian nodes to other nodes and back are possible. These links also incorporate the spatiotemporal information of the timetable, i.e., they only exist from near the time of a scheduled stop of a vehicle (maximum desired waiting time) until the scheduled departure of that vehicle.

In addition to the above graph representation we need to model the physical environment in a way that makes locomotion possible. This is accomplished by modeling the world according to Lynch's five elements: paths, nodes, barrier, district, and landmarks. Instead of using a graph representation for paths and nodes, for the locomotion we have to model a "walkable surface" (Fig. 3, see also Thomas and Donikian 2008) or conversely, all barriers to locomotion. This encompasses all pedestrian sidewalks and walkways, and all waiting areas at stops. Treating all walkable surfaces as potential scenes allows integrating the ODEON model for wayfinding.

When orienting herself, the traveler makes use of knowledge about landmarks. In classical spatial cognition studies, landmarks are considered to be point-like objects that are prominent and can be perceived from a large distance. In newer studies (Presson & Montello, 2003; Caduff 2007), landmarks are considered to encompass all memorable visible urban objects, be they point-like, linear or area features. In this case study there are not many landmarks - however the bridge crossing the railway tracks qualifies as landmark as well as the railway tracks (linear feature) themselves. Equally qualifying are a fountain, a kiosk, and a DVD shop (Fig. 2).

Figure 1. Transportation network in case study Zurich

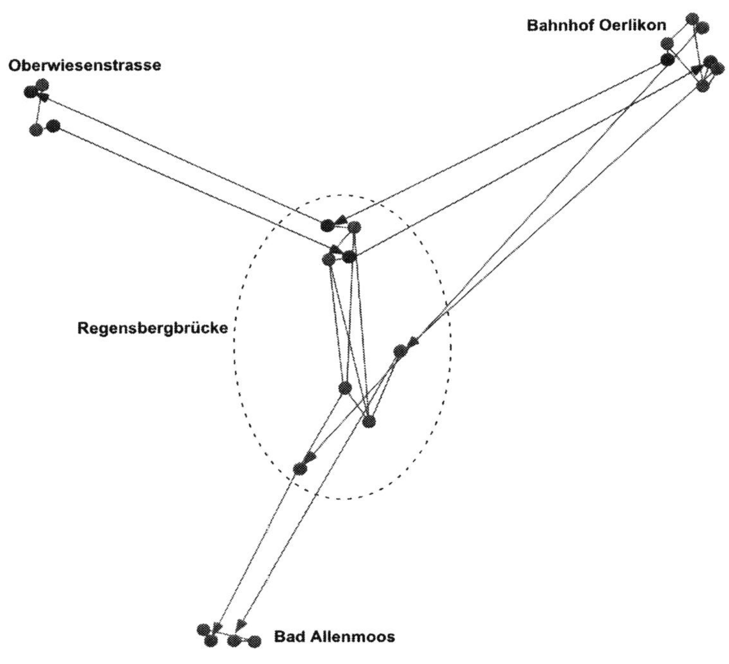

Figure 2. Walkable surfaces for locomotion and landmarks (map: GIS Viewer Zürich)

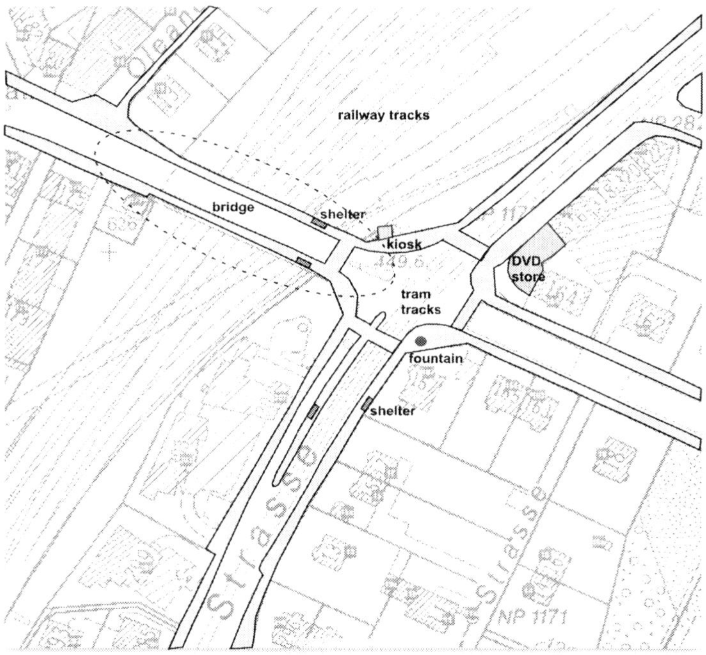

4.2 Perceiving and Mentally Mapping the Environment

During locomotion, the agent perceives the environment and stores the perceived information in a mental map. Perceivable features are landmarks and all stops in the public transportation network (we assume that all stops have some kind of sign bearing the name of the stop). Signalized crosswalks could also be considered as landmarks.

We cannot assume that an agent can perceive all these qualitative differences in the typification of environmental features. However, in a first shot, we will assume that the agent can perceive the features as we intended her to perceive them. This means that once our first agent has carried out the simulated wayfinding process, she should have a complete mental map of the route - no errors and no miss-classifications of features. For other agents we can change the perception using several filters, e.g., the attention filter focuses the attention on a single feature or feature type and only this feature will be encoded in the mental map. Similarly an error filter, a mis-classification filter etc. can be modeled. For such a simple physical situation only the attention filter makes sense and will be used in our simulation.

A filter that is associated with attention is the visibility filter. If a feature is not visible then it cannot be added to the mental map. However, calculating visibility within the representation at our disposal is not possible. Thus, an additional data structure for representing visibility is needed. There are several potential data structures for this purpose: one is the use of isovists, which are viewshed polygons that capture spatial properties by describing the visible area from a given observation point (Wiener, 2004). A second possibility is the use of a visibility graph (Raubal & Worboys, 1999), where for each node the visibility to each feature is encoded. For our model, the second option is preferable, since we already have a graph data structure at our disposal and the number of features is relatively small.

There is a consensus that the general structure of a mental map can rather be compared to a cognitive collage (Tversky, 1993) than to a map. However, a collage is not suited as a data model for a computer representation of mental maps. The most common underlying data structure for a mental map is considered to be a (usually planar) graph. An extension to a hierarchical form, encompassing differently detailed linked representations in the form of an information collage, has also been proposed (Timpf, 2005). Other representations are route graphs (Werner et al., 2000) or scene graphs for scene spaces (Rüetschi, 2007). We will therefore model a mental map as a graph, combining different spatial information fragments. Each information fragment is associated with a physical feature (nodes or landmarks in Lynch's sense). For our case study we will neglect the Lynch elements district and barriers. Barriers are implicitly modeled with the definition of the walkable areas. Districts could be modeled using a hierarchical graph structure, however, unless some specific reasoning about districts is going on, this does not seem to be a compulsory element in our model.

4.3 Wayfinding and Locomotion

The wayfinding process consists of first an orientation phase, then defining the goal in relation to the current location, determining the exact route (at least the next step and with the constraints given by the timetable of the means of transport), and finally, starting the journey by walking. At each transfer station the same sequence needs to be repeated at a smaller resolution, i.e., it is necessary to determine the location of the goal stop from the current position within the transfer station. Our agent also needs to understand concepts such as boarding a bus, signaling a stop and getting off the bus, although these processes are subtly different in each city. The thought processes can be modeled using an activity framework. Each phase is modeled as a separate action with the activity wayfinding - the sequence of actions leads to the locomotion phase.

The locomotion phase only seems to be straightforward: once the agent starts walking towards a goal, she will avoid barriers and obstacles, such as other agents, in order to reach the goal. However, an avoidance behavior for outdoor areas that looks "natural" is hard to come by. The usual solution, e.g. path following using predefined locations should be avoided because of being too deterministic. The social force approach (Helbing & Molnar, 1995) requires careful calibration of the model and even of parts of the model in order to yield a natural behavior. A study on the suitability of different locomotion models will be required.

5. CONCLUSION AND FUTURE WORK

A simulation allows for experimenting with models. It is a means of gaining insight into some topic. Simulations are frequently used when an experiment with the real system is not feasible for some reason. This is clearly the case with navigation, where many people are involved and the environment cannot easily be changed. Simulation models of wayfinding create their own environment, probably matching some real environment, and have agents representing human wayfinders, interact with this virtual environment and each other. For the simulation of realistic agent behavior complex and valid environmental models have to be modeled. In simulations, the modeled environment should always be a first order object that is as carefully developed as the agents themselves (Klügl, 2005). The result is a cognitively plausible simulation of multimodal navigation, which will provide insights into this complex task and the underlying cognitive processes.

As mentioned above, the goal of this research is to use a multi-agent system to model and simulate the wayfinding process as well as the locomotion process of navigation. The wayfinding process describes the information processing going on while navigating. As we have shown, many different representations and models of space and spatial relations are necessary to computationally model the navigation process. This is the main reason why we have not been able (yet) to implement and test our model within a multi-agent system. However, efforts in this direction are ongoing with the multi-agent system SeSAm (Klügl et al., 2006).

In addition to the breadth of representations, each representation has (usually) more than one variation that seems as plausible as the selected one. The only sensible way to distinguish between those variations lies in consistent experimentation with the different variations. Our design of this system helps with experimentation by providing a common and consistent framework for simulation. This requires even more representations.

Our future work then consists of the following points: finish the realization of the models for wayfinding and locomotion within a multi-agent system and start the experimentation with a single model; In parallel develop alternative models and alternate those in the experimental setup. Many models are only made for one such representation and the integration within a computational framework is a challenge in itself.

REFERENCES

Allen, G. L. (1999). Spatial abilities, cognitive maps, and wayfinding: Bases for individual differences in spatial cognition and behavior. In R. Golledge (Ed.), *Wayfinding behavior: Cognitive mapping and other spatial processes* (pp. 46-80). Baltimore: John Hopkins University Press.

Arthur, P., & Passini, R. (1992). *Wayfinding: People, signs, and architecture.* New York: McGraw-Hill Book Co.

Barkowsky, T. (2002). Mental representation and processing of geographic knowledge - a computational approach. *Künstliche Intelligenz,* (4), 42.

Bovy, P. H. L., & Stern, E. (1990). *Route choice: Wayfinding in transport networks.* Dordrecht; Boston: Kluwer Academic Publishers.

Caduff, D. (2007). *Assessing landmark salience for human navigation.* PhD thesis, University of Zurich, Zurich.

Denis, M. (1997). The description of routes: A cognitive approach to the production of spatial discourse. *Current Psychology of Cognition, 16,* 409-458.

Elias, B. (2003). Extracting Landmarks with Data Mining Methods. In W. Kuhn, M. Worboys, & S. Timpf (Eds.), *Spatial Information Theory: Foundations of Geographic Information Science* (375-389). Berlin: Springer Verlag.

Frank, A. U., Bittner, S., & Raubal, M. (2001). *Spatial and cognitive simulation with multi-agent systems.* Paper presented at the Spatial Information Theory - Foundations of Geographic Information Science (Int. Conference COSIT, September 2001), Morro Bay, U.S.A.

Freksa, C. (1991). Qualitative spatial reasoning. In D. M. Mark & A. U. Frank (Eds.), *Cognitive and linguistic aspects of geographic space* (pp. 361-372). Dordrecht, The Netherlands: Kluwer Academic Press.

Gärling, T., Böök, A., & Linderg, E. (1986). Spatial orientation and wayfinding in the designed environment - a conceptual analysis and some suggestions for post-occupancy evaluation. *Journal of Architectural and Planning Research, 3*, 55-64.

Gibson, J. J. (1986). *The ecological approach to visual perception.* Hillsdale, NJ: Lawrence Erlbaum.

Golledge, R. (1999). Human wayfinding and cognitive maps. In R. Golledge (Ed.), *Wayfinding behavior: Cognitive mapping and other spatial processes* (pp. 5-45). Baltimore: John Hopkins University Press.

Helbing, D., & Molnár, P. (1995). Social force model for pedestrian dynamics. *Physical Review E51*, 4282-4287.

Heye, C. (2002). *Deskriptive Beschreibung der Ergebnisse der Internetbefragung: Umsteigen an Haltestellen der VBZ (Technical Report No. KoMoNa02-03).* University of Zurich: Department of Geography, Zurich, Switzerland.

Heye, C., & Timpf, S. (2003). *Factors influencing the physical complexity of routes in public transportation networks.* Paper presented at the 10th International Conference on Travel Behaviour Research, Lucerne.

Hirtle, S. C., & Heidorn, P. B. (1993). The structure of cognitive maps: Representations and processes. In T. Gärling & R. G. Golledge (Eds.), *Behavior and environment: Psychological and geographical approaches* (pp. 1-29, Chapter 27).

Infopolis2. (1999). *Needs of Travellers: An Analysis Based on the Study of Their Tasks and Activities* (pp. 62). Brussels: Commission of the European Communities - DG XIII. WP3.2/DEL3/1999.

Johnson-Laird, P. N. (1992). Mental models. In S. C. Shapiro (Ed.), *Encyclopedia of artificial intelligence* (2nd ed.) (pp. 932-939). New York: John Wiley & Sons.

Kaptelinin, V., Nardi, B., & Macaulay, C. (1999). The activity checklist: A tool for representing the 'space' of context. *Interactions*(july/august), 29-39.

Klügl, F., Herrler, R., & Fehler, M. (2006). *Sesam: Implementation of agent-based simulation using visual programming.* Paper presented at the AAMAS 2006, Hakodate.

Krieg-Brückner, B., Frese, U., Lüttich, K., Mandel, C., Mossakowski, T., & Ross, R. J. (2005). Specification of an ontology for route graphs. In C. Freksa, M. Knauff, B. Krieg-Brückner, B. Nebel & T. Barkowsky (Eds.), *Spatial cognition IV. Reasoning, action, and interaction* (Vol. LNAI 3343, pp. 390-412). Berlin: Springer.

Kuipers, B. (1982). The 'map in the head' metaphor. *Environment and Behaviour, 14*(2), 202-220.

Kuipers, B., Tecuci, D. G., & Stankiewicz, B. J. (2003). The skeleton in the cognitive map: A computational and empirical exploration. *Environment and Behavior, 35*(1), 81-106.

Kuipers, B. (2000). The spatial semantic hierarchy. *Artificial Intelligence, 119*, 191-233.

Leiser, D., & A. Zilbershatz (1989). The Traveller—A Computational Model of Spatial Network Learning. *Environment and Behavior 21*(4), 435-463.

Lynch, K. (1960). *The image of the city.* Cambridge: MIT Press.

McDermott, D., & Davis, E. (1984). Planning Routes through Uncertain Territory. *Artificial Intelligence, 22*(1), 107-156.

Meier, E. (1997). Verkehrsnachfragemodelle: Hilf- oder Allheilmittel? In A. Müller (Ed.), *Wege und Umwege in der Verkehrsplanung.* Zürich: vdf Hochschulverlag.

Miller, H. J. (1992). Human Wayfinding, Environment-Behavior Relationships, and Artificial Intelligence. *Journal of Planning Literature, 7*(2), 139-150.

Montello, D. R. (1993). Scale and multiple psychologies of space. In A. U. Frank & I. Campari (Eds.), *Spatial information theory: Theoretical basis for gis, 716*, 312-321. Heidelberg-Berlin: Springer Verlag.

Montello, D. R. (2005). Navigation. In P. Shah & A. Miyake (Eds.), *Handbook of visuospatial cognition.* (pp. 257-294): Cambridge, Cambridge University Press.

Nagel, K. (2001). *Multi-modal traffic in TRANSIMS.* Paper presented at the ED'01 (Conference on Pedestrian and Evacuation Dynamics).

Nagel, K. (2002). Traffic networks. In S. Bornholdt & H. G. Schuster (Eds.), *Handbook of graphs and networks* (pp. 248-272): Wiley.

Nardi, B. (Ed.). (1996). *Context and consciousness - activity theory and human-computer interaction.* Cambridge, MA: The MIT Press.

Norman, D. A. (1988). *The design of everyday things.* Doubleday.

Ortuzar, J. D. D., & Willumsen, L. G. (2001). *Modeling transport.* Chichester, England: John Wiley & Sons.

Pontikakis, E. (2006). *Wayfinding in GIS: Formalization of basic needs of a passenger when using public transportation.* PhD thesis, TU Vienna, Vienna.

Portugali, J. (1996). *The Construction of Cognitive Maps.* Kluwer Publishers.

Presson, C. C., & Montello, D. R. (1988). Points of reference in spatial cognition: Stalking the elusive landmark. *British Journal of Developmental Psychology, 6*, 378-381.

Raney, B., Cetin, N., Völlmy, A., & Nagel, K. (2002a). *Large scale multi-agent transportation simulations.* Paper presented at the 42nd ERSA Congress (European Regional Science Association), Dortmund.

Raney, B., Cetin, N., Völlmy, A., Vrtic, M., Axhausen, K., & Nagel, K. (2002b). *Towards an activity-based microscopic simulation of all of Switzerland.* Zürich: Institut für Verkehrsplanung, Transporttechnik, Strassen- und Eisenbahnbau (IVT), ETHZ.

Raney, B., & Nagel, K. (in press). An improved framework for large-scale multi-agent simulations of travel behavior. In P. Rietveld, B. Jourquin & K. Westin (Eds.), *Towards better performing european transportation systems.*

Raubal, M., & Worboys, M. (1999). *A formal model of the process of wayfinding in built environments.* Paper presented at the Spatial Information Theory - cognitive and computational foundations of geographic information science, Stade, Germany.

Raubal, M. (2001). Human wayfinding in unfamiliar buildings: A simulation with a cognizing agent. *Cognitive Processing, 2-3*, 363-388.

Rüetschi, U. J., & Timpf, S. (2004). Modelling wayfinding in public transport: Network space and scene space. In C. Freksa, M. Knauff, B. Krieg-Brückner, B. Nebel & T. Barkowsky (Eds.), *Spatial cognition iv: Reasoning, action, interaction; international conference frauenchiemsee* (Vol. LNCS 3343, pp. 24-41). Heidelberg, Berlin: Springer.

Rüetschi, U.-J. (2007). *Wayfinding in science space - modelling transfers in public transport.* PhD thesis, University of Zurich, Zurich.

Schoggen, P. (1989). *Behavior Settings.* Stanford, CA: Stanford University Press.

Siegel, A. W., & White, S. H. (1975). The development of spatial representations of large-scale environments. In H. W. Reese (Ed.), *Advances in child development and behavior, 10*, 9-55. New York: Academic Press.

Thomas, R., & Donikian, S. (2008). A spatial cognitive map and a human-like memory model dedicated to pedestrian navigation in virtual urban environments. In T. Barkowsky, M. Knauff, G. Ligozat & D. R. Montello (Eds.), *Spatial cognition V: reasoning, action, interaction, LNCS,* 421-438. Berlin, Heidelberg: Springer.

Timpf, S. (2002). Ontologies of Wayfinding: a traveler's perspective. *Networks and Spatial Economics, 2*(1), 9-33.

Timpf, S. (2005). Wayfinding with mobile devices: Decision support for the mobile citizen. In S. Rana & J. Sharma (Eds.), *Frontiers of geographic information technology.* London, Berlin: Springer.

Timpf, S., & Kuhn, W. (2003). Granularity transformations in wayfinding. In C. Freksa, W. Brauer, C. Habel & K. F. Wender (Eds.), *Spatial cognition iii* (pp. 77-88). Berlin: Springer.

Timpf, S., Volta, G. S., Pollock, D. W., & Egenhofer, M. J. (1992). A conceptual model of wayfinding using multiple levels of abstractions. In A. U. Frank, I. Campari & U. Formentini (Eds.), *Theories and methods of spatio-temporal reasoning in geographic space, 639*, 348-367. Heidelberg-Berlin: Springer Verlag.

Tversky, B. (1993). Cognitive maps, cognitive collages, and spatial mental models. In A. U. Frank & I. Campari (Eds.), *Spatial information theory: A theoretical basis for GIS, 716*, 14-24. Heidelberg-Berlin: Springer Verlag.

Weisman, J. (1981). Evaluating architectural legibility: Wayfinding in the built environment. *Environment and Behavior, 13*(2), 189-204.

Werner, S., Krieg-Brückner, B., & Herrmann, T. (2000). Modelling navigational knowledge by route graphs. In C. Freksa, W. Brauer, C. Habel & K. F. Wender (Eds.), *Spatial cognition II, LNCS 1849*, 295-316) Berlin: Springer.

Widmer, P., & Axhausen, K. W. (2001). *Aktivitäten-orientierte Personenverkehrsmodelle: Vorstudie.* Bern, Schweiz: Eidgenössisches Departement für Umwelt, Verkehr, Energie und Kommunikation / Bundesamt für Strassen.

Wiener, J. M., & Franz, G. (2005). Isovists as a means to predict spatial experience and behavior. *Lecture notes in artificial intelligence, 3343*, 42-57.

Winter, S. (2003). Route Adaptive Selection of Salient Features. In W. Kuhn, M. Worboys, & S. Timpf (Eds.), *Spatial Information Theory: Foundations of Geographic Information Science* (pp. 349-361). Berlin: Springer Verlag.

Section II
Intelligent Traffic Management and Control

Chapter IX
An Unmanaged Intersection Protocol and Improved Intersection Safety for Autonomous Vehicles

Kurt Dresner
University of Texas at Austin, USA

Peter Stone
University of Texas at Austin, USA

Mark Van Middlesworth
Harvard University, USA

ABSTRACT

Fully autonomous vehicles promise enormous gains in safety, efficiency, and economy for transportation. In previous work, the authors of this chapter have introduced a system for managing autonomous vehicles at intersections that is capable of handling more vehicles and causing fewer delays than modern-day mechanisms such as traffic lights and stop signs [Dresner & Stone 2005]. This system makes two assumptions about the problem domain: that special infrastructure is present at each intersection, and that vehicles do not experience catastrophic physical malfunctions. In this chapter, they explore two separate extensions to their original work, each of which relaxes one of these assumptions. They demonstrate that for certain types of intersections—namely those with moderate to low amounts of traffic—a completely decentralized, peer-to-peer intersection management system can reap many of the benefits of a centralized system without the need for special infrastructure at the intersection. In the second half of the chapter, they show that their previously proposed intersection control mechanism can dramatically mitigate the effects of catastrophic physical malfunctions in vehicles such that in addition to being more efficient, autonomous intersections will be far safer than traditional intersections are today.

INTRODUCTION

Recent advances in technology have made it possible to construct a fully autonomous, computer-controlled vehicle capable of navigating a closed obstacle course. The DARPA Urban Challenge [DARPA 2007], at the forefront of this research, aims to create a full-sized driverless car capable of navigating alongside human drivers in heavy urban traffic. It is feasible that, in the near future, many vehicles will be controlled without direct human involvement. Our current traffic control mechanisms, designed for human drivers, will be upgraded to more efficient mechanisms, taking advantage of cutting-edge research in the field of Multiagent Systems (MAS).

Intersections are one aspect of traffic control that are particularly compelling multiagent systems. Often a source of great frustration for drivers, intersections represent both a sensitive point of failure as well as a major bottleneck in automobile travel. While fully autonomous open-road driving was demonstrated over ten years ago, events such as the DARPA Urban Challenge prove that city driving, including intersections, still pose substantial difficulty to AI and intelligent transportation systems (ITS) researchers.

Managed Intersection Control

Previously, we proposed an intersection control mechanism to direct autonomous agents safely through an intersection [Dresner & Stone 2005]. This system is based on the interaction of two classes of agents: *intersection managers* and *driver agents*. Driver agents "call ahead" to an intersection manager at the intersection, reserving the time and space needed to cross. Specifically, when approaching an intersection, a driver agent sends a request message containing a predicted arrival time and velocity, along with basic information about the vehicle it is controlling. The intersection manager responds with either a confirmation message containing details of the approved reservation, or a denial message, signaling that the parameters sent by the driver agent are unacceptable. In the case of confirmation, the driver agent will attempt to meet the parameters of the reservation, and will cancel the reservation if it cannot. In the case of denial, the driver agent must try to make a different reservation.

Intersection managers base their decisions on the supplied parameters and an *intersection control policy*. The most efficient policies, including FCFS or "first come, first served", simulate the trajectory of the vehicle through the intersection. At each stage in the simulation, the intersection manager checks whether the vehicle is within a certain buffer distance of any other vehicle in the intersection. If the requesting vehicle can cross the intersection without entering any space-time reserved by another vehicle, the policy creates the reservation, and the intersection manager approves the request. Otherwise, the policy does not create a reservation, and the intersection manager denies the request. By integrating these policies with traditional traffic light systems, we have also demonstrated that the system can accommodate human traffic [Dresner & Stone 2007]. This multiagent approach offers substantial efficiency benefits as compared to existing mechanisms, such as traffic lights and stop signs. Vehicles pass through the intersection faster, and congestion at intersections is significantly reduced.

Although at the city level this system is mostly decentralized, at each individual intersection, traffic is coordinated by a single arbiter agent, the intersection manager. We therefore designate this system a *managed* intersection control mechanism. An intersection controlled by a traffic light is also a managed intersection—the traffic light being the arbiter agent. Conversely, we designate intersection control mechanisms without an arbiter agent, such as stop signs and traffic circles, *unmanaged* intersection control mechanisms.

Managed intersection control mechanisms have a major drawback: cost. An arbiter agent of some sort must be stationed at the intersection, and our previously proposed managed system, this agent must have sufficient computational resources and communications bandwidth to rapidly negotiate a high volume of requests. Although the throughput benefits in large intersections would certainly warrant this expense, the system would be uneconomical for small intersections.

Stop signs are a low-overhead, unmanaged system designed for low-traffic intersections, complementing larger intersections managed by traffic lights. In the first half of this chapter, we propose an unmanaged intersection control mechanism for autonomous vehicles, designed specifically for low-traffic intersections. Our system—based on peer-to-peer communication and requiring no specialized infrastructure—is a similar complement to the managed intersection we previously proposed [Dresner & Stone 2005]. We make similar assumptions about the driver agent, such that a driver agent capable of using the managed system can be modified to use both systems seamlessly. We also present empirical data comparing our system to both traffic lights and stop signs. We focus our analysis primarily on the comparison between our system and the class of intersections that would currently be managed by a stop sign (low-traffic intersections), as these are the intersections for which our system is intended.

Safety Questions

In addition to gains in efficiency and economy, autonomous vehicles also promise vastly increased safety for automobile transportation. By taking the responsibility of driving away from humans, autonomous vehicles will completely eliminate driver error from the complicated equation of automobile traffic. By some estimates, driver error can be blamed for as much as 96% of all automobile accidents [Wierwille et al. 2002]. Thus, even if each accident were substantially worse, overall autonomous vehicles would represent an improvement in safety over the current situation. With automobile collisions costing the U.S. economy over $230 billion annually, any significant decrease would be a major triumph for artificial intelligence [National Highway Traffic Safety Administration 2002].

By coordinating the actions of many autonomous vehicles, our reservation-based system dramatically decreases time spent stopped or slowing down due to intersections. Because the system heavily exploits the precision sensory and control capabilities of computerized drivers, it offers dramatic improvements in efficiency. However, this increased efficiency is quite precarious. The system orchestrates what can only be described as "extremely close calls", with vehicles missing each other by the smallest (albeit adjustable) margins[1]. Figure 1 contains a screenshot depicting a particularly busy intersection.

While the system is safe in the face of communication failures, we have not addressed the possibility or effects of mechanical failures or unlikely "freak" accidents. In a world without vehicle malfunctions, this would be little cause for concern. However, one can easily imagine an otherwise ordinary problem, such as a flat tire or a slippery patch of road, quickly becoming a nightmare.

Even though the vast majority of automobile accidents can be blamed on driver error (or in some cases, the limitations of human drivers), if individual incidents are a hundred times more deadly, no reasonably achievable reduction in incident frequency will effect an overall improvement. However, if in the rare event of an accident, the total damage can be kept under control—perhaps at most a few times as many as normal—then, as a whole, riding in automobiles will be a safer experience than it is today.

Figure 1. A screenshot showing a busy intersection with a lot of "close calls"

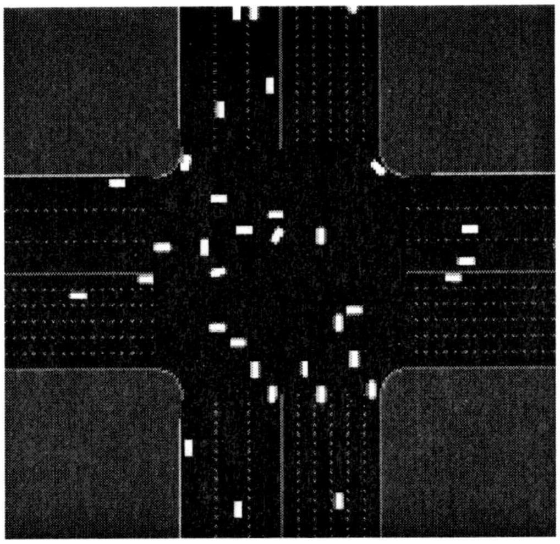

In the second half of this chapter, we describe safety features of the system designed to deal with these types of failures. We perform a basic failure mode analysis demonstrating the necessity of such features, and give extensive empirical evidence suggesting that these features are not only effective, but also robust to poor communications.

REMOVING THE INTERSECTION MANAGER

To address the issue of high cost associated with managed autonomous intersections, we have created a low-cost alternative for low-traffic intersections. In this section, we introduce our unmanaged autonomous intersection control mechanism. First, we specify the goals of our system. Next, we describe our assumptions about the driver agents. We then outline the protocol for communication between vehicles, and describe the rules that each vehicle must follow.

Goals of the System

For an unmanaged intersection control mechanism for autonomous vehicles to be both effective and economically viable, we believe it should have the following properties:

- Vehicles using the system should get through the intersection more quickly than they do using current mechanisms (i.e. stop signs).
- The protocol should have minimal (ideally none) per-intersection infrastructure costs.
- The protocol should guarantee the safety of the vehicles using it. Specifically, if all vehicles follow the protocol correctly, no collisions should result.

Vehicle-to-Vehicle (V2V) Protocol

Unlike the protocol for our managed intersection [Dresner & Stone 2004], our protocol for unmanaged autonomous intersection control is designed for communication among only one type of agent: driver agents. In our system, each agent sends and receives information to and from each other agent, maintaining up-to-date information about every vehicle approaching the intersection. Dropped packets and limited transmission distance may cause agents to have outdated or inconsistent information. Because data transmission is largely asynchronous in an ad-hoc wireless network of mobile agents, this protocol cannot rely on a dialogue between agents. As such, the protocol is simple, consisting only of broadcast messages. There are two types of messages: CLAIM and CLEAR.

Claim

A CLAIM message is sent by an agent in order to announce its intentions to use a specific space and time in the intersection. CLAIM contains information describing both the vehicle's intended path through the intersection, as well as when it believes its traversal will take place. Once the agent has chosen these parameters, it broadcasts its CLAIM repeatedly. The message contains seven fields:

- `vehicle_id`—The vehicle's unique Vehicle Identification Number (VIN).
- `message_id`—A monotonically increasing counter specific to this message. Other agents will use `message_id` to identify the most recent message from this vehicle. This number is not changed when a specific message is rebroadcast; it is incremented only when a vehicle generates a *new* message to broadcast.
- `stopped_at_intersection`—A boolean value representing whether the vehicle is stopped at the intersection.
- `arrival_lane`—The lane in which the vehicle will be when it arrives at the intersection. Each lane incident to the intersection has an absolute index available as part of the intersection's layout information.
- `arrival_time`—The time at which this vehicle will enter the intersection.
- `exit_lane`—The lane in which the vehicle will exit the intersection.
- `exit_time`—The time at which this vehicle will exit the intersection.

Clear

An agent sends a CLEAR message to release any currently held reservation. This message cancels any pending reservation; even if other agents have differing or outdated information about an agent's reservation, the agent can still cancel. The CLEAR message is broadcast repeatedly, with the same period as CLEAR, to ensure it is received by all other agents. This message has two fields:

- `vehicle_id`—This vehicle's VIN.
- `message_id`—A monotonically increasing number specific to this message. This is the same as the `message_id` field in CLAIM.

Message Broadcast

Because each message contains all the latest relevant information about the sending vehicle, agents need only pay attention to the most recent message from any other vehicle. Each message is also broadcast repeatedly with a set period to ensure its eventual delivery, should a new vehicle enter the transmission range of the sender. As a result, although occasional dropped messages may increase the delay in communications between vehicles, they should not pose a significant threat to the safety of vehicles in our system. In situations with higher rates of packet loss, messages may need to be broadcast more frequently to compensate. Conversely, in low-latency, high-reliability scenarios, messages can be sent less frequently.

For security purposes, we also assume that each message is digitally signed, ensuring that driver agents cannot falsify the `vehicle_id` parameter. Messages that do not conform to the protocol or are not digitally signed are ignored.

Conflict, Priority, and Dominance

In order to facilitate the discussion of agent behavior and protocol analysis, we define the following relations on CLAIM messages.

Two CLAIM messages are said to *conflict* if all of the following are true:

- The paths determined by the `lane` and `turn` parameters of the CLAIM messages are not compatible
- The time intervals specified in the CLAIM messages are not disjoint

We define the relative *priority* of two CLAIM messages based on the following rules, presented in order from most significant to least significant:

1. If neither CLAIM specifies that the sending vehicle is stopped at the intersection, the CLAIM with the earliest `exit_time` has priority.
2. If both CLAIM messages specify that the respective sending vehicles are stopped at the intersection, the CLAIM whose `lane` is "on the right" has priority. Here, "on the right" is defined similarly to current traffic laws regarding four-way stop signs. This binary relation on the incident lanes is globally available as a characteristic of the intersection.
3. If neither message's `lane` can be established as being "on the right," the CLAIM whose `turn` parameter indicates the sending vehicle is not turning has priority.
4. If priority cannot be established by the previous rules, the CLAIM with the lowest `vehicle_id` has priority.

Finally, given two claims 1 and 2, we say that 1 *dominates* 2 if either of the following rules is true:

- The `stopped_at_intersection` field of 1 is `true` and the `stopped_at_intersection` field of 2 is false.
- The `stopped_at_intersection` fields of 1 and 2 are identical, 1 and 2 conflict, and 1 has priority over 2.

Required Agent Actions

The consequences of failure in a traffic management system can be disastrous. As such, in addition to a communication protocol, a rigid set of rules must govern the interaction of agents within the system. With human drivers, traffic laws serve this purpose: if every driver obeys traffic laws, there is little or no potential for automobile accidents. Our multiagent system relies on an analogous set of rules. While there is nothing physically preventing an agent from ignoring them, the safety of each agent's vehicle can only be guaranteed if that agent follows the rules. Note that the rules restrict only how the agent behaves while in the intersection; driver agents have full autonomy everywhere else. The rules are as follows:

1. A vehicle may not enter the intersection if its own CLAIM is dominated by any other current CLAIM.
2. A vehicle may not enter the intersection without first broadcasting a CLAIM for at least time p. In our implementation, $p = 0.4$ seconds.
3. A vehicle must vacate the intersection at or before the `exit _ time` specified in its most recent CLAIM message.
4. If a vehicle is going to traverse the intersection, it must follow a reasonable path from the point of entry to the point of departure. This means, for example, that a vehicle going straight through the intersection must remain within its lane, and that a vehicle turning right must not enter any other lanes.
5. The `stopped _ at _ intersection` field of an agent's CLAIM must be set to `true` if and only if the agent is stopped at the intersection.
6. The agent may not broadcast unless it is within a certain distance of the intersection. This distance is called the *lurk distance*. In our implementation, the lurk distance is 75 meters.

Selfish and Malicious Agents

Agents in our system are assumed to be self-interested—they may take any possible legal action in order to ensure they traverse the intersection in as little time possible. Agents have little incentive to lie about their lane, path, or exit time, because lying about any of these puts the vehicle at risk for collision. However, an agent may have an incentive to falsely claim that it is stopped at the intersection. While there is a chance this may slow down the traffic in front of the offending vehicle, if there is no such traffic exists, an agent may gain some advantage by falsely claiming that it is stopped at the intersection, allowing its CLAIM to dominate the CLAIMS of other moving vehicles. This may result in the vehicle crossing the intersection earlier. This type of behavior is not currently disincentivized by our protocol, but if it were to become a problem, could be tested at random intersections to ensure compliance. This is analogous to current traffic enforcement, which relies on sporadic monitoring and associated penalties to decrease rule violations.

As with any multiagent system, malicious agents are a potential problem. In current traffic scenarios, nothing prevents someone from deliberately crashing into another vehicle, or disabling traffic signals. Similarly, a malicious driver agent could flood the network with useless traffic, preventing the system from operating properly. While nothing can be done to stop a determined saboteur, the fact that all messages are signed makes it impossible for vehicles to conceal their identity while using the protocol.

Driver Agent Behavior

Our proposed unmanaged intersection control mechanism relies not only on a communication protocol, but also on the existence of driver agents that can abide by the protocol. Our prototype driver agent's behavior is comprised of three phases: lurking, making a reservation, and intersection traversal.

Lurking

As the vehicle approaches the intersection, it begins to receive messages from other agents. However, it may not broadcast a reservation until it is within the *lurk distance*. The lurk distance is calculated to ensure that an agent is within transmission range of other vehicles long enough to be reasonably sure that it is aware of every pending CLAIM. CLAIMS are broadcast repeatedly at a set frequency; more frequent broadcasts reduce the amount of time an agent must spend within transmission range to assemble all pending CLAIMS. Therefore, lurk distance depends on both transmission range and broadcast frequency. In our simulations, we set lurk distance to 75 meters—a reasonable approximation given current communication technology.

Making a Reservation

The most important part of our driver agent behavior starts when vehicle reaches the lurk distance. At this point, it needs to let the other driver agents know how it intends to cross the intersection. We call this part of the process "making a reservation," as an analogue to our managed system, which also uses a reservation paradigm [Dresner & Stone 2005]. During this time, the vehicle needs to compute its expected arrival time, arrival velocity, departure time, and, given the messages it has accumulated from other vehicles, determine the soonest time at which the intersection will be available. This behavior is shown in Algorithm 1.

As an agent approaches the intersection, it generates a CLAIM based on predictions of its arrival time, arrival velocity, and path through the intersection (line 3). To predict the time required to cross the intersection, the agent must know its arrival velocity. Initially, the agent calculates the earliest possible arrival time, and the predicted velocity of the vehicle at this time based on the speed limit and its own acceleration constraints (the physical constraints of the vehicle, in addition to the constraints imposed by traffic front of it) Based on this arrival velocity, the agent predicts the time at which it will exit the intersection, assuming that it can accelerate as needed within the intersection. If the agent has received no CLAIMS from other vehicles that dominate this CLAIM, the agent will begin to broadcast this CLAIM (line 11).

Otherwise, the agent generates a new CLAIM at the earliest possible time such that it will not be dominated by any existing CLAIM of another vehicle (line 9). To do so, the agent searches through existing CLAIMS to find the next block of time that it could potentially dominate, assuming it can arrive at the highest legal velocity. After finding a suitable block, the agent predicts its arrival velocity based on arrival time (which is generally lower than the maximum legal velocity), which it uses to determine the actual time required to cross the intersection. If the agent can traverse the intersection in the available time, it begins broadcasting a CLAIM; if not, it searches for the next suitable block and repeats these calculations.

Algorithm 1. Behavior of the driver agent from coming within lurk distance of the intersection to entering the intersection

```
1: loop
2:  if do not have a current CLAIM
3:        generate a new CLAIM
4:  end if
5:  if not at the intersection and another vehicle is then
6:        broadcast CLEAR
7:  else
8:  if arrival estimate changes or CLAIM is dominated then
9:              generate new CLAIM
10:             end if
11:             broadcast the CLAIM
12:       end if
13: end loop
```

Intersection Traversal

Once a vehicle has made a reservation, it needs only to broadcast the CLAIM continually and to arrive at the intersection in accordance with its reservation. However, sometimes the vehicle may want to change an existing claim in order to take advantage of an unexpected early arrival (line 8). On the other hand, traffic patterns may occasionally cause a vehicle to arrive late. If a vehicle predicts that it cannot fulfill the parameters of its CLAIM message, it must either send a CLEAR or a new CLAIM. Similarly, if a new CLAIM message arrives that dominates the driver agent's CLAIM, the driver agent must also make a new reservation.

Once the vehicle reaches the intersection, it crosses in accordance with its CLAIM. While in the intersection, for safety purposes, the vehicle continues to broadcast its CLAIM, however this CLAIM cannot be dominated, as the vehicle is already executing the intersection traversal, which is clear from the fact that the current time is after the `arrival _ time` in the CLAIM. After a vehicle has vacated the intersection, it stops transmitting its CLAIM.

Vehicle Control

The driving actions taken by a vehicle to complete its reservation are very similar to those of the driver agent in our managed mechanism [Dresner & Stone 2004]. If a vehicle predicts that it will arrive late, it accelerates. If a vehicle predicts that it will arrive early, it slows down (unless it believes it can make an earlier CLAIM). The vehicle must also ensure that it arrives with sufficient velocity to traverse the intersection within the constraints of its reservation.

Canceling "Bad" Reservations

In some situations, a vehicle is unable to reach the intersection at the proper time and velocity. To detect these situations, the vehicle is constantly predicting its arrival time. As with the driver agent presented in our work on managed intersections, this agent calculates its arrival time and velocity either *optimistically* or *pessimistically* [Dresner & Stone 2005]. If a vehicle detects no vehicles in front of it, it will make an optimistic projection of arrival time, assuming it can accelerate as needed before it arrives. However, if a vehicle is obstructed by traffic, it will make a pessimistic projection of arrival time based on the assumption that it cannot accelerate before it arrives at the intersection. If the vehicle's predicted arrival time is later than that of its reservation, the vehicle will cancel its current reservation and attempt to make a reservation for a later time.

Improving Reservations

If a driver agent predicts that it will arrive at the intersection before the time specified in its reservation, it may be able to improve its reservation before reaching the intersection. To accomplish this, the agent looks for blocks of intersection time between its predicted arrival time and the arrival time specified in its reservation. If the vehicle determines that it can broadcast a suitably large CLAIM that will not be dominated, it will immediately begin broadcasting this CLAIM. As specified by the communication protocol, this implicitly cancels any previous reservation held by the vehicle.

If a vehicle arrives at the intersection before the time specified in its reservation, it changes its CLAIM to reflect that it is stopped and waiting to cross (as required by the protocol). As a result, this agent's CLAIM will now dominate the CLAIM of any vehicle not stopped at the intersection. The stopped agent will then begin broadcasting the earliest possible non-dominated CLAIM. If no other vehicles are stopped, this will be p seconds from the current time, as the vehicle must broadcast its claim for at least this amount of time before entering the intersection. If other vehicles are stopped at the intersection, the agent will broadcast a CLAIM for the earliest block of time not dominated by the CLAIM of any stopped vehicles.

Empirical Results

Here we present empirical results comparing our unmanaged autonomous intersection to intersections outfitted with four-way stop signs and traffic lights. After describing our metrics and experimental setup, we compare the average delay induced by each of these control policies. We then use these results to estimate the amounts of traffic for which a stop sign outperforms a traffic light. This range is the primary focus of the analysis of our system, as we consider it to be the range over which an unmanaged policy is more appropriate than a managed policy. We also compare the relative fuel consumption associated with the stop sign and unmanaged autonomous policies. Finally, we discuss the effects of dropped messages on our unmanaged autonomous control policy.

Metrics

In our analysis, we examine two key metrics: *average delay* and *average cumulative acceleration*. The primary metric is the average of the delay experienced by each vehicle as it crosses the intersection. The baseline for delay is the time it would take a vehicle to traverse a completely empty intersection. Because a vehicle must slow down to turn, the baseline is different for left turns, right turns, and straight passages through the intersection. We measured the trip time for an unobstructed vehicle following these three paths, giving us an accurate baseline for comparison. Delay is measured as actual trip time minus baseline trip time, which isolates the effect of the intersection control policies and allows us to accurately compare the among them.

The second metric we use is the average of the cumulative acceleration of each vehicle during its trip through the intersection. We define the *cumulative acceleration* of a vehicle, denoted α, as:

$$\alpha = \sum_{i=0}^{s} |a_i|$$

where s is the trip length of the vehicle measured in simulator steps, and i is the acceleration of the vehicle at simulator step i. Note that the baseline for α is nonzero in turning vehicles, as vehicles must slow down to turn and accelerate again to the speed limit afterwards. We chose to compare the average cumulative acceleration to examine the relative fuel efficiency of each system. Although not a direct measure of fuel efficiency, a vehicle's cumulative acceleration provides a reasonable approximation of gasoline usage, because substantially more fuel is required to accelerate than to maintain a constant velocity. Average delay is also an indicator of fuel efficiency, as the delay experienced by a vehicle correlates with the amount of fuel consumed while the vehicle was not accelerating (either idling at the intersection or traveling at a constant velocity). Thus, we can compare the relative fuel efficiency of each system by comparing both average delay and average cumulative acceleration.

Experimental Setup

To test these policies, we use a custom simulator which simulates a four-way intersection with one lane of traffic in each direction (see Figure 2). This small, symmetrical intersection is representative of those intersections currently configured as a four-way stop, and thus provides the best test case for unmanaged control mechanisms. We control traffic levels via a Poisson process governed by the probability of creating a new vehicle in a given lane at each time step. We simulate traffic levels between 0 and 0.5 vehicles per second, with 15% of vehicles turning left and 15% turning right. Each data point represents

Figure 2. A screenshot of the simulator

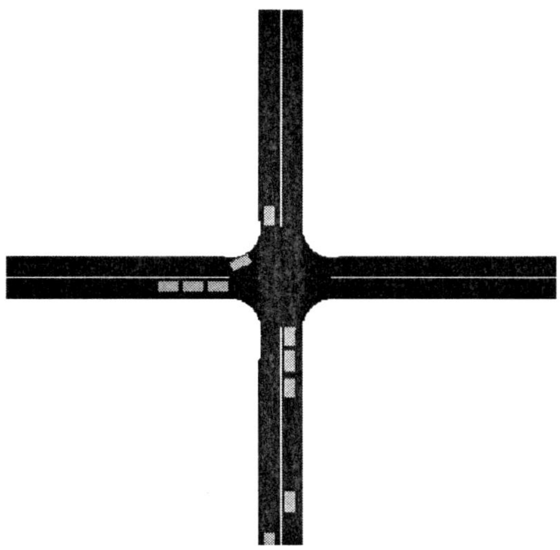

the average of 20 simulations, with each run consisting of 30 minutes of simulated time. All data are shown with error bars indicating a 95% confidence interval.

The traffic light timing is configured such that, in succession, each direction receives a green light for 10 seconds, followed by 3 seconds of yellow. There is a large body of theory and empirical evidence concerning the timing of traffic lights, but this work is largely irrelevant to our simulated scenario for two reasons. First, much of the theory deals with the timing of lights across multiple intersections, whereas we are examining one intersection in isolation. Second, our simulator generates symmetric traffic, which greatly simplifies light timing by eliminating the need to account for higher traffic levels in a particular direction or lane. For these reasons, we established a reasonable timing pattern experimentally by evaluating 10 different candidate patterns and selecting the one that led to the lowest average delay.

It should be noted that our four-way stop sign policy does not allow multiple vehicles to inhabit the intersection simultaneously. In the real world, stop signs can allow a limited sharing of the intersection. This is most apparent in intersections with multiple lanes of traffic in each direction: in this situation, cars traveling parallel to one another can cross the intersection at the same time. There is significantly less potential for sharing the intersection when there is only one lane of traffic in each direction. A human driver may observe the vehicle currently crossing the intersection and predict the vehicle's actions for the remainder of its journey (although this prediction is not always accurate!). If the other vehicle's path does not conflict with the intended path of the human driver, he or she may enter the intersection slightly before the other vehicle has exited. However, the benefits of this behavior are significantly reduced in small intersections. Therefore, we believe that our four-way stop sign policy is a reasonable approximation of a real-world four-way stop.

Delay

As shown in Figure 3, our system significantly reduces the average delay experienced by each vehicle. When traffic flow is below 0.35 vehicles per second, the four-way stop is a more effective policy than the traffic light.

Figure 3. A comparison of average delay of the traffic light, four-way stop, and our unmanaged mechanism

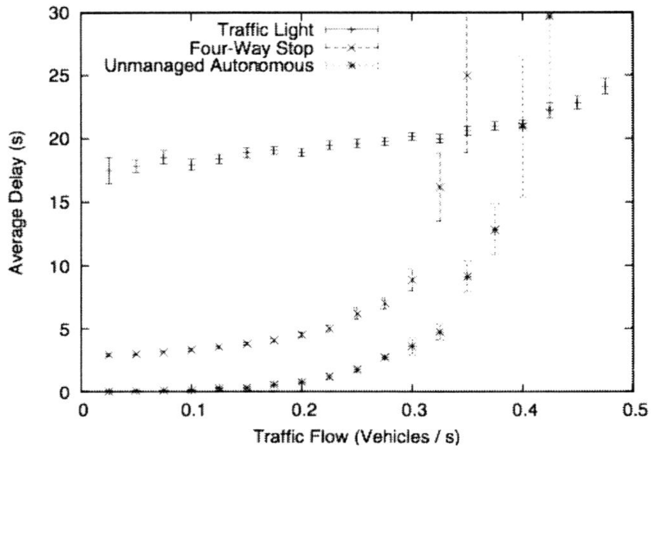

Our unmanaged system results in near-zero delay at traffic levels below 0.2 vehicles per second. In these situations, most agents are able to cross the intersection without slowing down to wait for other vehicles. With the four-way stop sign, each vehicle must stop even if no others are present, resulting in a baseline average delay of approximately 3 seconds. The traffic light system has a higher baseline average delay, around 18 seconds.

When traffic flow is between 0.2 and 0.35 vehicles per second, our system shows a somewhat increased delay. In these cases, cars must often slow down to accommodate other vehicles, but but only rarely will a vehicle need to make a complete stop. With the stop sign policy, vehicles begin to queue at the intersection, and must often wait for vehicles in front of them to cross. The traffic light policy shows almost no increase in delay at these levels.

At traffic levels above 0.35 vehicles per second, the stop sign policy deadlocks. At these traffic levels, our system is similar to a four-way stop: because there is almost always at least one vehicle waiting to cross, agents must wait until they are stopped at the intersection to make a reservation. However, the intersection sharing in our system (allowing four simultaneous right turns, for example) provides a noticeable benefit at these traffic levels. Our unmanaged system can safely handle traffic levels up to approximately 0.4 vehicles per second, at which point traffic begins to back up. The traffic light shows only a slight increase in delay at these traffic levels. In these situations, our data suggest that a managed mechanism is more appropriate.

Average Acceleration

Another benefit of our system is reduced average acceleration, as shown in Figure 4. With the stop sign policy, every vehicle must come to a complete stop at the intersection and accelerate to the speed limit after crossing. If vehicles are queued at the intersection, each vehicle must stop at the back of the queue. As the queue moves forward, each vehicle accelerates for a brief period of time, then decelerates to a stop until another car leaves the front of the queue. This behavior results in a very high average acceleration for the stop sign policy.

Figure 4. A comparison of average acceleration of the four-way stop and our unmanaged mechanism

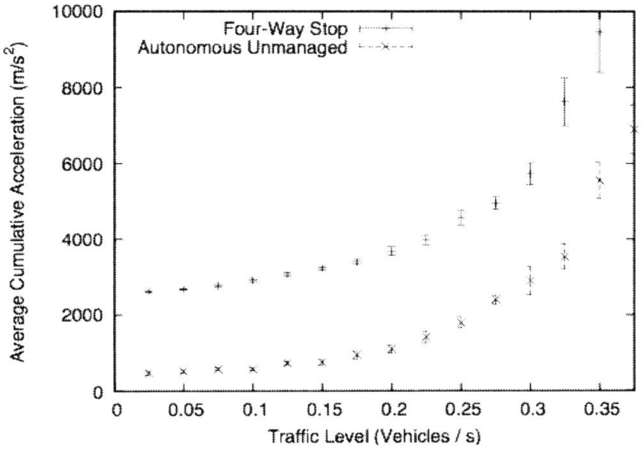

For low levels of traffic, our system allows most vehicles to pass directly through the intersection without slowing or stopping. Even at high traffic levels, when our system is essentially a modified four-way stop, our system results in lower average acceleration than a four-way stop. This is because our system causes shorter queues than a stop sign, reducing the amount of acceleration and braking required for each vehicle to reach the front of the queue. Combined with the data on average delay, these results suggest that our unmanaged autonomous system would allow significantly reduced fuel consumption.

Dropped Messages

We designed our system to be resistant to occasional communication failures such as dropped messages. In our previously proposed managed intersection, the vehicles must wait for a response from the intersection manager before entering the intersection [Dresner & Stone 2005]. Because of this, dropped packets may increase the delay of the system, but will not cause a collision. In our system, we have found no statistically significant correlation between dropped packets and delay. Rather, dropped packets introduce a possibility of failure that increases with the percentage of packets dropped.

To quantify this effect, we varied the proportion of dropped messages between 0 and 0.7 at intervals of 0.1, running 400 thirty-minute simulations at each level. The traffic level in these simulations was 0.3 vehicles per second. When fewer than 40% of messages were dropped, the system behaved normally. Between 40% and 60% packet loss, the system began to experience safety failures—five of the 1200 simulations in this range resulted in collisions. At 70% packet loss, the frequency of collisions is significantly higher, with collisions occurring in seven of 200 simulations.

These results suggest that, as proposed, our peer-to-peer protocol can tolerate moderate levels of packet loss with no ill effects, but that serious communication issues might make it unsafe. While a thorough analysis of communication failures is beyond the scope of this chapter, research in distributed systems has shown that fast and reliable information dissemination in ad-hoc wireless networks such as the kind we are simulating is possible [Drabkin et al. 2007]. We thus leave further communication analysis to future work.

MITIGATING CATASTROPHIC FAILURES

A collision in purely autonomous traffic can have any number of causes, including software errors in the driver agent, a physical malfunction in the vehicle, or even meteorological phenomena. In modern-day traffic, such factors are largely ignored for two reasons. First, the exclusively human-populated system, with its generous margins for error, is not as sensitive to small or moderate aberrations. Second, none of these causes are significant with respect to driver error as causes of accidents. In fact, according to a study from the 1980s, vehicle and road issues alone were responsible for fewer than 5% of accidents [Wierwille et al. 2002]. However, in the future of infallible autonomous driver agents, it is exactly these issues which will be the prevalent causes of automobile collisions. The safety buffers in our mechanism are adjustable—given some maximum allowable error in vehicle positioning, the buffers can be extended to handle that error—but no reasonable adjustment can account for gross mechanical malfunction like a blowout or failed brakes. Because these types of issues are infrequent, we believe the safety of the intersection control mechanism will be acceptable even if individual occurrences are slightly worse than accidents today. As we will show, without the safety measures presented in this section, the system is prone to spectacular failure modes, sometimes involving dozens of vehicles.

Responding to an Incident

When a vehicle deviates significantly from its planned course through the intersection, resulting in physical harm to the vehicle or its presumed occupants, we refer to the situation as an *incident*. Once an incident has occurred, the first priority is to ensure the safety of all persons and vehicles nearby. Because we expect these incidents to be very infrequent occurrences, re-establishing normal operation of the intersection is a lower priority and the optimization of that process is left to future work.

Assumptions

In order to reduce the average number of vehicles involved in a crash from dozens to one or two, we must make one assumption beyond those previously stated. We assume that once an accident has occurred in the intersection, the intersection manager can detect it. There are two basic ways by which the intersection manager could detect that a vehicle has encountered some sort of problem: the vehicle can inform the intersection manager, or the intersection manager can detect the vehicle directly. For instance, in the event of a collision, a device similar to that which triggers an airbag can send a signal to the intersection manager. Devices such as this already exist in vehicles equipped with General Motors' OnStar system, which automatically calls for help when an accident has happened. The intersection manager itself might notice a less severe problem, such as a vehicle that is not where it is supposed to be, using cameras or sensors at the intersection. However, this method of detection is likely to be much slower to react to a problem. Each has advantages and disadvantages, and a combination of the two would most likely be the safest. What is important is that whenever a vehicle violates its reservation in any way, the intersection manager should become aware as soon as possible.

Intersection Manager Response

As soon as the intersection manager detects or is notified of an incident, it immediately stops granting reservations. All subsequent received requests are rejected without consideration. Due to the nature of the protocol, the intersection manager cannot revoke reservations, as driver agents would have no incentive to acknowledge their receipt. However, the intersection manager can send a message to the vehicles that an incident has occurred. This message is the special EMERGENCY-STOP message, which the intersection manager may only send in an emergency situation, and which (as with the rest of the protocol) it must assume has not been received.

The EMERGENCY-STOP message lets vehicles know that an event has taken place in the intersection such that:

- no further reservations will be accepted
- vehicles able to come to a stop before entering the intersection should do so
- vehicles in the intersection should no longer assume that "near misses" will not result in collisions

Vehicle Response

For the Emergency-Stop message to be useful in any way, driver agents must react to it. Here we explain the specific actions our implementation of the driver agent takes when it receives this message. Normally, when approaching the intersection, our driver agent ignores any vehicles sensed in the intersection. This is because what might otherwise appear to be an imminent collision on the open road is almost certainly a precisely coordinated "near-miss" in the intersection. However, once the driver agent receives the Emergency-Stop message from the intersection manager, it disables this behavior. If the vehicle is in the intersection, the driver agent will not blindly drive into another vehicle if it can help it. If the vehicle is not in the intersection and can stop in time, it will not enter, even if it has a reservation.

While our first inclination was to make the driver agent immediately decelerate to a stop, we quickly realized that this is not the safest behavior. If all vehicles that receive the message come to a stop, vehicles that would otherwise have cleared the intersection without colliding may find themselves stuck in the intersection—another object for other vehicles to run into. This is especially true if the vehicle that caused the incident is on the edge of the intersection where it is unlikely to be hit. Trying to stop all the other vehicles in the intersection just makes the situation worse.

If a driver agent does detect an impending collision, however, it is allowed to take evasive actions or apply the brakes. Since this is a true multiagent system with self-interested agents, we cannot prevent the driver agents from doing so. In our experiments, our driver agent brakes if it believes a collision is imminent.

Experimental Results

In order to evaluate the effects of our reactive safety measures, we performed several experiments in which various components were intentionally disabled. The various configurations can be separated into three classes. An *oblivious* intersection manager takes no action at all upon detecting an incident. An intersection manager utilizing *passive* safety measures stops accepting reservations, but does not send any Emergency-Stop messages to nearby driver agents. Finally, the *active* configuration of the intersection manager has all safety features in place. In addition to considering these three incarnations of the intersection manager, we also study the effects of unreliable communication in the active case. Note that when no vehicles receive the Emergency-Stop message, the active and passive configurations are identical.

Experimental Setup

With the great efficiency of the reservation-based system comes an extreme sensitivity to error. While buffering can protect against the more minute discrepancies, it cannot hope to cover gross mechanical malfunctions. To determine just how much of an effect such a malfunction would have, we created a simulation in which individual vehicles could be "crashed", causing them to immediately stop and remain stopped. Whenever a vehicle that is not crashed comes into contact with one that is, it becomes crashed as well. While this does not model the specifics of individual impacts, it does allow us to estimate how a malfunction might lead to collisions.

In order to ensure that we included malfunctions in all different parts of the intersection, we triggered each incident by choosing a random (x,y) coordinate pair inside the intersection, and crashing the first vehicle to cross either the x or y coordinate. This is akin to creating two infinitesimally thin walls, one horizontal and the other vertical, that intersect at (x,y). Figure 5 provides a visual depiction of this process.

After initiating an incident, we ran the simulator for an additional 60 seconds, recording any additional collisions and when they occurred. Using this information, we constructed a *crash log*, which is essentially a histogram of crashed vehicles. For each step of the remaining simulation, the crash log indicates how many vehicles were crashed by that step. By averaging over many such crash logs for each configuration, we were able to construct an "average" crash log, which gives a picture of what a typical incident would produce.

For these experiments, we ran our simulator with scenarios of 3, 4, 5, and 6 lanes in each of the four cardinal directions, although we will discuss results only for the 3- and 6-lane cases (other results were similar, but space is limited). Vehicles are spawned equally likely in all directions, and are generated via a Poisson process which is controlled by the probability that a vehicle will be generated at each step. Vehicles are generated with a set destination—15% of vehicles turn left, 15% turn right, and the remaining 70% go straight. The leftmost lane is always a left turn lane, while the right lane is always a right turn lane. Turning vehicles are always spawned in the correct lane, and non-turning vehicles are not spawned in the turn lanes. In scenarios involving only autonomous vehicles, we set the traffic level at an average of 1.667 vehicles per second per lane in each direction. This equates to 5 total vehicles per second for 3 lanes, and 10 total vehicles per second for 6 lanes. While we wanted traffic to be flowing smoothly, we also wanted the intersection to be full of vehicles to test situations that likely lead to the most destructive possible collisions.

How Bad is it?

As we suspected, the average crash log of the oblivious intersection manager is quite grisly. Normally, driver agents must ignore their sensors while in the intersection, because many of the "close calls"

Figure 5. Triggering an incident in the intersection simulator. The dark vehicle turning left is crashed because it has crossed the randomly chosen x coordinate. If a different vehicle had crossed that x coordinate or the randomly chosen y coordinate earlier, it would be crashed instead.

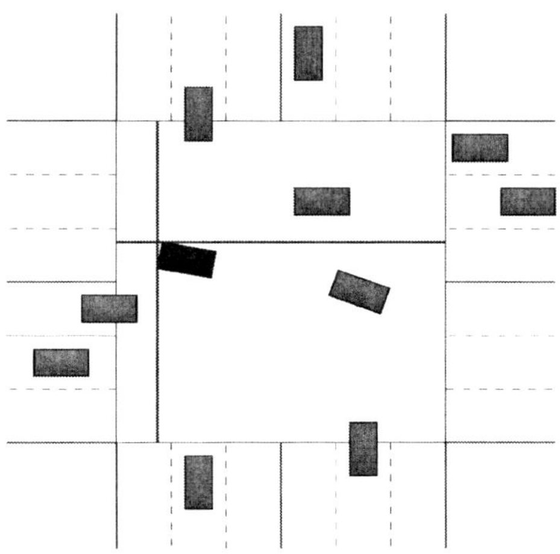

would appear to be impending collisions. Without any way to react the situation going awry, vehicles careen into the intersection, piling up until the entire intersection is filled and crashed vehicles protrude into the incoming lanes. Figure 6 shows that the rate of collisions does not abate until over 70 vehicles have crashed. Even a full 60 seconds after the incident begins, vehicles are still colliding. In the 3-lane case, the intersection is much smaller and thus fills much more rapidly; by 50 seconds, the number of collided vehicles levels off.

Reducing the Number of Collisions

There are two main components to the safety mechanism we introduced. First, the intersection manager stops accepting reservations. Second, the intersection manager sends messages informing the driver agents that an incident has taken place. There is a possibility that this second part might not always work perfectly; some vehicles might not receive the message. To investigate the effects of these potential communication failures, we intentionally disabled some of the vehicles' ability to receive the EMERGENCY-STOP message. A parameter in our simulator controls the fraction of vehicles created with this property, and by varying this parameter, we could observe its subsequent effect on the average number of vehicles involved in incidents.

As compared to the oblivious intersection manager, the number of vehicles involved in the average incident for an active intersection manager decreases dramatically. Table 1 shows the numerical results for both the 3- and 6-lane intersections, along with a 95% confidence interval. The average crash logs for these runs are not shown in Figure 6, as they would be indistinguishable from one another at that scale. Instead, we present them in Figure 7.

Figure 7 shows the effect of our safety system on intersections with 6 lanes, with the proportion of receiving vehicles varying from 0% (passive) to 100% in increments of 20%. Even in the passive case, the overall number of vehicles involved in the average incident declines by a factor of almost 30. As expected, when more vehicles receive the emergency signal (in the active case), fewer vehicles crash. Figure 7 shows only the first 15 seconds of the crash log, because in no case did a collision occur more than 15 seconds after the incident started.

Figure 6. Average crash logs (with 95% confidence interval) for 3- and 6-lane oblivious intersection managers

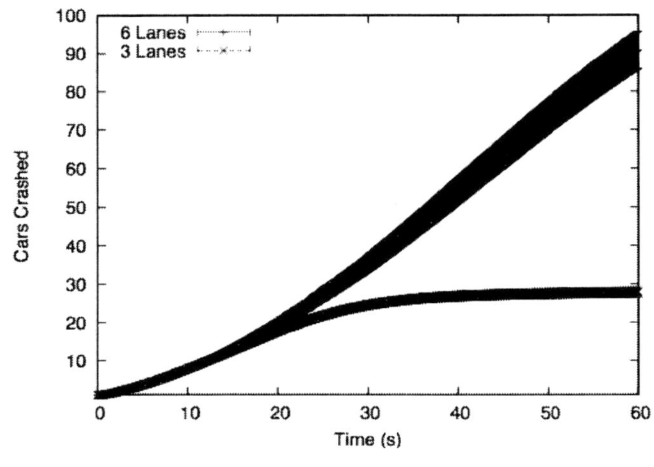

Table 1. Average number of vehicles involved in incidents for 3- and 6-lane intersections with various percentages of vehicles receiving the Emergency-Stop *message. Even in the passive case, the number of crashed vehicles decreases dramatically.*

		3 Lanes	6 Lanes
Oblivious		**27.9**±1.3	**90.9**±4.9
Passive		**2.63**±.13	**3.23**±.16
Active	20%	2.44±.13	3.15±.17
	40%	2.28±.12	2.90±.16
	60%	1.89±.10	2.69±.15
	80%	1.71±.08	2.30±.13
	100%	**1.36**±.06	**1.77**±.10

Reducing the Severity of Collisions

While it is reassuring to know that the number of vehicles involved in the average incident can be kept fairly low, these data do not give the entire picture. For example, compare an incident in which 30 vehicles each lose a hubcap to one in which two vehicles are completely destroyed and all occupants killed. While we do not currently have any plans to model the intricate physics of each individual collision with high fidelity, our simulations do allow us to observe the velocity at which the collisions occur. In the previous example, we might notice that the 30 vehicles all bumped into one another at low velocities, while the two vehicles were traveling at full speed. To quantify this information, we record not only when a collision happens, but the velocity at which it happens. In a collision, the amount of damage done is approximately proportional to the amount of kinetic energy that is lost. Because kinetic energy is proportional to the square of velocity, we can use a running total of the squares of these crash velocities to create a rough estimate of the amount of damage caused by the incident. Figure 8 shows

Figure 7. Average crash logs for the passive (0% receiving) and active (20%-100% receiving) intersection managers. Only the first 15 seconds of the 6-lane scenario are shown.

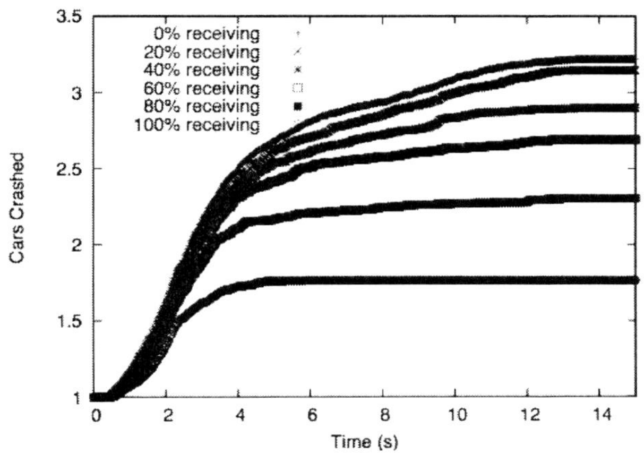

an average "damage log" of a 6-lane intersection of autonomous vehicles. Qualitatively similar results were found for the other intersection types.

As Figure 8 shows, the effect of our safety measures under this metric is quite dramatic as well. In the passive case the total accumulated squared velocity decreases by a factor of over 25. In the active case, with all vehicles receiving the signal, it decreases by another factor of 2. Of particular note is the zoomed-in graph in Figure 8(b). In the passive configuration, the total squared velocity accumulates as if the intersection manager were oblivious, until the first vehicles stop short of the intersection at around 3 seconds; without a reservation, they may not enter. In the active scenario, when all the vehicles receive the message, the improvement is almost immediate.

Figure 8. Average total squared velocity of crashed vehicles for a 6-lane intersection with only autonomous vehicles. Sending the emergency message to vehicles not only causes fewer collisions, but also makes the collisions that do happen less dangerous.

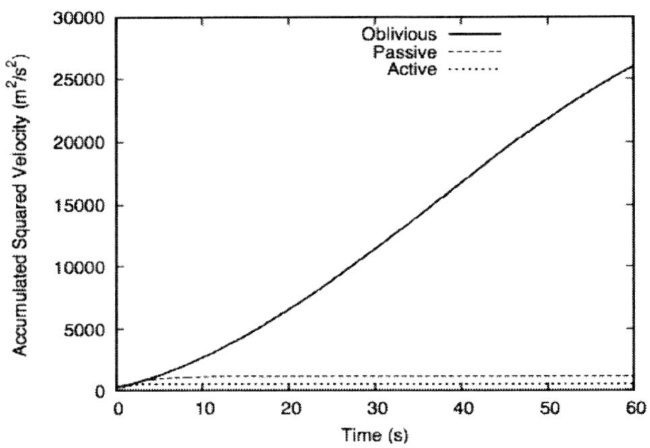

(a) The average incident

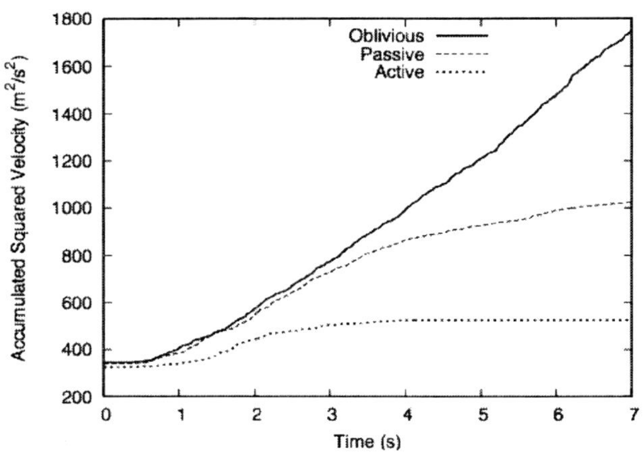

(b) Zoomed in

Delayed Incident Detection

Implicit in these results is the assumption that intersection managers become aware of incidents instantaneously. While this could be the case in many collisions—vehicles should communicate when they have collided—if a vehicle's communications are faulty, or if the vehicle does not realize it has collided, the intersection may not discover the problem for a few seconds, when another vehicle or sensor will detect the problem. To assess the effects of delayed incident detection, we artificially delayed the intersection manager's response in some of our simulations. Figure 9 shows the results from these experiments.

In Figure 9, the intersection manager's reaction was delayed 0, 1, 3, and 5 seconds. Note that the total number of crashed vehicles with a delay of 5 seconds is on par with the number in the experiment in which the intersection manager reacts immediately, but none of the vehicles receive the message, shown in Figure 7. Figure 9(b) shows what happens with both delayed detection and faulty communication. This graph, along with the earlier results, suggests that for small values, each second of delay is approximately equivalent to 20% of vehicles not receiving the EMERGENCY-STOP message, and that when combined, delayed detection and faulty communication have an additive effect. For larger delays, the

Figure 9. Crash logs showing the effects of delayed incident detection

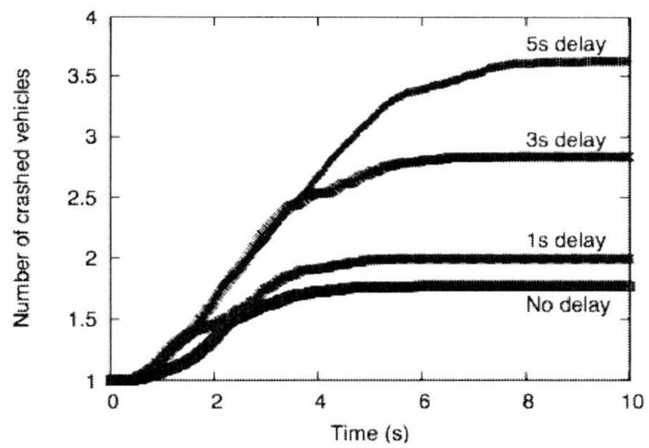

(a) Delaying detection

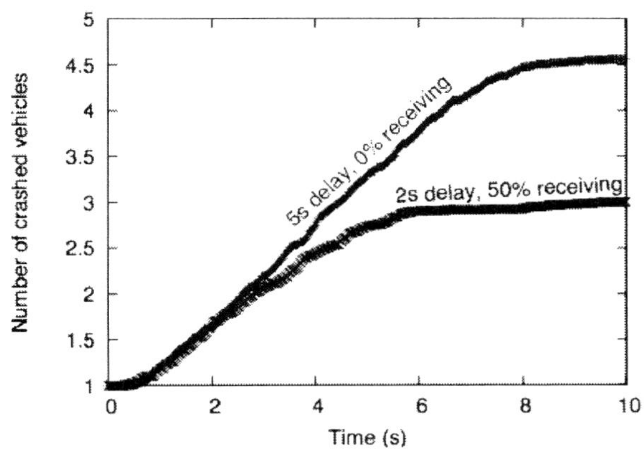

(b) Delays and faulty communication

number of vehicles involved can be approximated using the data shown in Figure 7, because in these cases, the number of vehicles that crash after the intersection reacts is much smaller than the number that crash before it reacts.

A Safer System Overall

In our experiments, we showed that the number of vehicles involved in individual incidents can be drastically reduced by virtue of some of the safety properties built into our intersection control mechanism. In fact, when all vehicles received the warning, a large portion of the incidents involved only one vehicle: the one we intentionally crashed. Even in the worst case—6 lanes of traffic and no vehicles receiving the warning signals—an average of only 3.23 vehicles were involved. But how does this compare with current systems? If we conservatively assume that accidents in traffic today involve only one vehicle, this represents a 223% increase per occurrence. Thus, all other things being equal, if the frequency of accidents can be reduced by 70%, the the autonomous intersection management system will be safer overall. A 2002 report for the Federal Highway Administration blamed over 95% of all accidents on driver error [Wierwille et al. 2002]. The report blamed 2% of accidents on vehicle failures and another 2% on problems with roads. It is important to note that these numbers are for all driving, not just intersection driving. Accidents in intersections are even more likely to be caused by driver error, sometimes by drivers willfully disobeying the law: running red lights and stop signs or making illegal "U"-turns.

Even if we make overly conservative assumptions—that all driving is as dangerous as intersection driving, and that driver error is no more accountable for intersection crashes than it is in other types of driving—our data suggest that automobile traffic with autonomous driver agents and an intersection control mechanism like ours will reduce collisions in intersections by over 80%. We believe that in reality, the improvement will be staggeringly greater.

The technique presented here is just one method for improving the safety of this system's failure modes. More sophisticated methods involving explicit cooperation amongst vehicles may create an even safer system. The main thrust of our discussion is not that this particular safety mechanism is by any means the best possible. Rather, it is that even with this fairly simple response to accidents, the overall safety of the system can be strengthened well beyond that of current automobile traffic—all without sacrificing the benefit of vastly improved efficiency.

RELATED WORK

Intersection management—especially for intersections of autonomous vehicles—is an exciting and promising area of research for autonomous agents and multiagent systems. Many projects in AI and intelligent transportation systems address this increasingly important problem. Using techniques from computer networking, Naumann and Rasche created an algorithm in which drivers attempt to obtain *tokens* for contested parts of the intersection, without which they cannot cross [Naumann & Rasche 1997]. While this allows vehicles to cross unimpeded in very light traffic, the system has no notion of "planning ahead"; only one vehicle may hold a token at any given time, no agent can plan to have the token in the future if another agent has it currently. Kolodko and Vlacic have created a system very similar to ours on golf cart–like IMARA vehicles [Kolodko & Vlacic 2003]. However, their system requires all vehicles to come to a stop, irrespective of traffic conditions.

In the context of video games and animation, Reynolds has developed autonomous steering algorithms that attempt to avoid collisions in intersections that do not have any signaling mechanisms [Reynolds1999]. While such a system does have the enormous advantage of not requiring any special infrastructure or agent at the intersection, it has two fatal drawbacks that make it unsuitable for use with real-life traffic. First, the algorithm does not let driver agents choose which path they will take out of the intersection; a vehicle may even find itself exiting the intersection the same way it came in, due to efforts to avoid colliding with other vehicles. Second, the algorithm only *attempts* to avoid collisions—it does not make any guarantees about safety.

To the best of our knowledge, our work on collision mitigation represents the first study of the impact of an efficient, multiagent intersection control protocol for fully autonomous vehicles on driver safety. However, there is an enormous body of work regarding safety properties of traditional intersections. This includes the general—correlating traffic level and accident frequency [Sayed & Zein 1999] and analyses of particular types of intersections [Bonneson & McCoy 1993, Harwood et al. 2003, Persaud et al. 2001]—as well as plenty of the esoteric, such as characterizing the role of Alzheimer's Disease in intersection collisions [Rizzo et al. 2001]. However, because it concerns only human-operated vehicles, none of this work is particularly applicable to the setting we are concerned with here.

CONCLUSION

After introducing the reservation-based protocol for managed intersections based on the assumption that all cars are autonomous, we later presented a policy which allows both computer- and human-controlled vehicles to safely interact at the same intersection [Dresner & Stone 2007]. Our protocol for unmanaged intersections can be similarly adapted to accommodate human drivers using traffic signs. The human drivers would be directed to behave as if they were stopped at a two-way stop, yielding to all approaching vehicles (this also assumes that the computer-controlled vehicles have some signal identifying them as autonomous). Because our system is designed for low-traffic intersections, human drivers could generally expect to wait for no more than a few seconds. Our proposed system for accommodating human drivers and the corresponding managed system both put human-controlled vehicles at somewhat of a disadvantage–an incentive for human drivers to transition to fully computer-controlled vehicles. Future research could formalize and optimize a policy for accommodating human drivers in our unmanaged autonomous intersection.

Another potential area for future research in unmanaged autonomous intersections is allowing the system to adapt to asymmetric traffic flow. Many intersections consistently receive higher traffic in some lanes than others. In these intersections, a two-way stop is often more efficient than a four-way stop. In our current system, all agents stopped at the intersection are given equal priority, regardless of the number of vehicles queued behind them. This approximates the behavior at a four-way stop. However, by granting priority to lanes with longer queues, our system could alleviate congestion in high-traffic lanes. This would allow our system to function like a two-way stop in situations with asymmetric traffic flow, while functioning like a four-way stop in situations with more symmetrical traffic.

Our work on accident mitigation still leaves some unanswered questions. For example, we have examined only one method of disabling vehicles. In the future, we would like to explore other possibilities such as locking a vehicle's steering, simulating a blowout, sticking the accelerator, or disabling the brakes. For this work, our aim was to initiate incidents that would test the limits of the intersection control mechanism by disrupting as much of the traffic flow as possible. A truly comprehensive failure mode analysis must include a much wider array of potential hazards. While our very conservative estimates indicate that this intersection control mechanism will be vastly safer than current systems with human drivers, we would like to conduct a more detailed study comparing the two, to quantify the improvement more precisely.

Recent research has already produced fully autonomous, computer-controlled vehicles. As these vehicles become more common, we will be able to phase out human-centric traffic control mechanisms in favor of vastly more efficient computer-controlled systems. This will be especially beneficial at intersections, which are a major cause of delays. For a transition of this magnitude, infrastructure cost will be a central, if not primary, concern. This chapter presents a novel, unmanaged intersection control mechanism requiring no specialized infrastructure at the intersection. We have described in detail a protocol for our unmanaged autonomous intersection, and created a prototype driver agent capable of utilizing this protocol. As illustrated by our empirical results, our protocol can significantly reduce both delay and fuel consumption as compared to a four-way stop. Unsignalized intersections far outnumber those that are sufficiently large or busy to warrant the cost of a managed solution. Whereas busier intersections may need to wait for the funding and installation of requisite infrastructure, our proposed mechanism has the potential to open every one of these unsignalized intersections to be used safely and efficiently by autonomous vehicles.

Autonomous vehicles, and the promise of easier, safer, and more efficient travel that they offer are a fascinating and exciting development. Before the benefits of this technology can be realized, much more work must be done to ensure that they are as safe as possible for the hundreds of millions of passengers that will use it on a daily basis. Our failure mode analysis calls attention to the need for keeping an eye toward safety throughout the development of the algorithms and protocols that will control the transportation systems of the future. In this way, we believe we have accomplished a portion of this important work. Further analysis will of course be necessary, first in simulation, and ultimately with real physical vehicles.

ACKNOWLEDGMENT

This research is supported in part by NSF CAREER award IIS-0237699 and by the United States Federal Highway Administration under cooperative agreement DTFH61-07-H-00030. Computational resources were provided in large part by NSF grant EIA-0303609.

REFERENCES

Bonneson, J. A., & McCoy, P. T. (1993). Estimation of safety at two-way stop–controlled intersections on rural highways. *Transportation Research Record, 1401*, 83–89.

DARPA 2007. The DARPA urban challenge. http://www.darpa.mil/grandchallenge.

Drabkin, V., Friedman, R., Kliot, G., & Segal, M. (2007). Rapid: Reliable probabilistic dissemination in wireless ad-hoc networks. *In The 26th IEEE International Symposium on Reliable Distributed Systems*, Beijing, China.

Dresner, K. & Stone, P. (2004). *Multiagent traffic management: A protocol for defining intersection control policies UT-AI-TR-04-315*, The University of Texas at Austin, Department of Computer Sciences, AI Laboratory.

Dresner, K. & Stone, P. (2005). Multiagent traffic management: An improved intersection control mechanism. *In The Fourth International Joint Conference on Autonomous Agents and Multiagent Systems*, (pp. 471–477), Utrecht, The Netherlands.

Dresner, K. & Stone, P. (2007). Sharing the road: Autonomous vehicles meet human drivers In Proceedings of the Twentieth International Joint Conference on Artificial Intelligence, (pp. 1263–68), Hyderabad, India.

Harwood, D. W., Bauer, K. M., Potts, I. B., Torbic, D. J., Richard, K. R., Rabbani, E. R. K., Hauer, E., Elefteriadou, L., & Griffith, M. S. (2003). Safety effectiveness of intersection left- and right-turn lanes. *Transportation Research Record, 1840*, 131–139.

Kolodko, J. & Vlacic, L. (2003). Cooperative autonomous driving at the intelligent control systems laboratory. *IEEE Intelligent Systems, 18*(4), 8–11.

National Highway Traffic Safety Administration (2002). *Economic impact of U.S. motor vehicle crashes reaches $230.6 billion*, new NHTSA study shows NHTSA Press Release 38-02. http://www.nhtsa.dot.gov.

Naumann, R. & Rasche, R. (1997). Intersection collision avoidance by means of decentralized security and communication management of autonomous vehicles. *In Proceedings of the 30th ISATA - ATT/IST Conference.*

Persaud, B. N., Retting, R. A., Gardner, P. E., & Lord, D. (2001). Safety effect of roundabout conversions in the united states: Empirical bayes observational before-after study. *Transportation Research Record, 1751*, 1–8.

Reynolds, C. W. (1999). Steering behaviors for autonomous characters. *In Proceedings of the Game Developers Conference,* (pp. 763–782).

Rizzo, M., McGehee, D. V., Dawson, J. D., & Anderson, S. N. (2001). Simulated car crashes at intersections in drivers with Alzheimer disease. *Alzheimer Disease and Associated Disorders, 15*(1), 10–20.

Sayed, T. & Zein, S. (1999). Traffic conflict standards for intersections. *Transportation Planning and Technology, 22*(4), 309–323.

Wierwille, W. W., Hanowski, R. J., Hankey, J. M., Kieliszewski, C. A., Lee, S. E., Medina, A., Keisler, A. S., & Dingus, T. A. (2002). *Identification and evaluation of driver errors: Overview and recommendations FHWA-RD-02-003*, Virginia Tech Transportation Institute, Blacksburg, Virginia, USA. Sponsored by the Federal Highway Administration.

Chapter X
Valuation-Aware Traffic Control:
The Notion and the Issues

Heiko Schepperle
Universität Karlsruhe (TH), Germany

Klemens Böhm
Universität Karlsruhe (TH), Germany

ABSTRACT

Current intersection-control systems lack one important feature: They are unaware of the different valuations of reduced waiting time of the drivers. Drivers with high valuations may be willing to pay to be prioritized at intersections. In this chapter, the authors describe an agent-based valuation-aware intelligent traffic-control system for road intersections which increases the overall driver satisfaction. It combines valuation-aware intersection-control mechanisms with sophisticated driver-assistance features, subsequently referred to as adaptive cruise and crossing control (A3C). Driver-assistance agents and intersection agents negotiate so-called time slots to cross the intersection. The driver-assistance agent adapts the speed of the vehicle in line with the time slot obtained. Various obstacles are in the way of realizing such a system and making it operational. The authors discuss these challenges and present ideas for solutions. They examine the intersection-control and the driver-assistance perspective of the intelligent traffic-control system. After a brief evaluation, they finally describe application scenarios where agent-based valuation-aware intersection control may become operational in the near future.

INTRODUCTION

The growing need for mobility makes it more difficult for cities to cope with the increasing number of vehicles and to provide the necessary infrastructure. Optimizing the use of existing traffic resources may be much cheaper than building new ones. Thus, cities introduce more sophisticated *intelligent traffic-control (ITC) systems* for road intersections. Current ITC systems do not take the driver valuations of reduced waiting time into account. This valuation may be different for each driver and for trip types, e.g., a driver who is in danger of missing a flight may have a higher valuation than one on a weekend trip. Taking these valuations into account can increase the overall driver satisfaction.

The ongoing progress in vehicle technology allows for more sophisticated driver-assistance systems. The automotive industry currently offers various driver-assistance systems which increase safety or comfort of drivers and passengers. A new application area for such systems is intersection control, i.e., traffic control at road intersections. The more information is available to an ITC system, the better it can deal with the current traffic. While current ITC systems for intersection control are already context-aware, i.e., they rely on historic traffic data or on actual traffic data collected with stationary sensor technology, driver-assistance systems may provide access to information which stationary sensors cannot collect. The driver valuation of reduced waiting time is a prominent example. We call systems which also consider this valuation of the drivers *valuation-aware*.

This chapter describes a valuation-aware ITC system for road intersections. Such a system requires negotiation between vehicles and infrastructure. To avoid unnecessary distraction of the drivers, we propose to use autonomous software agents for the negotiation. Next, the system should be able to adapt to changes in the traffic in real time. Further, it must be more effective than existing traffic-control systems. The ITC system combines valuation-aware intersection-control mechanisms with sophisticated driver-assistance features: Driver-assistance systems play a central role for valuation-aware intersection control. Using valuation-aware mechanisms, driver-assistance systems which are located in vehicles negotiate the right to cross an intersection with the intersection-control unit located at the intersection. The driver-assistance systems report the valuation of reduced waiting time of their drivers. The intersection-control unit assigns time slots to the road users taking the valuations into account. After having been notified about its time slot the driver-assistance system knows when to cross the intersection. Thus, it recommends an appropriate speed to the driver. If the driver-assistance system can also adapt the speed autonomously, we can extend the functionality of an ACC system with crossing control features. This means that the driver-assistance system does not only adapt the speed autonomously to keep a time gap to preceding vehicles, but also to reach the intersection in time. We call this *adaptive cruise and crossing control (A3C)*. We examine the two perspectives of the system, the intersection-control perspective, i.e., valuation-aware intersection-control mechanisms, and the driver-assistance perspective, i.e., the A3C system.

The outline of this chapter is as follows: We discuss other approaches for intersection control in Section "Background". Section "Definitions" features some definitions. Various obstacles are in the way of realizing such a system and making it operational. We provide a survey of the various challenges related to traffic engineering, information technology, economy, road users, and law in Section "Challenges". Since our research project now is in its third year, and we have had interaction with researchers from other disciplines, our survey addresses all severe issues which have been raised. Sections "Intersection-Control Perspective" and "Driver-Assistance Perspective" examine the two perspectives of our system in detail. Section "Evaluation" contains the evaluation. Finally, we describe settings where our system can become operational soon in Section "Application Scenarios".

BACKGROUND

There exist various approaches for traffic control at intersections. Some of them are agent-based. Only a few consider valuation-awareness. We only discuss the approaches we deem relevant for our valuation-aware ITC system. Because the ITC system comprises certain driver-assistance features, this section also discusses adaptive cruise-control systems.

Intersection Control Using Vehicular Sensors

Current approaches for intersection control rely on traffic data which can be detected using stationary sensors. But traffic data which such sensors cannot easily detect – like the turning direction or the valuation of reduced waiting time – could also be used for traffic control. One of the first approaches using traffic data observed by vehicular sensors is Dresner and Stone (2004). The authors propose a reservation-based system for autonomous vehicles which uses a first-in first-out strategy. They show that their approach performs better than traffic lights. Vehicles are equipped with agents. They report the time when the vehicle will arrive at the intersection, the arrival speed, the turning direction, and specifics of the vehicle type like the maximum speed, maximum and minimum acceleration, and the width and length of the vehicle to a so-called intersection manager (Dresner & Stone, 2004). Stationary sensors cannot detect this. While this approach shows the benefits of using vehicular sensor data, it is not valuation-aware.

Valuation-Aware Intersection Control

The traditional goal of a traffic-control system is to reduce the delay respectively the waiting time for each incoming vehicle stream (Garber & Hoel, 1988). Clearly, taking the waiting time into account is important. But overall satisfaction can increase further if we do not only consider the average waiting time of all road users. For example, if the intersection-control strategy also considers the valuation of reduced waiting time, at least for road users with high valuations, this should further increase overall satisfaction. We discuss such valuation-aware approaches in the following.

For public transportation there already exist so-called traffic-signal-priority systems which prioritize public transportation vehicles like buses or trams (e.g., Greschner & Gerland, 2000). Dresner and Stone (2008) enhance their reservation-based system by giving priority to emergency vehicles like ambulance or police cars. Both approaches have the disadvantage that they prioritize only few and specialized road users. The first valuation-aware approaches for traffic control at road intersections are the ones of Schepperle, Böhm, and Forster (2008) and Schepperle and Böhm (2007; 2008). They represent driver-assistance systems and intersections by agents. In Schepperle and Böhm (2007; 2008) driver-assistance agents (DAA) additionally report their valuation of reduced waiting time to the intersection agent (IA). The IA uses auctions to identify the next vehicle to cross the intersection. Schepperle et al. (2008) allow DAAs to exchange their rights to cross the intersection. They show that these approaches are more effective than a plain first-in first-out system. However, the various challenges which one must address before such an approach becomes operational are not discussed. Further, it is not mentioned how to combine these mechanisms with an A3C system.

Adaptive Cruise Control

Vehicles can approach an intersection in different ways, e.g., with maximum speed and – if necessary – a full brake just before the intersection or with constant speed to arrive just in time. In general, it is impossible to arrive at a certain time with a certain speed using a constant acceleration or deceleration. In this case, several acceleration and deceleration steps may be necessary. We call the sequence of acceleration and deceleration steps required by an intersection-control system respectively planned by a vehicle *acceleration profile*. Driving according to an acceleration profile may be difficult for human drivers. However, features of adaptive-cruise control (ACC) systems should be helpful in this context.

An ACC system can control throttle and brake of a vehicle in order to maintain a time-gap relationship between a vehicle and the preceding one. The driver can activate the ACC providing a desired speed v_{desire} and a desired time gap g_{desire}. The ACC system adapts the vehicle speed to maintain g_{desire} and – if possible – v_{desire}. The driver only has to steer (Ioannou & Chien, 1993).

ACC systems are already available for several years. Early generations could only be used for speeds between 30 km/h and 200 km/h (Bosch, 2008) because in slower traffic the characteristics of traffic are too diverse (Marsden, McDonald, & Brackstone, 2001). The latest generation also allows speeds lower than 30 km/h (ACCplus) (Bosch, 2008; Persson, Botling, Hesslow, & Johansson, 1999). Because we want to use ACC for intersection control as well, we have to extend it to an adaptive cruise and crossing-control (A3C) system. This means that it adapts the speed not only to keep a time-gap relationship to preceding vehicles, but also to arrive at the intersection in time. The project Travolution examines a similar approach (GEVAS, 2008): The traffic lights inform the driver about the next green phase. Because the traffic light does not know the number of vehicles approaching it cannot – in contrast to our intersection-control system – propose an appropriate speed, but only a maximum speed which is sufficient to arrive in time. The system also lacks valuation-awareness. Dresner and Stone (2004; 2008) present a similar system, again without valuation-awareness.

DEFINITIONS

We now introduce definitions which ease the presentation of our intersection-control system. We borrow some definitions from Schepperle and Böhm (2007).

Scenario

Our focus is on isolated intersections. An *intersection* consists of several *intersection lanes*. An intersection lane connects one *incoming lane* with one *outgoing lane*. Thus, the intersection lanes emanating from the same incoming lane determine the directions a vehicle can choose when approaching the intersection on this incoming lane. The *neighborhood* of an intersection consists of the incoming and outgoing lanes within communication range and the intersection lanes. Vehicles must not overtake in the neighborhood of the intersection.

The intersection lanes correspond to the traffic streams allowed. Conflicts can occur when traffic streams interfere with each other (Garber & Hoel, 1988). We define a *conflict area* as the intersection area of two intersection lanes. We distinguish between diverging, crossing and merging conflict areas. A *diverging conflict area* occurs where two intersection lanes emanate from one incoming lane. A *merging conflict area* occurs where two intersection lanes merge to one outgoing lane. We call all other conflict areas *crossing conflict areas*. Because an intersection lane connects an incoming to an outgoing lane, we also refer to it as *connector*. A conflict area consists of two overlapping *connector-conflict areas*, i.e., one connector-conflict area for each conflicting intersection lane.

We define a *time slot* as the right for a certain vehicle to cross an intersection on a certain intersection lane in a certain period of time. Two time slots are *conflicting* if the use of one slot excludes the use of the other one. The duration of a time slot may depend on the preferences and abilities of the vehicle respectively the driver. The *entrance time* of a time slot of a vehicle is the time when it can enter the intersection. The *leaving time* is the time when it must have left the intersection at the latest. The time slot may not only determine the time but also the speed to enter or leave the intersection. The *entrance* respectively *leaving speed* is the speed required while entering or leaving the intersection. For instance, a vehicle which has already stopped at the intersection may receive a lower entrance speed than a vehicle which is approaching the intersection with maximum speed. In any case, entrance and leaving speed on the one hand and the duration of a time slot on the other hand are related.

Architecture

We suppose that vehicles are equipped with a driver-assistance system which hosts a so-called *driver-assistance agent* (DAA). The DAA represents the driver. The driver instructs the DAA before the trip, and the DAA then acts according to these instructions autonomously. During the trip the DAA gives recommendations to the driver. Depending on the instructions, the DAA may also control the vehicle but the driver can always overrule the DAA. On the other hand, the intersection-control unit hosts an *intersection agent* (IA) which represents the traffic planner. IAs and DAAs negotiate appropriate time slots. Figure 1 illustrates this architecture.

IA and DAA may have different goals. The goals of the DAA depend on the driver. We expect the DAA to be self-interested. This means that it aims to increase the local utility of the driver, e.g., by reducing the waiting time or just to arrive at the destination in time. The goals of the IA depend on the intersection-control mechanisms used. We present different mechanisms later in this chapter.

Figure 1. Architecture of the valuation-aware ITC system

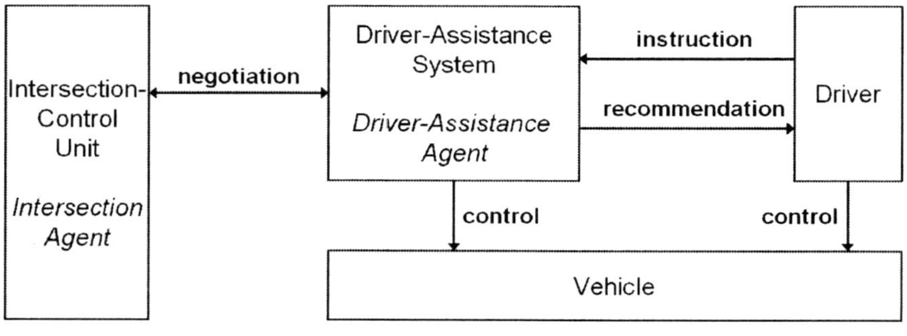

Measures

To evaluate valuation-aware mechanisms we propose the following measures. To ease presentation we use the term 'vehicle' synonymously for vehicle, driver and DAA. The *travel time* T_t^j *of Vehicle j* is the time from its first appearance in the neighborhood until it leaves the neighborhood. The *minimal travel time* $T_{t,\min}^j$ *of j* is the travel time if j was the only vehicle at the intersection, observed any constraints regulating the driving of an isolated vehicle like speed limits or one-way streets, but ignored all rules concerning the right of way (i.e., crossed red lights, did not stop at stop signs, etc.). This corresponds to the overpass strategy of Dresner and Stone (2004). The *waiting time* T_w^j of j is the difference of the travel time T_t^j and the minimal travel time $T_{t,\min}^j$. Thus, the waiting time is different from the standstill time and corresponds to the delay in Dresner and Stone (2004).

We use the waiting time to compute the average waiting time $\overline{T_w}$. Let V be the set of all vehicles observed, then

$$\overline{T_w} = \frac{\sum_{j \in V} T_w^j}{|V|}.$$

The average waiting time is not appropriate to evaluate valuation-aware mechanisms because it does not reflect if mechanisms let vehicles with higher valuations cross the intersection earlier. Nevertheless, it is a common measure to compare mechanisms for intersection control (Garber & Hoel, 1988). A more meaningful measure in our context is the waiting time weighted by the driver valuations of reduced waiting time. The *valuation* $v^j(t)$ *of Vehicle j* is the price j is willing to pay if it waits t seconds less. The *weighted waiting time of Vehicle j* is $v^j(T_w^j)$. Let $v^j = v^j(1)$ be the valuation of Vehicle j of its waiting time reduced by one second. If we confine ourselves to linear valuations, v^j determines the valuation per second. Then we can also write the weighted waiting time as $v^j(T_w^j) = v^j(1 \cdot T_w^j) = v^j(1) \cdot T_w^j = v^j \cdot T_w^j$. Thus, the average weighted waiting time is

$$\overline{vT_w} = \frac{\sum_{j \in V} v^j(T_w^j)}{|V|} \text{ respectively } \overline{vT_w} = \frac{\sum_{j \in V} v^j \cdot T_w^j}{|V|}.$$

Each vehicle has a limited travel budget, i.e., a certain amount of money to spend for the entire trip. We refer to the travel budget of Vehicle j as b^j. If valuation-aware intersection-control mechanisms are used, a driver may earn or spend money by negotiating time slots. We denote the expenses e^j as expenditure minus income of Vehicle j. Using the initial travel budget b^j and the expenses e^j we define the *utility* u^j *of Vehicle j* as travel budget minus expenses minus weighted waiting time $u^j = b^j - e^j - v^j(T_w^j)$.

Example: *Let the valuation of Vehicle j be linear and the valuation of its waiting time reduced by one second v^j=3. Let the travel budget be b^j=100. If the vehicle has a time slot which makes the vehicle wait T_w^j=20 seconds, the utility will be $u^j = b^j - e^j - v^j \cdot T_w^j = 100 - 0 - 3 \cdot 20 = 40$. If it paid 5 to receive* *an earlier time slot and therefore reduced its waiting time from 20 seconds to 10 seconds, it would have the utility* $u^j = 100 - 5 - 3 \cdot 10 = 65$. *Thus, the utility would increase by 25.*

Therefore, minimizing the average weighted waiting time increases total utility. This is identical with social welfare respectively the overall driver satisfaction mentioned earlier.

CHALLENGES

In this section we discuss challenges of a valuation-aware ITC system for road intersections and potential solutions. As mentioned, a respective system combines valuation-aware intersection-control mechanisms with sophisticated driver-assistance features. Making valuation-aware intersection control operational poses various challenges. In the following we group these challenges in the categories traffic engineering (Figure 2, CT1 – CT4), information technology (CI1, CI2), economy (CE1, CE2), road users (CU1, CU2), and law (CL1 – CL3) and describe ideas for solutions.

Traffic Engineering

Important challenges related to traffic engineering are physical constraints, traffic safety, heterogeneous environment, and effectiveness.

Physical Constraints (CT1)

In traffic we have to deal with *physical constraints*. Acceleration and deceleration of drivers and vehicles is naturally bounded and may vary depending on the road surface or on the weather conditions. In the neighborhood of an intersection, vehicles must not overtake. Both valuation-aware mechanisms computing time slots and A3C systems adapting the vehicle speed must consider these constraints.

Traffic Safety (CT2)

Any intersection-control system is safety-relevant. Injuries or loss of lives must be avoided in any case. An intersection-control system is *safe* if intersection-control mechanisms never send conflicting time

Figure 2. Challenges of a valuation-aware ITC system

	Traffic Engineering		**Information Technology**
CT1	Physical constraints	CI1	Inter-vehicle communication
CT2	Traffic safety	CI2	Security
CT3	Heterogeneous environment		
CT4	Effectiveness		**Economy**
		CE1	Mechanism design
	Law	CE2	Market penetration
CL1	Liability		
CL2	Traffic regulations		**Road users**
CL3	Privacy and anonymity	CU1	User acceptance
		CU2	Impact on driving behavior

slots to different vehicles, and if an A3C system always prevents vehicles from entering the intersection without a valid time slot. Note that the driver can still override the A3C system and enter the intersection without a valid time slot, as he can with other intersection-control mechanisms, e.g., by ignoring red lights.

Intersection-control mechanisms can be designed *fail-safe*, i.e., no physical damage in case of a failure: Before sending a time slot to a DAA, the IA has to reserve the slot. The IA uses the reservations to compute new time slots which are not in conflict with time slots previously sent. If the mechanism allows DAAs to return time slots, the IA reserves the time slot even in the case that the DAA is likely to return the time slot to the IA. The IA may clear reservations of time slots only after the owning DAA has definitely returned it. This may lead to unnecessarily reserved time slots but prevents conflicting time slots to be used by different DAAs.

An A3C system can also be designed fail-safe. This is because it is an extension of a state-of-the-art ACC system using similar techniques.

Heterogeneous Environment (CT3)

An intersection-control system always acts in a *heterogeneous environment* with different kinds of road users. This means that it has to be capable to deal with pedestrians, cyclists and drivers of vehicles of different types, age and equipment. This is notoriously difficult.

We propose three different solutions. One possibility is to use stationary detectors for pedestrians like in the Puffin pedestrian facilities (County Surveyors' Society & Department for Transport, 2006) and to feed the detector information to a pedestrian-assistance agent which negotiates with other agents on behalf of the pedestrians. This solution may be costly, but the necessary technology exists. Another possible solution for pedestrians and cyclists which do not have an assistance system as powerful as the ones in vehicles could be to use a mobile device, e.g., an advanced mobile cellular phone or a so-called personal-travel assistant (PTA). These devices can host the necessary assistance agents. This solution depends on the progress and the market penetration of mobile devices and may only be applicable in some years. Further, there may also be vehicles without the necessary capabilities, e.g., older vehicles which are not equipped at all or only with older assistance systems or vehicles whose assistance system is broken. Assuming a high degree of dissemination of such assistance systems these vehicles are an exception. Thus, we can use vehicles waiting behind to inform the traffic-control mechanism about 'blind' preceding vehicles. As mentioned, this solution is applicable only with a high market penetration. In the transition period with a lower market penetration we can apply the solutions suggested for pedestrians to vehicles: stationary detectors and assistance agents, which negotiate on behalf of the vehicles equipped insufficiently, or mobile devices of the drivers.

We assume a more homogeneous environment in certain application scenarios which we discuss in Section "Application Scenarios".

Effectiveness (CT4)

A new intersection-control system must outperform state-of-the-art intersection-control systems. In the context of valuation-aware intersection control, we define a traffic-control system to be *effective* if it reduces the average weighted waiting time compared to other state-of-the-art intersection-control systems. Section "Evaluation" shows that valuation-aware mechanisms can be effective.

Information Technology

Inter-vehicle communication (CI1) and security (CI2) are challenges related to information technology. A valuation-aware intersection-control system relies on robust communication between vehicles and infrastructure. Although the automotive industry pushes inter-vehicle communication (Vehicle Infrastructure Integration (VII) Initiative, http://www.vehicle-infrastructure.org/; CAR 2 CAR Communication Consortium, http://www.car-to-car.org/) the state-of-the-art currently is not robust enough (e.g., Torrent-Moreno, Killat, & Hartenstein, 2005) for safety-related systems. The degree of robustness also depends on the number of vehicles involved. Further, not only the communication, but also the driver-assistance systems and the intersection-control units must be secure. This means that they have to prevent attacks of any kind. Because of the efforts of the automotive industry mentioned we expect robust and secure communication between vehicles and infrastructure to be available in the future.

Economy

The design of a valuation-aware ITC system poses several economic challenges. We briefly discuss mechanism design and market penetration.

Mechanism Design (CE1)

To design mechanisms for valuation-aware intersection control, we have to take the inherent incentives into account. In order to be *efficient*, the mechanism should be *incentive-compatible* (Krishna, 2002). This means that drivers should have an incentive to reveal their valuations truthfully. This is because in this case drivers with the highest valuations obtain the next time slot. Next to incentive compatibility other desiderata are Pareto optimality, maximized social welfare, budget balance and individual rationality (Dash, Jennings, & Parkes, 2003) etc. There exist several impossibility theorems stating that certain desiderata are mutually exclusive. E. g., the Myerson-Satterthwaite theorem shows that for bilateral exchange no mechanism can exist which is efficient, incentive-compatible, individually rational, and budget balanced at the same time (Krishna, 2002). A thorough examination of these desiderata depends on the mechanism actually used. If drivers also receive payments, e.g., by letting other vehicles pass, we further have to *eliminate incentives to use traffic infrastructure more than necessary* just to earn money. Experiments with human drivers may be necessary to understand sophisticated auction mechanisms. Investigating strategies for DAAs to report their valuations is a promising research direction.

Market Penetration (CE2)

Intersection control relying on driver-assistance systems, like A3C systems, presumes that all road users are equipped with such a system. Vehicles have a life expectation of more than 10 years. Thus, it is difficult to achieve and maintain a high market penetration of state-of-the-art vehicular hardware soon. The life expectation of mobile devices in turn is shorter. Thus, the necessary *market penetration* is easier to achieve for *mobile devices* sold independently from vehicles. This is why potential solutions for the heterogeneity problem also help to achieve the necessary market penetration.

Road Users

Introducing a valuation-aware ITC system also poses challenges concerning the road users: user acceptance and impact on driving behavior.

User Acceptance (CU1)

Given the current rapid increase of the oil price, traffic-management systems which seem to further increase mobility cost have difficulties to achieve high *user acceptance*. Even in countries where toll systems are common resistance may be high. Therefore, we suggest the following steps to increase user acceptance of our approach: A mechanism for intersection control is *budget-balanced* if it allows only payments among road users or at least returns the revenue to road users completely, e.g., if the money earned by an ITC system is used to reduce traffic taxes, or if vehicles waiting at an intersection are refunded immediately. We expect users to accept a system which features a budget-balanced mechanism much easier. Further, in the context of intersection control, waiting times at intersections should be bounded. In other words, the system has to *avoid starvation*. We can avoid starvation if we suspend a mechanism when the waiting time exceeds an upper bound. In this case we let the vehicle cross the intersection and resume the mechanism thereafter.

Impact on Driving Behavior (CU2)

The impact of automated technology in vehicles on the driving behavior has been subject of several studies: One aspect is *situation awareness*. This means the awareness of environmental information relevant for driving and the ability to predict future traffic states, e.g., recognizing a dangerous situation several vehicles ahead and expecting preceding vehicles to brake abruptly (Ward, 2000). When drivers get used to advanced technology in their vehicles, this may reduce the situation awareness (or mode awareness) of the driver or even lead to a complete loss of situation awareness in hazardous situations (Ward, 2000; Furukawa, Shiraishi, Inagaki, & Watanabe, 2003). This negative effect is sometimes called driving without awareness. On the other hand automated technology may free limited mental resources. In this case drivers could pay more attention to the driving environment, and situation awareness would increase (Ma & Kaber, 2005). Closely related is the aspect of *risk compensation*. If drivers trust technology designed to increase safety more and more, drivers tend to drive riskier (Itoh, Sakami, & Tanaka, 2000). In our context this could mean that drivers reduce the time gap of their A3C system too much, or that they approach the intersection faster than allowed. One must take both situation awareness and risk compensation into account when designing valuation-aware ITC systems. But the actual impact on driver behavior can only be evaluated using a full-functional prototype.

If vehicles are equipped with more automated systems which demand interaction with the driver, his workload may be too high. E.g., simultaneous interaction with different driver-assistance systems like route guidance or A3C while driving may overburden the drivers. In other words, to achieve a *moderate workload*, interaction with systems in vehicles has to be minimal. This avoids unnecessary distraction of the driver. It is obvious that agent technology can avoid such distraction. DAAs are able to negotiate time slots with IAs autonomously, in line with the goals specified by the driver.

Law

The proposed intersection-control system also faces legal challenges like liability, traffic regulations, and privacy and anonymity.

Liability (CL1)

If an ITC system as described in this chapter does not work properly, the automotive industry may be liable. But vendors will only launch systems for which the risk is limited. Because we assume a safe design of such a system to be possible, we also expect liability risks to be limited.

Traffic Regulations (CL2)

Some of today's national and international traffic regulations seem to prohibit technical systems which the driver cannot overrule (e.g., UNECE, 2006, Article 8(5)). But advances in automotive technologies increase the pressure on lawmakers to adapt such traffic regulations. In the Section "Application Scenarios" we also examine closed traffic areas where such traffic regulations do not have to hold.

Privacy and Anonymity (CL3)

Because DAAs have to interact with IAs, drivers may loose their anonymity. If DAAs pay money to the IA, traceability on the one hand and anonymity of the driver on the other hand, which is wanted as well, are in conflict. Further, valuation-aware intersection control leaves traces which could be collected to generate travel profiles of road users. The negotiation may also include private data like account data, which should not be accessible for third parties.

Absolute anonymity is difficult and typically not desired by governments. Vehicles are equipped with license plates just to make them traceable. It is also challenging to exclude the possibility of generating travel profiles in any case. Governments may even want to use such data for law enforcement. However, 'random' individuals or organizations should not be able to identify a driver and access his private data. It is difficult to ensure both privacy and anonymity as well as traceability of drivers, e.g., for law enforcement. But each country can decide on an appropriate trade-off.

Discussion

In this section we have described various challenges posed by valuation-aware ITC systems. While none of them seems to prevent the start of operations of such systems in any case, some challenges may only be solved in the future. We show in Section "Application Scenarios" how this can be achieved earlier in certain application scenarios.

INTERSECTION-CONTROL PERSPECTIVE

In the following we describe the intersection-control perspective of the ITC system. The intersection-control unit is the part of the ITC system which is located at the intersection. It uses valuation-aware mechanisms. Such a mechanism consists of a contact step, a reservation and notification step, and an optional modification step.

Contact Step

The driver chooses the initial route before the trip. The route determines all intersection lanes to cross. At each intersection along the route, all DAAs entering the neighborhood of the intersection must establish contact with the corresponding IA. In order to determine an appropriate time slot the IA needs the following parameters: a unique id of the vehicle, the intersection lane desired, the entrance time desired, the length of the vehicle, the crossing speed desired, the acceleration and deceleration preferences, and the unique id of the preceding vehicle. The valuation is not mandatory to compute a time slot for a vehicle. In the mechanisms we present, the DAA reveals its valuation or proposes an offer for a certain time slot in the following reservation and notification step.

The unique id of the vehicle is necessary to communicate and to distinguish vehicles. The intersection lane to use is necessary to reserve the relevant zones of the intersection for the vehicle. The IA may offer the DAA a time slot for an alternative intersection lane, i.e., an intersection lane which connects the same incoming direction to the same outgoing direction. The desired entrance time is the earliest possible point of time when the vehicle can reach the intersection. Time slots with an earlier entrance time are not appropriate. The IA uses the length of the vehicle, the desired crossing speed and the acceleration and deceleration preferences to compute the time slot to offer. Longer and slower vehicles need more time to cross an intersection than shorter and faster ones. The acceleration and deceleration preferences allow altering the time slot further to meet the needs of the drivers. This could mean that the IA computes an acceleration profile for a vehicle. This profile influences the duration of the time slot. The vehicle has to follow the profile in order to cross the intersection.

For each incoming lane the IA maintains a queue of vehicles approaching the intersection. Because it cannot be guaranteed that the contact messages arrive in the order the vehicles approach the intersection, the IA has to use further information to maintain the queues. This is because vehicles are not allowed to overtake in the neighborhood of an intersection. We propose using the unique id of the preceding vehicle to check the order for each incoming lane.

Each DAA knows the specifics of the type of its vehicle. It might know the intersection lane to use from the route-guidance system. It knows the earliest entrance time and the desired crossing speed if the driver-assistance system has the capabilities of an A3C system. The driver can choose all other preferences before the trip.

Reservation and Notification

After DAAs and IA have made contact, different valuation-aware mechanisms to reserve time slots and notify DAAs about their time slots are conceivable. In the following we give an overview of such mechanisms and say how time slots can actually be reserved to avoid the assignment of conflicting time slots to different vehicles.

Notification

Different mechanisms to determine the time slot for each vehicle are conceivable. Examples are the valuation-aware mechanism Initial Time-Slot Auction (Schepperle & Böhm, 2007) and its variants Free Choice and Clocked (Schepperle & Böhm, 2008). After DAAs have made contact with the IA, the IA initiates an auction for the next free time slot periodically. The time when an auction is initiated influences its outcome. If it is too early, late arriving vehicles with high valuations cannot participate in the auction. If it is too late, vehicles may not be able to adapt their speed avoiding standstill. In this case, vehicles which have to stop before the intersection may not be able to enter the intersection in time because they need more time to accelerate from standstill. A more thorough discussion of this issue is beyond the scope of this chapter. We refer to all vehicles which are in the neighborhood, which have no time slot so far, and whose preceding vehicle has already received a time slot or has crossed the intersection as *candidates*. The IA initiates an auction by sending a call for bids to candidates. Each call for bids contains the next free time slot for the candidate. The time slot offered may be different for each candidate. It must not conflict with any time slot already reserved. The candidates reply with their bid. Finally, the IA reserves the time slot offered to the candidate with the highest bid and notifies this candidate.

Using this auction, subsequent vehicles with a higher valuation of reduced waiting time driving behind another vehicle with a very low valuation may have to wait a long time because the preceding vehicle looses several auctions in a row. To overcome this problem, the IA can allow these vehicles to subsidize the candidate in front of them. It does not only send a call for bids to the candidates, but also a call for subsidy to all DAAs waiting behind candidates. They can decide to subsidize the candidate in front. The IA accumulates the bids and subsidies for each incoming lane and returns the time slot to the candidate with the highest accumulated bid.

Note that candidates may offer less than their true valuation. I.e., the mechanism may lack incentive compatibility and therefore not optimize allocative efficiency. A simple second-price sealed-bid (Vickrey) auction is incentive-compatible (Dash et al., 2003). But even in a second-price auction, candidates might hope for subsidies by subsequent vehicles and bid less. Further, the IA executes sequential auctions. This means that candidates not winning in an auction have a chance to win the same time slot in the next auction. Thus, even without subsidies, candidates are tempted to offer less than their true valuation (Krishna, 2002).

Reservation

A time slot can only be sent to a vehicle if it does not conflict with time slots which have already been sent to other vehicles. Therefore, the IA always reserves a time slot before giving it to a DAA. It only offers time slots to a DAA which do not conflict with any time slot reserved. The actual reservation of a time slot depends on the degree of concurrency used by the IA. We distinguish four degrees of concurrency: intersection exclusive, lane exclusive, lane shared, and conflict-area exclusive.

The reservation of a slot demands that some zones of the intersection are *allocated* for a certain period of time. The vehicle which owns the slot can use the zones allocated exclusively. Other zones may be *blocked*. Blocked zones cannot be used by any vehicle. A zone can be blocked several times due to reservations of different slots. A zone already allocated cannot be blocked. All zones which are neither

Figure 3. IE

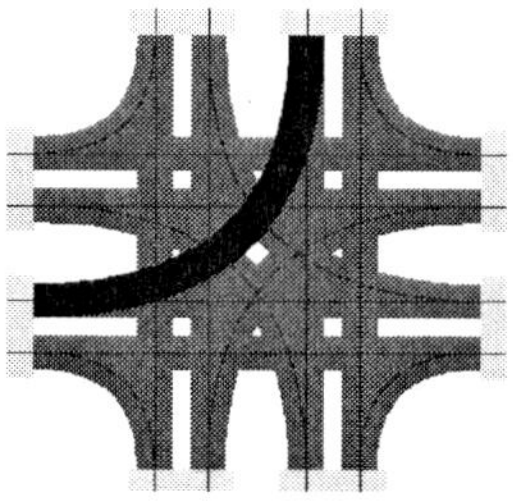

Figure 4. LE

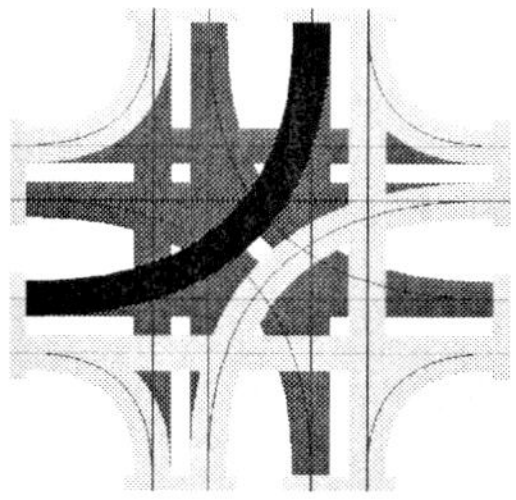

Figure 5. LS

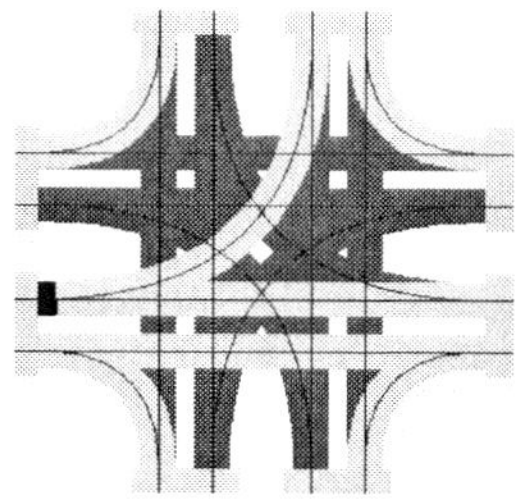

Figure 6. CAE

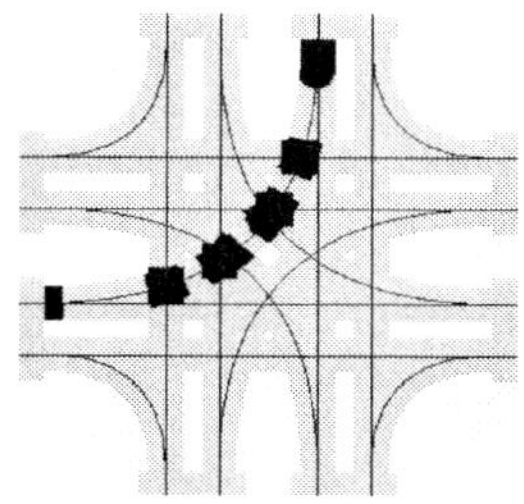

allocated nor blocked are free and can still be allocated or blocked. Figures 3-6 show an example for different degrees of concurrency where a time slot is allocated to a left-turning vehicle arriving from the left. We use black to indicate allocated zones, dark grey for blocked zones and light grey for free zones.

Intersection exclusive (IE) does not allow for any concurrency. Only one vehicle may cross the intersection per time. While the vehicle is crossing the intersection, the intersection lane used is allocated, and all other intersection lanes are blocked (Figure 3).

Lane exclusive (LE) blocks intersection lanes which have conflict areas with the intersection lane used while the vehicle is crossing the intersection. During this time, the whole intersection lane used is allocated. Non-conflicting intersection lanes remain free (Figure 4).

For the time a vehicle takes to cross the intersection, *lane shared* (LS) blocks all intersection lanes which have crossing or merging conflict areas with the intersection lane used. These are all conflicting intersection lanes with an incoming lane different from the intersection lane used. Note that this is different from lane exclusive where intersection lanes with diverging conflict areas are blocked as well, and subsequent vehicles have to wait until the preceding vehicle has left the intersection. The idea is that vehicles approaching on the same lane do not block each other unnecessarily. However, to avoid that vehicles following each other receive time slots with the same entrance time, the first part of the intersection lane used must be allocated for the minimum time gap of successive vehicles. If the first conflict area on the intersection lane is a diverging conflict area, the connector-conflict area on the intersection lane used is allocated, and the corresponding connector conflict area on the conflicting intersection lane is blocked for this minimum time gap of successive vehicles (Figure 5).

Lane shared (LS) is similar to the way how traffic lights are organized. While one traffic stream (from one incoming lane) has a green light, all other (conflicting) traffic streams have a red light. It depends on the actual design of the traffic light if non-conflicting traffic streams have a green light in the meantime or not, and how the green phase switches among the different traffic streams.

Conflict-area exclusive (CAE) allows not more than one vehicle to cross a conflict area. For each conflict area on the intersection lane used, the connector-conflict area is allocated for the time the vehicle needs to pass it. The corresponding connector-conflict area on the conflicting intersection lane is blocked for the same time. All other parts of the intersection remain free. This means that the vehicle uses only conflict areas exclusively (Figure 6).

From IE to CAE, concurrency and throughput increase, but safety decreases. Clearly, a high degree of concurrency is desirable. But the ITC system must take the limited capabilities of human drivers like reaction times or deviations from the speed proposed into account. Deviations from the entrance time and speed could cause accidents. The higher the degree of concurrency, the higher is the necessity not to deviate from the entrance time and speed. If the DAA only gives recommendations, this may be difficult for drivers with conflict-area exclusive (CAE) in particular. This is because the allocation and blocking times for conflict areas are short in this case. Thus, we either have to restrict the degree of concurrency to, say, lane shared (LS), or we have to use driver-assistance systems which can at least adapt the speed of the vehicle autonomously. This is why we propose to have an adaptive cruise and crossing control (A3C) system which can adapt the speed of the vehicle to the time slot even with conflict-area exclusive (CAE). We use intersection exclusive (IE), lane exclusive (LE), and lane shared (LS) to compare mechanisms in more restricted but safer environments.

Modification Step

After DAAs have received a time slot they may try to receive a better one. Different mechanisms to facilitate this are conceivable, e.g., sale, purchase, or again auctions (Schepperle et al., 2006). Another example is Time-Slot Exchange (Schepperle et al., 2008). An exchange is different from sale or purchase because all participating vehicles own a valid time slot both before and after the exchange. If a vehicle approaches the intersection, the DAA may look for time slots earlier than the one already received. In this case, the DAA asks a so-called exchange agent to arrange an exchange of time slots with another DAA. Like the IA, the exchange agent belongs to the intersection-control unit.

The DAA a_e informs the exchange agent about its valuation, i.e., the price it would be willing to pay for an earlier time slot. The exchange agent contacts the IA and looks for DAAs a_p which have a time slot with the following characteristics:

1. the time slot of a_p is earlier than the time slot of a_e but not too early,
2. the time slot of a_p is later than the time slot of the preceding vehicle of a_e,
3. the time slot of a_e is earlier than the time slot of the vehicle following a_p, and
4. an exchange of both time slots does not conflict with time slots of other vehicles already reserved which may cross the intersection at the same time.

If Constraint 1 is not fulfilled, an exchange does not increase the utility of a_e. If the latter constraints are not fulfilled, the exchange is not possible. This is because at least one vehicle could not use its time slot because a preceding vehicle would have a later time slot (Constraint 2, 3), or because the reservation of the time slots exchanged would be impossible (Constraint 4).

Example: *Let the IA use lane exclusive, and Vehicles j and k cross the intersection simultaneously going straight coming from opposite directions. Then Vehicle l passes the intersection, crossing the intersection lanes used by Vehicles j and k before. An exchange of time slots between Vehicles k and l is not possible. This is because the time slot of Vehicle l conflicts with the one of Vehicle j, which does not take part in the exchange.*

If there are DAAs a_p for which all constraints are fulfilled, the exchange agent contacts these agents beginning with the one with the earliest slot. If it accepts the exchange price offered, the exchange is executed. Otherwise, the exchange agent contacts the next DAA. If there are no more agents which fulfill the constraints, the exchange fails, and the requesting vehicle has to stick to its slot.

Clearly, the constraints restrict the number of exchanges. But Schepperle et al. (2008) have shown that, depending on the scenario, Time-Slot Exchange can reduce the average weighted waiting time by up to 15.7% for a heterogeneous intersection layout, compared to a state-of-the-art mechanism which is not valuation-aware.

DRIVER-ASSISTANCE PERSPECTIVE

As discussed, a valuation-aware ITC system can achieve a high degree of concurrency if the speed of the vehicles can be adapted autonomously (see also Section "Reservation"). Thus, we introduce a driver-assistance system as part of the ITC system. We refer to it as *adaptive cruise and crossing control* (A3C) system. We examine how the system computes the appropriate speed in different states.

A3C States

In order to describe the behavior of an A3C system we use three states: The A3C system can be completely switched off (*off*), only provide ACC functionality (*ACC only*) or include the crossing control features as well (*A3C*). Crossing control means that the system also adapts the speed to enter the intersection in time. If the A3C system is in state *off*, it is switched off and does not control the speed of the vehicle. In state *ACC only*, the A3C system only controls the speed to keep the desired time gap, but not the entrance time of the time slot. If the A3C system is in state *A3C*, it controls both the desired time gap and the crossing time of an intersection. We distinguish the following four substates (see Figure 7) of State *A3C*: If the vehicle is not in the neighborhood of an intersection, we call this state *unaffected (U)* because the system is unaffected by the intersection control. After the vehicle has entered the neighborhood passing a traffic sign indicating an agent-controlled intersection it is *approaching without a time slot (A-)*. The vehicle is *approaching with a time slot (A+)* as soon as it has received one. After the vehicle has entered the intersection, the state is *crossing (C)*. After leaving the intersection, the vehicle is unaffected.

Figure 7. The substates of the state A3C

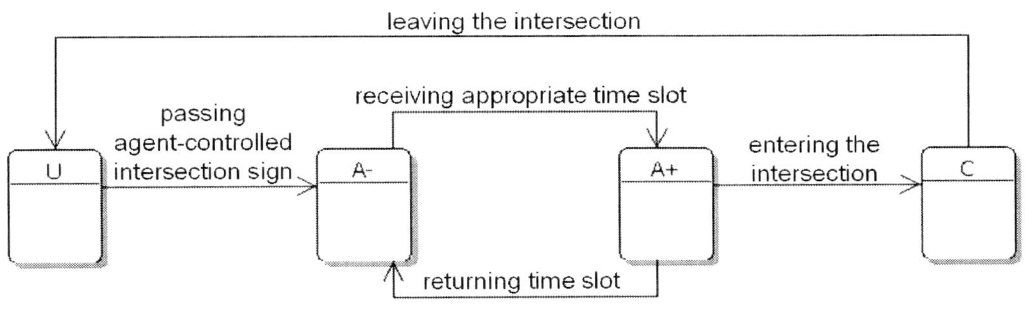

Parameters

We describe the preferences of the driver using the following parameters. They can be different in each state. The driver has a *desired speed* v_{desire} and a *desired time gap* g_{desire}. He prefers a certain (smooth) acceleration a_s and deceleration d_s for comfortable driving, but also accepts accelerations and decelerations up to a higher (hard) acceleration a_h and deceleration d_h. The A3C system computes the *adaptive cruise and crossing control speed* v_{A3C} computing the minimum of the *adaptive cruise-control speed* v_{ACC} and the *adaptive crossing-control speed* v_{CC}. v_{ACC} is the necessary speed to keep the time-gap relationship to the preceding vehicle. v_{CC} is the speed to reach the intersection in time. In any case the speed recommended by the A3C system must not exceed the *speed limit* v_{limit} and the desired speed v_{desire}. Thus, $v_{A3C} = \min(v_{ACC}, v_{CC}, v_{limit}, v_{desire})$.

Driving Strategies

In order to achieve a safe and efficient flow of traffic, we propose the following driving strategies for the A3C system. These strategies define how to compute the crossing-control speed v_{CC} in each substate of State *A3C*. Suppliers of the automotive industry may implement different strategies for their own A3C systems.

Unaffected

If the vehicle is in the state *unaffected* (*U*) it should drive – if possible – with the desired speed, i.e., $v_{CC} = \min(v_{limit}, v_{desire})$.

Approaching without a Time Slot

By passing a certain traffic sign indicating an agent-controlled intersection, the A3C system switches to State *approaching without a time slot* (*A-*). In this state the vehicle stops before the intersection by all means. As long as the stop is still possible, it should drive with the desired speed of the driver, i.e., $v_{CC} = \min(v_{limit}, v_{desire})$. The necessary deceleration depends on the current distance to the intersection. If the deceleration necessary is lower than the preferred smooth deceleration d_s, the A3C system should still drive with the desired speed. As soon as the necessary deceleration is greater than or equal to the preferred smooth deceleration d_s and less than or equal to the hard deceleration d_h, the system should decelerate with the necessary deceleration. We call this driving strategy *late deceleration*.

Approaching with a Time Slot

As soon as the vehicle receives a time slot, it does not immediately switch to *approaching with a time slot* (A+), but computes an acceleration profile to reach the intersection at the entrance time and with the entrance speed required by the time slot. This means that the vehicle does not necessarily drive with the desired speed of the driver any more. The acceleration profile should always use the smooth acceleration and deceleration. If this is not possible, it should use an acceleration/deceleration which is as smooth as possible, and which does not exceed the hard acceleration and deceleration. The acceleration profile also considers the desired speed of the driver and the actual speed limit. If the A3C system cannot compute such a valid profile, it refuses respectively returns the time slot and stays in State *approaching without a time slot (A-)*. Otherwise it accepts the time slot and switches to State *approaching with a time slot* (*A+*). We refer to this driving strategy as *early deceleration*.

If the DAA wants to receive an earlier time slot, e.g., by exchanging the time slot with another DAA, it uses a different driving strategy: The vehicle approaches the intersection with maximum speed $v_C = \min(v_{limit}, v_{desire})$ as long as it can still stop before the intersection. If necessary, it stops and waits until it can use its time slot previously received. This corresponds to the late deceleration driving strategy. If the vehicle drives slower, it risks missing some exchange opportunities because it reaches the intersection later than possible.

The DAA may not be able to accelerate and decelerate as computed in the profile, e.g., because the desired time gap to the preceding vehicle may prevent the vehicle from driving as computed. In this case the vehicle recomputes the acceleration profile. It does so periodically. If the A3C system cannot compute a valid profile only using accelerations and decelerations lower than the hard acceleration and deceleration, it returns the time slot to the intersection and switches to State *approaching without a time slot* (*A-*) again (see also Figure 7). This means that the DAA waits again for time slots and stops at the intersection if necessary.

Crossing

When entering the intersection, the state changes to *crossing* (C). In this state the vehicle behaves analogously to State *approaching with a time slot* (A+). The only difference is that the vehicle computes its acceleration profile not according to the entrance but to the leaving time and leaving speed of its time slot. After leaving the intersection the state switches back to *unaffected* (U).

EVALUATION

Our evaluation consists of simulations. This is in line with other research on intersection control (e.g., Dresner & Stone, 2004; Fürstenberg & Lages, 2005). Figure 8 shows the simulation results of the auction mechanism Free Choice (FC) described in Schepperle and Böhm (2008) compared to the valuation-unaware mechanism Time-Slot Request (TSR) for the three degrees of concurrency

Figure 8. Simulation results of the auction mechanism Free Choice

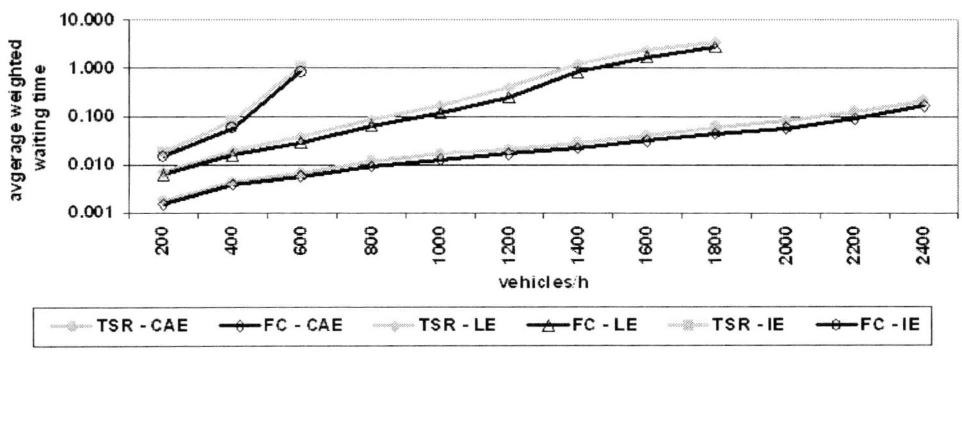

intersection exclusive (IE), lane exclusive (LE) and conflict-area exclusive (CAE). Time-Slot Request is a reimplementation of the reservation-based system by Dresner and Stone (2004). We compare Free Choice to Time-Slot Request because Dresner and Stone (2004) have shown that it already outperforms traffic lights in certain scenarios.

The degree of concurrency limits the throughput of the intersection. Therefore, IE is only evaluated up to 600 vehicles/h, LE up to 1800 vehicles/h and CAE up to 2400 vehicles/h. The results show that the degree of concurrency used mainly determines the average weighted waiting time. The higher the degree of concurrency, the lower the average weighted waiting time. Schepperle and Böhm (2008) show further that Free Choice always outperforms Time-Slot Request. Thus, Free Choice is always effective. The biggest reduction of the average-weighted waiting time of Free Choice for CAE is with 2000 vehicles/h (27.8%), for LE with 1200 vehicles/h (38.1%) and for IE with 480 vehicles/h (34.5%).

APPLICATION SCENARIOS

The range of application for valuation-aware ITC systems is broad. Next to road intersections both in urban and in highway traffic there are also settings where some of the challenges discussed are much easier to overcome. In the following we present some examples.

Closed Traffic Areas

Most of the challenges discussed do not occur in closed traffic areas like company premises, or transship areas of harbors or airports. Other challenges can be resolved much easier. In those traffic areas vehicles belong to one organization. The organization can decide to equip all of its vehicles and its premises with the necessary technology for valuation-aware intersection control. In this case, the organization can avoid heterogeneous environments (CT3), and market penetration is not an issue any more (CE2). The number of vehicles is limited and known in advance. Thus, the necessary robustness level of inter-vehicle communication (CI1) can be achieved much easier. Only members of the organization move in closed traffic areas, thus security attacks (CI2) are less likely. Closed traffic areas also resolve several legal challenges. There is no need for privacy and anonymity (CL3) among vehicles of the same organization. Closed traffic areas are not necessarily covered by road traffic regulations (CL2). This means that exceptions from such regulations may ease the introduction of valuation-aware A3C systems.

Example: *In the HHLA Container Terminal Altenwerder at the Port of Hamburg, Germany, containers are already handled using automated guided vehicles (AGV) (HHLA, 2008). Our system for valuation-aware intersection control is applicable to the intersections crossed by these vehicles. In this case, IAs could prioritize urgent containers with higher valuations.*

In an application scenario with vehicles guided automatically traffic safety (CT2) is still important, but accidents of automatically guided vehicles do not lead to injured human drivers. We do not have to deal with issues like user acceptance (CU1) and impact on driving behavior (CU2). Because humans cannot be injured, liability (CL1) has to cover only damages and production downtimes.

In closed traffic areas in particular, fully centralized approaches may be also applicable and even lead to more efficient solutions. In this case an optimal schedule is planned in advance. But this is not always feasible: The valuation of vehicles may not be known in advance or may change over time. Further, the failure of a centralized component could disrupt the whole system. In our decentralized approach the failure of an IA disrupts only the intersections handled by this IA. All other intersections will still operate.

Alternative Routes

So far, we discussed only scenarios where an agent-based driver-assistance system was mandatory for all road users because vehicles had no choice to use another route. If at least one alternative exists, vehicles can choose between an agent-controlled intersection and an intersection for which an agent-based driver-assistance system is not mandatory. Using these scenarios we can easily deal with a heterogeneous environment (CT3) and only need a lower level of market penetration (CE2). This also is an incentive for drivers to upgrade their vehicle or to buy a vehicle with an agent-based driver-assistance system.

Example: *We restrict the access to a tunnel only to vehicles which are equipped with an appropriate driver-assistance system. All other vehicles have to use the alternative mountain pass. The number of vehicles may even be limited for the tunnel. This would also make travel times more reliable for vehicles using the tunnel. This means that vehicles with an agent-based driver-assistance system could negotiate time slots to use the tunnel, analogously to crossing an intersection. Vehicles which do not receive appropriate time slots have to use – together with the vehicles without an agent-based driver-assistance system – the mountain pass.*

CONCLUSION

In this chapter we have proposed a novel approach for agent-based valuation-aware intersection control. Driver-assistance agents and intersection agents negotiate the right to cross an intersection. Valuation-aware systems consider the valuations of the road users of reduced waiting time and give priority to those with high valuations. Such systems can increase overall satisfaction of road users. We have discussed challenges and potential solutions related to traffic engineering, information technology, economy, road users, and law. We have shown how to combine valuation-aware mechanisms with novel adaptive cruise and crossing control (A3C) systems. This combination allows for a higher degree of concurrent usage of an intersection and leads to a more effective outcome than state-of-the-art intersection-control systems. Next to road intersections, we have examined closed traffic areas and traffic areas with alternatives where some challenges are easier to deal with.

FUTURE RESEARCH DIRECTIONS

The field of valuation-aware intersection control is new and provides many opportunities for further research. So far, it is unclear which strategies driver-assistance agents should use to bid in auctions for time slots. The bids should depend on the intersection, on the time of the auction and probably also on the intersection lane to use and on the necessary time to cross the intersection. Earlier slots are more useful than later ones. Thus, we expect bids for later time slots to be lower. On the other hand, each lost auction increases the waiting time of a vehicle and increases the chance to miss a given deadline. Therefore, we also expect bids for later time slots to increase under certain circumstances.

In general, a vehicle crosses several intersections on a trip. A driver-assistance agent has to keep budgets and time constraints under control when bidding to cross an intersection. Otherwise it risks running out of money or time at the next intersections. Thus, bidding strategies should also take the next intersections on the trip into account.

ACKNOWLEDGMENT

This work is part of the project DAMAST (Driver Assistance using Multi-Agent Systems in Traffic) (http://www.ipd.uni-karlsruhe.de/~damast/) which is partially funded by init innovation in traffic systems AG (http://www.initag.com/).

REFERENCES

Bosch (2008). *ACC for more room.* Retrieved August 4, 2008, from http://rb-k.bosch.de/en/safety_comfort/driving_comfort/driverassistancesystems/adaptivecruisecontrolacc/application_range/index.html

County Surveyors' Society, & Department for Transport (2006). *Puffin crossings. Good practice guide – Release 1.* Retrieved August 4, 2008, from http://www.dft.gov.uk/pgr/roads/tss/gpg/puffingoodpracticeguide01

Dash, R. K., Jennings, N. R., & Parkes, D. C. (2003). Computational-mechanism design: A call to arms. *IEEE Intelligent Systems 18*(6), 40-47.

Dresner, K., & Stone, P. (2004). Multiagent traffic management: A reservation-based intersection control mechanism. In *Proceedings of the Third International Joint Conference on Autonomous Agents and Multiagent Systems* (pp. 530-537). Washington, DC, USA: IEEE Computer Society.

Dresner, K., & Stone, P. (2008). A multiagent approach to autonomous intersection management. *Journal of Artificial Intelligence Research, 31*, 591-656.

Fürstenberg, K. C., & Lages, U. (2005). New European approach for intersection safety - The EC-project INTERSAFE. In *2005 IEEE Intelligent Vehicles Symposium* (pp. 177-180). IEEE.

Furukawa, H., Shiraishi, Y., Inagaki, T., & Watanabe, T. (2003). Mode awareness of a dual-mode adaptive cruise control system. In *IEEE International Conference on Systems, Man, and Cybernetics, 2005, 1*, 832-837. IEEE.

Garber, N. J., & Hoel, L. A. (1988). *Traffic and highway engineering.* St. Paul, USA: West Publishing Company.

GEVAS (2008). Travolution. Retrieved August 4, 2008 from http://www.gevas.eu/index.php?id=65&L=1

Greschner, J., & Gerland, H. E. (2000). Traffic signal priority: Tool to increase service quality and efficiency. In *Proceedings APTA Bus & Paratransit Conference* (pp. 138-143). American Public Transportation Association.

HHLA (2008). HHLA Container Terminals GmbH: A division of Hamburger Hafen und Logistik AG. Retrieved August 4, 2008, from http://www.hhla.de/fileadmin/download/HHLA_Container_Broschuere_ENG.pdf

Ioannou, P. A., & Chien, C. C. (1993). Autonomous intelligent cruise control. *IEEE Transactions on Vehicular Technology, 42*(4), 657-672.

Itoh, M., Sakami, D., & Tanaka, K. (2000). Dependence of human adaptation and risk compensation on modification in level of automation for system safety. In *IEEE International Conference on Systems, Man, and Cybernetics, 2000, 2*, 1295-1300. IEEE.

Krishna, V. (2002). *Auction theory.* London: Academic Press.

Ma, R., & Kaber, D. B. (2005). Situation awareness and workload in driving while using adaptive cruise control and a cell phone. *International Journal of Industrial Ergonomics, 35*(10), 939-953.

Marsden, G., McDonald, M., & Brackstone, M. (2001). Towards an understanding of adaptive cruise control. *Transportation Research Part C: Emerging Technologies 9*(1), 33-51.

Persson, M., Botling, F., Hesslow, E., & Johansson, R. (1999). Stop & go controller for adaptive cruise control. In *Proceedings of the 1999 IEEE International Conference on Control Applications* (pp. 1692-1697), IEEE.

Schepperle, H., Barz, C., Böhm, K., Kunze, J., Laborde, C. M., Seifert, S., & Stockmar, K. (2006). *Auction mechanisms for traffic management.* In *Group Decision and Negotiation (GDN) 2006* (pp. 214-217). Universitätsverlag Karlsruhe.

Schepperle, H., & Böhm, K. (2007). Agent-based traffic control using auctions. In *Cooperative Information Agents XI* (pp. 119-133). Berlin/Heidelberg, Germany: Springer.

Schepperle, H., & Böhm, K. (2008). Auction-based traffic management: Towards effective concurrent utilization of road intersections. In *The 10th IEEE Conference on E-Commerce Technology and the 5th International Conference on Enterprise Computing, E-Commerce and E-Services* (pp. 105-112). IEEE.

Schepperle, H., Böhm, K., & Forster, S. (2008). Traffic management based on negotiations between vehicles – a feasibility demonstration using agents. In *Agent-Mediated Electronic Commerce IX/Trading Agent Design and Analysis*, (pp. 90-104). Berlin/Heidelberg, Germany: Springer.

Torrent-Moreno, M., Killat, M., & Hartenstein, H. (2005). The challenges of robust inter-vehicle communications. In *2005 IEEE 62nd Vehicular Technology Conference, 1*, 319-323. IEEE.

UNECE (2006). Convention on road traffic done at Vienna on 8 November 1968 (2006 consolidated version). Retrieved August 4, 2008, from http://www.unece.org/trans/conventn/Conv_road_traffic_EN.pdf

Ward, N. J. (2000). Automation of task processes: An example of intelligent transportation systems. *Human Factors and Ergonomics in Manufacturing, 10*(4), 395-408.

Chapter XI
Learning Agents for Collaborative Driving

Charles Desjardins
Laval University, Canada

Julien Laumônier
Laval University, Canada

Brahim Chaib-draa
Laval University, Canada

ABSTRACT

This chapter studies the use of agent technology in the domain of vehicle control. More specifically, it illustrates how agents can address the problem of collaborative driving. First, the authors briefly survey the related work in the field of intelligent vehicle control and inter-vehicle cooperation that is part of Intelligent Transportation Systems (ITS) research. Next, they detail how these technologies are especially adapted to the integration, for decision-making, of autonomous agents. In particular, they describe an agent-based cooperative architecture that aims at controlling and coordinating vehicles. In this context, the authors show how reinforcement learning can be used for the design of collaborative driving agents, and they explain why this learning approach is well-suited for the resolution of this problem.

INTRODUCTION

Modern automotive transportation technologies have faced, in recent years, numerous issues resulting from the increase of vehicular traffic and having important consequences on passenger safety, on the environment and on the efficiency of the traffic flow.

In response, both manufacturers and public institutions have focused on such issues through research and development efforts, and have come up with many solutions. Among them, as mentioned in the introductory chapters, the field of Intelligent Transportation Systems (ITS) has gathered particular interest in the past twenty years. This chapter concerns a specific domain of ITS, which aims at designing fully autonomous vehicle controllers.

Many terms have been used to describe this field and its related technologies, such as Collaborative Driving Systems (CDS), Advanced Vehicle Control and Safety Systems (AVCSS) and Automated Vehicle Control Systems (AVCS). According to Bishop (2005), these systems could be defined as Intelligent Vehicle (IV) technology. Bishop characterized IV systems by their use of sensors to perceive their environment and by the fact that they are designed to give assistance to the driver in the operation of the vehicle. This definition of Intelligent Vehicles describes both Autonomous Vehicle Control and Collaborative Driving systems that we consider in this chapter.

Of course, the agent abstraction can be directly adapted to the definition of IV, as agents have the ability to sense their environment and make autonomous decisions to take the right actions. In the past, work related to the problem of autonomous vehicle control has already considered using intelligent agents. What we propose in this chapter is to show how agent technology can be used to design intelligent driving systems. More precisely, we will detail the design of an agent architecture for autonomous and collaborative driving based on the use of reinforcement learning techniques. We intend to show that reinforcement learning can be an efficient technique for learning both low-level vehicle control and high-level vehicle coordination as it enables the design of a controller that can efficiently manage the complexity of the application, i.e. the number of possible vehicle states and the number of coordination situations.

The next section of this chapter surveys the field of autonomous vehicle control and collaborative driving. It also details what has been done in this field in relation to agent technology. The third section briefly explains agent learning techniques while the fourth and final section describes how reinforcement learning can be used to build agents that can drive and coordinate themselves with others autonomously.

SURVEY OF COLLABORATIVE DRIVING SYSTEMS BASED ON AGENT TECHNOLOGY

This section first surveys what has been done in the field of autonomous vehicle control and collaborative driving systems. Then, it describes how the software agent abstraction and machine learning algorithms have already been used in the design of such systems.

Autonomous Vehicle Control and Collaborative Driving Systems

In response to the problems related to the increase of vehicular traffic, most industrialized countries have decided in recent years to adopt a road-map detailing the future of their investments in Intelligent Transportation Systems (ITS) research. Starting in the early '90s, this resulted in the fact that many research projects, often in the form of partnerships between academia and industry, began addressing the design of autonomous vehicle control systems. Research has rapidly led to the development of various applications, as detailed in Table 1.

Already, vehicle manufacturers have integrated some of these technologies in vehicles. For example, many luxury cars are now equipped with Adaptive Cruise Control (ACC) systems, automated parking technologies and even lane-keeping assistance systems. Technologies that have been included in vehicles are, for the moment, used in the form of driving-assistance systems where most of the driving task still belongs to the driver.

Of course, a great amount of research is still being done in this field in order to implement these technologies in consumer products. Clearly, it seems inevitable that the industry will, in a few decades, move towards fully automated vehicles. However, a couple of technological hurdles must be addressed before such sophisticated systems can become a reality.

Currently, the next step towards the implementation of fully autonomous collaborative driving systems is the development of efficient communication technology. Clearly, a robust communication protocol, for both vehicle-to-vehicle (V2V) and road-to-vehicle (R2V) communication, is a pre-requisite for collaboration. As a result, many research institutions have already been working on the development and on the implementation of a standardized communication protocol named DSRC (Dedicated Short-Range Communication). Evidently, a lot of research is still being done in that field, and we refer to Tsugawa (2005) for more details on the state of the art of inter-vehicle communications.

Table 1. Autonomous vehicle control technologies and their description (Bishop, 2005)

Technology	Description
Collision Detection and Avoidance	This technology uses sensors to monitor the surroundings of the vehicle and to detect possible collisions. The driver is alerted of possible accidents. In the future, these systems could even take action directly on the vehicle to avoid collision.
Lane-Keeping Assistance	This technology uses computer vision systems to detect the curvature of the highway. It can react accordingly, with small adjustments to steering, in order to keep the center of the current lane.
Adaptive Cruise Control (ACC)	This technology uses a laser sensor to detect the presence of a front vehicle. The system adapts the vehicle's cruising velocity in order to avoid collision. Once the obstacle is gone, the vehicle goes back to its initial, desired velocity
Cooperative Adaptive Cruise Control (CACC)	This technology adds a communication layer to ACC systems. Information about the acceleration of a front vehicle is shared and is used to reduce the distance between vehicles.
Platooning	This technology takes CACC to the next level by using communication to exchange acceleration data of an important number of vehicles travelling in a platoon formation.
Automated Longitudinal and Lateral Vehicle Control	This technology uses fully automated controllers to act on a vehicle's longitudinal and lateral components.
Collaborative Driving	This technology is the ultimate goal of autonomous vehicle control. It uses inter-vehicle communication in order to share sensor information and driving intentions with surrounding vehicles (not necessarily a platoon) and select an optimal driving action.

Research Projects

Many research projects have been active in the development and design of autonomous vehicle control, collaborative driving and related technologies.

Perhaps the most famous and influential program in this field is the program of the University of California at Berkeley called PATH (Partners for Advanced Transit and Highway). This program regroups numerous research projects that share the ultimate goal of solving the issues of transportation systems through the use of modern technologies. PATH projects have designed and tested an important range of solutions related to vehicle control. They have studied solutions to complex problems such as automated longitudinal (Raza & Ioannou, 1997; Lu et al., 2000) and lateral vehicle control (Peng et al., 1992), cooperative collision warning systems (Sengupta et al., 2007) and platooning (Godbole & Lygeros, 1994; Sheikholeslam & Desoer, 1990). Bana (2001), has also worked on the use of vehicle communications for advanced vehicle coordination. For more details about the history of PATH and its future research directions, we refer to Shladover (2007). Finally, PATH is also famous in part because it has implemented and demonstrated an autonomous platooning control system as early as in 1997, as part of the Demo '97 event (NAHSC, 1998).

Another important research program has been Japan's Advanced Cruise-Assist Highway Systems Research Association (AHSRA). Similarly to PATH, this program has focused on the development of intelligent systems for the infrastructure, but has also worked on Advanced Security Vehicle (ASV) systems which promote the development and integration of intelligent systems in vehicles. The next step of their project consists in linking both types of systems using a communication architecture. Their program is also well-known for its implementation and demonstration of ASV technologies in the Demo2000 (Tsugawa et al., 2000) event. Moreover, since AHSRA regroups many manufacturers, a large number of the technologies developed through this program have rapidly been integrated in Japanese vehicles.

Many European countries have also been active in this field. For instance, recent work at the TNO Automotive (a research institute of The Netherlands) through the CarTalk2000 project, with partners DaimlerChrysler and Siemens, has focused on the development of communication systems and their application to autonomous vehicle control (de Bruin et al., 2004; Hallouzi et al., 2004).

Of course, the projects described here only offer a glimpse of all the research that has been done on this topic. A lot of other research organizations have also financed projects in this field, such as Italy's ARGO (Broggi et al., 1999), Canada's Auto21 (Auto21, 2007) and European's CHAUFFEUR projects (Schulze, 2007), just to name a few.

Design of Intelligent Vehicles Using Agents and Machine Learning

As we have described earlier, the agent abstraction is especially adapted to the problem of automated vehicle control and collaborative driving. It is not surprising to see that a number of research projects have considered using agents to design such control systems. Moreover, since agents need a decision-making mechanism, the use of agents has often been in conjunction with machine learning techniques. This section overviews previous work on the use of agents and machine learning techniques for autonomous vehicle control and coordination.

Machine Learning for Vehicle Control

One of the first interesting applications of machine learning to the problem of vehicle control was Pomerleau's ALVINN (Pomerleau, 1995). Pomerleau has designed a supervised learning system based on computer vision that featured a neural network which received, as inputs from the vision system, patterns representing the road ahead. The task of the network was to learn to match vision patterns to an accurate driving action. Examples were given by watching a real person driving.

The PATH program, through its Bayesian Automated Taxi (BAT) (Forbes et al., 1995) project has also studied the use of agents and machine learning for autonomous driving in traffic. They have shown that the use of a decision theoretic architecture and of dynamic Bayesian networks has produced a good solution to the problems of sensor noise and uncertainty about the other vehicles' behavior.

Later, Forbes also introduced a longitudinal agent controller (Forbes, 2002) based on reinforcement learning. This controller has been compared to a hand-coded controller, and results showed that the hand-coded controller was generally more precise than the learned controller, but was less adaptable in some situations.

Another interesting approach to longitudinal vehicle control was developed by Naranjo et al. (2003) as part of Spain's AUTOPIA project. Naranjo and his colleagues designed a longitudinal controller based on fuzzy logic. Their controller used inter-vehicle communication to share positioning information of a lead vehicle. It was even embedded in a vehicle and tested in demo sessions of the IEEE (Institute of Electrical and Electronics Engineers) Intelligent Vehicles Conference of 2002.

Machine Learning for Vehicle Coordination

The problem of coordination between vehicles has also received much interest from many researchers as this problem is especially adapted to multi-agent learning algorithms.

Among the numerous examples is work by Ünsal et al. (1999). These researchers have tackled the problem by using multiple stochastic learning automata as a mean to control the longitudinal and lateral motion of a single vehicle. Using reinforcement learning, these automata were able to learn to act in order to avoid collisions. The interactions between the automata have been modeled using game theory, with the objective of optimizing the traffic flow.

In his work, Pendrith (2000) presented a distributed variant of the Q-learning algorithm and applied it to a lane change advisory system. The author considered using a local perspective to gather state information, by considering the relative velocities of the surrounding vehicles. Whereas the solution provided by the algorithm increases the traffic efficiency, the problem of this algorithm is the lack of learning stability.

Moriarty & Langley (1998) proposed a traffic management approach where vehicles select by themselves the lane which optimizes the performance of the traffic flow. The authors have used a combination of reinforcement learning and neuro-evolution methods to keep a set of possible strategies for the vehicles. They have shown that their approach optimizes the velocities of the cars while reducing the number of lane changes.

Finally, Blumer et al. (1995) have used a neural network and an expert system to control vehicles from a coordination point of view (changing lanes, joining a platoon, etc.). The neural network was used to classify traffic situations and a reinforcement learning algorithm was used to evaluate the risk of the situation observed in order to choose the adequate action.

Agent Abstraction

Agents are autonomous software entities that try to achieve their goals by interacting with their environment and with other agents (Russell & Norvig, 2002). With their ability for autonomy and social interactions, agents are a logical choice of mechanism to rely on in order to embed in vehicles a deliberative engine adapted for control and collaboration. Indeed, this abstraction is especially adapted to the problem of collaborative driving that we address here, as vehicle controllers must autonomously make decisions in a decentralized manner while interacting with other vehicles in order to reach their goals of optimizing safety and traffic flow efficiency.

The agent abstraction can also be used to model the driving task using a deliberative architecture, and many different approaches have already been considered. For example, Rosenblatt (1995) has proposed a framework based on a centralized arbitration of votes from distributed, independent, asynchronous decision making processes. This framework has been used for obstacle avoidance by vehicles. In related work, Sukthankar et al. (1998) have focused on tactical driving using several agents that are specialized on one particular task (e.g. change lane agent or velocity agent). A voting arbiter aggregates the recommendation of all agents to choose the best vehicle action. Similarly, work by Ehlert (2001) describes tactical driving agents based on the subsumption approach (Brooks, 1991) and uses behavioral robotics to consider the real-time aspects of the driving task.

A different agent architecture has been proposed by Hallé & Chaib-draa (2005). Their work features a deliberative architecture based on team work (Tambe & Zhang, 2000) and is used for platoon management. This approach relies on a three level architecture (Guidance, Management and Traffic) as in PATH's architecture. In Hallé and Chaib-draa's approach, each vehicle is assigned a specific role in the platoon (Leader, Follower, Splitter, etc.) according to its current task. They have also compared their approach to a centralized and a decentralized platoon and they have given advantages and disadvantages of each type of platoon organization.

Of course, the papers we have presented here on the use of machine learning and of the agent abstraction applied to vehicle control and coordination only represent an overview of what has been done in this field. Nonetheless, it clearly illustrates what can be done when applying agent abstraction and machine learning to vehicle control.

LEARNING AND AGENTS

The resolution of the problems of autonomous vehicle control and of collaborative driving using intelligent agents requires the use of methods that are adapted to making decisions in a complex environment. One important problem that agents must face is the presence, in most environments, of uncertainty. In recent years, reinforcement learning has gathered much interest for the resolution of such problems as it can be used in this context to obtain efficient control policies.

Thus, this section will briefly present the Markov Decision Processes (MDP) model and the corresponding reinforcement learning algorithms classically used to find an optimal solution for a single agent. Afterwards, we introduce multi-agent models and describe algorithms that can learn in situations where interaction and coordination between agents is possible.

Markov Decision Processes

To take action, autonomous agents rely on a deliberation mechanism to select the appropriate action to take according to the current perception of the environment. Since driving can be considered as a sequential task where decisions need to be taken at fixed intervals of time, the framework of Markov Decision Processes (MDPs) is an efficient candidate to model this problem. More precisely, MDPs are sequential decision problems in which the goal is to find the best actions to take to maximize the agent's utility (Sutton & Barto, 1998).

The Markov property is needed to find the optimal solution of an MDP via classic dynamic programming or reinforcement learning approaches. This property is satisfied if the current state of the agent encapsulates all knowledge required to make a decision. More precisely, an environment is said to be markovian if its evolution can be described only by the current state and by the current action of the agent.

The resolution of an MDP yields a policy, which is a function that maps states to actions and which actually represents the behaviour of the agents. When the dynamics of the system (represented by the probabilities of going from current state *s* to next state *s'* when taking action *a*) are known, it is possible to use the Value Iteration algorithm (Russell & Norvig, 2002) to obtain a policy that maximizes the expected reward that the agent can obtain when executing it from starting state *s* (this policy is called the optimal policy).

The Q-Learning algorithm is also particularly interesting for the resolution of an MDP. It is a model-free approach that enables an agent to learn to maximize its expected reward without the availability of the transition and the reward functions that both characterize knowledge of the environment. With this algorithm, the agent learns an optimal action policy simply by trying actions in the environment and by observing their results. This algorithm is based on the notion of Q-value *Q(s,a)* which represents the reward an agent can expect to obtain when it is in state *s* and selects action *a*.

The downside of these algorithms is that they face the "curse of dimensionality". This curse refers to the fact that the size of the state space (the number of *Q(s, a)* pairs) can grow exponentially with the number of variables contained in the states and with the number of possible actions. This renders convergence nearly impossible for complex problems. Moreover, the use of a Q-values table means that continuous environments cannot be treated and need to be discretized.

Policy-gradient algorithms can address some of these issues. Instead of updating a value function in order to obtain the optimal function, these algorithms work by updating directly a parameterized stochastic policy according to the gradient of a policy's performance with respect to the parameters (the performance of a policy is generally defined as the expected reward one can get by following this policy). The advantages of these methods are that they can easily treat continuous state variables and that there is no problem related to the growth of the state space. For more details, we refer the reader to both Baxter & Bartlett (2001) and Williams (1992), as these authors make a good overview of this family of learning algorithms.

Multiple Agents

When multiple agents are involved, their interactions need to be handled since each agent needs to take into account the actions of others for efficient action selection. Usually, we can distinguish two cases: cooperative interactions, where all agents share the same goals, and non-cooperative interactions, where agents may have different or even opposite goals. In this section, we will only focus on the cooperative case.

When several cooperative agents act in the same environment, a decentralized MDP (DEC-MDP) can be used to describe the interaction of these agents (Bernstein et al., 2002). DEC-MDPs adapt some concepts of MDPs to deal with multiple agents and partially observable domains. In DEC-MDPs, observations have a special property: each agent can observe only a part of the current system state and each joint observation corresponds to a unique system state. Note that in this model, any optimal solution maximizes the social welfare, i.e. the sum of all agent rewards.

As far as we know, there exists no reinforcement learning algorithm that can find an optimal solution of a DEC-MDP without knowing the model of the environment. All working algorithms are based on dynamic programming (Bertsekas, 2000) and can only solve problems of small size because the DEC-MDP model is known as being an intractable problem (Bernstein et al., 2002). However, when agents are able to exactly observe the global state of the environment, the Friend Q-Learning algorithm introduced by Littman (2001) allows building an optimal policy for all agents.

Notice that even if this algorithm converges to the optimal joint policy, agents need some information about the others in order to achieve a good coordination. In general, individual states, individual actions and sometimes individual rewards need to be transmitted by communication between agents so that they can learn good policies. This multi-agent learning algorithm will be used later as part of the layer that manages vehicle coordination.

DESIGN OF COLLABORATIVE DRIVING AGENTS

In this section, we present how agents making decisions based on reinforcement learning algorithms can be used to design an autonomous vehicle controller and a collaborative driving system. First, we present our architecture and the different layers it relies on to manage vehicle control. Then, we detail the design of both a low-level vehicle controller and a high-level coordination module. Finally, we describe the results we obtained by executing the policies learned for both modules.

Architecture Design

For the past thirty years, manufacturers have integrated classic Cruise Control (CC) systems into vehicles to automatically maintain a driver's desired cruising velocity. More recently, constructors have introduced Adaptive Cruise Control (ACC) systems that make use of sensors to detect the presence of obstacles in front of a vehicle (Bishop, 2005). These systems are designed to react automatically to obstacles by taking direct action on the vehicle to adjust its current velocity in order to keep a safe distance behind the preceding vehicle.

Cooperative Adaptive Cruise Control systems (CACC), which integrate the use of inter-vehicle communication in the control loop, are often seen as the next step towards autonomous control systems (Bishop, 2005). These systems use wireless communication for the broadcast of positioning, velocity, acceleration and heading information to other vehicles nearby, to improve the receiver's awareness of the environment. By providing this extra information that would normally be out of the range of standard sensors, communication helps vehicles make better driving decisions and increase both traffic efficiency and safety.

In particular, CACC systems benefit from the use of communication to assure the string stability of a group of vehicles. This expression signifies that vehicles do not propagate and amplify perturbations of a front vehicle's velocity. Thus, for example, vehicles do not have to brake more than the preceding vehicle when observing changes in velocity. Non-stability eventually leads to vehicles needing to brake to a stand-still in order to avoid collision, which is often what causes traffic jams. Sheikholeslam and Desoer (1990) have showed that communicating acceleration actions of preceding vehicles through inter-vehicle communication is necessary to observe the stability of a stream of vehicles separated by constant spacing.

The Cooperative Adaptive Cruise Control (CACC) architecture presented here is thus based on this previous work of the automotive industry on vehicle control. The system, which is described in more detail in work by Desjardins et al. (2007), is actually an autonomous, intelligent agent that takes decisions in order to control the vehicle. This agent relies on two layers for decision-making and on a communication module to interact with other vehicles.

The two control layers work at different abstraction levels yet are complementary at coordinating interactions and at achieving cooperation between vehicles. First, the *Coordination Layer* is responsible for the selection of high-level driving actions. It uses information from other communicating vehicles to select an action that is the best response it can take according to the other vehicles' actions in order to maximize local and global security and traffic efficiency criteria. When such an action has been chosen, the low-level vehicle controller, also named the *Action Layer*, is responsible for selecting an action that has a direct effect on the vehicle's actuators. Figure 1 shows how our CACC architecture acts as part of the basic control loop of the navigational system of a vehicle.

When the current low-level action has terminated (either by success of by failure), the *Action Layer* notifies the *Coordination Layer*. This termination is then broadcast to the neighborhood to inform other vehicles. When all neighbors of a vehicle have finished their respective action, the *Coordination Layer* is

Figure 1. CACC system architecture

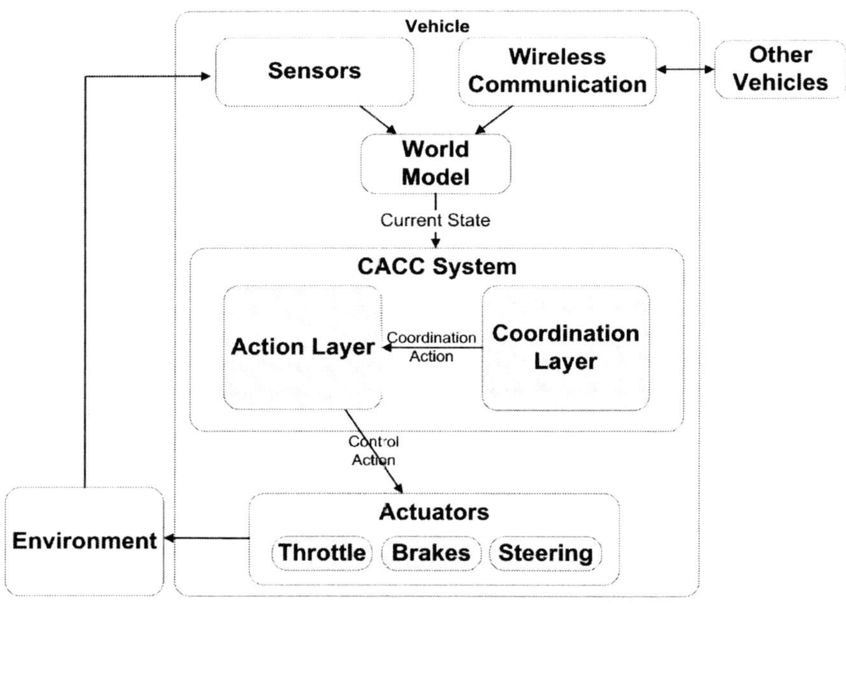

Figure 2. CACC system architecture interactions

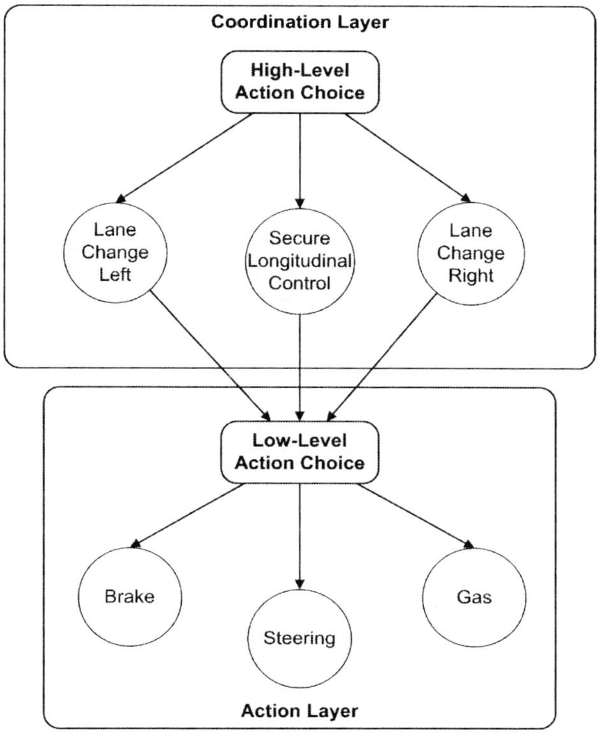

able to take another coordination action according to the current state. The state diagram of Figure 2 illustrates the possible transitions that can be triggered by the *Coordination Layer* for a single vehicle.

The exact behavior of each layer has been designed using reinforcement learning algorithms. This learning approach is particularly useful since it allows the agent vehicle to adapt to its environment even if it does not know its dynamics. More specifically, the *Action Layer* uses algorithms to learn the selection of the best low-level actions according to the environment's state in order to achieve the high-level action selected, while the *Coordination Layer* uses learning to optimize the agent interactions.

In the following subsections, we present the design of both layers in detail.

Design of the Action Layer

For the design of our system's *Action Layer*, we have focused on offering a control policy that enables secure longitudinal velocity control. In particular, instead of solving directly the complex problem of Cooperative Adaptive Cruise Control, the work we present here tries to solve a simpler problem by designing an Adaptive Cruise Control (ACC) system, as we intend to show that our approach based on reinforcement learning can lead to good results.

First, we have considered for the inputs (state variables in the MDP framework) of the system the time headway, which gives the distance in time from a front vehicle (as illustrated in Eq. 1), and its difference between two timesteps, which indicates whether the follower has been closing in or going farther from its front vehicle (as given by Eq. 2).

Headway = Hw , Position = Pt , Velocity = V

$$Hw = \frac{(Pt_{Leader} - Pt_{Follower})}{V_{Follower}} \quad (1)$$

$$\Delta Hw = Hw_t - Hw_{t-1} \quad (2)$$

The headway information is perceived by a laser sensor, and detects vehicles in front in a range of up to *120* meters. Through our experiments, we make the hypothesis that there are no delays in the sensory system (as sensor delay will be addressed in future work).

Of course, this ACC state definition can easily be extended by using the communication system to propagate information about the state of surrounding vehicles (position, velocity, acceleration and heading). More specifically, we would like to integrate information about a lead vehicle's acceleration as inputs of this process, so that our system becomes a fully-functional Cooperative Adaptive Cruise Control (CACC) system.

For both the ACC and CACC cases, we will compute the control policies using reinforcement learning. This kind of learning is advantageous and efficient since it enables us to make an abstraction of the vehicle physics but still learn a valuable control policy. This is particularly useful when learning a control policy in an environment containing complex vehicle physics similar to the one used for our experiments (which we briefly detail at the beginning of the Results section).

The reward function we use gives negative rewards when the vehicle is too far from or too close to a secure distance (*2* seconds, a common value in ACC systems (Bishop, 2005)). Positive rewards are given when the vehicle is in the desired range. To direct the exploration of the vehicle to interesting places of the state space, we also give a positive reward to the vehicle if it is too far from the goal but is closing up.

An interesting characteristic of this learning task is that the choice of these state variables was carefully considered. As a result, the behaviour learned does not depend on the current velocity of the vehicles and should generalize to any driving scenario. The only fixed aspect of the controller that would not change with different scenarios is the distance from which the vehicle is following, which depends on the goal region defined by the reward function.

Finally, we also design manually a basic lane change policy, which can be triggered whenever needed by the vehicle *Coordination Layer*. The design of this layer is described in detail in the next section.

Design of the Coordination Layer

The goal of vehicle coordination is to handle dynamically the interactions between cars on the road in order to obtain an intelligent collaborative driving system. To achieve this, the *Coordination Layer* uses policies defined by the *Action Layer* and chooses at each step which policy should be applied in order to improve the coordination. In this subsection, we describe the method we considered for the design of coordination policies. To solve this problem, we use multi-agent learning algorithms and DEC-MDP models, and we introduce the notion of distance of observation between vehicles. Basically, with communication and sensors, each vehicle only has a limited view of its surrounding environment, and can choose an action which will give good coordination results.

More formally, based on the DEC-MDP model described previously, we make assumptions about the observations of the vehicles, splitting these into two categories: observations over world states and observations over actions. Each observation is assumed to be perfect but only for a sub-part of the environment. Moreover, each agent only has a partial view of the other agents and cannot perceive the complete environment to learn the optimal actions. To define these partial views, we define a neighborhood function *neigh*, which returns the set of visible agents at a certain distance of observation from a central agent. Thus, observations are defined by the union of the exact states of visible agents. By this formulation, we assume that the need for coordination is higher when two agents are close than when they are far from each other. We also assume that every agent is in its own neighborhood and if an agent is in the neighborhood of another, the opposite is also true. Note that a maximal distance d_{max} is reached when each agent can observe all other existing agents.

Applied to the vehicle coordination problem, the functions calculating the partial state and the joint action are defined by the sensors and the communication of the vehicles. Figure 3 shows the partial view (state and action) for each vehicle where the global environment state is composed of 3 vehicles V_1, V_2 and V_3. In this figure, s_i^2 represents the partial vision of the vehicle *i*, which explains why the view s_3^2 is centered on Vehicle 3. Since the road is modeled as a ring, Vehicle 3 can observe Vehicle 2 in front of it and can observe Vehicle 1 behind it. At each step, the agent receives the information from other vehicles (velocities, positions) and the actions that have been chosen for the next step of the interaction.

Once all information needed to construct a partial state and joint action is received, the *Coordination Layer* decides to act by sending its command to the low-level vehicle controller. A vehicle can be ordered to follow the preceding vehicle, to keep a constant velocity or to change lanes to the right or to the left. All these actions correspond to the policies offered by the *Action Layer*.

Since the resolution of a DEC-MDP is known as an intractable problem, we will rather present an algorithm which finds an approximated joint policy using the distance of observation. Our algorithm,

Figure 3. Joint and partial states of a vehicle coordination scenario for a distance of observation of 2

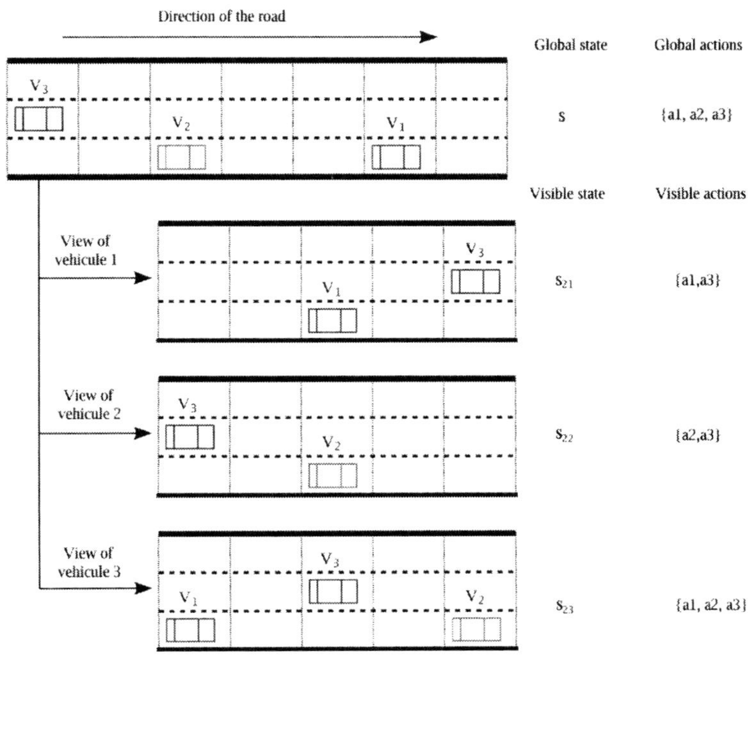

called Partial Friend Q-Learning (PFQ), is based on Friend Q-Learning (Littman, 2001), a multi-agent version of Q-Learning. The basic idea is to apply Friend Q-Learning on partial views and partial joint actions instead of on fully observable states and joint actions to limit the number of possible $Q(s,a)$ pairs. At each step, the agent chooses its action contained in the joint action that maximizes the Q-Value in the current state. Then, it observes partial states, partial joint actions and rewards, and updates the Q-Value as usual. In the end, the algorithm computes a policy π^d for each agent and for a fixed distance *d*. Further details of the coordination approach can be found in Laumônier & Chaib-draa (2006). From a vehicle coordination point of view, this algorithm allows us to take into account only a limited part of the environment by neglecting the influence of cars farther away. Thus, changes in the environment far away have no influence on the resulting policy.

Results

To test our architecture, we designed a microscopic traffic simulator in which vehicles are accurately modeled. It features vehicle physics and dynamics based on a single-track model (Kiencke & Nielsen, 2000). This model integrates both longitudinal and lateral vehicle movements and uses a wheel model that is complex enough to simulate with precision the behavior of a vehicle.

The simulator also includes an inter-vehicle communication system and a sensory system in order for vehicles to perceive their environment. The inter-vehicle communication module is a pre-requisite to an efficient CACC system as it makes possible extensive cooperation between vehicles. Both the *Action Layer* and the *Coordination Layer* rely on this module to share information and achieve good performance. The communication layer is loosely based on the DSRC protocol, which addresses many issues related to wireless inter-vehicle communications.

Actions of the vehicles in the simulator are controlled by acting directly on their actuators. This means that the longitudinal actions available to vehicles are to accelerate or to brake by pressing on the corresponding pedal. It is also possible not to take an action at the current time. As for the use of the steering wheel, it leads to the possible lateral actions of the vehicle. Before selecting a driving action, the *Action* and *Coordination Layers* can use sensors and communication to perceive the environment. We make the hypothesis that there are no delays or noise in the system whether it is from sensors, actuators or communication. As explained in the conclusion below, taking care of the issues of sensor delay and noise will be addressed in the following steps of the development of our architecture.

This simulation environment was used for learning control and coordination policies for both the *Action* and *Coordination Layers* of our system. How these experiments were done exactly for each layer is described in the following sections.

Vehicle Control

The *Action Layer* used for low-level vehicle control is designed using reinforcement learning. To obtain a control policy, we put the controller in "learning mode" in our simulated environment.

We tested a "Stop & Go" scenario where a leading vehicle accelerates to a velocity of *20 m/s*, slows down to *7 m/s* and then accelerates again, this time to a *20 m/s* cruising velocity. Our learning agent had to try actions in order to find the best longitudinal following policy. The goal was to reach a secure distance of *2* seconds behind a preceding vehicle, using only a front sensor, which effectively models the behavior of an ACC system.

The learning task definition corresponded exactly to what was presented in the "Design of the Action Layer" section. Experiments to learn an efficient control policy have been done using the OLPOMDP policy-gradient algorithm (Baxter & Bartlett, 2001). This reinforcement learning algorithm generates a stochastic parameterized policy (a policy that returns probabilities of selecting the actions in a particular state). To represent this policy, we have used a neural network, and the parameters of the policy are actually the weights of the network. As a result, the algorithm modifies the network's weights in order to increase the probability of selecting the actions that give us positive rewards.

Figure 4 illustrates data related to the execution of *10* learning simulations of *5,000* episodes. Since the algorithm is actually a stochastic gradient descent method, multiple learning simulations were needed in order to compare the resulting policies. Thus, the figure shows the worst, the average and the best policy obtained through the learning phase. Figure 4 also illustrates the fact that the learning algorithm did optimize the number of steps in which the vehicle is located in the desired "safe" region, as, by the end of the learning episodes, the vehicle is in the goal region for approximately *475* steps over *500*, which can be considered as a near-optimal behavior.

After the learning phase, we executed a "Stop & Go" scenario with two vehicles, the follower being controlled by using the learned ACC policy. Figure 5 illustrates the velocities of both vehicles during this execution scenario. This figure illustrates the fact that the learned policy was able to precisely match the velocity of the front vehicle, even when it did accelerate or brake.

Furthermore, Figure 6 shows the associated headway metric of the second vehicle during the execution scenario. It clearly shows that the learned policy resulted in an efficient behavior, with the headway oscillating closely around the desired value for the duration of the simulation.

Figure 4. ACC learning results

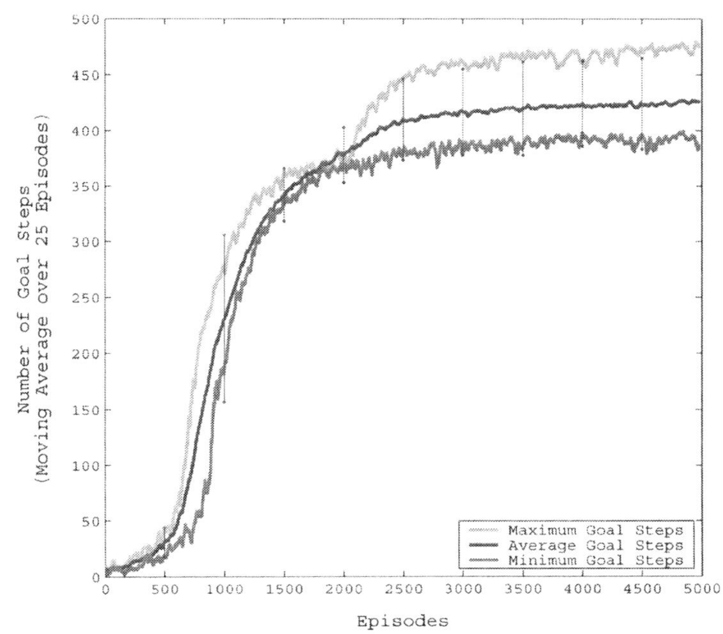

Work still needs to be done to achieve our goal of designing a complete longitudinal CACC controller but, for now, the results we have obtained with our Adaptive Cruise Control (ACC) system show that reinforcement learning can be used to provide efficient vehicle following controllers.

Vehicle Coordination

The PFQ algorithm has been tested on a simplified three vehicles scenario as described in Figure 3, where each vehicle had to choose the best lane in order to optimize the velocity of every vehicle. This coordination scenario uses simpler dynamics than the single track model. Moreover, we discretize the positions and velocities of the vehicles and, for each car, we note Y the longitudinal position (in meters, assuming that a car is a $1\ m^2$ square) and X the current lane. We discretize also the velocities to the set $V = 0, 4, 8, 12, 16, 20\ m/s$. Learning coordination allows us to design an efficient controller which can take into account the actions of the other vehicles situated at a close range. Here, we summarize these results to show that each vehicle only needs to observe a subset of the other vehicles, those that are close to itself, to learn a near-optimal coordination policy.

With the results of those simulations, we can compare empirically the performance of a coordination policy learned in a fully-observable environment (using Friend Q-Learning) with the performance of an approximated coordination policy learned using observations of a subset of the environment (using our approach, PFQ). Here, we compare the algorithms on two situations: the scenario S_1 is defined by size $X = 3$, $Y = 7$, by the set of velocities $V = 0, 4, 8, 12, 16, 20\ m/s$ and by the number of agents $N = 3$. In the second scenario S_2, we enlarge the number of lanes and the length of the road ($X = 5$, $Y = 20$, $V = 0, 4, 8, 12, 16, 20\ m/s$ and $N = 3$). Consequently, in these problems, the maximal distance that we can use to approximate the total problem is $d_{max} = 3$ for S_1 and $d_{max} = 10$ for S_2. In the initial state (Figure 3), ve-

Figure 5. ACC vehicle velocities

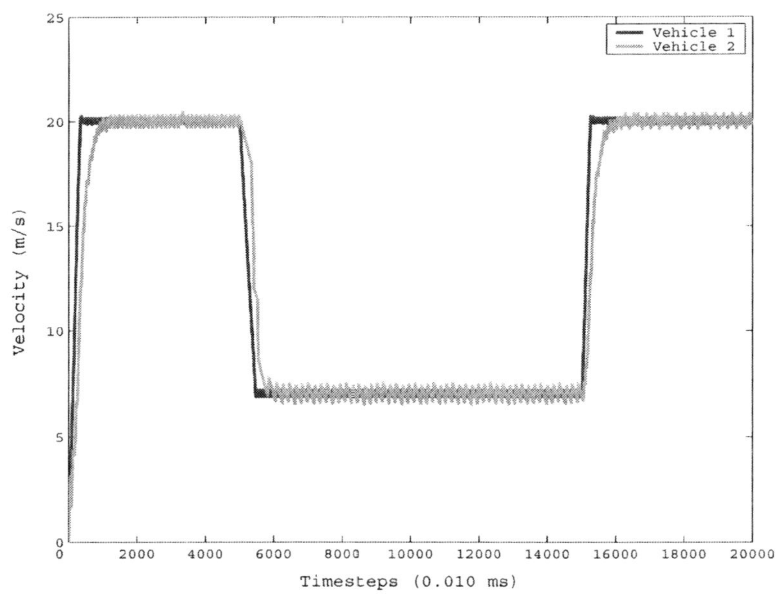

Figure 6. ACC headway results

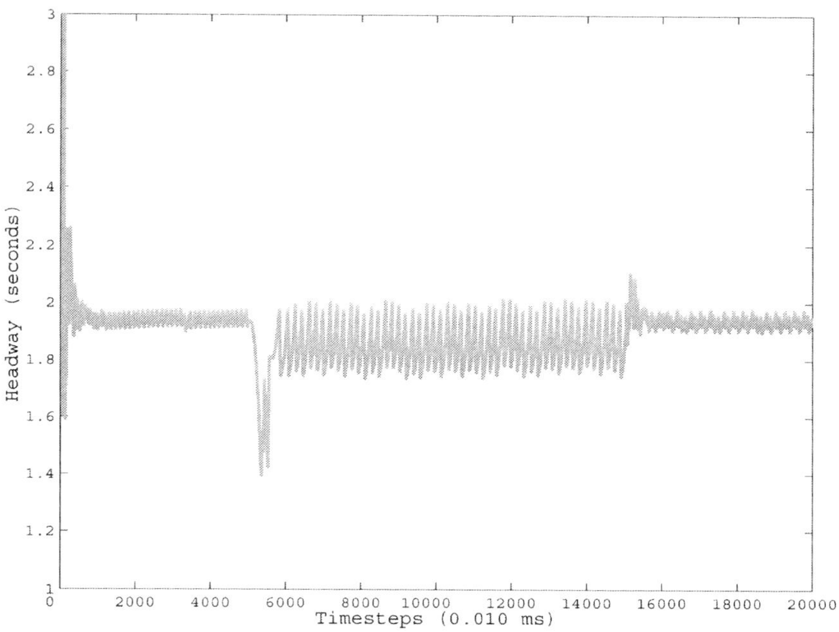

locities of the agents are $V_1 = 4\ m/s$, $V_2 = 8\ m/s$ and $V_3 = 12\ m/s$. We present, for all results, the average velocity of all vehicles, averaged over *25* learning simulations, with each episode lasting *10* steps.

Figure 7 shows the results of PFQ with distance from $d = 0$ to $d = 3$. This algorithm is compared to the total observation problem resolved by Friend Q-learning. For $d = 0$, $d = 1$ and $d = 2$, PFQ converges to a local maximum, which increases with *d*. In these cases, the approximated values are respectively of *76%*, *86%* and *97%* of the optimal velocity. When $d = 3$, which is when the local view is equivalent to the totally observable view, the average velocity converges to the optimal average velocity. Thus, without observing everything around them (distance $d = 2$) vehicles are able to coordinate themselves and learn a near optimal policy while reducing the number of vehicles taken into account in the coordination.

Practically, the observation distance is determined by the distance of communication between vehicles. In general, using communication protocol like DSRC, the distance of communication depends on the density of vehicles. Indeed, in order to keep the number of messages relatively low, vehicles can only send to their close neighbors if there are many vehicles around. By doing this, we limit the number of vehicles taken into account in our reinforcement learning algorithms. This limitation is also coherent with the fact that current communication and sensor systems are not designed to handle the perception of remote vehicles. Consequently, we are able to design a coordination layer with good efficiency limiting the number of states in which the optimal policy should be found. The collaborative driving policy learned using a total distance of observation ($d = 3$) is represented by Figure 8. We can observe that, with this near optimal policy, Vehicle 3 learned to pass Vehicle 2 and Vehicle 1 learned to let Vehicle 2 to pass.

Figure 7. Velocity for partial friend q-learning

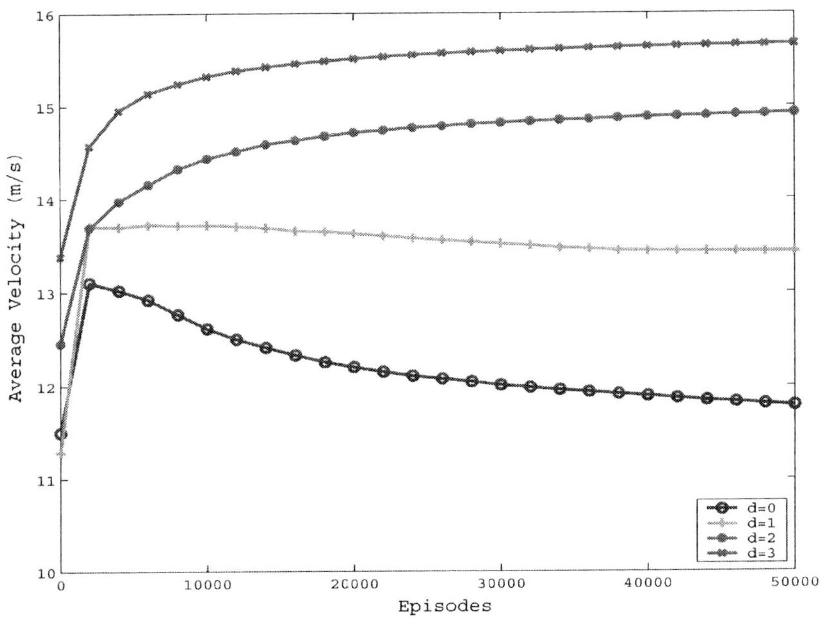

Figure 8. Coordination between 3 vehicles. Vehicle 3 learned to pass Vehicle 2 and Vehicle 1 learned to let Vehicle 2 pass.

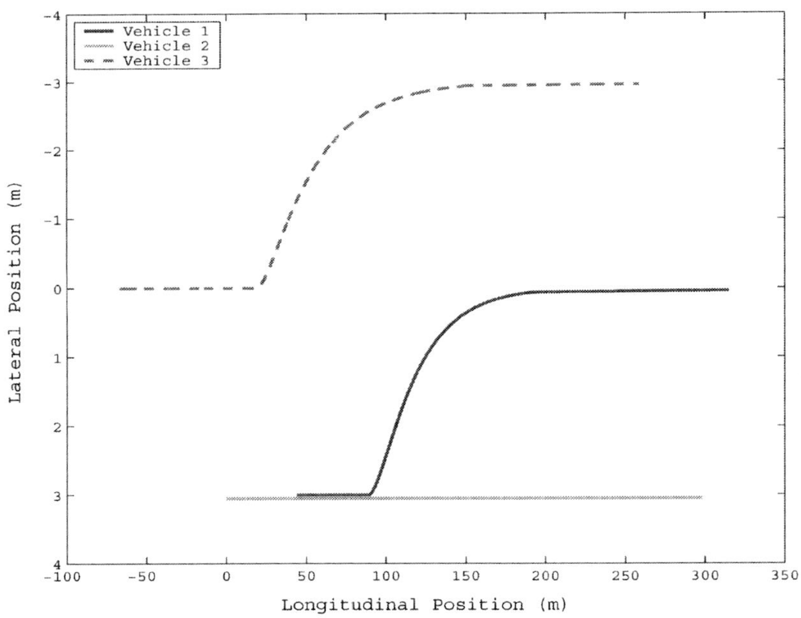

CONCLUSION

In this chapter, we proposed a system for autonomous vehicle control and collaborative driving based on the use of agent technology and of machine learning. More specifically, we presented a multi-layered architecture that relies on both an *Action Layer* and a *Coordination Layer*: the *Action Layer* is used to manage low-level vehicle control actions such as braking, accelerating or steering, while the *Coordination Layer* is responsible for high-level action choice by integrating cooperative decision-making between vehicles. These two layers were designed using agent and multi-agent reinforcement learning techniques. Finally, we showed that the integration of reinforcement learning techniques at all levels of our autonomous driving controller gives efficient results for vehicle control and coordination. This approach clearly facilitates the efforts of the system's designer, as the complex details related to vehicle control and related to the numerous possibilities of inter-vehicle interactions are automatically handled by the learning algorithm.

Unfortunately, even though our approach was tested on a realistic vehicle dynamics simulator, we obviously did not take into account all of the requirements needed for the implementation of our system in a real vehicle. For example, we assumed the sensors of the vehicle to be perfect and without noise. In practice, however, sensors like GPS and lasers have limited precision. Obviously, this can lead to a degradation of the efficiency of the *Action* and the *Coordination Layers's* policies. Therefore, future work could consider solving this particular problem, which could be done by using Partially Observable Markov Decision Processes (POMDPs). This framework generalizes MDPs and can be used to find control policies under uncertainty and partial observability of the environment. Moreover, the control of the *Action Layer* should consider continuous actions instead of discrete ones in order to improve the efficiency of the vehicle following behavior. As for the *Coordination Layer*, experiments should be done on more complex scenarios in order to improve performance in high-density vehicular traffic. In this case, some approximation techniques could be considered in order to find an efficient coordination policy for a large number of vehicles.

REFERENCES

Auto21 (2007). Retrieved July 17th, 2008, from: http://www.auto21.ca/

Bana, S. V. (2001). *Coordinating Automated Vehicles via Communication.* PhD thesis, University of California at Berkeley, Berkeley, CA.

Baxter, J. & Bartlett, P. L. (2001). Infinite-horizon policy-gradient estimation. *Journal of Artificial Intelligence Research, 15*, 319–350.

Bernstein, D., Givan, R., Immerman, N., & Zilberstein, S. (2002). The complexity of decentralized control of markov decision processes. *Mathematics of Operations Research, 27*(4), 819–840.

Bertsekas, D. P. (2000). *Dynamic Programming and Optimal Control, Vols. 1 & 2*, 2nd ed., Nashua, NH: Athena Scientific.

Bishop, W. R. (2005). *Intelligent Vehicle Technology and Trends*. Norwood, MA: Artech House.

Blumer, A., Noonan, J., & Schmolze, J. G. (1995). *Knowledge based systems and learning methods for automated highway systems.* (Technical Report), Waltham, MA: Raytheon Corp.

Broggi, A., Bertozzi, M., Fascioli, A., Lo, C., & Piazzi, B. (1999). The argo autonomous vehicle's vision and control systems. International *Journal of Intelligent Control and Systems, 3*(4), 409–441.

Brooks, R. A. (1991). Intelligence without representation. *Artificial Intelligence Journal, 47*, 139–159.

de Bruin, D., Kroon, J., van Klaveren, R., & Nelisse, M. (2004). Design and test of a cooperative adaptive cruise control system. *In Proceedings of IEEE Intelligent Vehicles Symposium* (pp. 392–396).

Desjardins, C., Grégoire, P.-L., Laumônier, J., & Chaib-draa, B. (2007). Architecture and design of a multi-layered cooperative cruise control system. *In Proceedings of the Society of Automobile Engineering World Congress (SAE'07).*

Ehlert, P. A. (2001). *The agent approach to tactical driving in autonomous vehicle and traffic simulation.* Master's thesis, Knowledge Based Systems Group, Delft University of Technology, Delft, The Netherlands.

Forbes, J., Huang, T., Kanazawa, K., & Russell, S. J. (1995). The batmobile: towards a bayesian automated taxi. *In Proceedings of International Joint Conference on Artificial Intelligence* (pp. 1878–1885), Morgan Kaufmann.

Forbes, J. R. (2002). *Reinforcement Learning for Autonomous Vehicles.* PhD thesis, University of California at Berkeley, Berkeley, CA.

Godbole, D. N., & Lygeros, J. (1994). Longitudinal control of a lead car of a platoon. *IEEE Transaction on Vehicular Technology, 43*(4), 1125–1135.

Hallé, S. & Chaib-draa, B. (2005). A Collaborative Driving System based on Multiagent Modelling and Simulations. *Journal of Transportation Research Part C (TRC-C): Emergent Technologies, 13*(4), 320–345.

Hallouzi, R., Verdult, V., Hellendorn, H., Morsink, P. L., & Ploeg, J. (2004). Communication based longitudinal vehicle control using an extended kalman filter. *In Proceedings of International Federation of Automatic Control Symposium on Advances in Automotive Control* (pp. 745-750).

Kiencke, U., & Nielsen, L. (2000). *Automotive control systems: for engine, driveline and vehicle.* Berlin, Germany: Springer-Verlag.

Laumônier, J., & Chaib-draa, B. (2006). Partial local FriendQ multiagent learning: application to team automobile coordination problem. In L. Lamontagne and M. Marchand (Ed.), *Canadian AI, Lecture Notes in Artificial Intelligence* (pp. 361–372). Berlin, Germany: Springer-Verlag.

Littman, M. (2001). Friend-or-Foe Q-learning in General-Sum Games. In C.E. Brodley and A. P. Danyluk (Ed.), *Proceedings of the Eighteenth International Conference on Machine Learning* (pp. 322–328), San Francisco, CA: Morgan Kaufmann.

Lu, X.-Y., Tan, H.-S., Empey, D., Shladover, S. E., & Hedrick, J. K. (2000). *Nonlinear longitudinal controller development and real-time implementation* (Technical Report UCB-ITS-PRR-2000-15).

Los Angeles, CA: University of Southern California, California Partners for Advanced Transit and Highways (PATH).

Moriarty, D., & Langley, P. (1998). Learning cooperative lane selection strategies for highways. *In Proceedings of the Fifteenth National Conference on Artificial Intelligence* (pp. 684–691), Menlo Park, CA: AAAI Press.

NAHSC (1998). *Technical Feasibility Demonstration Summary Report (Technical Report).* Troy, MI, USA: National Automated Highway System Consortium.

Naranjo, J., Gonzalez, C., Reviejo, J., Garcia, R., & de Pedro, T. (2003). Adaptive fuzzy control for inter-vehicle gap keeping. *IEEE Transactions on Intelligent Transportation Systems, 4*(3), 132–142.

Pendrith, M. D. (2000). Distributed reinforcement learning for a traffic engineering application. C. Sierra, M. Geni and J.S. Rosenschein (Ed.), *Proceedings of the Fourth International Conference on Autonomous Agents* (pp. 404-411), New York, NY: ACM Press.

Peng, H., bin Zhang, W., Arai, A., Lin, Y., Hessburg, T., Devlin, P., Tomizuka, M., & Shladover, S. (1992). *Experimental automatic lateral control system for an automobile (Technical Report UCB-ITS-PRR-92-11)*, Los Angeles, CA: University of Southern California, California Partners for Advanced Transit and Highways (PATH).

Pomerleau, D. (1995). Neural network vision for robot driving. In S. Nayar and T. Poggio (Eds.), *Early Visual Learning* (pp. 161-181), New York, NY: Oxford University Press.

Raza, H., & Ioannou, P. (1997). *Vehicle following control design for automated highway systems (Technical Report UCB-ITS-PRR-97-2)*, Los Angeles, CA: University of Southern California, California Partners for Advanced Transit and Highways (PATH).

Rosenblatt, J. K. (1995). DAMN: A distributed architecture for mobile navigation. In H. Hexmoor and D. Kortenkamp (Eds.), *Proceedings of the American Association of Artificial Intelligence Spring Symposium on Lessons Learned from Implemented Software Architectures for Physical Agents*, Menlo Park, CA: AAAI Press.

Russell, S. J., & Norvig, P. (2002). *Artificial Intelligence: A Modern Approach.* 2nd ed. Upper Saddle River, NJ: Prentice Hall.

Schulze, M. (2007). *Summary of chauffeur project.* Retrieved July 17th, 2008, from http://cordis.europa.eu/telematics/tap_transport/research/projects/chauffeur.html

Sengupta, R., Rezaei, S., Shladover, S. E., Cody, D., Dickey, S., & Krishnan, H. (2007). Cooperative collision warning systems: Concept definition and experimental implementation. *Journal of Intelligent Transportation Systems, 11*(3), 143–155.

Sheikholeslam, S. & Desoer, C. A. (1990). Longitudinal control of a platoon of vehicles. *In Proceedings of the American Control Conference, 1,* 291–297).

Shladover, S. E. (2007). Path at 20 – history and major milestones. *IEEE Transactions on Intelligent Transportation Systems, 8*(4), 584–592.

Sukthankar, R., Baluja, S., & Hancock, J. (1998). Multiple adaptive agents for tactical driving. *Applied Intelligence, 9*(1), 7–23.

Sutton, R. S., & Barto, A. G. (1998). *Reinforcement Learning: An Introduction.* Cambridge, MA: MIT Press.

Tambe, M., & Zhang, W. (2000). Toward flexible teamwork in persistent teams: extended report. *Journal of Autonomous Agents and Multi-agents Systems, 3*, 159–183.

Tsugawa, S. (2005). Issues and recent trends in vehicle safety communication systems. *International Association of Traffic and Safety Sciences Research, 29*, 7–15.

Tsugawa, S., Kato, S., Matsui, T., Naganawa, H., & Fujii, H. (2000). An architecture for cooperative driving of automated vehicles. *In Proceedings of IEEE Intelligent Transportation Systems* (pp. 422–427).

Ünsal, C., Kachroo, P., & Bay, J. S. (1999). Simulation study of multiple intelligent vehicle control using stochastic learning automata. *IEEE Transactions on Systems, Man and Cybernetics - Part A: Systems and Humans, 29*(1), 120–128.

Williams, R. J. (1992). Simple statistical gradient-following algorithms for connectionist reinforcement learning. *Machine Learning, 8*, 229–256.

Chapter XII
Traffic Congestion Management as a Learning Agent Coordination Problem

Kagan Tumer
Oregon State University, USA

Zachary T. Welch
Oregon State University, USA

Adrian Agogino
NASA Ames Research Center, USA

ABSTRACT

Traffic management problems provide a unique environment to study how multi-agent systems promote desired system level behavior. In particular, they represent a special class of problems where the individual actions of the agents are neither intrinsically "good" nor "bad" for the system. Instead, it is the combinations of actions among agents that lead to desirable or undesirable outcomes. As a consequence, agents need to learn how to coordinate their actions with those of other agents, rather than learn a particular set of "good" actions. In this chapter, the authors focus on problems where there is no communication among the drivers, which puts the burden of coordination on the principled selection of the agent reward functions. They explore the impact of agent reward functions on two types of traffic problems. In the first problem, the authors study how agents learn the best departure times in a daily commuting environment and how following those departure times alleviates congestion. In the second problem, the authors study how agents learn to select desirable lanes to improve traffic flow and minimize delays for all drivers. In both cases, they focus on having an agent select the most suitable action for each driver using reinforcement learning, and explore the impact of different reward functions on system behavior. Their results show that agent rewards that are both aligned with and sensitive to, the system reward lead to significantly better results than purely local or global agent rewards. They conclude this chapter by discussing how changing the way in which the system performance is measured affects the relative performance of these rewards functions, and how agent rewards derived for one setting (timely arrivals) can be modified to meet a new system setting (maximize throughput).

1. INTRODUCTION

This purpose of this chapter is to quantify how decisions of local agents in a traffic system (e.g., drivers) affect overall traffic patterns. In particular, this chapter explores the system coordination problem of how to configure and update the system so that individual decisions lead to good system level behavior. From a broader perspective, this chapter demonstrates how to measure the alignment between the local agents in a system and the system at large. Because the focus of this chapter is on multiagent coordination and reward analysis, we focus on abstract, mathematical models of traffic rather full fledged simulations. Our main purpose is to demonstrate the impact of reward design and extract the key properties rewards need to have to alleviate congestion in large agent coordination problems, such as traffic.

In this chapter we apply multi-agent learning algorithms to two separate congestion problems. First we investigate how to coordinate the departure times of a set of drivers so that they do not end up producing traffic "spikes" at certain times, both providing delays at those times and causing congestion for future departures. In this problem, different time slots have different desirabilities that reflect user preferences for particular time slots. The system objective is to maximize the overall system's satisfaction as a weighted average of those desirabilities. In the second problem we investigate lane selection, where a set of drivers need to select different lanes to a destination (Moriarty and Langley, 1998, Pendrith, 2000). In this problem, different lanes have different capacities and the problem is for the agents to minimize the total congestion. Both problems share the same underlying property that agents greedily pursuing the best interests of their own drivers cause traffic to worsen for everyone in the system, including themselves.

Figure 1. Reinforcement Learning for Congestion Management. A set of agents (cars) take actions. The result of each action is rewarded. Agents then modify their policy using reward.

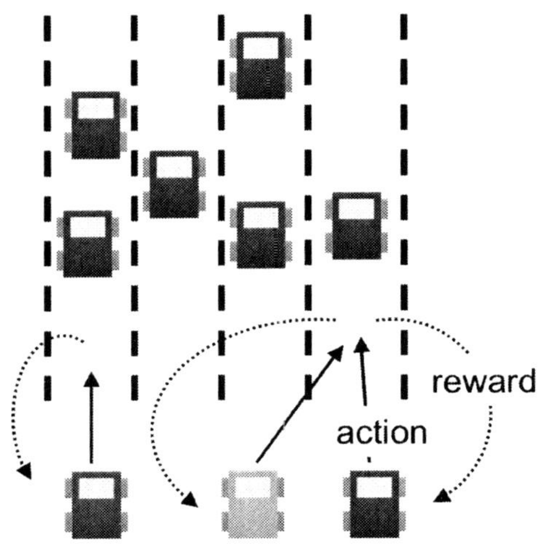

Indeed, multi-agent learning algorithms provide a natural approach to addressing congestion problems in traffic and transportation domains (Bazzan et al., 1999, Dresner and Stone, 2004, Klügl et al., 2005). Congestion problems are characterized by having the system performance depend on the number of agents that select a particular action, rather on the intrinsic value of those actions. Examples of such problems include lane/route selection in traffic flow (Kerner and Rehborn, 1996, Nagel, 1997), path selection in data routing (Lazar et al., 1997), and side selection in the minority game (Challet and Zhang, 1998, Jefferies et al., 2002). In those problems, the desirability of lanes, paths or sides depends solely on the number of agents having selected them. Hence, multi-agent approaches that focus on agent coordination are ideally suited for these domains where agent coordination is critical for achieving desirable system behavior.

The approach we present to alleviating congestion in traffic is based on assigning each driver an agent which determines the departure time/lane to select. The agents determine their actions based on a reinforcement learning algorithm (Littman, 1994, Sutton and Barto, 1998, Watkins and Dayan, 1992). In this reinforcement learning paradigm, agents go through a process of where they take actions and receive rewards evaluating the effect of those actions. Based on these rewards the agents try to improve their actions (see Figure 1). The key issue in this approach is to ensure that the agents receive rewards that promote good system level behavior. To that end, it is imperative that the agent rewards: (i) are aligned with the system reward[1], ensuring that when agents aim to maximize their own reward they also aim to maximize system reward; and (ii) are sensitive to the actions of the agents, so that the agents can determine the proper actions to select (i.e., they need to limit the impact of other agents in the reward functions of a particular agent).

The difficulty in agent reward selection stems from the fact that typically these two properties provide conflicting requirements. A reward that is aligned with the system reward usually accounts for the actions of other agents, and thus is likely to not be sensitive to the actions of one agent; on the other hand, a reward that is sensitive to the actions of one agent is likely not to be aligned with system reward. This issue is central to achieving coordination in a traffic congestion problem and has been investigated in various fields such as computational economics, mechanism design, computational ecologies and game theory (Boutilier, 1996, Sandholm and Crites, 1995, Huberman and Hogg, 1988, Parkes, 2001, Stone and Veloso, 2000). We address this reward design problem using the difference reward (Wolpert and Tumer, 2001, Tumer and Wolpert, 2004), which provides a good balance of alignedness and sensitivity. The difference reward has been applied to many domains, including rover coordination (Agogino and Tumer, 2004), faulty device selection problem (Tumer, 2005), packet routing over a data network (Tumer and Wolpert, 2000, Wolpert et al., 1999), and modeling nongenomic models of early life (Gupta et al., 2006).

The overall objective of this chapter is to show how agent reward design, coupled with reinforcement learning agents can be used to alleviate traffic congestion, and show experimental results illustrating that these methods are both effective and robust to non-compliance (i.e., drivers not following the suggestions of their agents). In Section 2 we discuss the properties agent rewards need to have and present a particular example of agent reward. In Sections 3.1 and 3.2 we present the departure coordination problem. The results in this domain show that total traffic delays can be improved significantly when agents use the difference reward. In Section 3.3 we present the lane selection problem. The results in this domain show that traffic congestion can be reduced by over 30% when agents use the difference reward. In Section 3.4, we investigate how the system performance degrades when the percentage of drivers who do not follow the advice of their agents increases from 0 to 100 %, a critical issue for the adoption of any new traffic algorithm. Finally, in Section 4 we discuss the implication of these results, discuss methods by which they can be applied in the traffic domain, and highlight future research directions.

2. BACKGROUND

In this chapter we propose modeling cars as agents that individually try to maximize a reward through a reinforcement learning process. The types of rewards the agents receive depend on our goals for the system and our approach to system. While in some cases all the agents will get the same reward, in general an agent will get a reward unique to the agent. Finding appropriate rewards that will entice agents take actions towards a collective goal is critical to the success of this method.

More formally we are modeling traffic congestion management as a multi-agent systems where each agent, *i*, tries to maximize its reward function *giz*, where *z* depends on the joint move of all agents. Furthermore, there is a system reward function, *G*(*z*) which rates the performance of the full system. To distinguish states that are impacted by actions of agent *i*, we decompose[2] *z* into $z=z_i+z_{-i}$, where z_i refers to the parts of *z* that are dependent on the actions of *i*, and z_{-i} refers to the components of *z* that do not depend on the actions of agent *i*.

2.1 Properties of Reward Functions

For learning agents to be able to perform effectively in a multi-agent system it is critical that the rewards have two properties:

- rewards are aligned with the overall goal.
- rewards are sensitive to the agent's actions.

First, the agent rewards have to be aligned with respect to *G*, quantifying the concept that an action taken by an agent that improves its own reward also improves the system reward. Formally, for systems with discrete states, the degree of **factoredness** for a given reward function g_i is defined as:

$$F_{g_i} = \frac{\sum_z \sum_{z'} u[(g_i(z)-g_i(z'))\,(G(z)-G(z'))]}{\sum_z \sum_{z'} 1} \quad (1)$$

for all *z*' such that $z_{-i}=z'_{-i}$ and where $u[x]$ is the unit step function, equal to 1 if $x>0$, and zero otherwise. Intuitively, the higher the degree of factoredness between two rewards, the more likely it is that a change of state will have the same impact on the two rewards. A system is fully factored when $F_{g_i}=1$.

Second, an agent's reward has to be sensitive to its own actions and insensitive to actions of others. Formally we can quantify the **learnability** of reward g_i, for agent *i* at *z*:

$$\lambda_{i,g_i}(z)=\frac{E_{z_i'}[|g_i(z)-g_i(z_{-i}+z_i')|]}{E_{z_{-i}'}[|g_i(z)-g_i(z_{-i}'+z_i)|]} \quad (2)$$

where $E[\cdot]$ is the expectation operator, z_i's are alternative actions of agent i at z, and z_{-i}'s are alternative joint actions of all agents other than i. Intuitively, learnability provides the ratio of the expected value of gi over variations in agent i's actions to the expected value of gi over variations in the actions of agents other than i. So at a given state z, the higher the learnability, the more giz depends on the move of agent i, i.e., the better the associated signal-to-noise ratio for i. Higher learnability means it is easier for i to achieve large values of its reward.

In the domain of congestion management a reward with the first property means that actions that help reduce the overall congestion are rewarded. This is in contrast to a greedy reward, which may reward actions taken that help an individual driver, but actually cause overall congestion to increase. However, in general, this property isn't sufficient since it does not concern itself with whether the agents can actually maximize their own reward. Consider the extreme example of a driver only being rewarded with a "good" or "bad" score depending on the traffic report at the end of the day summarizing the day's congestion. While this reward is aligned with our overall goal of reducing congestion, if there are thousands (if not millions) of drivers, a driver would not be able to see the effect of his/her individual actions on this reward. Instead we need rewards that balance being aligned with our goal while being sensitive to the driver's actions, so that the drivers can effectively learn to maximize their rewards.

2.2 Difference Reward Functions

Let us now focus on providing agent rewards that are both high factoredness and high learnability. Consider the **difference** reward (Wolpert and Tumer, 2001), which is of the form:

$$D_i \equiv G(z) - G(z_{-i} + c_i) \tag{3}$$

where z_{-i} contains all the states on which agent i has no effect, and c_i is a fixed vector. In other words, all the components of z that are affected by agent i are replaced with the fixed vector c_i. Such difference reward functions are fully factored no matter what the choice of c_i, because the second term does not depend on i's states (Wolpert and Tumer, 2001). Furthermore, they usually have far better learnability than does a system reward function, because the second term of D removes some of the effect of other agents (i.e., noise) from i's reward function. In many situations it is possible to use a c_i that is equivalent to taking agent i out of the system. Intuitively this causes the second term of the difference reward function to evaluate the value of the system without i and therefore D evaluates the agent's contribution to the system reward.

The difference reward can be applied to any linear or non-linear system reward function. However, its effectiveness is dependent on the domain and the interaction among the agent reward functions. At best, it fully cancels the effect of all other agents. At worst, it reduces to the system reward function, unable to remove any terms (e.g., when z_{-i} is empty, meaning that agent i effects all states). In most real world applications, it falls somewhere in between, and has been successfully used in many domains including agent coordination, satellite control, data routing, job scheduling and congestion games (Agogino and Tumer, 2004, Tumer and Wolpert, 2000, Wolpert and Tumer, 2001). Also note that computationally the difference reward is often easier to compute than the system reward function (Tumer and Wolpert, 2000). Indeed in the problem presented in this chapter, for agent i, D_i is easier to compute than G is (see details in Section 3.1.1).

2.3 Reward Maximization

In this chapter we assume that each agent maximize its own reward using its own reinforcement learner (though alternatives such as evolving neuro-controllers are also effective (Agogino and Tumer, 2004)). In this paradigm, an agent will take an action based on a policy and will then receive a reward evaluating its action. The agent will then use this reward to update its action policy. For complex delayed-reward problems, relatively sophisticated reinforcement learning systems such as temporal difference may have to be used. However, the traffic domain modeled in this chapter only needs to utilize immediate rewards, therefore a simple table-based immediate reward reinforcement learning is used. Our reinforcement learner is equivalent to an ε-greedy with a discount rate of 0. At every episode an agent takes an action and then receives a reward evaluating that action. After taking action a and receiving reward R a driver updates its table as follows:

$$Q(a) \leftarrow (1-\alpha)Q(a)+\alpha R$$

where α is the learning rate. At every time step the driver chooses the action with the highest table value with probability $1-\varepsilon$ and chooses a random action with probability ε. In the experiments described in the following section, α is equal to 0.5 and ε is equal to 0.05. The parameters were chosen experimentally, though system performance was not overly sensitive to these parameters.

3. EXPERIMENTS

To test the effectiveness of our rewards in the traffic congestion domain, we performed experiments using two abstract traffic models. In the first model each agent has to select a time slot to start its drive. In this model we explore both simple and cascading traffic flow. With non-cascading flow, drivers enter and exit the same time slot, while with cascading flow, drivers stuck in a time slot with too many other drivers stay on the road for future time slots.

In the second model, instead of choosing time slots, drivers choose lanes. This model differs from the time-slot model in that different lanes may also have different capacities (for example because of carpool restrictions). In this model we also use a slightly different objective function that seeks to avoid congestion, in contrast to maximizing throughput.

Between these models we performed six sets of experiments as follows:

1. Departure time selection for simple traffic flow model:
   (a) Single peak congestion - Heavy congestion peaks around a single time slot.
   (b) Double peak congestion - Heavy congestion peaks around two time slots.
   (c) Non-Symmetric congestion - Congestion progressively increases with time.
2. Departure time selection for cascading congestion for single peak congestion.
3. Lane Selection - Drivers reduce congestion using lane selection model.
4. Driver compliance - Test ability of learning agents to reduce congestion when some of the drivers are taking random actions instead of trying to reduce congestion.

3.1 Departure Time Selection

In the traffic congestion model we first explore, there is a fixed set of drivers, and the task of the agents is to find the time slot in which their drivers start their commutes. The system performance is measured from the perspective of a "city manager" (as opposed to a social welfare function based on the intrinsic rewards of the drivers) that directly measures a system wide performance criteria. To highlight this, we will denote the system level function of the City Manager by (dubbed G in the previous section) by $S(CM)$:

$$G=S(CM)=\sum_t w_t S(k_t)\,. \tag{4}$$

where weights wt model rush-hour scenarios where different time slots have different desirabilities, and $S(k)$ is a "time slot reward", depending on the number of agents that chose to depart in the time slot:

$$S(k)=\begin{cases} ke^{-1} & if\ k\leq c \\ k\ e^{-k/c} & \text{otherwise} \end{cases}, \tag{5}$$

The number of drivers in the time slot is given by k, and the optimal capacity of the time slot is given by c. Below an optimal capacity value c, the reward of the time slot increases linearly with the number of drivers. When the number of drivers is above the optimal capacity level, the value of the time slot decreases quickly (asymptotically exponential) with the number of drivers. This reward models how drivers do not particularly care how much traffic is on a road until it is congested. This function is shown in Figure 2. In this problem, the task of the system designer is to have the agents choose time slots that help maximize the system reward. To that end, agents have to balance the benefit of going at preferred time slots with the congestion at those time slots.

3.1.1 Driver Rewards

While as a system designer our goal is to maximize the system reward, we have each individual agent try to maximize a driver-specific reward that we select. The agents maximize their rewards through

Figure 2. Reward of time slot with c=30

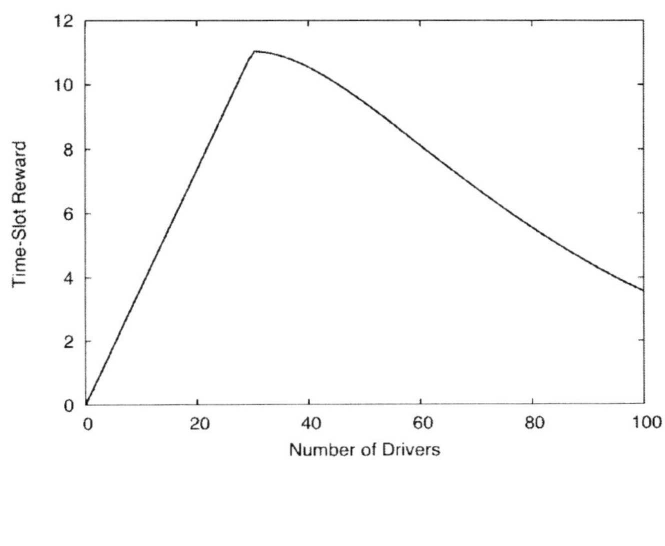

reinforcement learning, where they learn to choose time slots that have expected high reward. In these experiments, we evaluate the effectiveness of three different rewards. The first reward is simply the system reward $G=S(CM)$, where each agent tries to maximize the system reward directly. The second reward is a local reward, L-k/c_i where each agent tries to maximize a reward based on the time slot it selected:

$$L_i(k)=w_i S(k_i) \qquad (6)$$

where k_i is the number of drivers in the time slot chosen by driver i. The final reward is the difference reward, D:

$$\begin{aligned} D_i &= G(k)-G(k_{-i}) \\ &= \sum_j L_j(k)-\sum_j L_j(k_{-i}) \\ &= L_i(k)-L_i(k_{-i}) \\ &= w_i k_i S(k_i)-w_i(k_i-1)S(k_i-1), \end{aligned}$$

where k_{-i} represents the driver counts when driver i is taken out of the system. Note that since taking away driver i only affects one time slot, all of the terms but one cancel out, making the difference reward simpler to compute than the system reward.

3.1.2 Single Peak Congestion Results

In this set of experiments there were 500 drivers, and the optimal capacity of each time slot was 125. Furthermore, the weighting vector was centered at the most desirable time slot (e.g., 5 PM departures), simulating a single peak congestion:

$$w = [1\ 5\ 10\ 15\ 20\ 15\ 10\ 5\ 1]^T .$$

This weighting vector reflects a preference for starting a commute at the end of the workday with the desirability of a time slot decreasing for earlier and later times. All performance plots reflect agent daily agent learning (the "time step" is one day, in that each day the agents make new choices).

This experiment shows that drivers using the difference reward are able to quickly obtain near-optimal system performance (see Figure 3). In contrast, drivers that try to directly maximize the system reward do not learn at all and never achieve good performance during the time-frame of the experiment. This lack of learning is a result of the system reward having low learnability to the agents' actions. Even if a driver were to take a system wide coordinated action, it is likely that some of the 499 other drivers would take uncoordinated actions at the same time, lowering the value of the system reward. A driver using the system reward typically does not get proper credit assignment for its actions, since the reward is dominated by other drivers.

Figure 3. Performance on departure time selection problem with single peak congestion. In this and all subsequent figures, we present local (L), Difference (D) and System (S) rewards based on the City Manager (CM) perspective. Drivers using difference reward quickly learn to achieve near optimal performance (1.0). Drivers using system reward do not learn at all. Drivers using non-factored local reward slowly learn counterproductive actions.

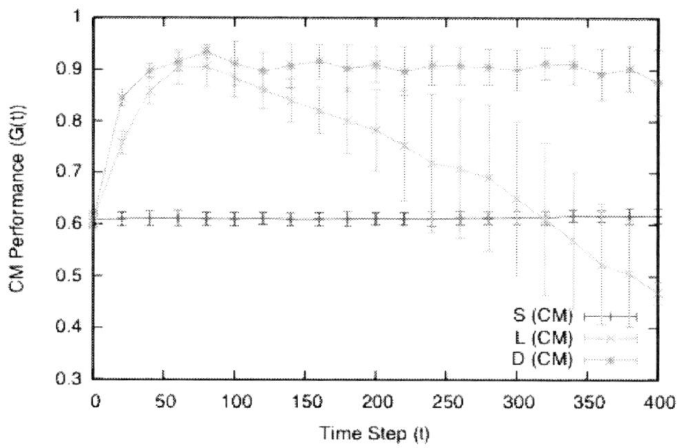

The experiment where drivers are using L (a non-factored local reward) exhibit some interesting performance properties. At first these drivers learn to improve the system reward. However, after about episode seventy their performance starts to decline. Figure 4 gives greater insight into this phenomenon. At the beginning of the experiment, the drivers are randomly distributed among time slots, resulting in a low reward. Later in training agents begin to learn to use the time slots that have the most benefit. When the number of drivers reach near optimal values for those time slots, the system reward is high. However, all agents in the system covet those time slots and more agents start to select the desirable time slots. This causes congestion and system reward starts to decline. This performance characteristics is typical of system with agent rewards of low factoredness. In such a case, agents attempting to maximize their own rewards lead to undesirable system behavior. In contrast, because their rewards are factored with the system reward, agents using the difference reward form a distribution that more closely matches the optimal distribution (Figure 4).

3.1.3 Double Peak Congestion Results

In many situations, there are multiple desirable departure times, resulting in multi-modal peak departure distribution. To verify that the performance obtained in the previous section was not due to the weight vector, we investigated the agent response to a weight profile that provided double peaks:

$w=[1\ 10\ 20\ 10\ 1\ 10\ 20\ 10\ 1]^T$

Figures 5 and 6 show performance for the double peak weight vector, along with the histograms of slot counts for agents using the local reward (over time) and all rewards (at the end of the simulation), respectively. In this case, because the problem was more difficult and required some degree of coordination from the starting point, the performance of the local reward never reached the performance

Figure 4. Slot distributions for single peak congestion: (a) Distribution of drivers using local reward. Early in training drivers learn good policies. Later in learning, the maximization of local reward causes drivers to over utilize high valued time slots. (b) Distribution of drivers at end of training for all three rewards. Drivers using difference reward form distribution that is closer to optimal than drivers using system of local rewards.

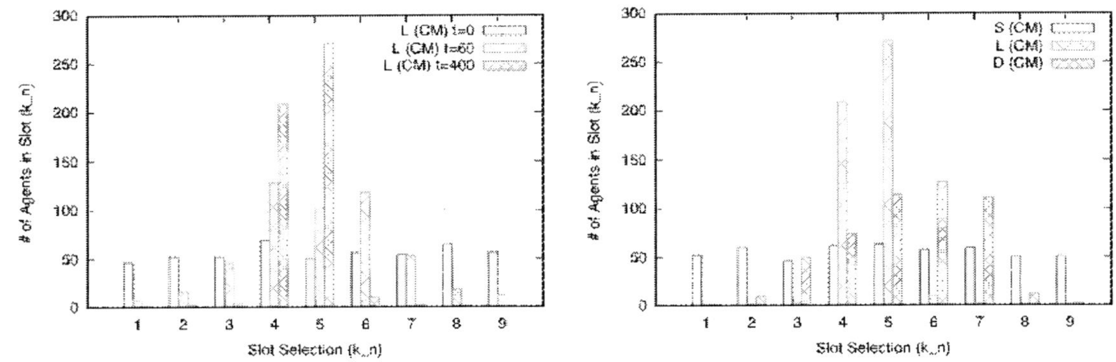

Figure 5. Performance on Departure Time Selection Problem with double peak congestion. Drivers using difference reward quickly learn to achieve near optimal performance (1.0). Drivers using system reward do not learn at all. Drivers using non-factored local reward quickly learn counterproductive actions.

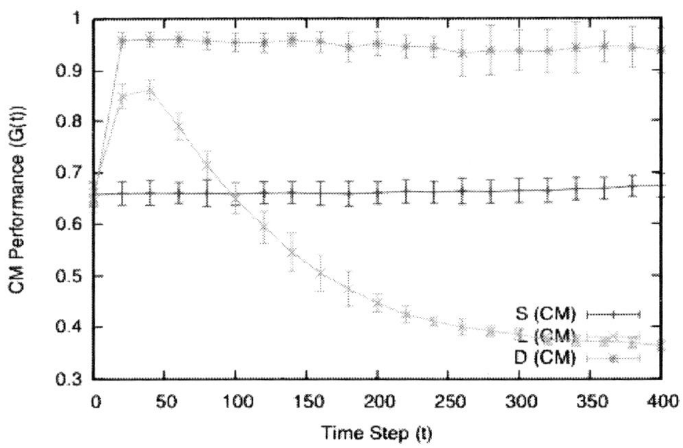

of the difference reward. However, the same performance drop is observed in this case, where agents pursuing the local reward start a decline that leads them to very poor solutions. This can be seen in Figure 6(a) where the local reward never finds the "good" distribution found by the difference reward in Figure 6(b). In contrast, the agents using the difference reward were not affected by the difficulty of the problem and reached a good solution in very few training steps.

3.1.4 Non-Symmetric Congestion Results

Finally, we explored the performance of the various rewards functions for a non-symmetric weight distribution:

Figure 6. Slot distributions for double peak congestion: (a) Distribution of Drivers using Local Reward. where The maximization of local reward causes drivers to quickly start to over utilize high valued time slots. (b) Distribution of Drivers at end of Training for all three rewards. Drivers using difference reward form near optimal distribution.

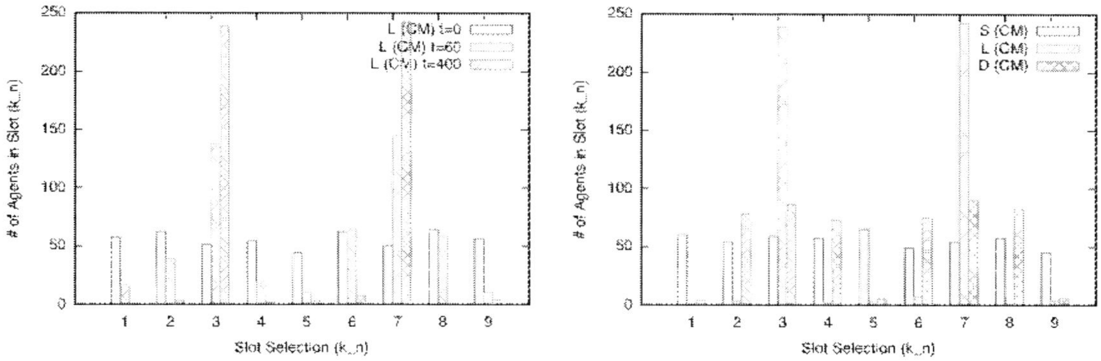

$w=[1\ 1\ 2\ 3\ 5\ 8\ 13\ 21\ 34]^T$

Figures 7 and 8 show performance for the non-symmetric weight vector, along with the histograms of slot counts for agents using the local reward (over time) and all rewards (at the end of the simulation), respectively. It is clear from this (and the double peak experiment) that the initial, single peak weights were more favorable to agents using the local reward than to agents using either the difference reward or the full system reward. In these two difficult cases, agents using the local reward never reach the performance of the difference reward, and their drop in performance begins almost immediately. In contrast, the original single peak environment had yielded improved performance for a longer time period before succumbing to clustering effects.

Figure 7. Performance on departure time selection problem with non-symmetric congestion. Drivers using difference reward quickly learn to achieve near optimal performance (1.0). Drivers using system reward do not learn at all. Drivers using non-factored local reward quickly learn counterproductive actions.

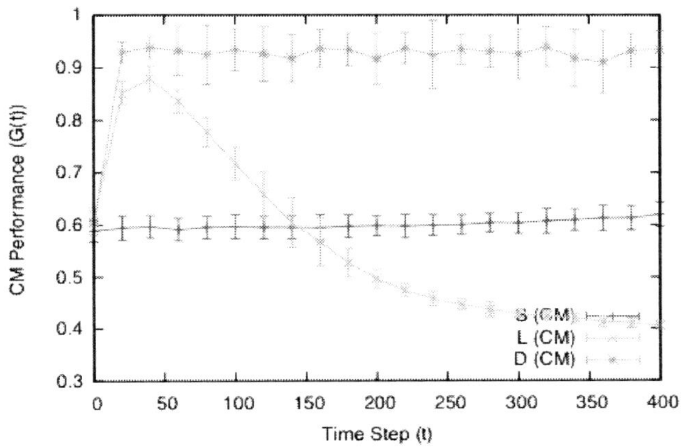

Figure 8. Slot distributions for non-symmetric congestion: (a) Distribution of Drivers using Local Reward. The maximization of local reward causes drivers to quickly start to over utilize high valued time slots. (b) Distribution of Drivers at end of Training for all three rewards. Drivers using difference reward form distribution that is closer to optimal than drivers using system of local rewards.

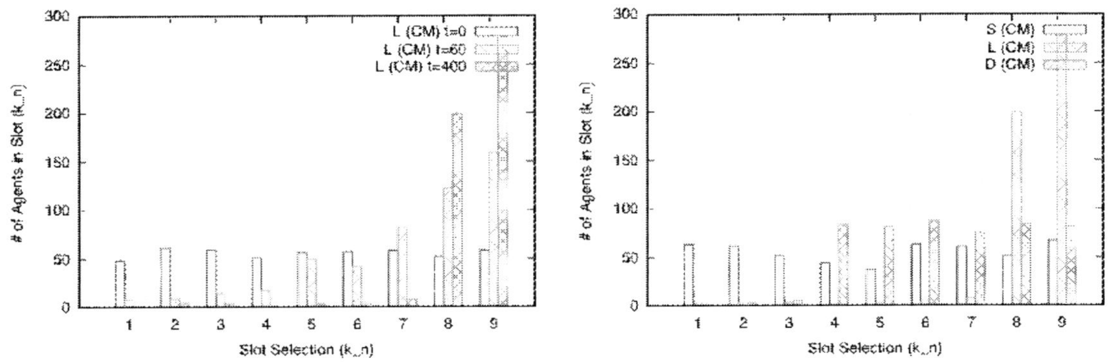

3.2 Cascading Traffic for Departure Time Selection

The previous model assumes that drivers enter and leave the same time slot. Here we introduce a more complex model, where drivers remain in the system longer when it is congested. This property is modeled by having drivers over the optimal capacity, c stay in the system until they reach a time slot with a traffic level below c. When the number of drivers in a time slot is less than c the reward for a time slot is the same as before. When the number of drivers is above c the linear term k is replaced with c:

$$S(k)=\begin{cases} ke^{-1} & if\ k\leq c \\ c\ e^{-k/c} & \text{otherwise} \end{cases} \qquad (7)$$

As before the system reward is a sum of the time slot rewards: $G=\sum_t S(k_t)$.

3.2.1 Driver Rewards

Again the local reward is the weighted time slot reward:

$$L_i=w_i S(k_i) \qquad (8)$$

where k_i is the number of drivers in the time slot chosen by driver i. However the difference reward is more difficult to simplify as the actions of a driver can have influence over several time slots:

$$\begin{aligned} D_i &= G(k)-G(k_{-i}) \\ &= \sum_j w_j S(k_j)-\sum_j w_j S(k_{-i_j}), \end{aligned}$$

where k_{-i_j} is the number of drivers there would have been in time slot j had driver i not been in the system.

3.2.2 Results

Figure 9 shows the results for cascading traffic model for the single peak weight vector given by w=[1 5 10 15 20 15 10 5 1]^T. As previously, there are 500 drivers and time slot capacities are 125. Drivers using the different rewards exhibit similar characteristics on this model than on the non-cascading one. Again drivers using the system reward are unable to improve their performance significantly beyond their initial random performance.

In this model drivers using the local reward perform worse than in the simple cascading model (Results in Figure 3) once they become proficient at maximizing their own reward. This is because bad choices have longer lasting impact in this model. As a result, when drivers using the local reward cause congestion for their time slots, the congestion cascades as drivers spill into future time slots causing a significant decrease in performance. The performance of the three different rewards for the double peak and non-symmetric weight vectors are similar to those obtained in Sections 3.1.3 and 3.1.4, in that the local rewards degrade faster than for the single peak vector. We omit the details of those experiments for brevity, as they do not provide additional insight into agent behavior.

3.3 Lane Selection Congestion Model

In this model instead of selecting time slots, drivers select lanes. The main difference in this model is the functional form of the reward for a lane as shown in Figure 10. In this model the objective is to keep the lanes uncongested. The system reward does not care how many drivers are on a particular lane as long as that lane is below its congestion point. Each lane has a different weight representing overall driver preference for a lane. Furthermore, each lane has its own capacity, modeling the realities that some lanes having more restrictions such as tolls and/or carpools.

Figure 9. Performance on cascading departure time selection problem. In this domain drivers above the capacity in one time slot remain in system in future time slots. Drivers using difference reward quickly learn to achieve near optimal performance (1.0).

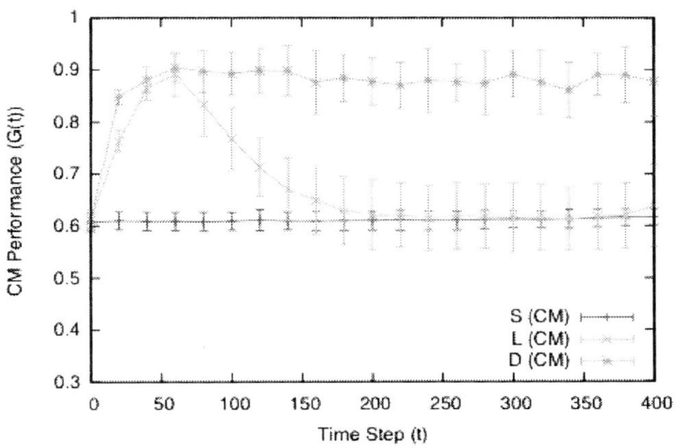

In this model the reward for an individual lane is:

$$S_{Lane}(k,c)=\begin{cases} e^{-1} & if k\leq c \\ e^{-k/c} & \text{otherwise} \end{cases} \quad (9)$$

The system reward is then the sum of all lane rewards weighted by the value of the lane.

$$G=\sum_i w_i S_{Lane}(k_i,c_i), \quad (10)$$

where wi is the weighting for lane i and c_i is the capacity for lane i.

3.3.1 Driver Rewards

Again three rewards were tested: the system reward, the local reward and the difference reward. The local reward is the weighted reward for a single lane:

$$L_i=w_i S_{Lane}(k_i,c_i). \quad (11)$$

The final reward is the difference reward, D:

$$\begin{aligned} Di &= G(k)-G(k_{-i}) \\ &= L_i(k)-L_i(k_{-i}) \\ &= w_i S_{Lane}(k_i,c_i)-w_i S_{Lane}(k_i-1,c_i), \end{aligned}$$

Figure 10. Reward of Road with c=30

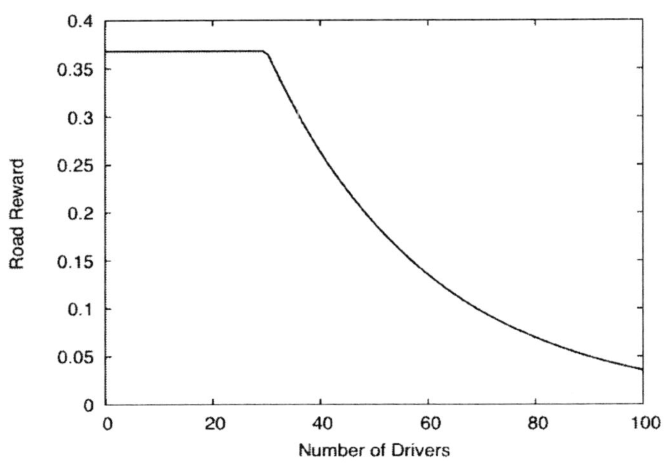

representing the difference between the actual system reward and what the system reward would have been if the driver had not been in the system.

3.3.2 Results

Here we show the results of experiments where we test performance of the three rewards in the multi-lane model, where different lanes have different value weightings and different capacities. There were 500 drivers in these experiments and the lane capacities were 167, 83, 33, 17, 9, 17, 33, 83, 167. Each lane is weighted with the weights 1, 5, 10, 1, 5, 10, 1, 5, 10. Figure 11 shows that drivers using the system reward perform poorly, and learn slowly. Again drivers using the difference reward perform the best, learning quickly to achieve an almost optimal solution. Drivers using the local reward learn more quickly early in training than drivers using the system reward, but never achieve as high as performance as those using the difference reward. However in this domain the drivers using the local reward do not degrade from their maximal performance, but instead enter a steady state that is significantly below that of the drivers using the difference reward.

3.4 Non-Compliant Drivers

All the pervious experiments presented in this chapter have assumed that all the drivers are actively participating in the learning system. However in most real-world situations this will not happen. In many traffic scenarios it may be only possible to convince some of the drivers to participate in a particular scheme. In addition even if all the drivers agree participate, due to various information/sensing limitations, some of the drivers may not be able to. To test this situation we conducted a set of experiments where a certain percentage of drivers did not participate in the learning paradigm. Instead they took random actions.

The results (Figure 12) show that the proposed paradigm is robust even when a moderate amount of drivers are non-compliant. As before, drivers using the system reward perform uniformly poorly. Interestingly drivers using the local reward actually can improve their performance when the number

Figure 11. Performance on domain with multiple lanes. Best observed performance = 1.0 (optimal not calculated)

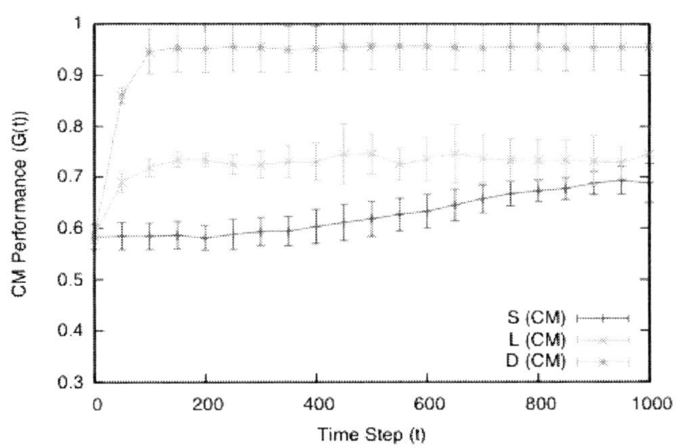

of non-compliant drivers increases. This is not surprising since the drivers using local rewards where actually learning how to make counter productive actions. When noise is added to the system in the form of non-compliant drivers, the drivers using the local reward were less able to learn these counter productive actions.

Finally, drivers using the difference reward perform better when more drivers conform, but their performance degrades gracefully with the number of non-compliant drivers. This is a key result that implies such a system can be implemented in stages with improvements acting as "advertising" to entice others to participate in the system.

4. CONCLUSION AND FUTURE RESEARCH DIRECTIONS

This chapter presented a method for improving congestion in two different traffic problems. First we presented a method by which agents can coordinate the departure times of drivers in order to alleviate spiking at peak traffic times, demonstrating its effectiveness in two similar congestion models. Second we showed that agents can manage effective lane selection and significantly reduce congestion by using a reward structure that penalizes greedily seeking the lanes with high capacity.

These results are based on agents receiving rewards that have high factoredness and high learnability (i.e., are both aligned with the system reward and are as sensitive as possible to changes in the reward of each agent). In these experiments, agents using difference rewards produced near optimal performance (93-96% of optimal). Agents using system rewards (63-68%) performed comparably to random action selection (62-64%), and agents using local rewards (48-72%) provided performance ranging from mediocre to worse than random in the instances when their own interests did not align with the system reward (i.e., city manager's reward).

Finally, one issue that arises in traffic problems that does not arise in many other domains (e.g., rover coordination) is in ensuring that drivers follow the advice of their agents. We showed that the system is

Figure 12. Performance on time selection problem with non-compliant drivers. With a moderate number of non-compliant drivers difference reward still performs well.

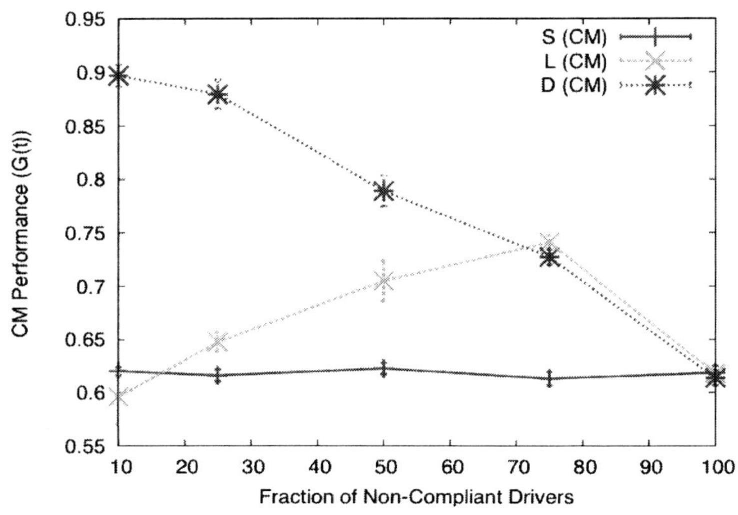

robust when a large number of drivers do not participate in the optimization system. A related problem also arises when the city manager's system reward is at odds with a social welfare function based on timeliness desires of the drivers. Determining what incentives to provide to the agents so that these two seemingly different objectives can be simultaneously maximized is a critical problem that has recently been investigated (Tumer et al., 2008), but bears further study.

However, in this chapter, we did not address the issue of what drivers do when it is not in their interest to follow the advice of their agents. The purpose of this chapter was to show that solutions to the difficult traffic congestion problem can be addressed in a distributed adaptive manner using intelligent agents. Ensuring that drivers follow the advice of their agents is a fundamentally different problem. One can expect that drivers will notice that the departure times/lanes suggested by their agents provide significant improvement over their regular patterns. However, as formulated, there are no mechanisms for ensuring that a driver does not gain an advantage by ignoring the advice of his or her agent. Future work includes investigating this issue, exploring the alignment/mismatch between a city manager's utility and a social welfare reward based on the agent's intrinsic rewards and verifying these results in a traffic simulator.

REFERENCES

Agogino, A., & Tumer, K. (2004). Efficient evaluation functions for multi-rover systems. In *The Genetic and Evolutionary Computation Conference*, (pp. 1–12), Seatle, WA.

Balmer, M., Cetin, N., Nagel, K., & Raney, B. (2004). Towards truly agent-based traffic and mobility simulations. In *Proceedings of the Third International Joint Conference on Autonomous Agents and Multi-Agent Systems*, (pp. 60–67), New York, NY.

Bazzan, A. L., & Klügl, F. (2005). Case studies on the Braess paradox: Simulating route recommendation and learning in abstract and microscopic models. *Transportation Research C*, *13*(4), 299–319.

Bazzan, A. L., Wahle, J., & Klügl, F. (1999). Agents in traffic modelling – from reactive to social behaviour. In *KI – Kunstliche Intelligenz*, (pp. 303–306).

Boutilier, C. (1996). Planning, learning and coordination in multiagent decision processes. In *Proceedings of the Sixth Conference on Theoretical Aspects of Rationality and Knowledge*, Holland.

Challet, D., & Zhang, Y. C. (1998). On the minority game: Analytical and numerical studies. *Physica A*, *256*, 514.

Dresner, K. and Stone, P. (2004). Multiagent traffic management: A reservation-based intersection control mechanism. In *Proceedings of the Third International Joint Conference on Autonomous Agents and Multi-Agent Systems*, (pp. 530–537), New York, NY.

Gupta, N., Agogino, A., & Tumer, K. (2006). Efficient agent-based models for non-genomic evolution. In *Proceedings of the Fifth International Joint Conference on Autonomous Agents and Multi-Agent Systems*, Hakodate, Japan.

Hall, S., & Draa, B. C. (2004). Collaborative driving system using teamwork for platoon formations. In *The third workshop on Agents in Traffic and Transportation*.

Huberman, B. A., & Hogg, T. (1988). The behavior of computational ecologies. In *The Ecology of Computation*, (pp. 77–115). North-Holland.

Jefferies, P., Hart, M. L., & Johnson, N. F. (2002). Deterministic dynamics in the minority game. *Physical Review E*, 65 (016105).

Kerner, B. S., & Rehborn, H. (1996). Experimental properties of complexity in traffic flow. *Physical Review E, 53*(5), R4275–4278.

Klügl, F., Bazzan, A., & Ossowski, S., editors (2005). *Applications of Agent Technology in Traffic and Transportation*. Springer.

Lazar, A. A., Orda, A., & Pendarakis, D. E. (1997). Capacity allocation under noncooperative routing. *IEEE Transactions on Networking*, 5(6), 861–871.

Littman, M. L. (1994). Markov games as a framework for multi-agent reinforcement learning. In *Proceedings of the 11th International Conference on Machine Learning*, (pp. 157–163).

Moriarty, D. E., & Langley, P. (1998). Learning cooperative lane selection strategies for highways. In *Proceedings of the Fifteenth National Conference on Artificial Intelligence*, (pp. 684–691), Madisson, WI.

Nagel, K. (1997). *Experiences with iterated traffic microsimulations in dallas*. pre-print adap-org/9712001.

Nagel, K. (2001). Multi-modal traffic in TRANSIMS. In *Pedestrian and Evacuation Dynamics*, (pp. 161–172). Springer, Berlin.

Parkes, D. C. (2001). *Iterative Combinatorial Auctions: Theory and Practice*. PhD thesis, University of Pennsylvania.

Pendrith, M. D. (2000). Distributed reinforcement learning for a traffic engineering application. In *Proceedings of the fourth international conference on Autonomous Agents*, Barcelona, Spain.

Sandholm, T., & Crites, R. (1995). Multiagent reinforcement learning in the iterated prisoner's dilemma. *Biosystems, 37*, 147–166.

Stone, P., & Veloso, M. (2000). Multiagent systems: A survey from a machine learning perspective. *Autonomous Robots*, *8*(3), 345–383.

Sutton, R. S., & Barto, A. G. (1998). *Reinforcement Learning: An Introduction*. MIT Press, Cambridge, MA.

Tumer, K. (2005). Designing agent utilities for coordinated, scalable and robust multi-agent systems. In P. Scerri, R. Mailler, & R. Vincent, (Eds.), *Challenges in the Coordination of Large Scale Multiagent Systems*. Springer.

Tumer, K., Welch, Z., & Agogino, A. (2008). Aligning social welfare and agent preferences to alleviate traffic congestion. In *Proceedings of the Seventh International Joint Conference on Autonomous Agents and Multi-Agent Systems*, Estoril, Portugal.

Tumer, K., & Wolpert, D. (Eds.) (2004). *Collectives and the Design of Complex Systems*. Springer, New York.

Tumer, K., & Wolpert, D. H. (2000). Collective intelligence and Braess' paradox. In *Proceedings of the Seventeenth National Conference on Artificial Intelligence*, (pp. 104–109), Austin, TX.

Watkins, C., & Dayan, P. (1992). Q-learning. *Machine Learning*, *8*(3/4), 279–292.

Wolpert, D. H., & Tumer, K. (2001). Optimal payoff functions for members of collectives. *Advances in Complex Systems*, *4*(2/3), 265–279.

Wolpert, D. H., Tumer, K., & Frank, J. (1999). Using collective intelligence to route internet traffic. In *Advances in Neural Information Processing Systems - 11*, (pp. 952–958). MIT Press.

Chapter XIII
Exploring the Potential of Multiagent Learning for Autonomous Intersection Control

Matteo Vasirani
University Rey Juan Carlos, Spain

Sascha Ossowski
University Rey Juan Carlos, Spain

ABSTRACT

The problem of advanced intersection control is being discovered as a promising application field for multiagent technology. In this context, drivers interact autonomously with a coordination facility that controls the traffic flow through an intersection, with the aim of avoiding collisions and minimizing delays. This is particularly interesting in the case of autonomous vehicles that are controlled entirely by agents, a scenario that will become possible in the near future. In this chapter, the authors seize the opportunities of multiagent learning offered by such a scenario, by introducing a coordination mechanism where teams of agents coordinate their velocities when approaching the intersection in a decentralized way. They show that this approach enables the agents to improve the intersection efficiency, by reducing the average travel time and so contributing to alleviate traffic congestions.

INTRODUCTION

Traffic congestion is a costly problem in all developed countries. Many human-centered instruments and solutions (e.g. message signs, temporary lane closings, speed limit changes), are deployed in highways and roads in order to speed up the traffic flow. Nevertheless, in line with the recent advances of computerized infrastructures, the problem of road traffic management is being discovered as a promising application field for multiagent technology (Klügl, 2005). Multiagent systems (MAS) are the ideal candidates for the implementation of road traffic management systems, due to the intrinsically distributed nature of traffic-related problems.

In this context, the problem of advanced intersection control, where drivers interact autonomously with a coordination facility that controls the traffic flow through an intersection so as to avoid collisions while minimizing delays, is receiving more and more attention.

In (Dresner, 2004) is introduced a reservation-based system in which vehicles request an intersection manager to reserve the necessary time slots during which they may pass through the intersection. This work opens many possibilities for multiagent learning, with the goal of improving the efficiency of intersections.

In this chapter, we present a coordination mechanism based on Probability Collectives (PC) (Wolpert, 2004). With such an approach, teams of agents coordinate their velocities during their approximation to the intersection in a decentralized way, with the aim of reducing the average travel time by making better, non-conflicting, reservations.

RESERVATION-BASED INTERSECTION CONTROL

In the chapter by Dresner et al. in this book, a reservation-based system for intersection control is proposed. In such system, an intersection manager is responsible for managing the vehicles that want to pass through the intersection, by assigning the necessary time slots, while the driver agents are responsible for controlling the vehicles to which they are assigned.

A driver agent, when approaching the intersection, "calls ahead" the intersection manager and requests a reservation of space and time in the intersection, providing all the necessary information to simulate the vehicle journey through the intersection (vehicle ID, vehicle size, arrival time, arrival velocity, type of turn, arrival lane, arrival road segment,...).

If the request is confirmed by the intersection manager, the driver agent stores the reservation details and tries to meet them. Otherwise, it slows down and makes another request at a later time.

The reservation system offers many opportunities for improving the efficiency of intersection, by incorporating learning mechanisms in the agents (Dresner, 2006). For example, since the intersection manager serves the requests in a "first-come-first-served" fashion, it is possible to relax this constraint and allow the intersection manager to respond to the requests at a later time. In this way the intersection manager can evaluate more competing requests at the same time and make a more well-informed decision.

While the learning opportunities for the intersection manager are of the form of *single agent learning*, the very *multiagent learning* opportunities reside in the driver agents. In the current implementation, driver agents must estimate the arrival time at the intersection, the arrival velocity, the arrival lane ... without communication nor coordination with the other driver agents; each agent makes its request on the basis of its actual velocity, and, if the request is rejected, the driver slows down and tries again. On the other hand, by letting the agents form teams and coordinate their actions, we provide them with more information that they use to make decisions.

Agent Model

We assume that every driver agent wants to keep a preferred velocity during its journey. We also assume that when a vehicle starts to commute in a lane of a road segment, it cannot change it during the approach to the intersection[1]. In this way, if the front vehicle proceeds at a lower velocity, the following vehicle is obliged to slow down. Furthermore, as demonstrated in (Dresner, 2005), it is not convenient that the driver agents could turn from any lane, so turning right (left) is only possible from the rightmost (leftmost) lane of a road segment.

The actions that a driver agent can autonomously take are related to the velocity at which it crosses the intersection. In particular, an agent could set its velocity to a value in the (discretized) interval *[1,preferredVelocity]*.

So, for the generic driver agent a_i, the variable x_i that defines its action is x_i = *<vehicleID, direction, lane, turn, arrivalTimeAtIntersection, arrivalVelocityAtIntersection>*. The field *arrivalTimeAtIntersection* is implicitly set by the specific *arrivalVelocityAtIntersection*, while the fields *vehicleID, direction, lane* and *turn* are constant parameters.

We assume that there are no misunderstandings regarding the ontology that describes the geometric configuration of the intersection, e.g. the lane 3 along the *North* direction corresponds to the same physical lane for every vehicle.

LEARNING TO COORDINATE

Global Objective

To improve the efficiency of the intersection, we take the perspective of a system designer, whose goal is minimizing the travel time of the vehicles. The agent decision making (i.e. when and at which velocity crossing the intersection) affects each other travel time, since they compete for the common resource (i.e. the space in the intersection). So the travel time for the generic driver agent a_i depends not only on its velocity, but also on the conflicts that may occur among different requests.

Let C be a set of driver agents, $C = \{a_1, a_2, ..., a_n\}$. Each agent can take an action of the form defined in the previous section. So, the vector $x = <x_1, x_2, ..., x_n>$ defines the joint action of this set of agents. A possible function[2] that rates "how good" a joint action is, from the system designer perspective, is

$$G(x) = \left(1 + P(x)\right) \cdot D(x)$$

Equation 1. Global objective

where $P(x)$ is the number of collisions resulting from the full joint action x, and $D(x)$ is the time spent by the agents to cross the intersection. We remark that a generic joint action x contains all the necessary information to simulate the agent journeys through the intersection, so that is possible to calculate the number of conflicts among them as well as the total travel time.

Agent Private Utility

The multiagent learning challenge here is making the agents learn to act in an environment that is not merely a black-box that produces a reward for every action taken by the agent, but it is actually composed of other learning agents, i.e. the reward that an agent receives for its actions depends also on the actions of other agents. So there is a strict relation between the private utility function of a single agent and the global objective of the system.

A recent advance in this direction is that proposed by the COllective INtelligence (COIN) (Wolpert, 1999a; Wolpert, 1999b; Wolpert, 2001) framework. The aim of COIN is studying the the relation between the global objective and the private utility functions of the learning agent situated in a multiagent environment. COIN introduced the concepts of *factoredness* and *learnability* of an agent private utility function. A private utility function g_i is meant to be *factored* if it is aligned with the global utility G, i.e. if the private utility increases, the global utility does the same. Furthermore, it has to be easily *learnable*, i.e. it must enable the agent to distinguish its contribution to the global utility from that of the other agents. For example, the *Team Games Utility*, $TGU_i(x) = G(x)$, is trivially *aligned*, but is poorly *learnable*. If for example agent a_i takes an action that actually improves the global utility, while all the other agents take actions that worsen the global utility, agent a_i wrongly believes that its action was bad.

Better results have been obtained (Wolpert, 2001) with the *Difference Utility* (*DU*), defined as follows:

$$DU_i(x) = G(x) - G(CL_i(x))$$

Equation 2. Difference utility

where x is the joint action of the collective, $G(x)$ is the global utility derived from such joint action, and $G(CL_i(x))$ "virtual" joint action formed by replacing with a constant factor c all the components of x affected by agent a_i. If this constant is $\vec{0}$, i.e. the null action, the DU is equivalent to the global utility minus the global utility that would have arisen if the agent a_i had been removed from the system.

Such an utility function is *aligned* with the global utility; in fact, since the second term in Equation 2 does not depend on the action taken by agent a_i, any action that improves DU_i also improves the global utility $G(x)$. Furthermore, it is more *learnable* than *TGU* because, by removing agent a_i from the dynamics of the system, it provides a clearer signal to agent a_i.

In the case of intersection control, the driver agent computes the $DU_i(x)$ as follows:

$$DU_i(x) = \left(1 + P(x)\right) \cdot D(x) - \left(1 + P\left(CL_i(x)\right) \cdot D\left(CL_i(x)\right)\right)$$

where $CL_i(x) = < x_1, ..., x_{i-1}, x_{i+1},...,x_n >$

Probability Collectives (PC)

Once the agents in a collective have been provided with "well-designed" private utility functions, many methods are available for supporting the agent decision making, such as reinforcement learning (Sutton, 1998). In this paper we draw upon a novel method called Probability Collectives (PC) (Wolpert, 2004), which has been developed within the COIN framework, for the agent decision making. PC replaces the search in the space of *actions* with the search in the space of *probability distributions* over those actions. In other words, PC aims at learning the agent decision strategies that maximize the global objective.

Formally, let $C = \{a_1, a_2, ..., a_n\}$ be a collective of n agents. Each agent a_i can take an action by setting its action variable x_i, which can take on finite number of values from the set X_i. So these $|X_i|$ possible values constitute the action space of the agent a_i. The variable of the joint set of n agents describing the collective action is $x = <x_1, x_2, ..., x_n> \in X$, with $X = X_1 \times X_2 \times ... \times X_n$.

Given that each agent has a probability distribution (i.e. mixed strategy in game theory sense) over its possible actions, $q_i(x_i)$, the goal of PC is to induce a product distribution $q = \Pi q_i(x_i)$ that is highly peaked around the x that maximize the objective function of the problem, and then obtaining the optimized solution x by sampling q.

The main result of PC is that the best estimation of the distribution q_i that generates the highest expected utility values is the minimizer[3] of the Maxent Lagrangian (one for each agent):

$$L_i(q_i) = E_q[g_i(x)] - T \cdot S(q_i)$$ *Equation 3. Maxent Lagrangian*

where q_i is the agent probability distribution over the actions of agent a_i; $g_i(x)$ is the agent a_i private utility function (e.g. the *Difference Utility* defined in equation 2), which maps a joint action into the real numbers; the term $E_q[g_i(x)]$ is the expected utility value for agent a_i, subjected to its action and the actions of all the agents other than a_i; $S(q_i)$ is the Shannon entropy associated with the distribution q_i, $S(q_i) = -\sum_{x_i} q_i(x_i) \cdot ln[q(x_i)]$; T is an inverse Lagrangian multiplier, which can be treated as a "temperature": high temperature implies high uncertainty, i.e. *exploration*, while low temperature implies low uncertainty, i.e. *exploitation*.

Since the Maxent Lagrangian is a real valued function of a real valued vector, it is possible to use gradient descent or Newton methods for its minimization. Using Newton methods, the following update rule is obtained:

$$q_i^{t+1} = q_i^t - \alpha \cdot q_i^t \times \left\{ \frac{E_q(g_i \mid x_i) - E_q(g_i)}{T} + S(q_i^t) + \ln[q_i^t] \right\}$$ *Equation 4. Nearest Newton update*

where $E_q[g_i]$ is the expected utility, $E_q[g_i \mid x_i]$ is the expected utility associated with each of the agent a_i's possible actions, and α is the update step. Equation 4 shows how the agents should modify their distributions in order to jointly implement a step in the steepest descent direction of the Maxent Lagrangian.

Since at any time step t, an agent might not know the other agents' distributions, in this case it wouldn't be able to evaluate any expected value of g_i, because they depend on the full probability distribution q. Those expectation values can be estimated by repeated Monte Carlo sampling of the distribution q to produce a set of $(x; g_i(x))$ pairs. Each agent a_i then uses these pairs to estimate the values $E_q[g_i \mid x_i]$, for example by uniform averaging of the g_i values in the samples associated with each possible action.

PC for Intersection Control

PC is a broad framework for the analysis, control and optimization of distributed systems that offers new approaches to problems. Nevertheless, in order to be actually instantiated in a particular domain, several design decisions must be made.

Since the entire framework is based on the Monte Carlo-based estimation of the product distribution that maximizes the global objective, it is necessary to have a communication structure that enables to build the set of sampled joint actions. For example in (Waldock, 2007) such a set is constructed using a token-ring message passing architecture. In this work, we opted for letting the agents asynchronously request the other agents in the collective to sample their distributions. Then each agent constructs locally its set of sampled joint actions and uses them to update its distribution with equation 4. We assume that the agents truthfully sample their distributions without manipulation, even if investigating how an agent can exploit the coordination mechanism for its purposes deserves a further analysis.

Another design decision is the setting of the initial temperature T and the initial probability distribution q_i. The initial temperature usually depends on the particular domain, because its order of magnitude is strictly related with the expected utility values (see Equation 4). In our experiments we set the initial temperature to *1*. On the other hand, the initial probability distribution q_i is usually initialized with the maximum entropy distribution, i.e. the uniform distribution over the action space X_i. In this way we don't make any assumptions about the desirability of a particular action and all the actions are equiprobable.

Usually, the Lagrangian minimization proceeds as follows: for a given temperature T, the agents jointly implement a step in the steepest descent direction of the Maxent Lagrangian using Equation 4. Then the temperature is slightly reduced, and the process continues, until a minimum temperature is reached. The annealing schedule we implemented was geometrically reducing the temperature T as long as a driver agent approaches the point after which it is obliged to send a request to the intersection manager. When a driver agent arrives at that point, it evaluates the action with the highest probability, sets its velocity accordingly and makes a reservation request with the given velocity.

Algorithm in table 1 sketches the algorithmic structure of an agent program that implements PC for the intersection control problem. The algorithm starts initializing the temperature T and the probability distribution q_i (line 01 and 02). The main loop controls the annealing schedule of the temperature T (line 09), until the driver agents reaches the minimum distance to the intersection (line 03).

The minimization of L_i for a fixed temperature is accomplished by repeatedly determining all the conditional expected values $E_q[g_i \mid x_i]$ (line 06) and then using these values to update the distribution (line 07). Such values are obtained by requesting samples to the agents in the collective (line 04) and storing them when they are received (line 10), in order to have an estimation of the entire distribution q. At the end of the algorithm, agent a_i selects its "best" action by sampling the distribution q_i or di-

Table 1. PC Algorithm for intersection control

```
01: T ← 1
02: qi ← uniformDistribution
03: while minimum distance not reached do
04:     requestMCsamples
05:     if m not empty then
06:         ce ← evalConditionalExpectations(m)
07:         qi ← updateQ(ce)
08:     end if
09:     T ← updateT
10:     m ← storeIncomingMCSamples
11: end while
12:
13: xi ←mostProbableAction
14: velocity ← xi.arrivalVelocityAtIntersection
15: store request R = < vehicleID, direction, lane, turn,
arrivalTimeAtIntersection, velocity >
```

rectly selecting the action with the highest probability, and then store the request that will be sent to the intersection manager.

From this point on, the driver agent starts to behave like in the reservation-based scenario (for more details, see the chapter by Dresner et al.). It sends reservation requests to the intersection manager, until it receives a confirmation or a refuse message. In the first case, the driver agent stores the reservation details and tries to meet them. Otherwise, it decreases its velocity and makes another request in the next step.

A driver agent is not allowed to cross the intersection with an out-of-date reservation or without reservation at all. A confirmed reservation goes out-of-date if the agent cannot be at the intersection at the time specified in the reservation, due to the traffic conditions. In this case, the driver agent must cancel the reservation with the intersection manager and make a new one, whose constraints it is able to meet.

If a driver agent arrives at the intersection without a confirmed and valid reservation, it is obliged to stop at the intersection. At this point, the driver agent is only allowed to propose reservations for the time slots in the near future.

EXPERIMENTAL RESULTS

In this section we present the results of the experiments made with a simulator of a *4-ways-3-lanes* intersection (see Figure 1). The metric we used to evaluate the efficiency of the intersection was the average travel time of a set of vehicles. During the simulation, a total of 100 vehicles were generated using a Poisson distribution $f(k,\mu) = \frac{\mu^k \cdot e^{-\mu}}{k!}$ where μ is the number of expected occurrences (i.e. vehicles) in a given interval. In all the experiments, the μ parameter is kept fixed, while we progressively reduce the interval, simulating in this way different (increasing) traffic densities. Each spawned vehicle has a preferred velocity, whose value is generated randomly using a gaussian distribution with mean 3 and variance 1, and the maximum allowed velocity was set to 10.

One challenge in the implementation of the coordination mechanism was coping with the extreme dynamic and asynchronous nature of the system, as well as with the constraints imposed by the real-time. Furthermore, while in multiagent reinforcement learning it is assumed that in every learning episode the set of agents remains the same, in this case this assumption does not hold, because the set of learning agents is created dynamically. Once a driver agent appears in the managed area, its ID is stored by the road infrastructure. Then the road infrastructure periodically communicates the set of collected IDs to the agents, in order to create collective of coordinating agents.

Figure 2 shows the average travel time for two different configurations. In one configuration, each driver agent communicates exclusively with the intersection manager by making reservation requests solely on the basis of its knowledge; in the other configuration, the driver agents implement the coordination mechanism before starting making reservation requests. If the traffic density is low, the average travel time of the two configurations is approximatively the same. This is reasonable, since when the traffic density is low, few reservation requests are rejected, so no previous coordination is needed. Similarly, with high traffic density the average travel time tends to be the same for the two configurations. Again this is reasonable, because the intersection tends to be saturated by vehicles stopped at the intersection, waiting for its reservation request to be confirmed. On the other hand, in case of medium

Figure 1. Simulator snapshot

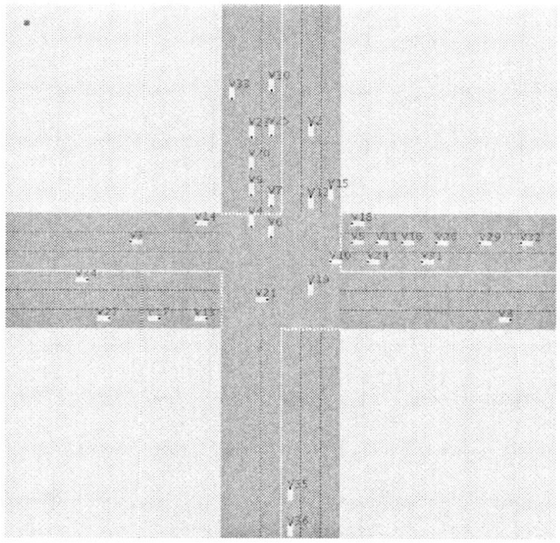

Figure 2. Average travel time

	coord.	stdev	no coord.	stdev	$\%\Delta$
Low density	143.95	5.3	143.9	4.9	+0.04%
Medium density	228.16	3.1	244.91	3.0	−6.84%
High density	269.34	4.3	269.19	4.5	+0.05%

traffic density, the coordination between drivers reduces the average travel time up to approximately the 7%, due to a lower number of refused reservations (see Figure 3).

The experimental results suggest that the fact that the intersection manager replies to each request in a *first-come-first-served* fashion shrinks the possibility of effective coordination among driver agents. Notwithstanding, there is space for further possible improvements of the agents learning capabilities. Firstly, the agent action space to act in the environment is quite reduced, since it can only set the velocity at which it intends to cross the intersection. For example, if there is a confirmed reservation of a very slow vehicle, which occupies the intersection for many time slots, it is reasonable to think that there is no way for an approaching agent to make a request that will not be refused, no matter the velocity it proposes. So, a possible improvement could derive from giving the agents the possibility of changing its lane.

Furthermore, with the current agent model, a collective of agent searches the product distribution q to maximize a global utility function $G(x)$. This is a function of the joint action x, and does not take in consideration external factors (i.e. noise). In the domain of the intersection control, for a given $x =$

Figure 3. Refused reservations

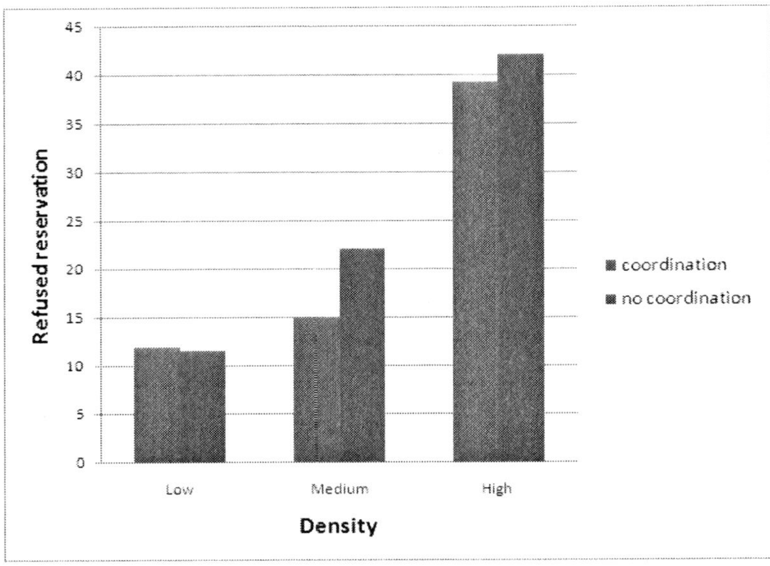

$<x_1, x_2, ..., x_n>$, an agent is only able to evaluate the number of conflicts that occurs among the x_is and their travel times, by simulating the journey of each agent a_i through the intersection. If for example the intersection is saturated due to a crash, or it has been reserved by very slow vehicles, the collective is not able to react to these events and adjust its collective behaviour, since it does not have such information.

A way to circumvent this problem is modifying the structure of the global utility as a 2-players game between the collective and the external world. At each time step, the collective sets its joint action x, while the world plays y. Such a vector y contains any external information not directly under control of the collective. Then the global objective $G(z)$ is calculated, as a function of the full vector $z = < x, y >$. In the domain of the intersection control, the vector y could contain information about the confirmed reservations that the intersection manager has in its database. Nevertheless, such noise could have side effects that eventually worsen the learning rather than improving it.

Another issue that it is worth mentioning is that the multiagent learning performed by the driver agents comes as a sort of coordination on-the-fly: an agent does not learn from the behaviours of the other agents that it observes, rather such information is explicitly provided by exchanging samples of the agents' distributions. This setting speeds up the learning, although it has an associated cost for the communication overhead. To evaluate the expected utility of its actions, $E_q[g_i \mid x_i]$ (see Equation 4), an agent uses the samples of the joint distribution q, by using the samples provided by all the agents. If we remove the communication between agents, the only way for an agent to evaluate the expected utility $E_q[g_i \mid x_i]$ is actually executing different actions in several episodes (i.e. crossing the intersection several times at different velocities), then using the utility values of the different actions it has executed in these episodes to compute $E_q[g_i \mid x_i]$ and adapt its mixed strategy q_i. Within this formalization, no communication is needed, although the learning process will take much more time.

CONCLUSION

This paper showed that the intersection control problem offers many opportunities for multiagent learning (Dresner, 2006; Bazzan, 1997; Bazzan, 2005). In particular, we started from the COIN framework for the definition of agent private utilities, and we applied Probability Collectives to make the agents learn to coordinate their action. The preliminary experiments showed some improvements of the intersection efficiency, with a reduction of the average travel time for a given traffic density interval.

Future works includes evaluating the model under different metrics (e.g. delay, congestion, number of refused reservation…), considering different private utility functions and global objectives, as well as modifying the model so that also external factors (i.e. noise) are taken into account in the decision making. More generally, the road traffic management scenario is open to a plethora of interesting research lines, from the study of "cooperative vs competitive" agent behaviour, to the impact of "malicious" agents that try to exploit the coordination mechanism.

REFERENCES

Bazzan, A. (1997). *An Evolutionary Game-Theoretic Approach for Coordination of Traffic Signal Agents.* PhD Thesis, University of Karlsruhe.

Bazzan, A. (2005). A Distributed Approach for Coordination of Traffic Signal Agents. *Autonomous Agents and Multi-Agent Systems, 10*(1), 131-164, Springer.

Dresner, K., & Stone, P. (2004). Multiagent traffic management: A reservation-based intersection control mechanism. *Proceedings of the 3rd International Joint Conference on Autonomous Agents and Multiagent Systems* (pp. 530-537). ACM Press.

Dresner, K., & Stone, P. (2005). Multiagent traffic management: an improved intersection control mechanism. *Proceedings of the 4th International Joint Conference on Autonomous Agents and Multiagent Systems* (pp. 471-477). ACM Press.

Dresner, K., & Stone, P. (2006). Multiagent Traffic Management: Opportunities for Multiagent Learning. *Lecture Notes in Computer Science* (pp. 129-138), volume 3898. Springer-Verlag.

Klügl, F., Bazzan, A., & Ossowski, S. (2005). *Application of agent Technology in Traffic and Transportation.* Birkhäuser.

Sutton, R. S., & Barto, A. G. (1998). *Reinforcement learning: An Introduction.* MIT Press.

Waldock, A., & Nicholson, D. (2007). Cooperative Decentralised Data Fusion Using Probability Collectives. *Proceedings of the 1st International Workshop on Agent Technology for Sensor Networks (ATSN-07), held at the 6th International Joint Conference on Autonomous Agents and Multiagent Systems.*

Wolpert, D., & Tumer, K. (1999). An introduction to COllective INtelligence. *Technical Report NASA-ARC-IC-99-63.* NASA Ames Research Center.

Wolpert, D., Wheeler, K., & Tumer, K. (1999). General principles of learning-based multi-agent systems. *Proceedings of the 3rd Annual Conference on Autonomous Agents* (pp. 77-83). ACM Press.

Wolpert, D., & Tumer, K. (2001). Optimal payoff functions for members of collectives. *Advances in Complex Systems, 4*(2/3), 265-279 .

Wolpert, D. (2004). Information theory - the bridge connecting bounded rational game theory and statistical physics. *Understanding Complex Systems* (pp. 262-290). Springer-Verlag.

ENDNOTES

1 This is a feature that we plan to remove from the model in the future.

2 It is possible to formulate other objective functions that take in consideration different relationship between collisions and time, as well as including other aspects, such as congestion or lane changes.

3 Without loss of generality, the global utility function is considered as a "cost" to be minimized, by simply flipping the sign of the utility value.

Chapter XIV
New Approach to Smooth Traffic Flow with Route Information Sharing

Tomohisa Yamashita
National Institute of Advanced Industrial Science and Technology (AIST), Japan

Koichi Kurumatani
National Institute of Advanced Industrial Science and Technology (AIST), Japan

ABSTRACT

With maturation of ubiquitous computing technology, it has become feasible to design new systems to improve our urban life. In this chapter, the authors introduce a new application for car navigation in a city. Every car navigation system in operation today has the current position of the vehicle, the destination, and the currently chosen route to the destination. If vehicles in a city could share this information, they could use traffic information to improve traffic efficiency for each vehicle and the whole system. Therefore, this chapter proposes a cooperative car navigation system with route information sharing (RIS). In the RIS system, each vehicle transmits route information (current position, destination, and route to the destination) to a route information server, which estimates future traffic congestion using current congestion information and this information and feeds its estimate back to each vehicle. Each vehicle uses the estimation to re-plan their route. This cycle is then repeated. The authors' purpose in this chapter is to confirm the effectiveness of the proposed cooperative car navigation system with multiagent simulation. To evaluate the effect of the RIS system, we introduce two indexes; individual incentive and social acceptability. In theor traffic simulation with three types of road networks, the authors observe that the average travel time of the drivers using the RIS system is substantially shorter than the time of other drivers. Moreover, as the number of the RIS drivers increases, the average travel time of all drivers decreases. As a result of simulation, this chapter confirms that a cooperative car navigation system with the RIS system generally satisfied individual incentive and social acceptability, and had a effect for the improvement of traffic efficiency.

1. INTRODUCTION

With maturation of ubiquitous computing technology, particularly with advances in positioning and telecommunications systems, we are now in a position to design advanced assist systems for many aspects of our lives. However, most of the research we have seen to date has focused on aspects of supporting a single person. We believe a mass user support system (Kurumatani, 2004; Kurumatani, 2003) would have a large impact on society. The new concept would benefit not only society as a whole but would also benefit individuals. In particular, Nakashima (Nakashima, 2003) and Noda (Noda, 2003) have focused on technologies that might enhance urban social life, especially transportation support systems. This chapter reports on our recent multiagent simulation demonstrating the effectiveness of a new kind of car navigation system.

Many researchers have been trying to design better navigation systems, by examining the variety of traffic information available (Bazzan, & Klügl, 2005; Bazzan, Fehler, & Klügl, 2006; Klügl, Bazzan, & Wahle, 2003; Shiose, Onitsuka, & Taura, 2001). However, previous research efforts have revealed that individually optimizing performance with only traffic congestion information is difficult (Mahmassani & Jayakrishnan, 1991; Tanahashi, Kitaoka, Baba, Mori, Terada, & Teramoto, 2002; Yoshii, Akahane, & Kuwahara, 1996). A navigation system recommends the route for the shortest estimated travel time based on the current state of traffic congestion. However, if other drivers, using the same information, simultaneously choose the same route, traffic would become concentrated on the new route.

Active queue management algorithms for TCP (Transmission Control Protocol) traffic, e.g., Random Early Detection (Floyd & Jacobson, 1993) are similar to city traffic management. TCP is one of the core protocols of the Internet protocol suite. Vehicles in road transport system are similar to IP packets in Internet. However, these algorithms are unsuitable for traffic flow in road transportation systems for two reasons. One is a physical constraint: dropping vehicles like packets in TCP traffic is impossible. The other is a social constraint: such algorithms are problematic from the standpoint of fairness because the utilities of the vehicles that are randomly dropped (or stopped) suffer a big loss.

Car navigation systems were originally designed as electronic enhancements of maps automatically indicating the current position of the vehicle and a route to the destination. Japan roads now support the second generation of car navigation systems connected to VICS (Vehicle Information and Communication System) (Vehicle Information and Communication System Center, 1995). This new system can download traffic information and display it on the map. The system uses the information to avoid congested routes when it plans a route. What we suggest in this chapter is yet another generation of car navigation systems. VICS measures traffic volume with sensors located on roadsides, e.g., radar, optical and ultrasonic vehicle detectors and CCTV (Closed Circuit Television) cameras. The gathered information is transmitted using infrared beacon, radio wave beacon, and FM multiplex broadcasting. Each car just receives information from VICS, but does not return any.

If a car could transmit information by using a mobile phone or other short-range communication, we believe that we could design a far better navigation system. Every car navigation system in operation today has the current position of the vehicle, the destination, and the currently chosen route to the destination. If vehicles in a city could share this information, they could use traffic information to improve traffic efficiency for each vehicle and the whole system. Our idea is thus a cooperative car navigation system with route information sharing.

The main purpose of this chapter is to report the results of simulations demonstrating the validity of our idea. In particular, the simulation showed the average travel time is substantially shorter when drivers use the RIS mechanism. As the number of the RIS users increases, the total amount of traffic congestion of the city decreases. Before we go into the details of the simulation, let us suggest a further capability of the idea presented here. Estimated traffic volume based on gathered route information can be used in many other ways. One of the simple usages is to reflect it in the timing of traffic signals. We can increase the green light time of the traffic signals that are expected to receive more traffic. Traffic lanes may also be dynamically changed. By connecting many systems in a city in a cooperative way, we can increase the physical capacity of the city's infrastructure.

2. TRAFFIC FLOW MODEL

We constructed a simple traffic flow model to examine the interdependence between traffic congestion as macro phenomena and route choice of individual drivers as micro behavior. Therefore, we did not consider the following factors: traffic signals (e.g., stopping at red lights), waiting for oncoming cars when turning at intersections, turn lanes, multiple lanes, passing, blind allies; and U-turns in lanes, not at intersections.

Our traffic flow model designates a road between intersections as a link. It is divided into several blocks based on block density method (Horiguchi, Katakura, Akahane, & Kuwahara, 1994). The block length is equal to the distance that a vehicle runs at the free flow speed of V_f of the link during one simulation step. After link division, an order is assigned to each block from downstream to upstream. Concerning the block assigned to be the *i*-th, we define K_i as the density of block *i*, *Li* as the length of block *i*, N_i as the number of the vehicles in block *i*, and *Vi* as the feasible speed of vehicles in block *i*. K_i is the division of *Ni* by *Li*. In block *i*, V_i is revised based on Greenshield's V-K relationship as follows:

$$V_i = \max(V_f(1 - \frac{K_i}{K_{jam}}), V_{\min}), \tag{1}$$

where K_{jam} is the traffic jam density. The density signifies the minimum density that prevents vehicles in a traffic jam from moving.

The process of the flow calculation between neighboring blocks *i* and *i*+1 is as follows. At every step, the speed of vehicles in each block is revised according to the *V-K* relationship. The vehicles then move forward based on this speed. The vehicles' movement is processed from downstream to upstream, as shown in Figure 1. Depending on V_i, vehicle *j* can move forward. When vehicle *j* moves from block *i* + 1 to block *i*, its speed changes from V_{i+1} to V_i. If K_i exceeds the jam density K_{jam}, no vehicles can move into block *i* from block *i* + 1. After *j*1 in front of vehicle j_2 moves, if j_1 is within a distance that allows j_2 to move forward at V_i, j_2 approaches j_1 to the minimum distance between them. Although j_2 has sufficient speed to advance, it must remain behind j_1. At the next step in block *i*, when V_i is revised based on K_i, vehicles can accelerate or slow down to V_i immediately, regardless of the speed in the last step.

3. ROUTE CHOICE MECHANISMS

In this section, we propose a new route choice mechanism with route information sharing (RIS), that is, cooperative car navigation system. To examine this proposed mechanism, we compared it with two basic

Figure 1. Direction of vehicle movement and revision of blocks

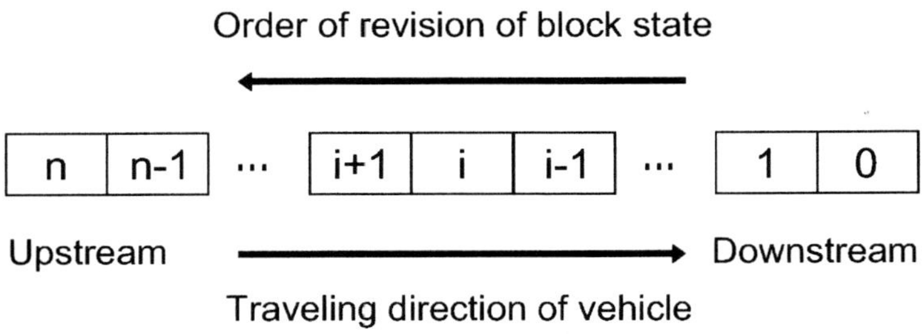

route choice mechanisms; the shortest distance route (SD) and shortest time route (ST). These mechanisms are well known and easy to understand because they seek routes minimizing the travel distance or travel time. At first, we define the mechanisms of the SD and ST. Next, we define the mechanism of the RIS based on that of the ST.

3.1 Shortest Distance Route

Drivers searching for the shortest distance route (SD drivers) select a route on a map without using information on traffic congestion. That is, SD drivers simply select the shortest distance route from their respective origin to their destination, and do not consider traffic congestion at all.

3.2 Shortest Time Route

Drivers searching for the shortest time route (ST drivers) decide a route with information on the current levels of traffic congestion. Their choice will thus vary based not only on map information, but also on current congestion information on the entire network, as would be obtained from a traffic information center (e.g., a VICS Center) via vehicle equipment. A traffic information center measures the current traffic density of all blocks, and calculates the expected travel time of each link by estimating the time spent on a link in light of the current traffic density. A traffic information center calculates expected travel time ETT_l of link l as follows.

1. Feasible speed $V_{i,l}$ on block i in l is calculated based on the V-K relationship with traffic density $K_{i,l}$.
2. Passage time $PT_{i,l}$ of block i in l is calculated based on length $L_{i,l}$ and speed $V_{i,l}$ on block i in l.
3. Expected travel time ETT_l of link l is calculated as

$$ETT_l = \sum\nolimits_{0 \le k \le n}^{n} PT_{k,l}, \qquad (2)$$

where n is the number of blocks in l.

The expected travel time is transmitted to all ST drivers at every simulation step. ST drivers search for the shortest route in terms of the expected travel times from their current position to their destination at every intersection.

3.3 Shortest Time Route with Route Information Sharing

Drivers searching for the shortest time route by using route information sharing (RIS drivers) base their selection on information sent from a route information server. Moreover, the RIS drivers transmit route information (current position, destination, and route to the destination) to the route information server. The route information server then estimates future traffic congestion levels based on this route information and transmits the estimate to the RIS drivers. The RIS drivers use the estimate to revise their route at every intersection. The route information server only provides traffic information to the RIS drivers, but does not plan the routes of drivers. Each RIS driver plans its route based on information sent from the route information server. Figure 2 shows the outline of route information sharing mechanism.

The route information sharing procedure between RIS drivers and the route information server is as follows.

1. RIS drivers search for the shortest route in terms of expected travel time from their origins to their destinations. RIS drivers decide a route using the current confusion information like ST drivers, only when departing from the destination. They transmit their route information to the route information server.
2. The route information server collects route information from all RIS drivers, and uses it to assign a passage weight for each RIS driver to a link. The passage weight indicates the degree of accuracy with which an RIS driver will pass through the link in the future. Passage weight $PW_{j,l}$ of RIS driver j to link l is calculated as follows.
    (a) If j's route passes through p links from the current position to a destination, the links are assigned numbers in ascending order from the destination to the driver's current position.
    (b) The order of each link is divided by p, and it is regarded as the passage weight of the link. (For example, $1/p$ is assigned to the link including the destination, and 1 ($=p/p$) is assigned to the link including the current position.)

Figure 2. Outline of route information sharing

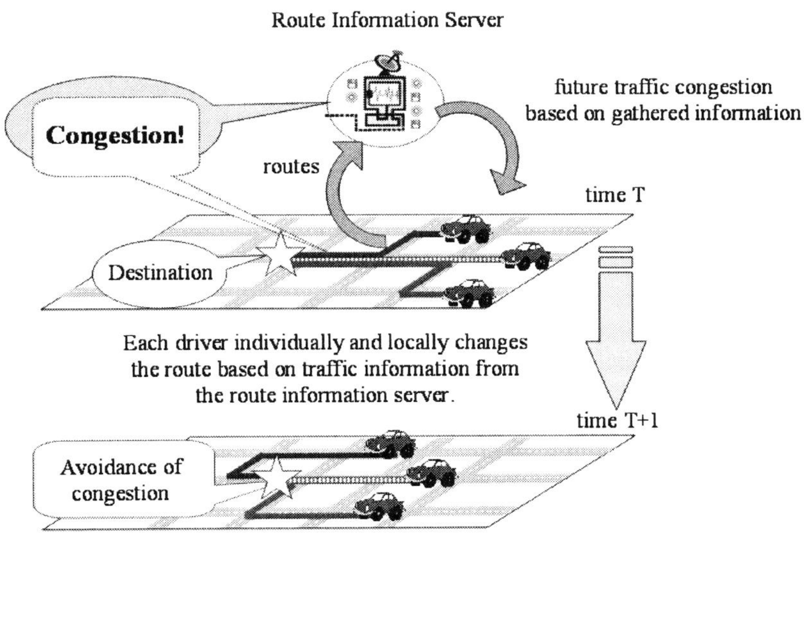

3. The route information server calculates the total passage weight of each link based on the passage weight of each link. Total passage weight means the sum of the passage weights of all RIS drivers. Total passage weight TPW_l of link l is calculated as

$$TPW_l = \sum_{k \in RIS}^{k} PW_{k,l}, \tag{3}$$

where RIS is the set of RIS drivers.

4. The route information server calculates the prospective traffic volume of each link based on the total passage weight and the expected travel time. Prospective traffic volume PTV_l of link l is calculated as

$$PTW_l = ETT_l \times (TPW_l + \alpha), \tag{4}$$

where α is a positive constant.

5. The prospective traffic volume is transmitted from the route information server to all RIS drivers. The RIS drivers revise the shortest route in the prospective traffic volume and again transmit route information to the route information server when they reach the next intersection.
6. Processes 2 to 5 are repeated.
7. The RIS drivers arriving in their destination stop transmitting their route information to the route information server.
8. The route information server stops calculating when all RIS drivers arrive in their destination.

Figure 3 shows an example of calculating total passage weight. Driver 1 has a route through six links 6, 5, 4, 3, 2, 1, from the current position on link 6 to the destination on link 1. Based on the current position, destination, and route of driver 1, the passage weights for links 1 to 7 of driver 1 are:

$$PW_{1,1} = 1/6, PW_{1,2} = 2/6, PW_{1,3} = 3/6, PW_{1,4} = 4/6,$$
$$PW_{1,5} = 5/6, PW_{1,6} = 6/6, PW_{1,7} = 0. \tag{5}$$

Figure 3. Sample calculation of total passage weight

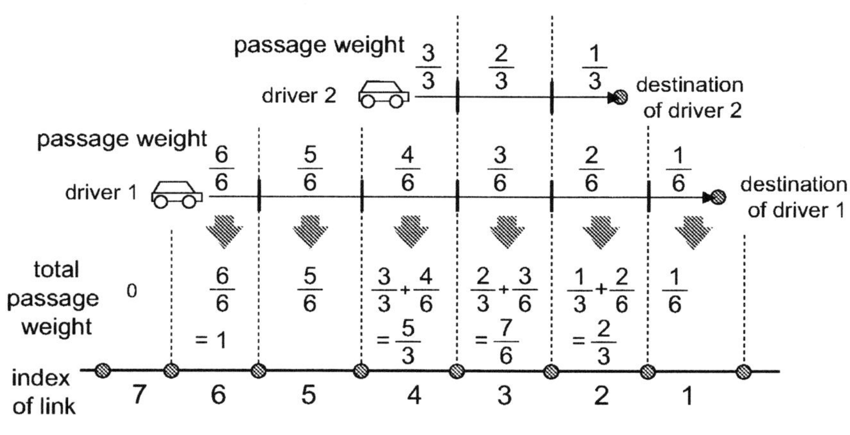

Driver 2 has a route through three links 4, 3, 2, from the current position on link 4 to the destination on link 2. Similarly, the passage weights of links 1 to 7 of driver 2 are:

$$PW_{2,1} = 0, PW_{2,2} = 1/3, PW_{2,3} = 2/3, PW_{2,4} = 3/3,$$
$$PW_{2,5} = PW_{2,6} = PW_{2,7} = 0. \quad (6)$$

Given the passage weights for links 1 to 7 of drivers 1 and 2, the total passage weights of link 1 to 7 are:

$$TPW_1 = 1/6, TPW_2 = 2/3, TPW_3 = 7/6, TPW_4 = 5/3,$$
$$TPW_5 = 5/6, TPW_6 = 1, TPW_7 = 0. \quad (7)$$

4. MULTIAGENT SIMULATION

4.1 Simulation Settings

To evaluate the RIS mechanism, we performed a multiagent simulation using the three route choice mechanisms for which the ratio of ST and RIS drivers varied from ST: RIS = 0.8:0 to ST: RIS = 0:0.8 and the ratio of the SD drivers was fixed at 0.2. This setting was based on estimation that car navigation systems and traffic information services will be more easily accessible for many drivers in the near future.

Furthermore, we evaluated the effectiveness of the RIS mechanism on three different road networks: a lattice network, a radial and ring network, the network around the city of Tokyo (see Figures 4, 5, and 6 and also Table 1). In particular, the Tokyo network matches the structure of the main trunk roads and expressways within the 8 km centered on the Imperial Palace in Tokyo, Japan. In these road networks, all links has only one lane and all blocks in a link have the same capacity. The origin and destination of a vehicle are assigned randomly to any block on any link. After reaching its destination, the vehicle is removed from the network.

The number of vehicles to a capacity of traffic systems influences the effect of traffic information systems for shorting travel time. It is preferable that the effect of the RIS mechanism is examined under general traffic conditions. In our simulation, we apply the number of vehicles in a traffic system belonging to following traffic environment; the number of vehicles is within range of the capacity of a traffic system. The concentration of certain vehicles causes traffic congestion, and their travel time should be lengthened. If vehicles can avoid traffic congestion, their travel time can be shortened.

Vehicles are generated every simulation step, until the amount of vehicles reaches 25,000. Table 2 lists the numbers of vehicles generated in one step N_{gen}. Based on our preliminary simulation results, it was confirmed that the number of vehicles generated in one step in Table 2 realizes a preferable traffic situation in which roads are not vacant, yet a deadlock does not occur.

Other parameters used in our traffic simulation are listed in Table 3.

Figure 4. Lattice network

Figure 5. Radial and ring network

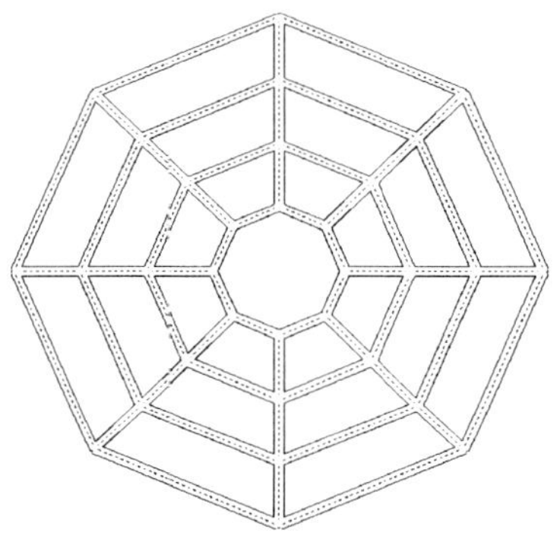

Figure 6. Tokyo network

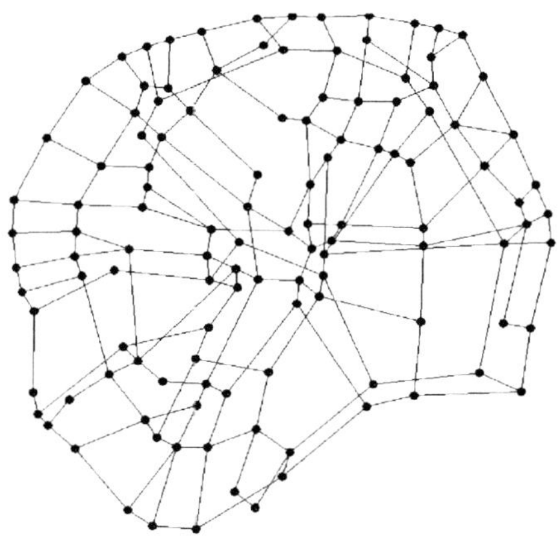

Table 1. Settings of three networks

	lattice	radial and ring	Tokyo
Number of nodes	36	32	120
Number of links	60	56	200
Number of blocks	1,200	1,168	4,034

Table 2. Number of vehicles generated in one step (total number of vehicles generated is 25,000)

⊠	lattice	radial and ring	Tokyo
N_{gen}	40, 45	30, 35	55, 65

Table 3. Parameters in traffic simulation

free flow speed	V_f	50km/h
minimum speed	V_{min}	6km/h
critical density	K_c	70veh/km
jam density	K_{jam}	140veh/km
minimum distance between two cars	D_{min}	3.5m
length of block	L_i	138.9m
traffic capacity of block		1750veh/h
1 simulation step		10sec
coefficient of prospective traffic volume	α	1.0

4.2 Simulation Results

We were particularly interested in the transition of the average travel time of each mechanism as the ratio of RIS drivers increased.

The travel time of each driver was normalized by the ideal travel time to compare the results the different road networks and different sets of vehicle origins and destinations. The ideal travel time is the time required from origin to destination when a driver passes through the shortest distance route at free flow speed. The travel time is thus defined as the ratio of the actual travel time to the ideal travel time.

Figures 7 to 12 show the simulation results averaging 30 trials. In these bar graphs with error bar, a bar graph represents the average travel time using each route choice mechanism. An error bar on the bar graph represents the upper and lower 95% confidence limits of the average travel time if the data are assumed to be normally distributed, i.e., 95% of all average travel times in 30 trials lies within the interval between the upper and lower limits. These limits are calculated as

$$mean \pm 1.96\sigma, \qquad (8)$$

Figure 7. Average travel time with N_{gen} = 40 in the lattice network (Ratio of SD drivers is fixed at 0.2)

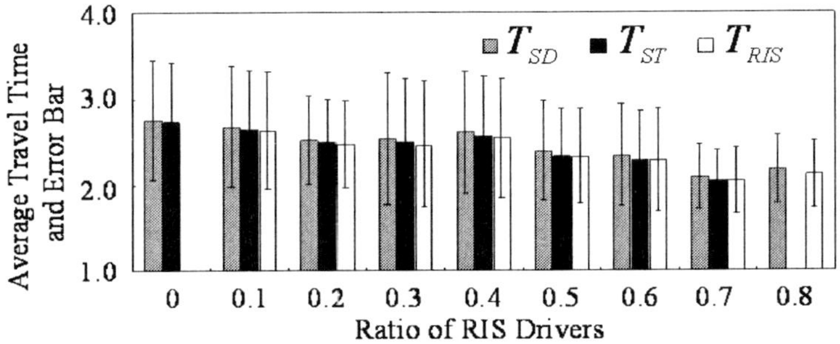

Figure 8. Average travel time with N_{gen} = 45 in the lattice network (Ratio of SD drivers is fixed at 0.2)

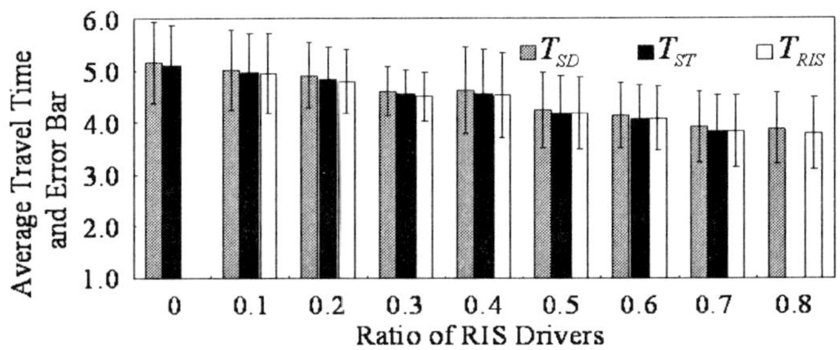

where *mean* is the average travel time in 30 trials and σ is standard deviation of the average travel time. The ratio of RIS drivers among all drivers is denoted as R_{RIS}. The average travel times of the SD, ST, and RIS drivers are denoted as T_{SD}, T_{ST}, and T_{RIS}.

Figure 7 shows the result with N_{gen} = 40 in the lattice network. The average travel times of the three types decreased irregularly as R_{RIS} increased. In particular, T_{SD}, T_{ST}, and T_{RIS} at R_{RIS} = 0.4 were longer than those at R_{RIS} = 0.3 as R_{RIS} increased. Similarly, T_{SD} and T_{RIS} at R_{RIS} = 0.8 was longer than those at R_{RIS} = 0.7. The average travel times were ranked in ascending order as T_{SD}, T_{ST}, and T_{RIS}. For R_{RIS} = 0.5.

Figure 8 shows the result with N_{gen} = 45 in the lattice network. The average times of all three types decreased monotonically as R_{RIS} increased, and they were always ranked in ascending order as T_{SD}, T_{ST}, and T_{RIS}. In all cases of R_{RIS}, there were only marginal differences among them.

In Figures 7 and 8, the error bars of the three types do not become short greatly although R_{RIS} increased.

Figures 9 and 10 show the the result with N_{gen} = 40 and with Ng_{en} = 45 in the radial and ring network. In both cases, the average times of all three types decreased monotonically as R_{RIS} increased and were ranked in ascending order as T_{SD}, T_{ST}, and T_{RIS}. There was only a marginal difference between T_{ST}, and T_{RIS}. Only in the case with N_{gen} = 35 at R_{RIS} = 0.7 was T_{ST} longer than T_{RIS}.

In Figures 9 and 10, the error bars of the three types decrease monotonically as R_{RIS} increased.

Figure 11 shows the result with N_{gen} = 55 in the Tokyo network. Except at R_{RIS} = 0.6 and 0.7, the average times of all three types decreased monotonically as R_{RIS} increased and were ranked in ascending order as T_{SD}, T_{ST}, and T_{RIS}. In all cases of R_{RIS}, there were only marginal differences between T_{ST} and T_{RIS}.

Figure 12 shows the result with N_{gen} = 65 in the Tokyo network. Except at R_{RIS} = 0.6, T_{ST}, and T_{RIS} decreased substantially as R_{RIS} increased. T_{SD} increased at R_{RIS} = 0.2, 0.3, and 0.6 as R_{RIS} increased. The average times were ranked in ascending order as T_{SD}, T_{ST}, and T_{RIS}. But, at R_{RIS} = 0.6 and 0.7, T_{ST} was shorter than T_{RIS}.

In Figures 11 and 12, the error bars of the ST and RIS become short as R_{RIS} increased. On the other hand, the error bar of the SD does not change.

5. DISCUSSION

5.1 Evaluation of RIS System

At first, based on our simulation result, we evaluate the effectiveness of the RIS mechanism from the viewpoint of whether it promotes individual incentive and social acceptability. Individual incentive

Figure 9. Average travel time with Ngen = 30 in the radial and ring network (Ratio of SD drivers is fixed at 0.2)

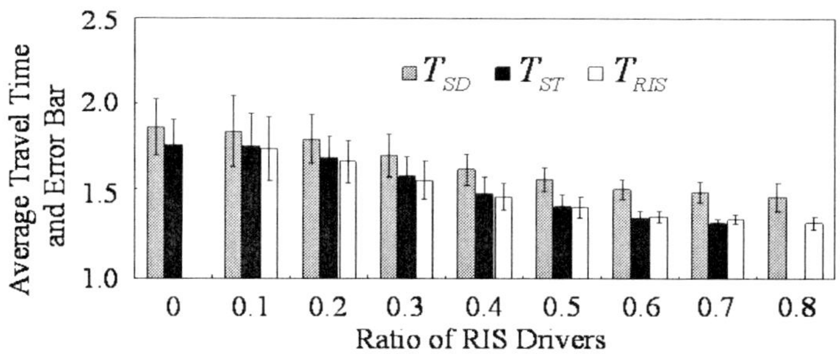

Figure 10. Average travel time with Ngen = 35 in the radial and ring network (Ratio of SD drivers is fixed at 0.2)

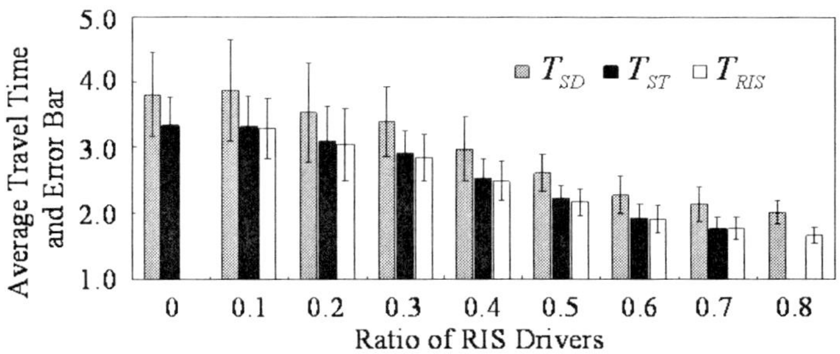

Figure 11. Average travel time with N_{gen} = 55 in the Tokyo network (Ratio of SD drivers is fixed at 0.2)

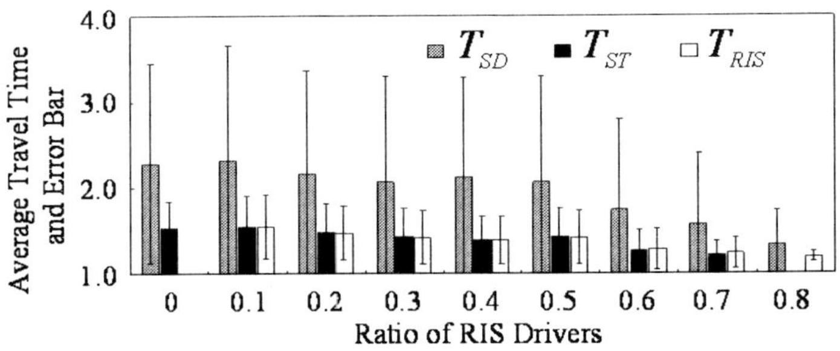

Figure 12. Average travel time with N_{gen} = 65 in the Tokyo network (Ratio of SD drivers is fixed at 0.2)

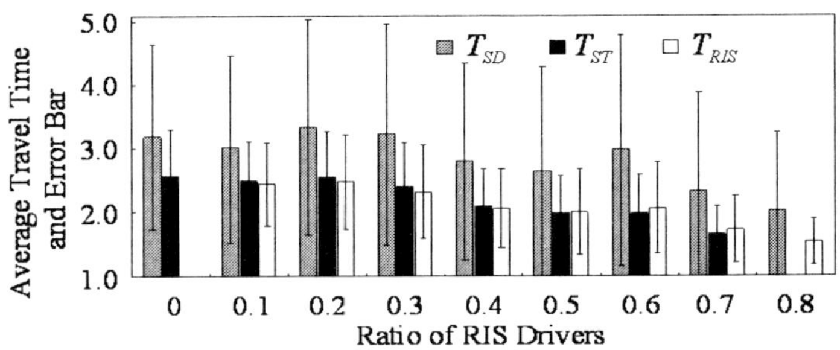

means an incentive by which a driver would switch from using the other navigation mechanisms to the RIS mechanism. Here it is significant that the traffic efficiency of the RIS drivers is always higher than that of drivers using the other mechanisms we simulated. Social acceptability means the acceptability of the RIS mechanism to promote its popularity. Here it is notable that as the number of RIS drivers increases, their traffic efficiency improves.

Our simulation showed that T_{RIS} was always shorter than the other average times. Therefore, of the RIS system seems to promote individual incentive in the lattice network. Individual incentive was also promoted in the radial and ring network and the Tokyo network, because T_{RIS} was shorter than the other times except that T_{RIS} was slightly longer than T_{ST} in the case of N_{gen} = 35 at R_{RIS} = 0.7 in the radial and ring network, and in the case of N_{gen} = 65 at R_{RIS} = 0.6 and 0.7 in the Tokyo network. In the lattice and the radial and ring networks, our method promoted social acceptability because T_{RIS} decreased monotonically as R_{RIS} increased. In the Tokyo network, social acceptability was substantially promoted because T_{RIS} increased slightly as R_{RIS} increased only in the case of N_{gen} = 65 at R_{RIS} = 0.6.

It follows from these results that the RIS mechanism can realize shorter travel time than other mechanisms, and that the travel time of the RIS drivers decreases as the number of RIS drivers increases. Moreover, the results confirm the RIS mechanism's effectiveness in promoting both individual incentive and social acceptability.

Next, we analyze the effectiveness of the RIS mechanism as follows. The reason of congestion based on current congestion information is that many ST drivers tend to choose the same vacant route simultaneously when current congestion information is broadcasted. After broadcasted, congestion is caused in the route because many ST drivers expecting to be vacant concentrate on one route. Therefore, to broadcast current congestion information may cause congestion. The route that is vacant now and will be crowded later can be detected based on the route information where each vehicle is and where it passes and is going. By broadcasting the route that is vacant now and will be crowded later, some RIS drivers change their routes and concentration can be prevented. This prevention of concentration is realized by cooperative car navigation system with route information sharing.

Many previous researches asserted only individual incentive of their proposed traffic information system at certain diffusion rate of them. However, the effect of traffic information systems significantly depends on its diffusion rate. In our research, we introduced social acceptability as another index estimating whether an information system can spread or not. Furthermore, we examine the effect of our proposed RIS system from the point of view of social acceptability, and confirm that the RIS system satisfied social acceptability. Previous research revealed that traffic Information systems providing current congestion status does not satisfy social acceptability (and partly satisfy individual incentive) (Mahmassani & Jayakrishnan, 1991; Tanahashi, Kitaoka, Baba, Mori, Terada, & Teramoto, 2002; Yoshii, Akahane, & Kuwahara, 1996). Therefore, the result that the RIS system satisfied both individual incentive and social acceptability is significantly valuable.

5.2 Influence of Network Structure

In this subsection, we discuss the different tendencies of effect of the RIS mechanism in three kinds of road networks.

5.2.1 Lattice Network

In the lattice network, the SD drivers did not seriously concentrate on the central links because they had a number of shortest routes to choose from randomly. Traffic congestion caused by the SD drivers tended to occur suddenly on any link. Therefore, under these circumstances, the ST and RIS drivers had more difficulty in preliminarily avoiding congested links, and were often caught in traffic congestion. Accordingly, the differences among T_{SD}, T_{ST}, and T_{RIS} were small. Because the ST and RIS drivers were often involved in the traffic congestion caused by the ST drivers, T_{SD}, T_{ST}, and T_{RIS} decreased slightly irregularly for $N_{gen} = 40$. Furthermore, when the congestion on a link was caused by the SD drivers, traffic congestion often occurred on other links because the ST drivers concentrated on links other than the ones affected by SD drivers. Therefore, T_{SD}, T_{ST}, and T_{RIS} decreased overall because this kind of traffic congestion decreased as R_{RIS} increased.

5.2.2 Radial and Ring Network

In the radial and ring network, the SD driver had only one or two shortest routes of the same distance. Because these shortest distance routes statistically tended to pass through the innermost ring when the origin and destination were assigned randomly on the entire map, the SD drivers tended to concentrate at the innermost ring. The ST and RIS drivers could avoid congestion occurring on the innermost ring.

Therefore, in the radial and ring network, T_{SD} was longer than T_{ST} and T_{RIS}. However, by avoiding the innermost ring and concentrating on vacant links, the ST drivers often caused congestion on the second and third innermost ring. Furthermore, the RIS drivers did not cause such traffic congestion on the second- and third-innermost rings. Therefore, T_{RIS} decreased monotonically as R_{RIS} increased.

5.2.3 Tokyo Network

In the Tokyo network, the SD drivers often concentrated on certain links in the center of the network because its structure is similar to the radial and ring network. The SD drivers often caused traffic congestion at the center of the network regardless of the ratio of the RIS and ST drivers. A certain distribution of origins and destinations of the SD drivers often caused traffic congestion on links besides those at the center because the asymmetric structure of the Tokyo network that naturally had bottlenecks. Therefore, in our simulation results, T_{SD} did not decrease monotonously and the error bar of the SD did not become short.

The congestion by only SD drivers is not based on the influence of traffic information and route information sharing, and is based on the setting of the SD driver, the network, and our traffic simulator. Removal of instability by the SD drivers is placed as our future work. Now, we have a plan to implement cooperative car navigation system on traffic simulator AIMSUN. The result with AIMSUN will be reported in next paper.

5.3 Realization of an RIS System

We discuss the RIS system as it might be implemented in actual services. To develop the RIS system, we must first consider its system architecture. For instance at the beginning of this chapter, we suggested direct communication between the RIS drivers and the route information server via long-distance communication, e.g., mobile phone. However, if we were to apply such an RIS system on a huge road network like the one in the Tokyo metropolitan area (with millions of cars on the roads), direct communication by phone would be impossible because the route information server could not deal with the heavy communication traffic. Instead of phones, we are considering using traffic signals as relay stations. In this architecture, traffic signals would collect route information from the RIS drivers, and transmit it to the route information server on a dedicated high-speed line.

Recently, DSRC (Dedicated Short Range Communication) (Inoue, 2004) and infrared beacons (Otakeguchi & Horiuchi, 2004) have been developed as short distance two-way communications for the intelligent transport system (ITS). These technologies have already been put to practical use. The RIS system could easily use them for communication between vehicles and traffic signals. Moreover, by connecting the traffic signal system with the RIS system (Bazzan, Oliveira, Klügl, & Nagel, 2008; Dresner & Stone, 2004), the prospective traffic volume could be used to control traffic signals.

6. CONCLUSION

We proposed a cooperative car navigation system with route information sharing (RIS). For the evaluation, we constructed a simple traffic flow model using multiagent modeling. Three types of route choice were compared in a simulation: the shortest distance route (SD), the shortest time route (ST) and the shortest time route with route information sharing (RIS). The simulations were of a lattice network, a radial and ring network, and the network around Tokyo. The simulation results confirmed that the RIS mechanism promoted i) drivers' individual incentive to switch to using it: the average travel time of the RIS drivers was always shorter than those of drivers using the other choice mechanisms, and ii) social acceptability: the travel time of RIS drivers became shorter as the percentage of RIS drivers increased. Moreover, the results showed that the network structure influenced the effectiveness of the RIS mechanism. Finally, the chapter discussed how the RIS mechanism might be implemented with traffic signals linked to a route information server.

7. REFERENCES

Bazzan, A. L. C., & Klügl, F. (2005). Case Studies on the Braess Paradox: Simulating Route Recommendation and Learning in Abstract and Microscopic Models. *Transportation Research, C 13*(4), 299-319.

Bazzan, A. L. C., Fehler, M., & Klügl, F. (2006). Learning To Coordinate In a Network of Social Drivers: The Role of Information. *Proceedings of International Workshop on Learning and Adaptation in MAS, 3898*, 115-128.

Bazzan, A. L. C., Oliveira, D., Klügl, F., & Nagel, K. (2008). Adapt or Not to Adapt – Consequences of Adapting Driver and Traffic Light Agents. *Conference proceedings of Adaptive Agents and Multi-Agent Systems III, 4865*, 1-14.

Dresner, K., & Stone, P. (2004). Multiagent Traffic Management: A Reservation-Based Intersection Control Mechanism. *Proceedings of The Third International Joint Conference on Autonomous Agents and Multi-Agent Systems*, (pp. 530-537).

Floyd, S., & Jacobson, V. (1993). Random Early Detection gateways for Congestion Avoidance. *IEEE/ACM Transactions on Networking, 1*(4), 397-413.

Horiguchi, R., Katakura, M., Akahane, H., & Kuwahara, M. (1994). A Development of A Traffic Simulator for Urban Road networks: AVENUE. *Conference proceedings of Vehicle Navigation and Information Systems Conference*, (pp. 245-250).

Inoue, M. (2004). Current Overview of ITS in Japan. *Conference proceedings of The 11th World Congress on Intelligent Transport Systems (CD-ROM).*

Klügl, F., Bazzan, A. L. C., & Wahle, J. (2003). Selection of Information Types Based on Personal Utility: A Testbed for Traffic Information Markets. *Conference proceedings of the Second International Joint Conference on Autonomous Agents and Multiagent Systems*, (pp. 377-384).

Kurumatani, K. (2004). Mass User Support by Social Coordination among Citizens in a Real Environment. *Multiagent for Mass User Support*, (pp. 1-19). LNAI 3012, Springer.

Kurumatani, K. (2003). Social Coordination with Architecture for Ubiquitous Agents: CONSORTS. *Conference proceedings of International Conference on Intelligent Agents, Web Technologies and Internet Commerce 2003* (CD-ROM).

Mahmassani, H. S., & Jayakrishnan, R. (1991). System Performance and User Response Under Real-Time Information in a Congested Traffic Corridor. *Transportation Research 25A*(5), 293-307.

Nakashima, H. (2003). Grounding to the Real World - Architecture for Ubiquitous Computing -. *Foundations of Intelligent Systems*, (pp. 7-11). Lecture Notes in Computer Science, 2871, Springer Berlin.

Noda, I., Ohta, M., Shinoda, K., Kumada, Y., & Nakashima, H. (2003). Evaluation of Usability of Dial-a-Ride Systems by Social Simulation. *Proceedings of Multi-Agent-Based Simulation III (4th International Workshop),* (pp. 167-181).

Otakeguchi, K., & Horiuchi, T. (2004). Conditions and Analysis of the Up-Link Information Gathered from Infrared Beacons in Japan. *Conference proceedings of The 11th World Congress on Intelligent Transport Systems (CD-ROM).*

Shiose, T., Onitsuka, T., & Taura, T. (2001). Effective Information Provision for Relieving Traffic congestion. *Conference proceedings of The 4th International Conference on Intelligence and Multimedia Applications*, (pp. 138-142).

Tanahashi, I., Kitaoka, H., Baba, M., H. Mori, H., Terada, S., & Teramoto, E. (2002). NETSTREAM, a Traffic Simulator for Large-scale Road networks. *R & D Review of Toyota CRDL, 37*(2), 47-53.

Yoshii, T., Akahane, H., & Kuwahara, M. (1996). Impacts of the Accuracy of Traffic Information in Dynamic Route Guidance Systems. *Conference proceedings of The 3rd Annual World Congress on Intelligent Transport Systems (CD-ROM).*

Vehicle Information and Communication System Center (1995). http://www.VICS.or.jp/english/index.html

Chapter XV
Multiagent Learning on Traffic Lights Control:
Effects of Using Shared Information

Denise de Oliveira
Universidade Federal do Rio Grande do Sul, Brazil

Ana L. C. Bazzan
Universidade Federal do Rio Grande do Sul, Brazil

ABSTRACT

In a complex multiagent system, agents may have different partial information about the system's state and the information held by other agents in the system. In a distributed urban traffic control, where each junction has an independent controller, agents that learn can benefit from exchanging information, but this exchange of information may not always be useful. In this chapter the authors analyze how agents can benefit from sharing information in an urban traffic control scenario and the consequences of this cooperation in the performance of the traffic system.

INTRODUCTION

Urban traffic control (UTC) is an important and challenging real-world problem. This problem has several important characteristics, related to its dynamics (changes in the environment are not only consequences of the agents actions, changes are beyond the agents control); to non-determinism (each action may have more than one possible effect); and to partial observability (each agent perceives a limited fraction of the current environment state).

Multiagent learning can be seen as a suitable tool for coping with the issues related to the dynamicity in this scenario. Formalizing the problem of control is an important part of the solution, and the theory of Markov Decision Processes (MDP) has shown to be particularly powerful in that context. Defining the traffic control problem as a single MDP, i.e. in a centralized way, would lead to an unsolvable problem, due the large number of possible states. For instance, consider a scenario where six traffic lights with five possible states each, according to the incoming links (streets) congestion: all links have the same number of stopped vehicles, North link has more waiting vehicles, South link has more waiting vehicles, East link has more waiting vehicles, and West link has more waiting vehicles. In this case, the number of possible states is 15,625 (56) and the number of possible joint actions is 729 (36), considering that each traffic light has three possible actions. The number of Q-values is 11,390,625 (729 x 15,625). In a decentralized solution, traffic controlling agents may have different partial information about the system state and the information held by other agents in the system.

The distributed urban traffic control (DUTC) problem has some important characteristics to be considered: a large number of possible traffic pattern configurations, limited communication, limited observation, limited action frequency and delayed reward information. Delayed reward, since the traffic flow takes some time to respond to the agent's actions, this time is, at least, the duration of one cycle time.

In this chapter we explore some questions about information shared among learning agents in a traffic scenario. We also discuss the multiagent reinforcement learning used as a solution to the traffic control problem, current limitations and further developments.

BACKGROUND

Reinforcement Learning

Reinforcement learning methods can be divided into two categories: model-free and model-based. Model-based methods assume that the transition function T and the reward function R are available or estimate those values. Model-free systems, such as Q-learning (Watkins & Dayan 1992), on the other hand, do not require the agent's access to the model of the environment.

Q-Learning works by estimating optimal state-action values, the Q-values, which are a numerical estimator of quality for a given pair of state and action. More precisely, a Q-value Q(s,a) represents the maximum discounted sum of future rewards an agent can expect to receive if it starts in *s*, chooses action *a* and then continues to follow an optimal policy.

Q-Learning algorithm approximates the Q-values Q(s,a) as the agent acts in a given environment. The algorithm uses an update rule, at line 7 in the algorithm description on Figure 1, for each experience tuple <s, a, s', r>, where α is the learning rate and γ is the discount rate for future rewards. As it can be seen in the update rule, the estimation of the Q-values does not rely on T or R, as Q-learning is model free.

The simplest action selection rule (line 5 on Figure 1) is to select the action with highest estimated action value at the current state (*greedy* solution). Although this selection method maximizes immediate reward, it does not explore other actions, (Sutton & Barto, 1998). An alternative is to choose greedily most of the time and select a random action, uniformly, independently of the action-value estimates, with a given small probability (ε). This selection method is called *ε -greedy*. Although *ε -greedy* is an

Figure 1. Q-learning algorithm

Initialize $Q(s,a)$ arbitrarily ;
foreach *episode* **do**
    Initialize s;
    repeat for each *step of episode*
        Choose a from s using a policy derived from Q;
        Take action a, observe s' and r;
        $Q(s,a) \leftarrow Q(s,a) + \alpha\,(r + \gamma max_{a'}\, Q(s',a') - Q(s,a))$;
        $s \leftarrow s'$;
    until *s is terminal*;
end

effective for balancing exploration and exploitation in reinforcement learning, it explores equally all actions. In tasks where taking the worst actions diverge significantly from the best action, this may be a problem. A solution may be in varying the action probabilities based on the estimated values. The *greedy* action remains with the highest selection probability and the others are weighted according to their value estimates. These are called *softmax* action selection rules. The most common *softmax* method uses a Boltzmann (Gibbs) distribution. It chooses action with probability given by Equation 1.

$$\frac{e^{\frac{Q_t(s,a)}{\tau}}}{\sum e^{\frac{Q_t(s,a)}{\tau}}} \qquad (1)$$

Where τ is a positive parameter called the *temperature*.

High temperatures cause the actions to have (almost) the same probability. Low temperatures cause a greater difference in selection probability for actions that differ in their value estimates. In the limit ($\tau \rightarrow 0$) as, *softmax* action selection becomes the same as *greedy* action selection. Both methods have only one parameter that must be set: ε or τ.

Independent vs. Cooperative Agents

In (Tan1993), three ways of agent cooperation are identified: information, episodic experience, and learned knowledge. The main thesis of Ming Tan is: "If cooperation is done intelligently, each agent can benefit from other agents instantaneous information, episodic experience, and learned knowledge."

Using a hunter-prey scenario, his work investigates where agents are seeking to capture a prey moving in a random direction in a 10x10 grid world. At each step, each agent has four possible actions: moving up, down, left, or right within the grid. More than one agent can occupy the same cell at the same time. There are three ways to capture the prey. One way is when the prey and one hunter occupy the same cell, another way is when two hunters and the prey are in the same cell, and the other way is when two hunters are next to the cell where the prey is. When capturing a prey, the hunters involved receive a reward of +1. One the other hand, on each move they do not capture a prey, they receive a negative reward of -0.1. Each agent has a limited visual field.

First the effect of the sensation from another agent is studied. To differentiate sensing from learning the experiments are performed on a one-prey/one-hunter scenario with a scouting agent that cannot capture prey. The scout makes random moves. At each step, the scout sends its action and sensation back to the hunter. The sensation inputs from the scout are used only if the hunter cannot sense any prey. As the scout's visual field depth increases, the difference in their performances becomes larger. After verifying that the scout information helps the hunter, this concept was extended to hunters that perform both scouting and hunting.

In the case where the agents share policies or episodes, is assumed that the agents do not share sensations. The question in this case is: "*If the agent can complete a task alone, is cooperation still useful?*" To answer this question, exchanging policies and exchanging episodes are studied. An episode is a sequence of sensation, action, and reward experienced by an agent.

An episode is exchanged between the agent that has accomplished the task and its partner, so the experience would be duplicated. The second possibility is to learn from an expert agent.

In the case with joint tasks, the hunter can capture a prey only with other agent. Cooperation can occur by passive observation or by active exchange of perceptions and location. The simulation result indicates that agents of the independent learning behavior tend to ignore the other and approach the prey directly. In the passive observation case (where the agent's perception is extended to include the partner location) the agents have a more effective learning. The mutual cooperation increases the state space without an increase in the state representation, having a slower but more effective learning process.

Cooperative reinforcement learning agents can learn faster and converge sooner than independent agents by sharing information about the environment. On the other hand, the experiments show that extra information can have a negative interference in the agent's learning if the information is unnecessary. These trade-offs must be considered for autonomous and cooperative learning agents.

Related MAS Approaches for Traffic Control

In order to have a configured traffic light, there must be a control that determines the *stage*, the *splits*, the *cycle time* and, in coordinated traffic lights, the *offset*. The offset time is a time delay between two successive intersections in order to allow vehicles to pass successive intersections without stopping. The stage specification determines the traffic movements in each part of the cycle time. The cycle time is the number of seconds needed for a complete sequence of phases. Each of these stages must have a relative green duration, this share of the cycle time is called split. We call a set of these specifications as a signal plan.

In (Bazzan2005), a MAS based approach is described in which each traffic light is modeled as an agent, each having a set of pre-defined signal plans to coordinate with neighbors. This approach uses techniques of evolutionary game theory: self-interested agents receive a reward or a penalty given by the environment. Moreover, each agent possesses only information about his or her local traffic states. However, payoff matrices (or at least the utilities and preferences of the agents) are required, i.e. these figures have to be explicitly formalized by the designer of the system. In (Oliveira et. al. 2005) an approach based on cooperative mediation is proposed, which is a compromise between totally autonomous coordination with implicit communication and the classical centralized solution. An algorithm which can deal with distributed constraint optimization problems (OptAPO) is used in a dynamic scenario, showing that the mediation is able to reduce the frequency of miscoordination between neighbor crossings. However, this mediation was not decentralized: group mediators communicate their decisions to the mediated agents in their groups and these agents just carry out the tasks. Also, the mediation process may take long in highly constrained scenarios, having a negative impact in the coordination mechanism. In (Oliveira et al. 2004) a decentralized, swarm-based approach was presented, but we have not collected and analyzed information about the group formation. Therefore, a decentralized, swarm-based model of task allocation was developed (Oliveira and Bazzan 2006, Oliveira and Bazzan 2007), in which the dynamic group formation without mediation combines the advantages of those two previous works (decentralization via swarm intelligence and dynamic group formation).

Camponogara and Kraus (2003) have studied a simple scenario with only two intersections, using stochastic game theory and reinforcement learning. Their results with this approach were better than a best effort (greedy), a random policy, and also better than Q-learning. In (Oliveira et al 2006) single agent reinforcement learning was applied on a traffic control scenario. The objective of this paper was to study if the single agent reinforcement learning methods ware capable of dealing with non-stationary traffic patterns in a microscopic simulator. The results showed that in non-stationary scenarios the learning mechanisms have more difficulty in recognizing the states. Also, it shows that independent learning mechanisms are not capable of dealing with over-saturated networks. Finally, approaches based on self-organization of traffic lights via thresholds (Gershenson 2005) or reservation-based systems (Dresner and Stone2004) have still to solve low-level abstraction issues in order to be adopted by traffic engineers and have a chance to be deployed.

The Microscopic Simulation Model

There are two main approaches to the simulation of traffic: macroscopic and microscopic. The microscopic allows the description of each road user as detailed as desired (given computational restrictions), thus permitting a model of the drivers' behaviors. Multi-agent simulation is a promising technique for microscopic traffic models as the drivers' behavior can be described incorporating complex and individual decision-making.

In general, microscopic traffic flow model describe the act of driving on a road i.e. the perception and reaction of a driver on a short time-scale. In these models, the drivers are the basic entities and their behavior is described using several different types of mathematical formulations, such as continuous models (e.g. car-following). Other models use cellular automata (CA). In particular, we use the Nagel-Schreckenberg model (Nagel and Schreckenberg 1992) because of its simplicity. The Nagel-Schreckenberg cellular automaton model represents a minimal model in the sense that it is capable to reproduce basic features of real traffic.

Next, the definition of the model for single lane traffic is briefly reviewed. The road is subdivided in cells with a length varying around 7.5 or 5 meters (for highways or urban traffic, respectively). Each cell is either empty or occupied by only one vehicle with an integer speed $v_i \in \{0, ..., v_{max}\}$, with vmax the maximum speed. The motion of the vehicles is described by the following rules (parallel dynamics):

R1: Acceleration: $v_i \leftarrow \min(v_i+1, v_{max})$;

R2: Deceleration to avoid accidents: $v_i' \leftarrow \min(v_i, gap)$;

R3: Randomizing: with a certain probability p do vi" ←max (vi-1, 0);
R4: Movement: xi ←xi-1+vi".

The variable gap denotes the number of empty cells in front of the vehicle at cell i. A time-step corresponds to $\Delta t = 1$ sec, the typical time a driver needs to react.

Every driver described by the Nagel-Schreckenberg model can be seen as a reactive agent: autonomous, situated in a discrete environment, and has individual characteristics such as its maximum speed vmax, and the deceleration probability p. During the process of driving, it perceives its distance to the predecessor (gap), its own current speed v, etc. This information is processed using the three rules (R1-R3) and changes in the environment are made using rule R4. The first rule describes one goal of the agent, as it wants to go by maximum speed vmax. The other goal is to drive safe i.e. not to collide with its predecessor (R2). In this rule the drivers assumes that its predecessor can brake to zero speed. However, this is a crude approximation of the perception of an agent. These first two rules describe deterministic behavior, i.e. the stationary state of the system is determined by the initial conditions. But drivers do not react in this optimal way: they vary their driving behavior without any obvious reasons. This uncertainty in the behavior is reflected by the braking noise p (R3). It mimics the complex interactions with the other agent s, i.e. the overreaction in braking and the delayed reaction during acceleration. Finally, the last rule is carried out, the agent acts on the environment and moves according to his current speed.

REINFORCEMENT LEARNING WITH SHARED INFORMATION IN TRAFFIC LIGHTS CONTROL

The aim of this chapter is to analyze the impact of the information exchange on the performance of learning agents in a traffic scenario, where an agent is a traffic light controller. We use the basic Q-Learning mechanism (using *softmax* as action-selection method) and the ideas about information exchange presented in (Tan, 1993) and discussed in the previous section. Each agent communication area is restricted according to the number of its incoming links. For instance, in Figure 2, the central agent (black filled circle), with 2 incoming and 2 outgoing links will have 2 partners (dash filled circles). A partner sends information for the agents directly connected to its output links and receives information from agents connected directly to its input links.

Figure 2. Example of partner's location

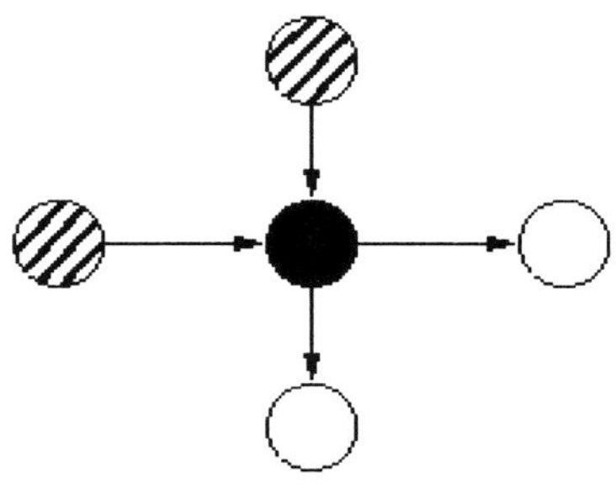

We consider three different agent types, according to the information used in the states composition. The states composition can be simple, when the agent uses only its local information (from its local sensors) or composed, when the agent uses information from its neighbors to compose the state information. The first type is the independent agent that has only the local information, having 3 possible states, according to the local traffic information: 0, 1 and 2. "State 0": streets from all directions have the approximate same number of stopped vehicles; "State 1": streets from North and South directions have a larger number of stopped vehicles than the streets from East and West; "State 2": streets from East and West directions have a larger number of stopped vehicles than the streets from North and South.

In some situations, each agent can profit from receiving information from its neighboring agents in having a wider perception of the traffic, but in others (i.e. when local links have low traffic), local information might be good enough. In these situations, the second agent type acts considering local information about the environment (when it considers the local state) indicates a direction with more stopped vehicles; otherwise it uses the information received from its neighbors. We call this process the *selective behavior*, since the agent selects the situation where to use the received information. This type of agent has 12 possible states: 3 local states plus 9 states from its neighbors.

The last type of agent is the one that uses complete information received by its partners with its local information to compose the states. This type of agent has 27 (3^3) possible states: 3 states of its own and 3 states from each partner. In this scenario we consider that the agents are always reachable if there are no changes in the topology of the traffic network (i.e.: removing one traffic light).

Another issue for DUTC is when we have coordinated traffic lights forming a "*green wave*". In this case, all the agents from a given street must have a synchronized timing (*offset*) in order to allow vehicles crossing through several junctions without stopping.

Scenario and Experiments

For the experiments we use the ITSUMO (Silva et al. 2006) microscopic traffic simulator to simulate a 9x9 Manhattan scenario with 81 intersections, each controlled by traffic lights. This simulator uses the Nagel-Schreckenberg model, previously described. Each agent controls one traffic light and all decisions are local. Figure 3 shows the graphical representation of the network. The 81 nodes correspond to RL controlled traffic lights and the street segments are represented as directed (one-way) links, with 300m, having a maximum allowed velocity set to 15m/s (3 cells/second). Each link has a capacity of 60 stopped vehicles, so the total amount of stopped vehicles in the scenario is 10,800. In all experiments we do not consider the input links in the total sum, since they are directly connected to the input mechanism and have a large number of stopped vehicles trying to get in the network. The maximum number of stopped vehicles is limited to 8,640. Each street has a *source* and a *sink*, located at the beginning and at the end of the street, respectively, and they are not shown in the network representation. Vehicles are inserted by *sources* and removed by *sinks* and do not change direction (i.e.: a vehicle inserted at the beginning of street "A" will be removed at the end of this same street, after crossing the junction "A9"). As for the insertion rates, we used a 4/10 probability of a vehicle being inserted on each *source* located on the South or in the North, at each time step, approximately 900 vehicles/hour. On *sources* located on nodes beginning at East and West (horizontal streets), there is a 1/10 probability of a vehicle being inserted at each time step, approximately 360 vehicles/hour. This means that there is a high insertion rate in the North and South sources and a low to medium insertion rate in the East and West sources.

Figure 3. Representation of the traffic network

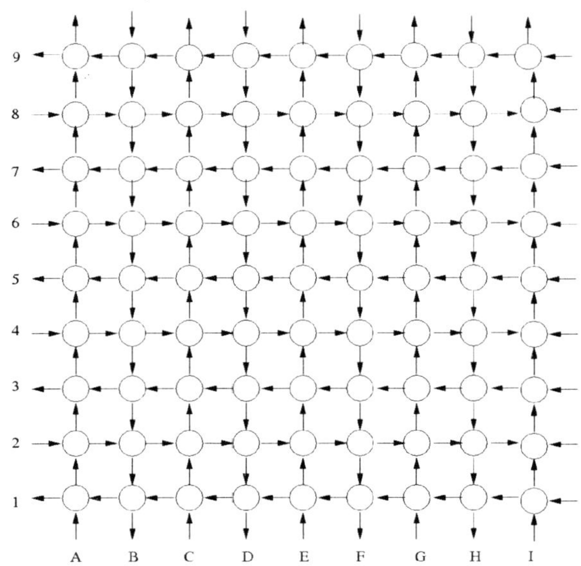

When arriving at the sinks vehicles are immediately removed. For instance, a vehicle inserted in the network by the source of "A" street with North direction will be removed at the end of the street. Vehicles may have a deceleration probability, which indicates a chance of reducing the speed without apparent reason (i.e.: no other vehicle blocking). This is aimed at modeling the phenomenon of "jams out of nothing", i.e. a source of noise in the network which causes traffic jams. Our experiments were made in three different scenarios considering the deceleration probability: no deceleration, and probability of deceleration of 5/100 or 1/10. Of course higher deceleration probabilities generate more noisy traffic patterns.

Even though decisions are local, we can assess how well the mechanism is performing by measuring global performance values as the traffic authority is normally interested in the welfare of the whole system. By using reinforcement learning to optimize isolated junctions, we implement decentralized controllers and avoid expensive processing of a central controller.

As a measure of effectiveness for the control systems, the performance is the total number of stopped vehicles. After the discretization of the traffic measure the state of each link can be empty, regular or full, based on the number of stopped vehicles (queue length). The state of an agent is given by the states of the links arriving in its corresponding traffic light. There are three possible local states for each agent, namely:

- "EQUAL": when the incoming links from the two directions are at the same state;
- "NORTH/SOUTH": when the incoming links from North and South directions have longer queues than the links from the other direction;
- "EAST/WEST": when the incoming links from East and West directions have longer queues than the links from the other direction.

The reward for each agent is based on the average number of waiting vehicles at the incoming links (average queue size). We consider three possible values for each link: 0 (if the average is lower of equal to 5 vehicles), 1 (from 6 to 10 vehicles) and 2 (11 or more waiting vehicles). The reward is higher when there are few stopped vehicles; if the queues are long (to many stopped vehicles) the reward is low. For this scenario, where the agents have only two input links, the reward function is calculated using Equation 2.

$$Reward = \frac{1.0\text{-}(2*(state\ of\ link\ 1 + state\ of\ link\ 2) - |(state\ of\ link\ 1\ \text{-}state\ of\ link\ 2)|)}{10.0} \qquad (2)$$

So, the maximum reward is 1 (no queue is longer than 5 vehicles at any link) and the lowest is 0.2 (when the queues are long in all incoming links).

The system performance is evaluated for the whole traffic network by adding up the number of stopped vehicles over all links (total stopped vehicles on the network), excluding links directly connected to sources.

Traffic lights usually have a set of signal plans. A signal plan is a set of timing and traffic movements (stage) specifications. We have two scenarios concerning the traffic plans in both scenarios. Each traffic agents has three plans with two phases: one allowing green time to direction north-south (NS), and other to direction east-west (EW). In the first scenario, the plans are not coordinated, so there is no offset time on those plans. Each of the three signal plans uses different green times for phases:

- Signal plan 1: gives equal green phase duration time for both phases;
- Signal plan 2: gives priority to the vertical direction;
- Signal plan 3: gives priority to the horizontal direction.

All signal plans have cycle time of 60 seconds and phases of either 42, 30 or 18 seconds (70% of cycle time for the preferential direction, 50% of cycle time and 30% of cycle time for non-preferential direction). The signal plan with equal phase times allows 30 seconds of green time for each direction (50% of the cycle time); the signal plan which prioritizes the vertical direction allows 42 seconds of green time to the phase N/S and 18 seconds to the phase E/W; and the signal plan which prioritizes the horizontal direction allows 42 seconds of green time to the phase E/W and 18 seconds to the phase N/S.

In the second scenario, all agents have 2 "pre-coordinated" plans, so a specific offset time was set to each plan, allowing a "green wave" formation among adjacent agents in a specific direction. All plans follow the same priority as the plans without offset. Here, instead of only giving priority to one direction, this direction has a specific offset for coordinating with the previous traffic light. In this way, we have created more dependent actions. The agent's action consists of selecting one of the three signal plans at each 120 seconds (simulation time steps). We use a 12-hour simulation time (43,200 steps) in all simulations; however, the figures show only 36,000 steps, since the first two hours are the traffic adaptation and learning period. We have made 10 simulations for each scenario and the presented graphics are relative to the average over those simulations.

Figure 4 shows a comparison of the performance of the three agent types in the same scenario with coordinated plans. In this scenario the agent with more information (27 states), which always considers the information received from its neighbors, has the better performance (lower number of stopped vehicles) compared to the others. The agent type with 3 states achieved the worst results in this scenario.

Figure 4. Performance in the scenario with no deceleration and coordinated plans

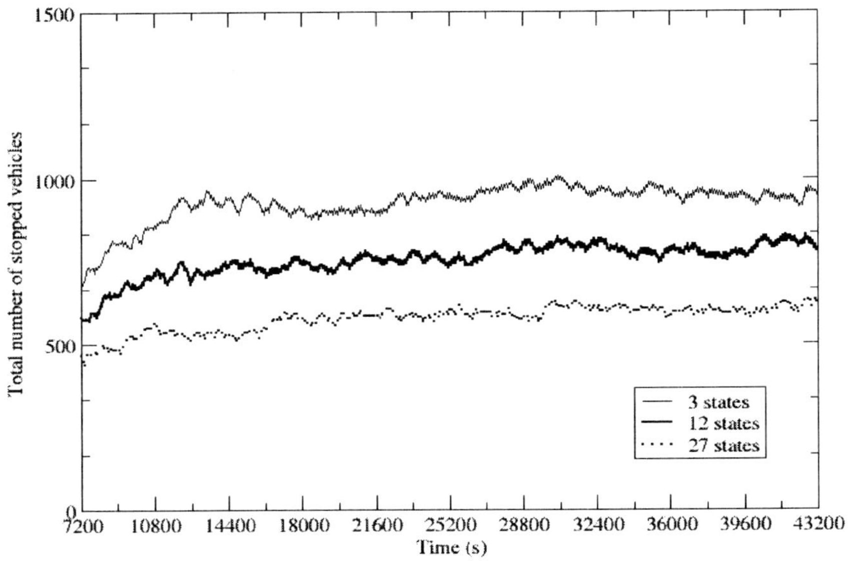

Figure 5. Performance in the scenario with no deceleration and without coordinated plans

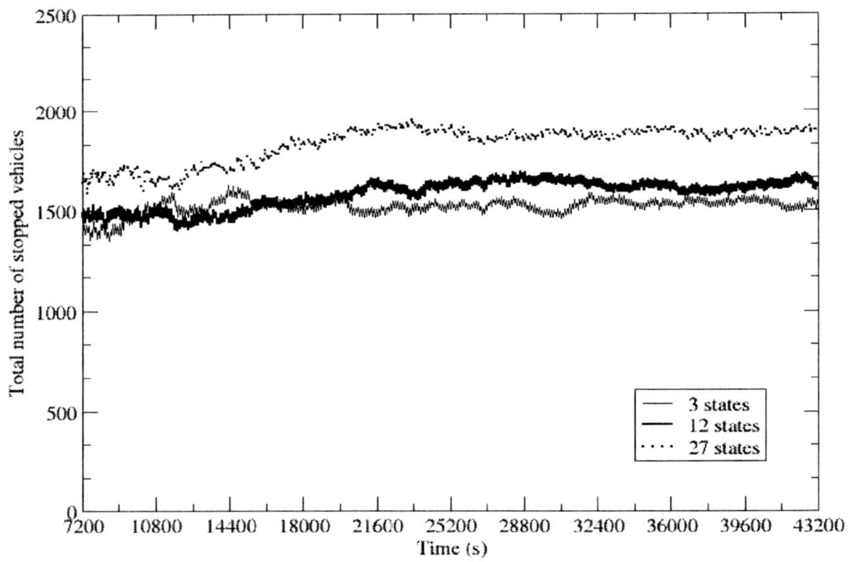

This result indicates that in this scenario, the agents with 27 states are using the information received to have a better view of the environment, meaning that the local information is not sufficient for an agent in this scenario to have a good performance.

Figure 5 shows the result for the three types of agents in a scenario with no deceleration and non-coordinated plans. The agent type with more information (27 states) has the lower performance compared with the others, it happened the contrary in the scenario with coordinated plans (Figure 4). This shows that having information is not sufficient f the other agents are not acting in a coordinated way.

Coordinated plans are pre-defined considering that the vehicles are all having an expected velocity but deceleration generates a different average velocity. In Figure 6 we perceive that the more informed agent has in average, the same performance as the other two kinds of agents. The difference seen here is in the stability of the result. The agents are trying to learn co-related actions (coordinated plans) in a too noisy scenario, leading to an unstable behavior. In the case where the actions are not directly related, Figure 7, the agents have a more stable performance and the more informed agent type is the agent with the best performance.

When we set the deceleration probability to 1/10, the noise in the scenario is too high for coordinated plans to work. Since the coordinated plans where made considering a given average velocity to cross the street, this velocity cannot be guaranteed with a 1/10-deceleration probability. Figure 8 shows that all three learning types have almost the same performance, similar to the performance seen on Figure 6. This result indicates that in some cases, increasing the information does not increase the agent's performance.

In Figure 9 we see that the agent that uses more information only when needed (12 states) adapts slightly better to a very noisy scenario. An interesting fact in this result is that the agent with 3 states has the same performance than the agent type with more information, confirming the results seen on Figure 8. This indicates that in a same scenario, the agent that only uses local information (with 3 possible states) might have a better performance compared with the agent with complete information (27 states).

Table 1 shows a summary of the presented results, including the values of the averages and standard deviations of all the experiments.

Figure 6. Results in the scenario with 5/100-deceleration and coordinated plans

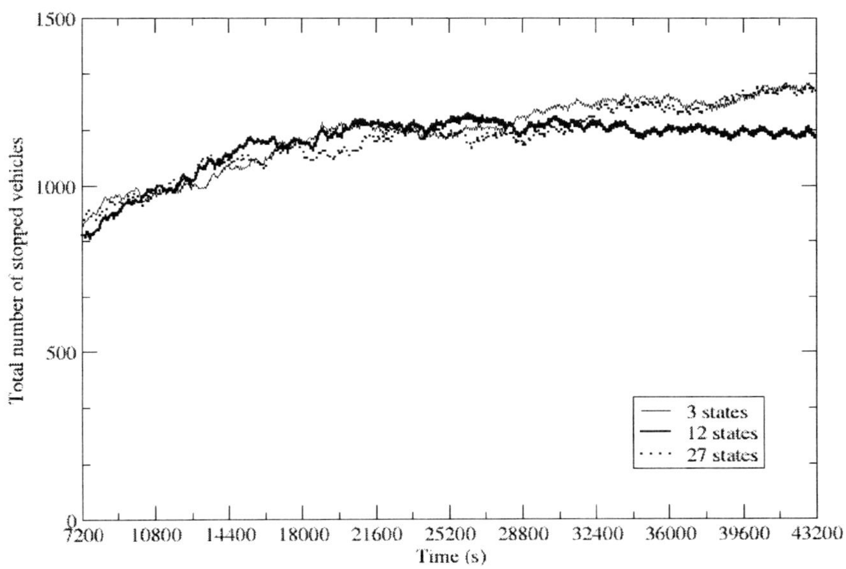

Figure 7. Performance in the scenario with 5/100- deceleration probability without coordinated plans

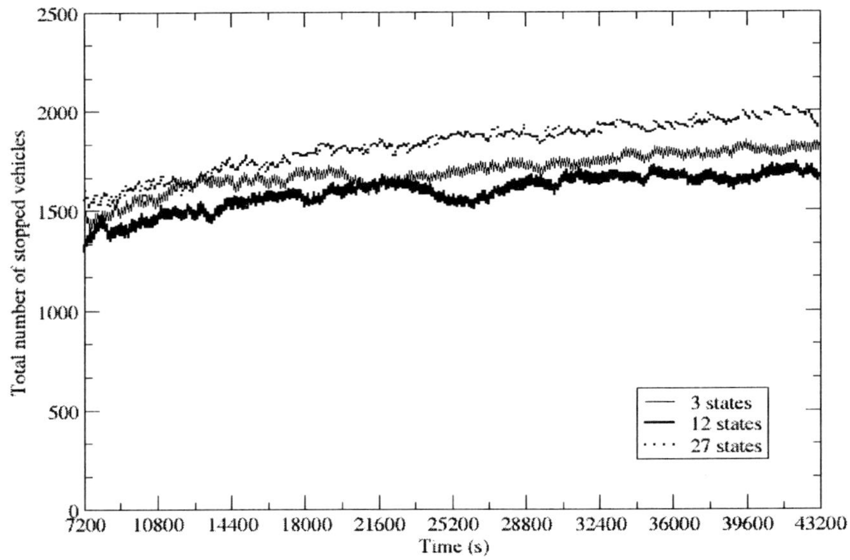

Figure 8. Results in the scenario with 1/10-deceleration probability and coordinated plans

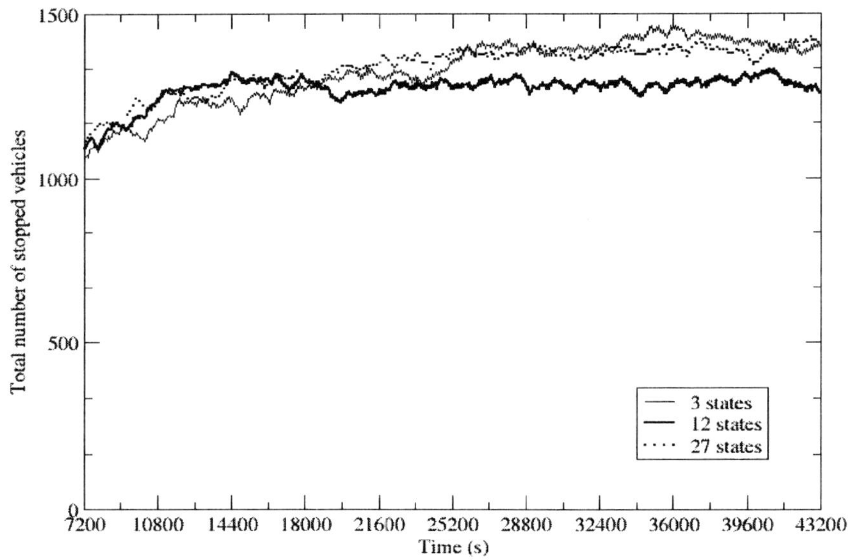

CONCLUSION

This chapter has presented a study of cooperative reinforcement learning applied on the traffic control scenario. Cooperative reinforcement learning agents can learn better than independent agents by sharing information about the environment. Our experiments show that extra information can have a negative

Figure 9. Performance in the scenario with 1/10-deceleration probability

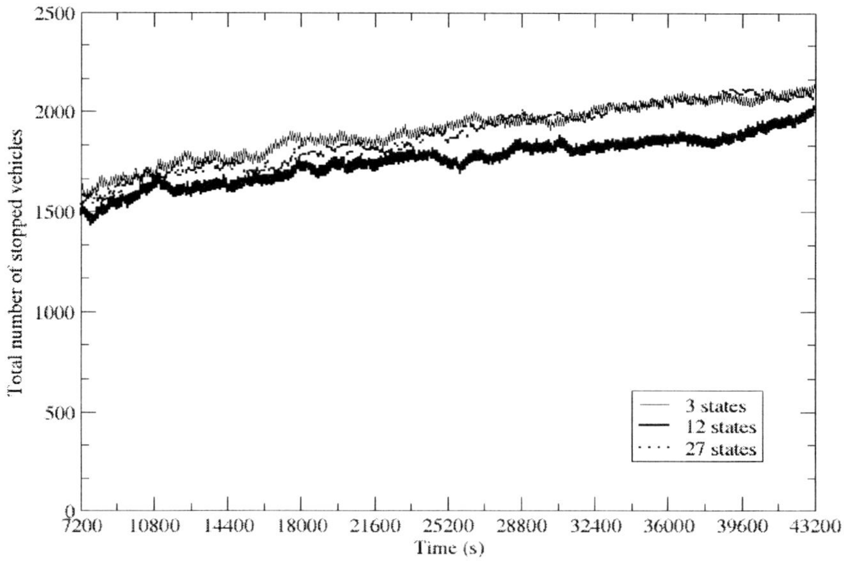

Table 1. Results summary

Deceleration	States	Coordinated Plans		Non-Coordinated Plans	
		Average	Std. Dev.	Average	Std. Dev.
No deceleration	3	924.72	68.80	1520.36	109.94
	12	751.34	63.33	1579.19	122.46
	27	575.59	64.76	1834.66	132.56
5/100	3	1156.34	112.50	1693.48	138.26
	12	1126.45	93.97	1589.92	132.22
	27	1138.00	109.37	1823.36	161.32
1/10	3	1324.95	106.30	1907.66	171.37
	12	1269.55	53.73	1757.15	156.55
	27	1335.13	79.45	1881.61	185.66

interference in the agent learning if the information is unnecessary. Also, noise in the traffic patterns and the inter-dependencies among the agent's actions are also very relevant factors to consider when there is a need to decide among three different possibilities: if we are going to use shared information, only local information, or if we are going to have a mid-term solution.

The applicability of this kind of the proposed distributed control on traffic regulation demands an infrastructure composed by: 1) traffic detectors, for the link state; 2) some kind of communication among neighboring junctions (wireless or wired), in the case of information exchange; and 3) a dedicated microprocessor at each junction, for running the Q-learning algorithm. In a future work, we intend to continue testing the three agent types viewed here on different scenarios (less regular flow, irregular topologies, etc) and also create more adaptive view of the states.

REFERENCES

Bazzan, A. L. C. (2005). A Distributed Approach for Coordination of Traffic Signal Agents. *Autonomous Agents and Multiagent Systems, 10*(1), 131–164, March.

Bernstein, D. S., Givan, R., Immerman, N., & Zilberstein, S. (2002). The complexity of decentralized control of Markov Decision Processes. *Mathematics of Operations Research, 27*(4), 819–840.

Camponogara, E., & Kraus Jr., W. (2003). Distributed Learning Agents in Urban Traffic Control. In: 11th Portuguese Conference on Artificial Intelligence, EPIA. (pp. 324–335). Lecture Notes in Computer Science 2902. Berlin: Springer.

Diakaki, C., Dinopoulou, V., Aboudolas, K., Papageorgiou, M., Benshabat, E., Seider, E., & Leibov, A. (2003). Extensions and New Applications of the Traffic Signal Control Strategy TUC. In: *82th Annual Meeting of the Transportation Research Board*, 2003. (pp. 12–16).

Dresner, K., & Stone, P. (2004). Multiagent Traffic Management: a reservation-based intersection control mechanism. In: *The Third International Joint Conference on Autonomous Agents and Multiagent Systems*, 2004, New York, USA. New York: IEEE Computer Society. (pp. 530–537).

Gershenson, C. (2005). Self-Organizing Traffic Lights. *Complex Systems, 16*(1), 29–53. Champaign, IL: Complex Systems Publications, Inc.

Nagel, K., & Schreckenberg, M. (1992). A cellular automaton model for freeway traffic. *Journal de Physique I, 2*(2221).

Oliveira, D., Bazzan, A. L. C., & Lesser, V. (2005). Using Cooperative Mediation to Coordinate Traffic Lights: a case study. In: *The Fourth International Joint Conference on Autonomous Agents and Multiagent System*, 2005. New York: IEEE Computer Society. (pp. 463–470).

Oliveira, D., & Bazzan, A. L. C. (2006). Traffic Lights Control with Adaptive Group Formation Based on Swarm Intelligence. In: *The Fifth International Workshop on Ant Colony Optimization and Swarm Intelligence,* ANTS 2006, 2006. (pp. 520–521). Lecture Notes in Computer Science. Berlin: Springer.

Oliveira, D., Bazzan, A. L. C., Silva, B. C., Basso, E. W., Nunes, L., Rossetti, R. J. F., Oliveira, E. C., Silva, R., & Lamb, L. C. (2006). Reinforcement learning based control of traffic lights in non-stationary environments: a case study in a microscopic simulator. In: *Fourth European Workshop On Multi-Agent Systems*, (EUMAS06). (pp. 31–42).

Panait, L., & Luke, S. (2005) Cooperative Multi-Agent Learning: the state of the art. *Autonomous Agents and Multi-Agent Systems, 11*(3), 387–434, Hingham, MA, USA.

Silva, B. C., Junges, R., Oliveira, D., & Bazzan, A. L. C.. (2006). *ITSUMO: an Intelligent Transportation System for Urban Mobility*. In: *The 5th International Joint Conference on Autonomous Agents and Multiagent Systems* (AAMAS 2006) - Demonstration Track. (pp. 1471-1472).

Sutton, R. S. & Barto, A. G. (1998). *Reinforcement Learning: An Introduction*. Cambridgem, MA: MIT Press.

Tan, M. (1993). Multi-agent reinforcement learning: Independent vs. cooperative agents. In *Proceedings of the Tenth International Conference on Machine Learning (ICML 1993)*, (pp. 330–337). Morgan Kaufmann.

Watkins, C. J. C. H., & Dayan, P. (1992). Q-learning. *Machine Learning*, *8*(3), 279–292.

Section III
Logistics and Air Traffic Management

Chapter XVI
The Merit of Agents in Freight Transport

Tamás Máhr
Almende/TU Delft, The Netherlands

F. Jordan Srour
Rotterdam School of Management, Erasmus University, The Netherlands

Mathijs de Weerdt
TU Delft, The Netherlands

Rob Zuidwijk
Rotterdam School of Management, Erasmus University, The Netherlands

ABSTRACT

While intermodal freight transport has the potential to introduce efficiency to the transport network, this transport method also suffers from uncertainty at the interface of modes. For example, trucks moving containers to and from a port terminal are often uncertain as to when exactly their container will be released from the ship, from the stack, or from customs. This leads to much difficulty and inefficiency in planning a profitable routing for multiple containers in one day. In this chapter, the authors examine agent-based solutions as a mechanism to handle job arrival uncertainty in the context of a drayage case at the Port of Rotterdam. They compare their agent-based solution approach to a well-known on-line optimization approach and study the comparative performance of both systems across four scenarios of varying job arrival uncertainty. The chapter concludes that when less than 50% of all jobs are known at the start of the day then an agent-based approach performs competitively with an on-line optimization approach.

INTRODUCTION

Scheduling the routes of trucks to pick-up and deliver containers is a complex problem. In general such Vehicle Routing Problems (VRPs) are known to be NP-complete and therefore inherently hard and time consuming to solve to optimality (Toth & Vigo, 2002). Fortunately, these problems have a structure that can facilitate efficient derivation of feasible (if not optimal) solutions. Specifically, the routes of different trucks are more or less independent. Such "locality" in a problem is a first sign that an agent-based approach may be viable.

Modeling and solving a VRP by coordinating a set of agents can bring a number of advantages over more established approaches in the field of operations research even when using state of-the-art mixed integer solvers such as CPLEX (ILOG, Inc., 1992). Agent advantages include the possibility for distributed computation, the ability to deal with proprietary data from multiple companies, the possibility to react quickly on local knowledge (Fischer et al., 1995), and the capacity for mixed-initiative planning (Bürckert et al., 2000).

In particular, agents have been shown to perform well in uncertain domains. That is, in domains where the problem is continually evolving (Fischer et al., 1995). In the VRP, for example, a very basic form of uncertainty is that of job arrivals over time. To the best of our knowledge, however, the effect of even this basic level of uncertainty on the performance of agent-based planning in a realistic logistics problem has never been shown.

We think it is safe to assume, based on its long history, that current practice in operations research (OR) outperforms agent-based approaches in settings where all information is known in advance (static settings). However, in situations with a high level of uncertainty, agent-based approaches are expected to outperform these traditional methods (Jennings & Bussmann, 2003).

In this chapter we investigate whether a distributed agent-based planning approach indeed suffers less from job arrival uncertainty than a centralized optimization-based approach. Our main contribution is to determine at which level of job arrival uncertainty agent-based planning outperforms on-line operations research methods. These results can help transportation companies decide when to adopt an agent-based approach, and when to use an on-line optimization tool, depending on the level of uncertainty job arrivals exhibit in their daily business.

In the next section we provide a survey of current work on agent-based approaches to logistics problems. We then introduce the case of a transportation company near the port of Rotterdam. Based on this literature review and the specific nature of our case study VRP, we propose a state-of-the-art agent-based approach where orders are auctioned among trucks in such a way that each order is assigned to the truck that can most efficiently transport the container. Moreover, these trucks continuously negotiate among each other to exchange orders as the routing situation evolves. This agent-based approach is fully described in this chapter. We follow this description with a description of the centralized on-line optimization approach used in comparison to our distributed agent-based system. The structure of our test problems and the computational results are covered in the next to last section. In the final section we discuss the consequences of our results, summarize our advice to transportation companies, and give directions for future work.

LITERATURE SURVEY

In their frequently cited 1995 paper, Fischer et al. argued that multi-agent models fit the transportation domain particularly well. Their main reasons were that (i) the domain is inherently distributed (trucks, customers, companies etc.); (ii) a distributed agent architecture can cope with multiple dynamic events; (iii) commercial companies may be reluctant to provide proprietary data needed for global optimization and agents can use local information; and (iv) inter-company cooperation can be more easily facilitated by agents. To illustrate the idea, the authors also provided a detailed agent architecture for transportation problems that evolve over time thereby exhibiting uncertainty over time. This architecture makes a distinction between a higher and a lower architectural level. At the higher level, company agents negotiate over transportation requests to eliminate ill-fitting orders. On the lower level, truck agents (clustered per company) participate in simulated market places, where they bid on offered transportation orders. Truck agents use simple insertion heuristics to calculate their costs and use those costs to bid on auctions implementing an extended contract net protocol (Smith, 1980). Although the heuristics that agents use to make decisions are rather crude, the authors suggested that in dynamic problems (problems with high uncertainty), such methods survive better than sophisticated optimization methods.

Fischer et al.'s (1995) bi-level approach recognizes that one shortcoming of a fully distributed system is that agents only have access to local information. The need to balance between the omniscience of a centralized model and the agility of a distributed model was similarly recognized by Mes et al. (2007). They also introduce a higher level of agents, but with a different role than the high-level agents of Fischer et al (1995). Mes et al.'s (2007) two high-level agents (the planner and the customer agent) gather information from and provide information to agents assigned beneath them. The role of the higher level agents is to centralize information essential for the lower level agents to make the right decisions.

Some researchers have gone even further in proposing centralized agent-based models. These researchers focused on centralizing the problem information to be able of make better distributed decisions. In one of the few models that has actually been applied in a commercial company, Dorer and Calisti (2005) cluster trucks geographically, using one agent per cluster. This way, one agent plans for multiple trucks. They use insertion heuristics to initially assign orders to trucks, and then use cyclic transfers (Thompson & Psaraftis, 1993) to enhance the solution. In an even more centralized model, Leong and Liu (2006) use a fully centralized optimizer to initialize the agents. The agents' role is then to change the plans as events are revealed. The authors analyze the performance of their model on a selection of Solomon benchmark sets, and show that it performs competitively.

As noted previously, however, the move towards centralization can hinder the ability of the agents to react quickly on local information. Given the uncertain environment of our problem, we are interested in the competitiveness of a system with fully distributed agents. One example of a fully distributed agent approach in the transportation domain is that of Bürckert et al. (2000). They proposed a more detailed (holonic) agent model. They distinguished truck, driver, chassis, and container agents that have to form groups (called holons) to serve orders. Already formed holons use the same techniques to allocate tasks as Fischer et al. (1995), but the higher agent level is omitted, since they model only a single company case. The main focus of their research is computer-human cooperative planning, and they do not test their model extensively against other models.

Generally, the decision to use a distributed approach is based on the expectation (included already in the reasons of Fischer et al., 1995) that distributed models handle uncertainty better. The agent architecture in these fully distributed models is completely flat, the models avoid centralizing information, and agents can use only local information when making decisions. Having lost the power of using (partial) global information, distributed agents need other ways to enhance their performance.

In the model of Fischer et al. (1995), as well as in the models of many of their followers, agents use simple approximation techniques to make decisions. In the related domain of production planning, Persson et al. (2005) embed optimization in the agents to improve local decisions. They show that optimizing agents outperform the approximating agents, but they also show that central optimization still outperforms the optimizing, but distributed agents.

While Persson et al. (2005) concentrated on making optimal decisions within agents there is still a need to coordinate between the distributed agents. For example, in the transport problem context, when orders are assigned to trucks sequentially, at every assignment the truck with the cheapest insertion gets the order. Later, however, it might turn out that it would be cheaper to assign the same order together with newly arrived orders to another truck. From the truck point of view this means that trucks that bid early and win assignments might not be able to bid later on more beneficial (better fitting) orders. This problem is called 'the eager bidder problem' (Schillo et al., 2002), and several researchers proposed alternative techniques as solutions. Kohout and Erol introduce an enhancement process that works between agents (1999). The process mimics a well known enhancement technique called 'swapping' or two-exchange (Cordeau et al., 2001). Kohout and Erol implement this swapping process in a fully distributed way, and show that it yields significant improvement.

Perugini et al. (2003) extend Fischer's contract-net protocol to allow trucks to place multiple possibly conflicting bids for partial routes. These bids are not binding; trucks are requested to commit to them only when one of the bids is accepted by an order agent. Since auctions are not necessarily cleared before other auctions are started, agents have a chance to "change their mind" if the situation changes. This extension helps to overcome the eager bidder problem to some extent and thereby produces better results. Another possible way to tackle the same problem is to use leveled commitment contracts introduced by Sandholm and Lesser (2001). Leveled commitment contracts represent agreements between agents that can be withdrawn. If a truck agent finds a new order that fits better, it can decommit an already committed order and take the new one. Hoen and La Poutré (2003) employ truck agents that bid for new orders considering decommitting already assigned ones. They show that decommitment yields more optimal plans in a single-company cooperative case.

Returning to Fischer's reasoning, however, the primary reason for using distributed agent models is that they are usually expected to outperform central optimization models in problem instances with high levels of uncertainty. Taking this for granted, researchers usually show that their distributed algorithm is better than the distributed algorithms of others. Experiments studying the behavior of distributed methods over varying levels of uncertainty in comparison to centralized optimization methods are generally absent from the literature (Davidsson et al., 2005).

If advanced swapping and decommitment techniques are used, can fully distributed agents perform competitively with (or better than) centralized optimization in highly uncertain settings? Can the time gained in doing local operations compensate for the loss of not considering crucial global information? In our opinion these questions have not been fully answered. In this chapter, we construct a distributed agent model using the most promising techniques as identified in the agent literature and compare this approach via experiments on a real data set to a state-of-the-art centralized on-line optimization approach. The lack of appropriate comparisons between agent-based approaches and existing techniques for transportation and logistics problems possibly indicates a belief on the part of agent researchers that agent-based systems outperform traditional methods (Davidsson et al., 2005). Our goal is to add credibility to this belief by studying a state-of-the-art agent-based system in comparison to a state-of-the-art centralized optimization approach for a real-world dynamic transportation problem. In the following section we define in detail the exact VRP that we use to study both the distributed agent-based and centralized optimization-based approaches.

VEHICLE ROUTING PROBLEM

Many of the agent-based approaches for vehicle routing problems are tested on generated data-sets. These data-sets are usually constructed to test specific features of the agent system – often focusing on the extreme ends of the performance spectrum. We, however, want to understand the potential of agent solutions in the highly uncertain real world. To that end we are fortunate to have access to operational data from a mid-sized Dutch logistics service provider (LSP) engaged in the road transport of sea containers. While the LSP that we study is active in several sectors, we focus only on the container division which has a fleet of around 40 trucks, handling an average of 65 customer orders each day.

The process of executing an order starts with receiving an order, generally one day before execution is required. While the orders are often called in one day early, the company does not generally use this information in planning routes or establishing schedules. This is due to the unreliable nature of the order information and the resulting uncertainty encountered during execution. An order is a customer request to the LSP for pickup and transport of a specific container from a container terminal (in the case of an import container) to the customer, with delivery within a certain time window. Arriving at the customer's requested location, the container is then unloaded, and the empty container is brought back to a container terminal or empty depot. This concludes the order, and the truck is ready for its next order. The process is reversed for export containers. What adds uncertainty to this process is that not all containers are available at the time indicated in the received order: either they have not physically left the ship at the expected time or they are delayed for administrative reasons; e.g. an unsettled payment or customs clearance. The LSP can only transport containers that have been released, and are allowed to leave the container terminal. For this reason it is hard to optimize the system in a traditional sense, since not all information is known beforehand, and will only become available at some point in time during the day.

The planning and control of operations is currently performed manually by a team of three human planners, who take care of order intake, arrange the proper amount of trucks based on the expected workload, and assign current orders to trucks. Given the primarily manual method of operations, the addition of a computerized decision support system may greatly enhance the profitability and scalability of the LSP's operations. To formalize the structure of this case study problem we make several formal assumptions:

- Each demand is available for scheduling at the time it is announced. The announcement of a demand includes all information on: the pick-up location (zip code), the customer location (zip code), return drop-off location (zip code), and the required time windows for arrival at each of these three locations.

- Loading and unloading at the terminals and customer takes time. Picking up a container requires 60 minutes; servicing the container at the customer requires 60 minutes; and returning a container to the final terminal takes 30 minutes.
- All travel times are measured according to data on the Benelux road network.
- No time window violations are allowed; if a job is going to violate time windows then it is rejected at a penalty.
- The penalty for rejecting a job is equal to the loaded time of the job. Given the problem structure defined here, loaded time serves as a proxy for revenue.
- Given the demand structure, the truckload nature of the problem, and the fact that the truck must remain with the container at the customer location, we bundle the pick-up, drop-off, and return activities into one job. The loaded time of a job is then the time spanning the arrival at the pick-up terminal through the completion of service at the return terminal -including all loading and unloading times.
- All trucks in the fleet are equivalent.

Given this context, the objective of this vehicle routing problem is to derive a schedule in real-time that serves as many jobs as possible at the least cost. Cost is defined here in terms of time, as the time spent traveling empty (i.e. non-revenue generating travel) to serve all jobs in addition to the loaded time penalty affiliated with rejecting jobs. By adding a penalty for rejecting jobs equal to the loaded distance (in terms of time) of each job, the obvious cost-minimizing solution of rejecting all jobs is avoided. In this regard, it is important to note that in our setting the loaded distance of an order is approximately four times as great as the empty distance incurred in serving that job.

AGENT-BASED APPROACH

Based on the agent-based modeling literature and the assumptions related to our problem as introduced in the previous section, our goal is to design, using selected techniques from the literature, a distributed agent model that can outperform a centralized optimization approach. Since we are primarily interested in distributed agent models, we use an uncompromisingly flat architecture: no agents can concentrate information from a multitude of other agents. The global idea of our agent-based planning system is to apply an advanced insertion heuristic in a distributed setting and combine this with two heuristics for making (local) improvements: substitution of orders, and random attempts for re-allocation of orders. The only two kinds of agents that participate in this planning system are truck agents and order (or container) agents.

Our order agents represent container orders. The particularity of container orders is identical to the real-world case of the previous section in that they are described by the three stops required: a pick up at a sea-terminal, a delivery at the customer's, and a drop-off return at a possibly different sea-terminal. With each of the three stops there is a time window and a service time associated, which are obeyed by the trucks. Truck agents represent trucks with a single chassis, which means that they can transport only one order at a time. They make plans in order to transport as many containers as they can.

Order agents hold auctions in order of their arrival, and truck agents bid in these auctions. This results in partially parallel sequential auctions. Trucks may bid on multiple orders at the same time; these bids are not binding. If a truck happens to win more than one order, it takes only the first one. All the other orders it won parallel to the first one are rejected, which results in the rejected order agents starting a new auction. Truck agents ultimately accept only one winning bid on parallel auctions as all bids submitted in parallel are highly dependent on the order of previously won and accepted bids. In this way, in the end, the orders are auctioned sequentially, even if they happen to arrive at the same time.

To clear an auction, order agents choose the best bid as winner, and respond positively to the winner and negatively to the others. For this we chose a one-shot auction (and more specifically, a Vickrey auction [Vickrey, 1961]) for its computational efficiency, as in the model of Hoen and La Poutré (2003). If the winner confirms the deal, a contract is made. These contracts are semi-binding, so truck agents might break it in order to achieve a better allocation.

At the heart of the agent model are the decisions truck agents make. The most important decision they have to make is the bid they submit for a given order. Every truck agent submits a bid that reflects its cost associated with transporting the given order. This cost is a quantity in the time domain. To calculate it, a truck considers inserting the new order into its plan, or alternatively substituting one of the already contracted orders by the new one.

To calculate the cost of insertion, the truck agent tries to insert the new order in-between every two adjacent orders in their plan (see Figure 1), plus at the beginning and the end. At every position, it calculates the amount of extra empty time it needs to drive if this order is inserted there. Suppose that an agent considers the position between container i and j, and calculates that the empty time the truck needs to travel to pick up j after returning i is d_{ij}. Here we use d_{ij} to represent the distance (in time) between the two jobs i and j, and d_{ii} to denote the loaded distance of job i. The amount of extra empty time the truck would need to drive for container l then equals $ins_{ij}^{l} = d_{il} + d_{lj} - d_{ij}$.

In addition to insertion, a truck agent also considers substitution (analogous to what others call decommitment). To calculate the cost of substituting one of the already contracted orders by the new

Figure 1. Example of insertion

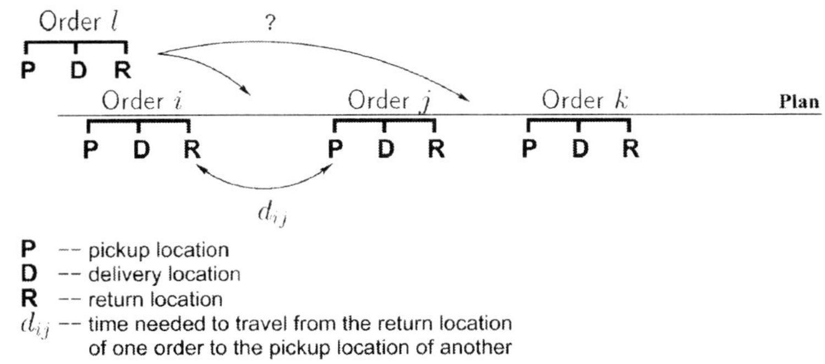

Figure 2. Example of substitution

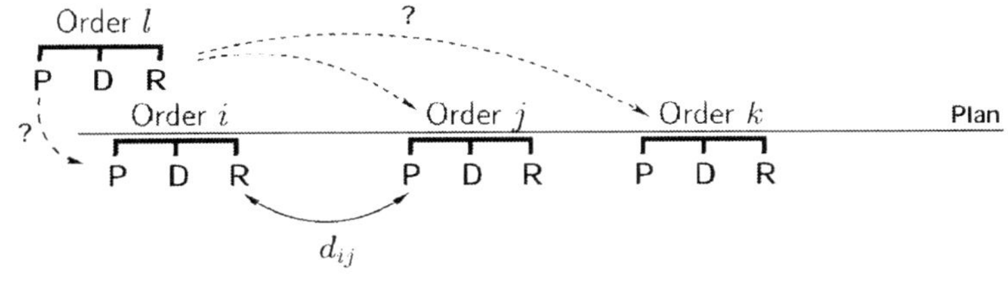

one, it sums up the cost components. The first component is the insertion cost of the new order at the place of the substituted order, the second component is the lost profit on the substituted order, and the third component is a penalty term. For example, we compute the cost of substituting order j with order l ($subs_j^l$) in Figure 2. Here $subs_j^l = ins_{ik}^l + profit_j + d_{jj}$. The insertion term ins_{ij}^l is the same as defined above. The value of $profit_j$ is the difference of the price received for order j and its insertion cost: $profit_j = price_j - ins_{ik}^j$. This term represents the market position of the substituted order in the bid. If the competition for order j is fierce, the profit on j would be low (since the second-best bid was hardly higher than the winning bid). This results in a low substitution cost, therefore such orders are more likely to be substituted. An order that is well suited for a specific truck is likely to produce a high profit for that truck, therefore it will have a high substitution cost. The last term in this expression, the amount of loaded time of order j, serves as a penalty on substituting that job. Using such a penalty discourages the substitution of long orders that may be harder to fit somewhere else. Additionally, the orders that are finally rejected (those that do not manage to make a contract with any truck agents) will be shorter, which will result in a better total cost. Algorithm 1 describes how new orders are dealt with.

In addition to bidding on auctions for new orders, truck agents have another way to enhance the overall solution. At random time intervals, every truck randomly selects an order in its plan and releases it. Trucks never select the order they are currently serving and also not one, for which the execution is about to begin (the pick-up time of the container is less than 10 seconds away – this small time buffer is selected to provide as much opportunity for route improvement as possible). Note, the same time limit is also applied to the insertion and substitution decisions explained earlier. An order agent that is released (just as those order agents that are substituted) initiates a new auction to find another place. In most cases, these auctions result in the very same allocation as before the release. Nevertheless, sometimes they do manage to find a better place and make a contract with another truck.

Whenever an order agent finalizes a contract with a truck agent, it sends a message to all other order agents to notify them about the changed plan of the given truck. This is important for order agents that do not have a contract yet. Any change in the trucks' plans may be their chance to find their place in a truck. Those order agents will start an auction in response to the notification message in the hope of finally making a contract. To summarize the agent-based approach, let us list the main techniques that characterize it:

Algorithm 1. Insertion and substitution of orders

1) Compute the extra costs for every possible insertion and every possible substitution.
2) Order the merged list of insertions and substitutions in increasing order of these costs.
3) Iterate over this list:
    a) If the new order's time windows are violated, continue with the next alternative.
    b) If a time window of an order after the new one is violated, continue with the next alternative.
    c) Else the cheapest feasible position is found. Return this position.

- Orders are allocated to trucks via second-price auctions sequentially, at the time they become known to the agent system.
- Truck agents consider insertion and substitution of new orders in their plan. Substituted orders are released from the truck. Released order agents hold a new auction to find another place. If a truck cannot deliver an order within the time windows, it rejects it.
- Truck agents randomly release contracted order agents. Randomly released order agents also hold a new auction to find a place.
- Order agents notify each other whenever they change the plan of a truck (make a contract). Rejected orders (without a contract) thereby get a chance to hold a new auction and find a truck.

To evaluate this approach, we implemented a real-time truck simulator that we connected to the agent system. Every truck agent assumes responsibility for a simulated truck. In the coupled agent-truck-simulator system, agents send plans to trucks for execution. Simulated trucks drive along the road network of the Benelux as the plans prescribe. They periodically report their position as well as their activities to the agents. This way truck agents can follow the execution of the plans and make decisions with the knowledge of what is happening in the (simulated) world.

Finally, we have a third element in the system, whose role is to monitor both the agents and the simulator, thereby gathering all information necessary to evaluate the performance of the agents, and to calculate the total cost of the routing. Just as described in the previous section, the ultimate objective of the agents is to minimize the total cost of the routing which is specified in terms of the time trucks travel empty plus the loaded-travel-time penalty associated with rejecting a container. The next section describes the on-line optimization approach that is used in comparison to the agent-based approach, based on this total cost.

ON-LINE OPTIMIZATION APPROACH

To estimate the value of the agent-based solution approach (described in the previous section), we study it in comparison to an optimization-based solution approach, reflective of those currently embedded in commercially available vehicle routing decision support software (DSS). We therefore examine two optimization based solution approaches: (i) a mixed-integer program for solving the static a priori case in order to provide a baseline benchmark, and (ii) an on-line optimization approach, comparable to the agent approach, and designed to represent current vehicle routing DSS.

At the core of both the static a priori solution and the on-line optimization is a mixed integer program (MIP) for a truck-load vehicle routing problem with time windows, which is passed to CPLEX for solving (ILOG, Inc., 1992). This MIP is based on the formulation put forth by Yang et al. (1999). The complete description of our modifications to Yang et al.'s MIP is the focus of this section. Before introducing the notation and mathematical formulation for this problem, we begin with a small example to illustrate exactly how Yang et al.'s MIP works to exploit the structure of this truckload pick-up and delivery problem with time windows. Imagine a scenario with three trucks and four jobs. The model of Yang et al. is constructed such that it will find a set of least cost cycles describing the order in which each truck should serve the jobs. For example, as depicted in Figure 3, the outcome may be a tour from truck 1 to job 1, then job 2, then truck 2, then job 3, then back to truck 1. This would indicate that truck

Figure 3. Example of cyles in MIP structure

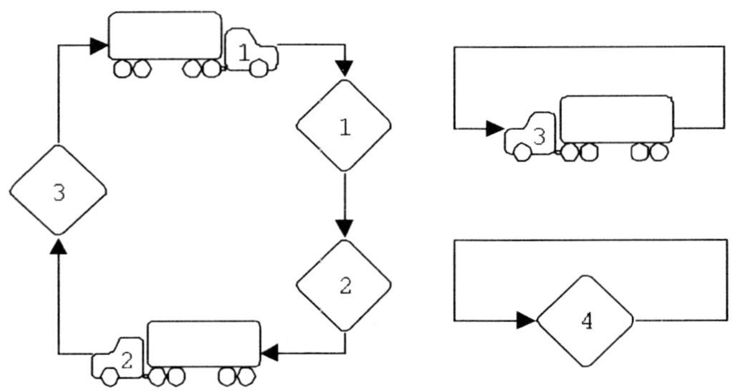

1 serves job 1 and 2, while truck 2 serves job 3. The cycle including only truck 3 indicates that truck 3 remains idle. Similarly, the cycle including only job 4 indicates that job 4 is rejected.

Given the assumptions in Section 3, we designate the following notation for the given information.

K	The total number of vehicles available in the fleet.
N	The total number of known demands.
d_{ij}	As introduced earlier, the travel time required to go from demand i's return terminal to the pick-up terminal of demand j. Note, if $i = j$ then the travel time d_{ii} represents the loaded distance of demand i.
d^k_{0i}	The travel time required to move from the location where truck k started to the pick-up terminal of demand i.
d^k_{iH}	The travel time from the return terminal of demand i to the home terminal of vehicle k.
v^k	The time vehicle k becomes available.
l_i	The loaded time required of job i (time from pick up at originating terminal to completion of service at the return terminal). Note, $l_i = d_{ii}$.
τ_i^-	The earliest possible arrival time at demand i's pick-up terminal.
τ_i^+	The latest possible arrival time at demand i's pick-up terminal.
M	A large number set to be $2\max\{d_{ij}\}$.

Note: τ_i^- and τ_i^+ are calculated to ensure that all subsequent time windows (at the customer location and return terminal) are respected. Given the problem of interest, we specify the following two decision variables.

x_{uv}	A binary variable indicating whether arc (u, v) is used in the final routing; $u, v = 1, \ldots, K+N$
δ_i	A continuous variable designating the time of arrival at the pick-up terminal of demand i.

Using the notation described above, we formulate a MIP that explicitly permits job rejections, based on the loaded distance of a job.

$$\min \sum_{k=1}^{K}\sum_{i=1}^{N} d_{0i}^{k} x_{k,K+i} + \sum_{i=1}^{N}\sum_{j=1}^{N} d_{ij} x_{K+i,K+j} + \sum_{i=1}^{N}\sum_{k=1}^{K} d_{iH}^{k} x_{K+i,k} \quad (1)$$

such that

$$\sum_{v=1}^{K+N} x_{uv} = 1 \qquad \forall\, u = 1, \ldots, K+N \quad (2)$$

$$\sum_{v=1}^{K+N} x_{vu} = 1 \qquad \forall\, u = 1, \ldots, K+N \quad (3)$$

$$\delta_i - \sum_{k=1}^{K} (d_{0i}^{k} + \nu^{k}) x_{k,K+i} \geq 0 \qquad \forall\, i = 1, \ldots, N \quad (4)$$

$$(l_i + d_{ij}) x_{K+i,K+i} - M x_{K+i,K+j} - \delta_i + \delta_j \geq -M + l_i + d_{ij} \qquad \forall\, i, j = 1, \ldots, N \quad (5)$$

$$\tau_i^{-} \leq \delta_i \leq \tau_i^{+} \qquad \forall\, i = 1, \ldots, N \quad (6)$$

$$\delta_i \in \Re^{+} \qquad \forall\, i = 1, \ldots, N \quad (7)$$

$$x_{uv} \in \{0,1\} \qquad \forall\, u, v = 1, \ldots, K+N \quad (8)$$

In words, the objective (1) of this model is to minimize the total amount of time spent traveling without a profit generating load. This objective is subject to the following seven constraints:

(2) Each demand and vehicle node must have one and only one arc entering.

(3) Each demand and vehicle node must have one and only one arc leaving.

(4) If demand i is the first demand assigned to vehicle k, then the start time of demand i (δ_i) must be later than the available time of vehicle k plus the time required to travel from the available location of vehicle k to the pick up location of demand i.

(5) If demand i follows demand j then the start time of demand j must be later than the start time of demand i plus the time required to serve demand i plus the time required to travel between demand i and demand j; if however, demand i is rejected, then the pick up time for job i is unconstrained.

(6) The arrival time at the pick up terminal of demand i must be within the specified time windows.

(7) δ_i is a positive real number.

(8) x_{uv} is binary.

Mathematically this model specification serves to find the least-cost (in terms of time) set of cycles that includes all nodes given in the set $\{1, \ldots, K, K+1, \ldots, K+N\}$. We define x_{uv}, ($u, v = 1, \ldots, K+N$) to indicate whether arc (u, v) is selected in one of the cycles. These tours require interpretation in terms of vehicle routing. This is done by noting that node k, ($1 \leq k \leq K$) represents the vehicle k and node $K + i$, ($1 \leq i \leq N$) corresponds to demand i. Thus, each tour that is formed may be seen as a sequential assignment of demands to vehicles respecting time window constraints.

The model described above is used to provide the optimal (yet realistically unattainable) lower bound for each day of data in each scenario. We denote this approach as the static a priori case. In this case, we obtain the route and schedule as if all the jobs are known and we have an unlimited amount of time to find the optimal solution. Thus, not only is this lower bound realistically unattainable due to a relaxation on the amount of information available, but also due to a relaxation on the amount of time available to CPLEX for obtaining the optimal solution. In this way, because the problem instances are relatively small (note, using this MIP structure CPLEX can handle a maximum of about 100 jobs and about 50 Trucks, yet our instances are only 34 trucks and 65 jobs) we are able to uncover the optimal solution for all 26 problem instances across all four uncertainty scenarios.

In order to provide a fair comparison with the agent-based approach, the MIP is then manipulated for use in on-line operations. In our on-line approach, this MIP is invoked at 30 second intervals. At each interval, the full and current state of the world is captured, and then encoded in the MIP. This "snapshot" of the world includes information of all jobs that are available and in need of scheduling, as well as the forecasted next available location and time of all trucks. The MIP is then solved via a call to CPLEX. The decision to use 30 second intervals was driven by the desire to be comparable to the agent-based approach while still providing CPLEX enough time to find a feasible solution for each snapshot problem. The solution given by CPLEX is parsed and any jobs that are within two intervals (i.e. 60 seconds) of being late, if travel is not commenced in the next interval, (i.e. missing the time specified by δ_i in the latest plan) are permanently assigned. Any jobs that were designated for rejection in the solution are rejected only if they are within two intervals of violating a time window; otherwise they are considered available for scheduling in a subsequent interval. The procedure continues in this fashion until the end of the working day at which point all jobs have been served or rejected.

The test problems and the results from the static a priori benchmark, the on-line optimization, and the agent-based solution approach as applied to these test problems are the topic of the next section.

COMPUTATIONAL EXPERIMENTS

In this section we report the computational results on the performance of the agent-based approach in comparison to the optimization-based approach. The first subsection describes how the test problems were generated and the second subsection presents the results of these tests.

Test Problems

The data we used for our experiments was based on data provided to us by the LSP described in Section 3. In all, we were given the execution data from January 2002 to October 2005 as well as the data from January 2006 through March 2006. We could not, however, simply use this data in its raw form. We first had to make multiple corrections to the customer address fields as many addresses referred to postal boxes and not to the physical terminal locations. After cleaning the address fields, we then extracted a random sample of jobs from the original data-set in order to generate a set of 26 days with 65 orders per day. The company from which these data are taken serves between 50 and 80 jobs per day, thus 65 jobs per day represents the average daily job load.

As discussed before, each order consists of a pickup location, customer location, and return location. To standardize the data for our experimental purposes we specified time windows at all locations as follows: for the terminals (the pickup and return locations) the time windows span a full twelve hour work day from 6am to 6pm and the time windows at the customer locations are always 2 hours. The start of the 65 customer time windows occurs throughout the working day in accordance with the data provided by the LSP, which roughly follows a uniform distribution. Given the variation in customer locations, the workload per day varies similarly. On average each job requires approximately 4.2 hours of loaded distance. When the routing is optimal in the case that all jobs are known at the start of the day the average empty time per job is approximately 25 minutes.

Given our interest in determining how the agent solution performs on this pick-up and delivery problem with time windows and order arrival uncertainty, we further rendered our 26 days of data into four separate scenarios with varying levels of order arrival uncertainty. This was done by altering the arrival times of the orders, i.e., the time at which the order data is revealed to the LSP. We generated these points in times over the day using a uniform distribution. We used such a uniform distribution as the original data did not show a fit with other distributions. The four different scenarios reflecting different levels of order arrival uncertainty were:

- ***Scenario A:*** All orders (100%) are known at the start of the working day, 6AM.
- ***Scenario B:*** About half of the orders (50%, selected randomly from the 65 jobs) are known at the start of the working day, 6AM. The other half of the orders arrive two hours before the start of the customer location time window (i.e., four hours before the end of the customer location time window, leaving slightly less than two hours on average before the latest departure time from the pickup location).
- ***Scenario C:*** Only seven of the jobs (10%, selected randomly from the 65 jobs) are known at the start of the working day, 6AM. The remaining 58 jobs arrive two hours before the start of the customer location time window.
- ***Scenario D:*** None of the jobs (0%) are known at the start of the working day. All 65 jobs arrive two hours before the start of the customer location time window.

If we classify these scenarios in terms of the effective degree of dynamism for vehicle routing problems with time windows as developed by Larsen et al. in 2002 then values of dynamism for Scenarios A, B, C, and D are .5, .7, .8, and .9, respectively. Noting that this form of measuring uncertainty may range from 0 to 1 with 1 being the most uncertain then we may say that our test problems range from partially uncertain to mostly uncertain.

Computational Results

All three solution approaches were applied to each of the 26 days of data in the four scenarios. The mean cost over the 26 days of these experiments may be seen in Figure 4. From this graphical depiction, the on-line optimization procedure clearly outperforms the agents only in Scenario A. In fact, in Scenario A in which all information is known at the start of the day, the on-line optimization performs at a level quite close to the realistically unattainable benchmark optimal. The on-line optimization does not, however, achieve optimal in Scenario A as the snapshot problem in the first 30 second interval represents the full problem size; a size for which finding the optimal solution in thirty seconds is quite difficult. In all cases, CPLEX does, however, provide a feasible solution which can then be improved in future intervals. In the remaining three scenarios, however, the agents perform at a level competitive to the on-line optimization.

To fully understand the competitive nature of the agents in the dynamic settings of Scenarios B, C, and D a t-test was performed to determine if the average total cost of the routing solutions were statistically equivalent. The results of these tests may be seen in Tables 1 and 2. From these results we

Figure 4. Mean over 26 days of the total cost for the three approaches across the four scenarios; bars indicate ± one standard deviation

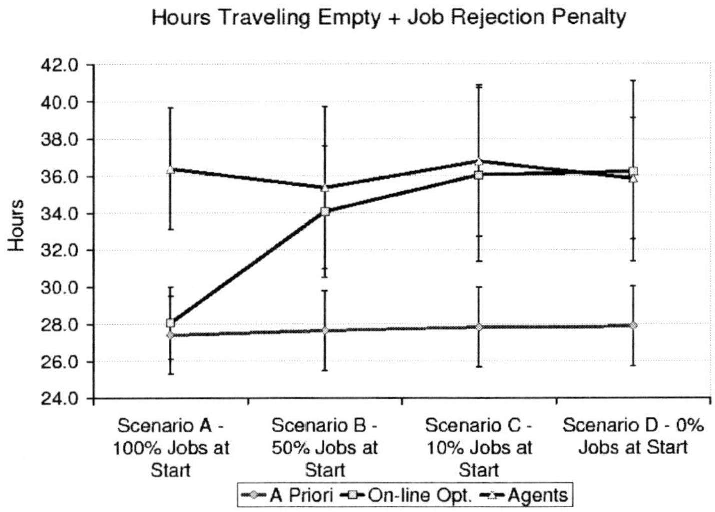

may conclude that for the reality-based datasets used in this study, agent-based solution approaches perform competitively with the on-line optimization when at least half of the jobs is unknown at the start of the day.

While the study of total cost and associated t-test results are promising for the agent approach, we must also look at the portion of this total cost due to the job rejection penalty and the portion of the cost due to empty travel time. Figure 5 depicts the penalty of rejected jobs on the left axis and the number of jobs rejected on the right axis. Note, we do not include the a priori optimal in this figure as no jobs were rejected using this approach. While the on-line optimization demonstrates a clear trend in the number of rejections (the more dynamic the setting the more jobs are rejected at a higher penalty), the agent approach does not demonstrate any trend. In comparing Figure 5 and Figure 6, it is clear that this irregular job rejection trend of the agent approach is having a significant impact on the trend in the total cost of the agent approach (see Figure 4).

Figure 6 depicts the average number of hours spent traveling empty in the routing solution provided by each approach in the four scenarios. From this figure, all three approaches show a general trend toward an increased level of empty travel with an increased level of uncertainty. Interestingly, however, the agent approach shows far more stability in this regard. In this sense we may conclude that despite

Table 1. Mean ± standard error over 26 days for on-line optimization and the agent-based approach on the total cost for scenarios A, B, C, and D

	A	B	C	D
On-line Opt.	28.07 ± .38	34.09 ± .70	36.06 ± .92	36.24 ± .95
Agents	36.4 ± .64	35.37 ± .86	36.81 ± .80	35.85 ± .64

Table 2. Results of the t-test on the null hypothesis that the means of the total cost of the two datasets are equal (with .05 significance)

	A	B	C	D
Calculated t-value	11.16	1.16	.61	.34
Tabulated t-value	2.01	2.01	2.01	2.01
Result	Reject	Fail to Reject	Fail to Reject	Fail to Reject

Figure 5. Job rejection in terms of penalty (left axis) and number of jobs rejected (right axis) for the on-line optimization and agent approaches across four scenarios

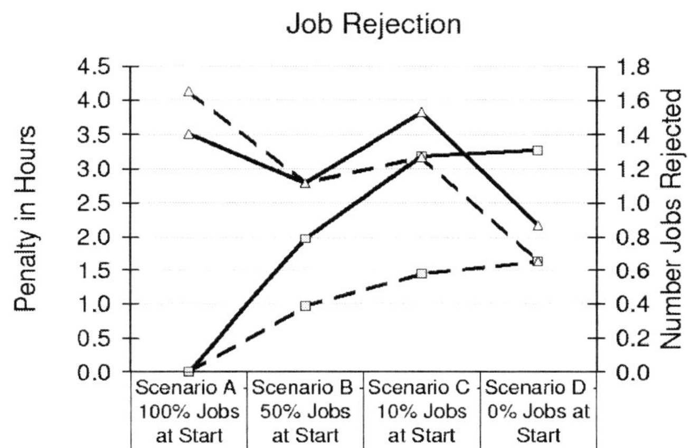

	Scenario A 100% Jobs at Start	Scenario B 50% Jobs at Start	Scenario C 10% Jobs at Start	Scenario D 0% Jobs at Start
On-line Opt. Penalty	0.00	1.98	3.18	3.28
Agents Penalty	3.51	2.79	3.82	2.18
On-line Opt. Number	0.00	0.38	0.58	0.65
Agents Number	1.65	1.12	1.27	0.65

the agents' poor performance in our less uncertain settings, they are, however, less susceptible than on-line optimization to the effects of high uncertainty. Yet, in the end, both systems perform comparatively well in the most uncertain setting.

DISCUSSION

In this chapter, we studied an on-line truckload vehicle routing problem arising from a real-world case study. We described a state-of-the-art agent-based solution approach and compared that approach to a well known on-line optimization approach. The computational results, from 26 days of data spanning four different scenarios representing various levels of job arrival uncertainty, indicate that the agent-based approach is highly competitive in cases where less than 50% of the jobs are known in advance.

Given these results, agents should be considered as a viable decision support mechanism for transportation planners that must cope with uncertain job arrivals. If, however, the job arrival environment

Figure 6. Time spent traveling empty for the three approaches across the four scenarios

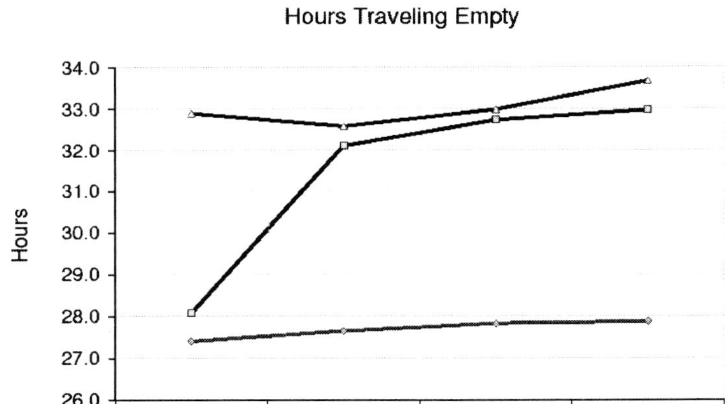

	Scenario A - 100% Jobs at Start	Scenario B - 50% Jobs at Start	Scenario C - 10% Jobs at Start	Scenario D - 0% Jobs at Start
A Priori	27.41	27.65	27.84	27.88
On-line Opt.	28.07	32.11	32.73	32.97
Agents	32.89	32.58	32.98	33.67

is relatively static, that is more than half of the jobs are known at the start of the day, then optimization should remain the tool of choice. Admittedly, this recommendation carries the following caveat. The agents do suffer a certain level of instability as reflected in the lack of a trend in job rejections relative to the level of uncertainty. The reason is that while job rejection is explicitly handled in the optimization model, it is implicit in the agent model. When an agent rejects an order, it has no way of knowing whether other agents will reject it too. In general, it is therefore more difficult to implement a global notion such as the number of rejected orders in an agent approach. In practice, a transportation provider must be very explicit about routing priorities. If a consistent or predictable level of job rejections is important then on-line optimization is a better choice.

One of the reasons that the agent-based solution performs consistently in terms of empty distance traveled is because of the sequential auction method used to handle jobs that arrive simultaneously. Thus, in Scenario A, in which the uncertainty is low, the agents must run many auctions at the start of the day; on-line optimization on the other hand may exploit all of this information at once to obtain a near optimal solution. In Scenario D, on the other hand, the agents approach the auctions in very much the same way as in Scenario A except that they are spread more evenly over time. In contrast the on-line optimization is forced to adopt job assignments that may preclude the assignment of jobs arriving late in the day.

In short, agent-based systems perform well in settings where less than half of all jobs are known in advance. Agents do, however, present issues concerning tractability in terms of rejected jobs. The number and penalty of rejected jobs is particularly variable with no clear trend across the four scenarios. Finally, in steep contrast to the online optimization, the agents used in this study are not well suited to exploit large batches of job arrivals; agents tend to perform better when a small number of jobs arrive evenly spaced through out the planning horizon.

Noting from these cases the impact of clumped job arrivals on the two approaches brings us to our first extension of this work. We recommend that both systems be tested across several problem sizes and a variety of uncertain job arrival patterns to truly understand the effect of clumped job arrivals.

Turning now to the theme of uncertainty, job arrival uncertainty as studied here represents only one narrow definition of uncertainty. A simple extension to this definition by including variable numbers of jobs across the days (i.e. each day would have a different number of jobs taken from the range 50 to 80) will provide additional insight on the strengths and weaknesses of agents in handling uncertainty. Furthermore examining other sources of uncertainty in the transportation domain, such as loading, unloading, and travel time variability, will not only add realism to the study, but will also yield a more robust view on the benefits and drawbacks of an agent approach as compared to centralized approaches.

Another extension of this work is the introduction of optimization into the agent approach. In this way, the agents may be able to capitalize on the benefit of optimization in less uncertain situations and the benefit of local heuristics in more uncertain situations. We conclude by stating that agent-based approaches may have even greater benefits when we consider modeling other forms of uncertainty such as travel time uncertainty, loading and unloading time uncertainty, and so forth. The field for agent-based approaches to the VRP is wide open, but should also be carefully explored to ensure that the practical everyday needs of real-world transport planners are met.

ACKNOWLEDGMENT

The authors would like to thank Hans Moonen for his help in retrieving and constructing the data set used for the experiments in this chapter. Furthermore, this work is partially supported by the Technology Foundation STW, applied science division of NWO, and the Ministry of Economic Affairs of the Netherlands. Finally, the authors gratefully acknowledge the comments provided by three anonymous reviewers.

REFERENCES

Bürckert, H.-J., Fischer, K., & Vierke, G. (2000). Holonic transport scheduling with Teletruck. *Applied Artificial Intelligence, 14*(7), 697–725.

Cordeau, J.-F., Desaulniers, G., Desrosiers, J., Solomon, M. M., & Soumis, F. (2001). VRP with time windows. In P. Toth & D. Vigo (Eds.), *The Vehicle Routing Problem* (pp. 157–193). Philadelphia, PA: Society for Industrial and Applied Mathematics Monographs on Discrete Mathematics and Applications.

Davidsson, P., Henesey, L., Ramstedt, L., Törnquist, J., & Wernstedt, F. (2005). An analysis of agent-based aproaches to transport logistics. *Transportation Research Part C, 13*(4), 255–271.

Dorer, K. & Calisti, M. (2005). An adaptive solution to dynamic transport optimization. *In Proc. of 4th Int. Joint Conf. on Autonomous Agents and Multiagent Systems (AAMAS 2005)*, (pp. 45–51). New York, NY: ACM Press.

Fischer, K., Muller, J. P., Pischel, M., & Schier, D. (1995). A model for cooperative transportation scheduling. *In Proc. of the 1st Int. Conf. on Multiagent Systems,* (pp. 109–116). Menlo park, California: AAAI Press / MIT Press.

Hoen, P. J., & La Poutré, J. A. (2003). A decommitment strategy in a competitive multi-agent transportation setting. *In Proc. of 2nd Int. Joint Conf. on Autonomous Agents and Multiagent Systems (AAMAS 2003),* (pp. 1010–1011), New York, NY: ACM Press.

ILOG, Inc. (1992). *Using the CPLEX Callable Library and CPLEX Mixed Integer Library.*

Jennings, N., & Bussmann, S. (2003). Agent-based control systems: Why are they suited to engineering complex systems? *Control Systems Magazine, IEEE, 23*(3), 61–73.

Kohout, R., & Erol, K. (1999). In-time agent-based vehicle routing with a stochastic improvement heuristic. *In Proc. of the 16th national conference on Artificial Intelligence and the 11th on Innovative Applications of Artificial Intelligence (AAAI '99/IAAI '99),* (pp. 864–869). Menlo Park, CA: American Association for Artificial Intelligence.

Larsen, A., Madsen, O., & Solomon, M. (2002). Partially dynamic vehicle routing-models and algorithms. *Journal of the Operational Research Society, 53,* 637–646.

Leong, H. W., & Liu, M. (2006). A multi-agent algorithm for vehicle routing problems with time window. *In Proc. of the ACM Symposium on Applied Computing (SAC 2006),* (pp. 106–111). New York, NY: ACM Press.

Mes, M., van der Heijden, M., & van Harten, A. (2007). Comparison of agent-based scheduling to look-ahead heuristics for real-time transportation problems. *European Journal of Operational Research, 181*(1), 59–75.

Persson, J. A., Davidsson, P., Johansson, S. J., & Wernstedt, F. (2005). Combining agent-based approaches and classical optimization techniques. *In Proc. of the European workshop on Multi-Agent Systems (EUMAS 2005),* (pp. 260–269). Koninklijke Vlaamse Academie van Belie voor Wetenschappen en Kunsten.

Perugini, D., Lambert, D., Sterling, L., & Pearce, A. (2003). A distributed agent approach to global transportation scheduling. *In IEEE/WIC Int. Conf. on Intelligent Agent Technology (IAT 2003),* (pp. 18–24).

Sandholm, T. W. & Lesser, V. R. (2001). Leveled commitment contracts and strategic breach. *Games and Economic Behaviour, 35,* 212–270.

Schillo, M., Kray, C., & Fischer, K. (2002). The eager bidder problem: a fundamental problem of DAI and selected solutions. *In Proc. of 1st Int. Joint Conf. on Autonomous Agents and Multiagent Systems (AAMAS 2002*), (pp. 599–606), New York, NY: ACM Press.

Smith, R. (1980). The contract net protocol: High-level communication and control in a distributed problem solver. *IEEE Transactions on Computers, C-29*(12), 1104–1113.

Thompson, P. & Psaraftis, H. (1993). Cyclic transfer algorithms for multivehicle routing and scheduling problems. *Operations Research, 41*(5), 935-946.

Toth, P. & Vigo, D., editors (2002). The Vehicle Routing Problem. *SIAM Monographs on Discrete Mathematics and Applications*. Society for Industrial and Applied Mathematics.

Vickrey, W. (1961). Counterspeculation, auctions, and competitive sealed tenders. *Journal of Finance, 16*, 8–37.

Yang, J., Jaillet, P., & Mahmassani, H. (1999). On-line algorithms for truck fleet assignment and scheduling under real-time information. *Transportation Research Record, 1667*, 107–113.

Chapter XVII
Analyzing Transactions Costs in Transport Corridors Using Multi Agent-Based Simulation

Lawrence Henesey
Blekinge Institute of Technology, Sweden

Jan A. Persson
Blekinge Institute of Technology, Sweden

ABSTRACT

In analyzing freight transportation systems, such as the intermodal transport of containers, often direct monetary costs associated with transportation are used to evaluate or determine choice of Transport Corridor. In forming decisions on Transport Corridor cooperation, this chapter proposes that transaction cost Simulation modelling can be considered as an additional determinant in conducting Transport Corridor analysis. The application of Transaction Costs theory in analyzing the organisational structures and the transactions that occur, assists in indicating as to which governance structure results in higher efficiencies. The use of Multi-Agent based Simulation for modelling the organisational structure and mechanisms provides a novel approach in understanding the organisational relationships in a regional Transport Corridor.

1. INTRODUCTION

The purpose of this chapter is to apply elements from Transaction Costs economic theory in the design of a conceptual computer Simulation model for analysing cooperation choice of Transport Corridor.

The Simulation model adopts a Multi-Agent approach in coordinating the intelligent behaviour among a collection of autonomous Agents representing actors involved in the transportation of goods. This technological approach implies that the Agents would be modelled to represent both users and providers in a Transport Corridor for Simulation and analysis. The Agents would be seeking to satisfy their own goals rather than searching an optimal organisational solution. Contracts and negotiations could be simulated and organisational structures analyzed, i.e., market, vertical and contract. The application of Transaction Costs theory would assist in explaining or predicting the behaviour of actors in a Transport Corridor. Additionally, Multi-Agent based Simulation (MABS) could assist in analysing the decisions that are influenced by the different levels of Transaction Costs, such as whether shipping lines should purchase or build their own Terminals as opposed to using Terminals of others (make or buy). The research question that is studied is: "how can agent-based technology be used in analyzing the Transaction Costs and organisational structures in a Transport Corridor?"

A market has to exist first before governance structures can be formed (Klos 2000). Therefore the objective of the research presented in this chapter is to analyze how the real organisations represented as Agents and their transactions, which are incorporated in a model, influence the choice of a suitable structure for organisation in a Transport Corridor. In order to achieve this objective, we study goods transferred through the entire transport chain, from origin to final destination, in the most efficient manner, i.e., cost- and time-effective. Some examples of cooperation in transport chains are:

- the use of a common standard, e.g. an ISO container, may create strong interconnectivity with other actors in the organisation of shipping.
- improve the operations and utilization of resources. For instance, it is important that time tables meet customer requirements.
- use of new technologies which may help to bind firms closer, settle claims and develop trust.

The chapter is structured as follows: In Section 2 a description of transaction cost theory is presented. The components that are to be represented by Agents, in a generic Transport Corridor are described in Section 3. A Simulation architecture based on a MABS approach is presented in Section 4. The model and design of the simulator is outlined in Section 5. Finally, in Section 6, we discuss our conclusions and provide an outlook onto future work.

2. DESCRIPTION OF TRANSACTION COST THEORY

In the book "*The Nature of the Firm*" (Coase, 1937), Ronald Coase observed that market prices often govern the relationships between firms, known as transactions. Ronald Coase noted that if transactions are not governed by the price system then an organisational structure must exist. The transaction cost approach was developed by Ronald Coase to identify what are the costs of providing for some transaction through the market rather than having it provided from within the firm (Klos, 2000). Some Transaction Costs types are: searching costs, negotiation costs, and monitoring or policing costs.

In further developing transaction cost economics, Williamson (1979) and Williamson (1995) has studied the organization of transactions and "governance structures" that occur whenever or wherever a good or service is transferred from a provider to a user. As one transaction occurs when a good or service is transferred, a stage of activity is terminating and another is beginning (Williamson,1979).

Transaction Costs economics focuses on the transactions between the stages of activity where the firm is one type of organisational structure. Transaction cost economics can be seen as the mapping of forms of organisations into transactions. The existence of low Transaction Costs in global trade has been a leading element in globalization.

Transaction can be either internal or external to organizations. The transactions that occur within the organization are internal and may include such costs as managing and monitoring staff, products, or services. The external transactions costs when buying from an external provider may consider the source selection, performance measurement, and managing the contract. Transaction cost economics tries to answer such questions as: shall we make or buy? Is the market structure the best method to organize purchasing? When is cooperation beneficial? In a paper by Kylaheiko et al., (2000) provides some examples of transactions costs that can be considered to be related to trading partners located in a Transport Corridor:

- *Searching costs* – Caused by the search for transaction partners or alternative actions (examples are: the amount of time needed for the search at special organisations or institutions, costs which are caused by the use of telecommunication, online services or special publications or management consultants).
- *Information costs* – Due to lack of information in the process of interaction. This covers costs that are caused by the use of different languages (e.g. translation costs) or by technical problems that disturb the exchange of information (costs of technical equipment to overcome this disturbance).
- *Decision costs* – Arises from the participation of a group in the decision process. Due to different aims and motives of participants of decision groups, coming to an (shared) agreement is a very time-consuming process. Moreover, decision costs are caused by contracts that were not fulfilled in the way they were negotiated or by contracts that were not closed in the intended meaning.
- *Bargaining costs* – Caused by the process of negotiation (examples: costs of lawyers and consultants, costs of the required resources like costs of travelling and travelling time).
- *Control costs* – Emerge from the adaptation and supervision of transaction results (examples: costs controlling payments or arranged technical standards or quality).
- *Handling costs* – Emerge from the management of converging action cooperation (examples: costs involving human resources, costs which are caused by the definition of business processes).
- *Adjustment costs* – Caused by the change of transaction conditions can be defined as costs of adjustment (examples: costs which are caused by the implementation of new laws or new IT-standards).
- *Disincentive costs* – Emerge by an opportunistic behaviour of the transaction partners or employees, i.e. every partner tries to interpret the contract to his own advantage (examples: unannounced high increase of prices by a supplier of products which have a very high level of specificity).
- *Execution costs* – Arise from the collection of overdue performances or payments. A possible example is the collection of proceedings.

Williamson (1991) lists six key elements of which two are assumptions, fixed factors and four are variables, used to characterise a transaction. According to the theory, the variables can determine whether the Transaction Costs will be lowest in a market or in a hierarchy that can affect Transaction Costs *assumptions:*

- *Opportunism* – A situation in which one partner in a relationships exploits the dependence of another partner, i.e. increasing prices or reducing quality.
- *Bounded rationality* – Not possessing perfect information due to limited time or span of control. It is difficult to locate the best solution or know what alternatives may exist.

Transaction cost *variables*:

- *Asset specificity* – These investments are made by the trading partners who are specific, such as the tools, routines, knowledge or machines to serve a certain trade partner.
- *Uncertainty* – The plethora of new technologies and the increasing complexity that characterizes many systems impacts the decisions that are made.
- *Information Asymmetry* – Information or quality is not disseminated among all partners evenly. Typically characterized in many transportation networks are the number of "islands of information" which generate, release or retrieve information that is useful for a specific trading partner.
- *Frequency* – The number or volume of orders.

A major concentration of Transaction Costs theory has been on *governance structures* that seek to maximize the value net of production and Transaction Costs. Most transactions are carried out through a market governance structure. There are three main types of governance structures: *market, contracts, and vertical integration*. Markets are seen the most preferred solution to organize activities, when uncertainty and knowledge is imperfect. Contracts provide protection for transaction specific assets by binding both the provider and the user together for a certain time period. Vertical integration is employed in order to internalize the values of transaction specific assets. In Table 1 we compare the advantages and disadvantages listed by RAND (2002) on the three main governance structures resulting from Transaction Costs.

3. COMPONENTS OF A TRANSPORT CORRIDOR

A main objective of the European Union's (E.U.) *Motorways of the Sea* initiative and especially in the *BalticGateway (*BalticGateway, 2008) and *EastWest* (EastWest Transport Corridor, 2008) projects is to increase the use of intermodal freight, seaports and Terminals in order to take more freight traffic off the road and rail systems. The enlargement of the European Union, especially in the East Baltic region offers many tantalizing opportunities and uncertainties for policy makers regarding to the choice of

Table 1. Three types of governance structures

Governance structure	Advantages	Disadvantages
Market	Incentive on maximizing net value	Can't protect transaction-specific investments
Contracts	Some protection on investments	Not all possible contingencies can be contracted
Vertical Integration	Internalize values of transaction-specific investments	Can't control costs as well as markets

freight transportation systems and Transport Corridors. The investments and business decisions on seaports, rail networks, and roads in moving cargo between the new members states in the Baltic incites many questions that require further analysis. In particular, the Terminals (seaports) require much attention and need to be studied since they are the "nodal point" between the land-based transport networks and marine transport networks. The Terminals are often not explicitly taken into account when cargo transportation flows are analyzed at a regional level (Kondratowicz, 1992).

Shipping can be viewed as a network coupled with land-based transport networks (by trucks or railway), marine transport networks (ships) and seaports or Terminals. As network organisations, *shipping* can be considered to be virtual organizations linked by supplier-customer relationships. Such relationships are often modelled as markets where goods are bought and sold between actors in the network. Transportation costs include physical movement costs and the non-monetary Transaction Costs between the organizations in the Transport Corridor. The use of market mechanisms in coordination or control has assisted in eliminating much of the administrative overhead, meaning that the fall in Transaction Costs significantly decreases the whole transportation costs.

In introducing a generic Transport Corridor, we have concentrated on a few actors that are involved in the transport of goods, e.g. the transport activity between Karlshamn, Sweden with Klaipeda, Lithuania (EastWest Transport Corridor, 2008). Actors that we consider in modelling and simulating are the following: *terminal, freight forwarder, inland transportation providers, governmental legal authorities, shipper* and *ship line*. Some decisions made by the actors that could be modelled and simulated are for example, shippers decision of whether to use rail, ship or road, or the shippers decisions of whether to use a hierarchy (freight forwarder) or just contact the market (inland transportation providers and shipping lines) directly. By evaluating the agent's decisions we aim to identify the most cost-effective governance structure for moving goods between two ports, given for example current asset specificity and switching costs. We describe in more detail the following modelled types of actors in alphabetical order:

1. **Freight Forwarder:** The business of transporting goods involves many various activities. The use of sales contracts between the exporter and importer are the starting phase, where intermediaries may intervene such as freight forwarders. If the exporter or importer does not have their own shipping department, they will contact a freight forwarder. The freight forwarder will have contacts and contracts with various road haulers and steamship lines. The freight forwarder makes the necessary arrangements in taking responsibility of transporting a good from place of origin to the destination. In practice, this means that the freight forwarder will check with the government-legal authorities (e.g. customs), insurance companies, and the banks to insure the transport activity is cleared.
2. **Governmental-legal authorities:** Customs and governmental agencies from regional, national, and international make policies that effect shipping across borders, either by taxation or subsidizing. The inspections and clearance of goods and the way this activity is carried out can influence the transportation of goods and choice of Transport Corridor. The importance of fast clearance and transparency of the process is paramount as can be see on from the example of many shippers choosing Finnish ports over Russian ports in moving cargo to Russia (Mivitrans, 1998). The choice of Transport Corridor is influenced by such policies.
3. **Inland Transportation Provider (Road and Rail):** As road transport and rail cargo transport are becoming more and more effective competitors of sea transport, it is no longer possible to look at maritime transport, including port economics, separately from the total transport system. This

explains why traditional modal split issues are reconsidered in a so-called system split model: "the choice will not primarily be a modal choice; it will really be a choice between different transport systems, some of which will contain a combination of several modes and some of which will depend on only one mode" (Ljungstrom, 1985). Consequently, shippers do not necessarily choose a seaport, but they select a transport chain in which a seaport is merely a node.

With road and rail networks connecting many Terminals to their shippers and with vessels calling at multiple Terminals, the seaport or terminal is sensitive to freight variations and to competition. The seaport must develop a strategic plan and coordinate with its stakeholders on a path that will support the seaport and develop more mutual business in order to compete. The notion that a terminal will be competing with other Terminals is now being redefined.

4. **Shipper:** Often a shipper is the person or organisation that initially decides to transport a good. Shippers are either seen as the exporter or importer, which depends on the contract, and in general are responsible for influencing the transport activity. The shipper can be a manufacturer in which it may ship parts to its factories- in this case it is taking an importer role. When the manufacturer ships the finished autos to its markets- it is taking an exporter role. In both examples the manufacturer was taking the shipping role. In other situations, a shipper may represent a large group of small firms, e.g. the Swedish log industry. By having such an organisation represent the thousands of small log companies, it can assist in negotiating better rates and contracts with shipping lines and Terminals.
5. **Shipping lines:** Often shipping lines are only associated with transporting goods between ports on ships. The emergence of logistics has propelled many shipping lines, such as Maersk or DFDS lines, to develop integrated logistics systems where the ships are one component to a total transport system. In many cases shipping lines can take competing or cooperating roles. The "foot-loose" characteristics of the shipping lines influences the decisions on which Transport Corridors should be taken. The example of TEAM lines (a shipping line) moving its container operations from Karlshamn, Sweden to Åhus, Sweden has severely impacted the flow of containers in Karlshamn.
6. **Terminal:** The terminal is has an important position in Transport Corridors as the intermediaries in helping to reduce the number of transactions, which then leads to lower transport costs. A part of the Transaction Costs in a Transport Corridor can be seen as handling costs at railroad stations, seaport and Terminals. Seaports and Terminals are used by customers to reach the hinterlands or markets that they serve customers by accessing through Transport Corridors, which try to achieve overall transportation system performance by having lower costs and wider access to markets (Henesey et al., 2003).

   Modern seaport and Terminals are no longer passive points of interface between sea and land transport, used by ships and cargo as the natural point of intermodal interchange (Henesey & Tornquist, 2002). They have become logistic centres acting as 'nodal points' in a global transport system. The emergence of integrated freight transport system leads to new challenges in the field of efficiency, equity and sustainability. In order to meet the new requirements, active forms of inter-governmental co-operation, on the sub-regional and even global level, are indispensable.

The importance of operational integration among the actors in a Transport Corridor is generated by the need for greater efficiency. Operational intermodal integration in Transport Corridors is influenced by such forces as, trends in out-sourcing, more focus on supply-chain management concepts and liberalisation of new markets, i.e. Lithuania, Latvia, Estonia, Poland, etc. Governance structure of the market facilitates the exchange for goods to be transported and is considered as an input for decisions that influence the cooperation in the Transport Corridor.

By understanding the most efficient system for organising this integration (such as between actors) may be achieved by applying the transaction cost economics approach (Williamson,1979). The choice of governance systems in the Transaction Costs approach seeks to understand how economic efficiencies can be created in a Transport Corridor. This choice of governance structure is dependent upon cost difference between a market, contract, or vertical (hierarchy).[1] In the case that asset specificity is high, such as a terminal buying a new crane or building a ramp to serve RoRo ships, a vertical form of governance structure is preferred by a terminal actor. If the asset specificity is low, such as locating a truck to move cargo to a terminal, then a market organization would be preferred. Often, market structure is characterized as being preferred in terms of incentives and ability to aggregate demand for exploiting economies of scale. Hierarchy is preferred for adaptive sequential decision making.

In Figure 1, we present the six actors and illustrate their relationships within a port or terminal community. This community can be seen as a subset of a larger Transport Corridor community. In transporting cargo from one port to another port, such as Karlshamn to Klaipeda in Figure 1, both ports have terminal communities with similar groupings of entities. In Henesey et al. (2003), the relationships in a port community are identified to be either physical, implying that the relationships between the entities are more optional in nature, or incorporeal suggesting that these type of relationships are not material. Incorporeal relationships include for example, behaviour that may have impacts on the efficiency objective (Henesey et al., 2003). The solid lines in Figure 1 indicate where a physical relationship

Figure 1. Illustration of the transport corridor simulation

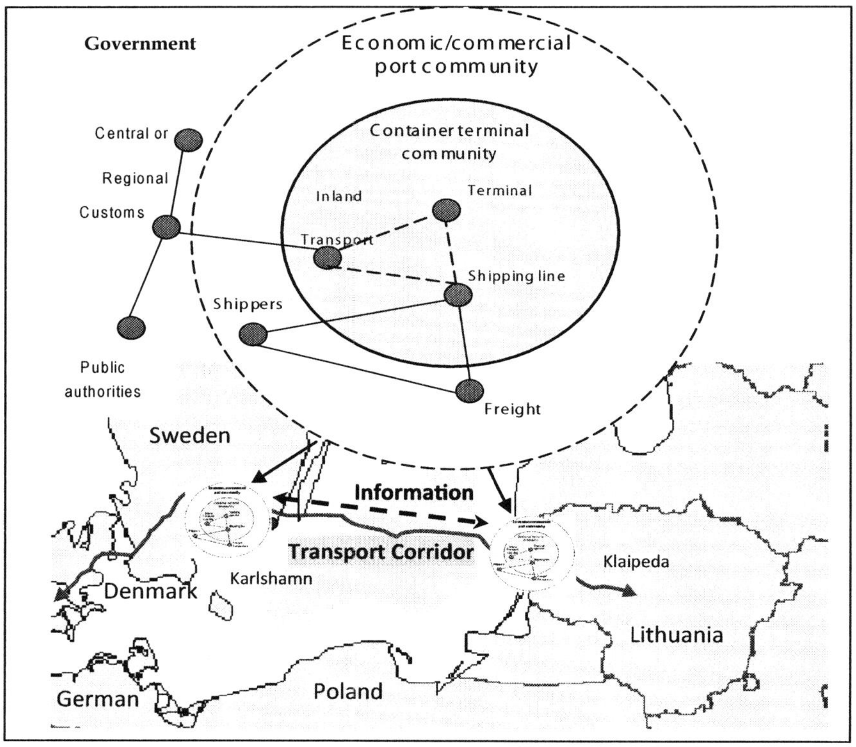

exists the incorporeal relationships are identified as a broken line that suggest incomplete information transmitting between the actors. The Transport Corridor between the two ports is represented as a red line that flows through both ports. The information flow is seen as being communicated between two port communities represented in the Transport Corridor.

4. ARCHITECTURE FOR SIMULATING TRANSPORT CORRIDOR CHOICES

A computer based simulator model is suggested to model the actors that are involved in a Transport Corridor as Agents. The following actors are modelled: *Freight forwarder Agents, Governmental legal authority's Agents, Inland transportation provider Agents, Shipper Agents, Ship line Agents, Terminal Agents.*

In order to simplify the model the transaction cost types that are considered in the model are the handling costs, information costs and switching costs. For the transaction cost assumption we use bounded rationality. Both asset specificity and frequency are transaction cost variables that we analyse. The Agents are considered to be bounded rationally and through their interactions with other Agents organisational patterns will emerge. The output, such as types of governance structures, could be useful in future decision making for evaluating the total transport costs that include both the cost for transporting and the Transaction Costs.

Economic models incorporating MABS have been developed in investigating the theory of transactions cost economics. Klos (2000) has developed an agent-based model for simulating and analyzing Transaction Costs economics. In our proposed model, Agents can act autonomously on deciding preferences of which particular agent(s) to work with. The Agents develop different preferences for other Agents representing trading partners. The Agents and three types of Transaction Costs are considered in the model, where the Agents adaptively search suitable structural forms for organizing in order to satisfy transport demands.

Different polices and strategies for integrating terminal, shipping and logistics operations in Transport Corridor could be analyzed and compared through extending the work on Simulation proposed by Klos (2000). The simulator is expected to generate results that would offer decision makers the ability to view the structure of a Transport Corridor system and the functions that the stakeholders have under various "what if" analyses. Different type of Transaction Costs questions that could be evaluated are for example:

- How can seaports, transport operators (land and sea based) and Terminals improve performance by selecting a suitable governance structure?
- Which actors are working together and how they are cooperating in the Transport Corridor?

5. SIMULATION DESIGN AND MODEL

The proposed simulator model extends the work by Klos (2000). The extensions that we consider into the proposed model are;

- Employing the Beliefs Desires and Intentions (BDI) model for developing the individual Agents and their behaviours.
- Introducing real transportation costs by considering the transport costs.
- Finally, we consider the agent's abilities to satisfy other agent's demands by considering the actual tasks required for satisfying transport demands.

5.1 BDI Model

The BDI architecture model is suggested to capture some of the characteristics of real stakeholders in the Transport Corridor. The Agents will be representing the stakeholders in the system and would have incomplete beliefs – bounded rationality. The desires of the Agents could be considered the individual goals that could be achieved by each of the Agents, whether executing a task alone or with other Agents. Intentions are similar to plans, which may be tightly integrated with other agent plans, to satisfy a transport demand.

One motivation for using the MABS approach is that it has been useful when applied to other areas of policymaking (Downing et al., 2000). In particular to the Transport Corridor choices and Transaction Costs that influence those decisions, different forms of organisation could be investigated. Scenarios representing different levels of transactions costs and various forms of organisation could also be generated and analyzed. These analyses would help to assess what are the factors influencing performance in a systems perspective and give indication on what are proper governance structures. In order to achieve an objective such as intermodality, intensive cooperation and coordination amongst trading partners in the Transport Corridor are essential.

Further motivation in suggesting the BDI model is that given when bounded rationality exists and opportunism exists, transaction cost economics includes a rational analysis component that searches for the best organisational structure for various types of transactions. The proposed agent methodology would deal with the complexity in modelling the behaviour of the individual actors in the system.

5.2 Model Design

To satisfy a demand for transport, a specified set of tasks must be conducted. For instance, a task, *m*, could be the moving of a product or the handling of a container in a terminal. In executing the tasks, Agents will represent different actors able to execute a particular task. We suggest that the model includes both a MABS, that would update input parameters to the Simulation, and a matching algorithm formulated by Klos (2000). See the diagram illustrating the proposed simulator in Figure 2. The matching algorithm formulated in Klos (2000) is based on Tesfatsion's (1997) Deferred Choice and Refusal (DCR) algorithm, which extends Gale and Shapley's deferred acceptance algorithm (Gale & Shapley, 1962). The matching algorithm will compute weights of working with other Agents on a task based on dynamically updated input parameters from the MABS, and use the weights for deciding which agent should work with which agent for a particular task.

Our suggested approach is similar to an approach described by Robert Axtell's set-up for a variable effort model of firm formation described in (Axtell, 1999), in that each agent should posses preferences for profit and past-experience with more of either preferred to less, *ceteris paribus*. Agent *i*'s profit is monotonically increasing with additional tasks, which implies that adding more tasks to satisfy a transport demand never decreases the profit. The assignment of preference weights is extended from Klos

(2000) by extending the Cobb-Douglas functional form. We suggest the following general formula for computing weights:

$$s_{ij}^{m} = f^{cost}\left(t_{ij}^{m}\right) + f^{transaction\,costs}\left(o_{ij}^{m}\right) + \left(p_{ij}^{\alpha_i} \cdot r_{ij}^{1-\alpha_i}\right) \quad (1)$$

where we adopt from Klos (2000):

sijm = weight assigned for Agent *i* to cooperate with Agent *j* in order to perform task, *m*,
pij = *estimated* profit that Agent *i* calculates from coordinating with Agent *j*,
rij = preference based upon past experience of Agent *i* coordinating with Agent *j* (we substitute trust in Klos (2000), with preference),
αi ∈ [0,1] = weight Agent *i* assigns to *pij* relative *rij,*
our extension to the model:
tijm = transport cost parameter for agent *i* to perform task *m* for agent *j,*
oijm = transaction cost parameter in which a task *m* is associated with a demand for a specific transaction cost,
fcost(tijm) and *ftransactioncosts(oijm)* are functions (to be detailed in future work) for influencing the weights to a suitable degree due to transport cost and Transaction Costs, respectively.

The parameters (p_{ij}, r_{ij} *and* t_{ij}^{m} and o_{ij}^{m}) are used as input to the Simulation and will dynamically change during the Simulation as the Agents in the MAS update their preferences.

Each agent attaches a profitability weight, p_{ij} based on Agents *i* general estimate of profit for agent *i* working with agent *j.* The estimate is partly based on asset specificity. We assume that a specialized provider of transport, e.g., a terminal offering cranes to lift cargo on or off a ship, can enjoy efficiency advantages. This interpretation of efficiency is applicable for the providers of transport. See Klos (2000) for details on handling the general asset specificity.

Provided that asset specificity is proportional to differences in transport demand, it is possible that the assets required to satisfy a transport demand may not be easily switched to another provider. Since the asset specificity is connected to the actual tasks of the transport demand, we use o_{ij}^{m} to consider this. The more differentiated an agent's transport task is, the more specialized to that agent's are the assets which an agent that provides transport service, Agents *j,* will be. We suggest to use parameter o_{ij}^{m} for primarily modelling Transaction Costs connected to handling cost, switching cost and information cost.

Initially, some of the Agents in the Transport Corridor would possess predefined preferences, r_{ij}, of agent *i* working with agent *j* based on historical experience of past profit made and Transaction Costs incurred. The ability of an agent *i* to perform a task *m* with agent *j* is modelled through cost parameter, t_{ij}^{m}.

The calculated weights s_{ij}^{m} are used in the DCR for identifying which Agents should be cooperating. The DCR Algorithm will conduct one transport demand (and all its associated tasks) per time step and one task at a time. The result of the calculations performed by the DCR Algorithm will lead to governance structures being formed, i.e. a matching of Agents. Note that the DCR Algorithm is capable to identify the situation where an agent should carry out the work himself, i.e. make instead of buying. The operation in Transport Corridor can be viewed in (at least) three hierarchical levels. The first level is occupied by the shipper Agents, which receive transport demands. The transport demands may be sent to the freight forwarder Agents, which are defined in the second level. Alternatively, the shipper agent or freight forwarder Agents may contact Agents directly in the third level for determining cooperation for satisfying transport demands. In order to identify the structure in the result of the DCR Algorithm, the identification starts at the highest level, with continuation at the nearest level below which has an agent allocated for the task. Hence, it can be identified whether a shipper agent employs a forwarder (second level) or employs an agent at the third level directly. The use of the input from the MAS coupled with the selection process conducted by the DCR Algorithm implies that organisational forms may emerge such as; alliances, coalitions, groups, networks and unions.

In the MAS diagram presented in Figure 2 we illustrate how the proposed simulator can realise the transportation costs by considering both the transport costs and Transaction Costs. For example, in the diagram we view the shipper Agents processing transport demands and coordinating with other Agents by calculating profitability and assigning a weight for a Transaction Costs. The government legal authority agent(s) may influence the environment of the system by levying taxes or offering incentives or subsides.

Figure 2. Input to the Simulation and MAS parts; Agents' interactions leading to input to the DCR Algorithm, which results in the formation of organisation structures

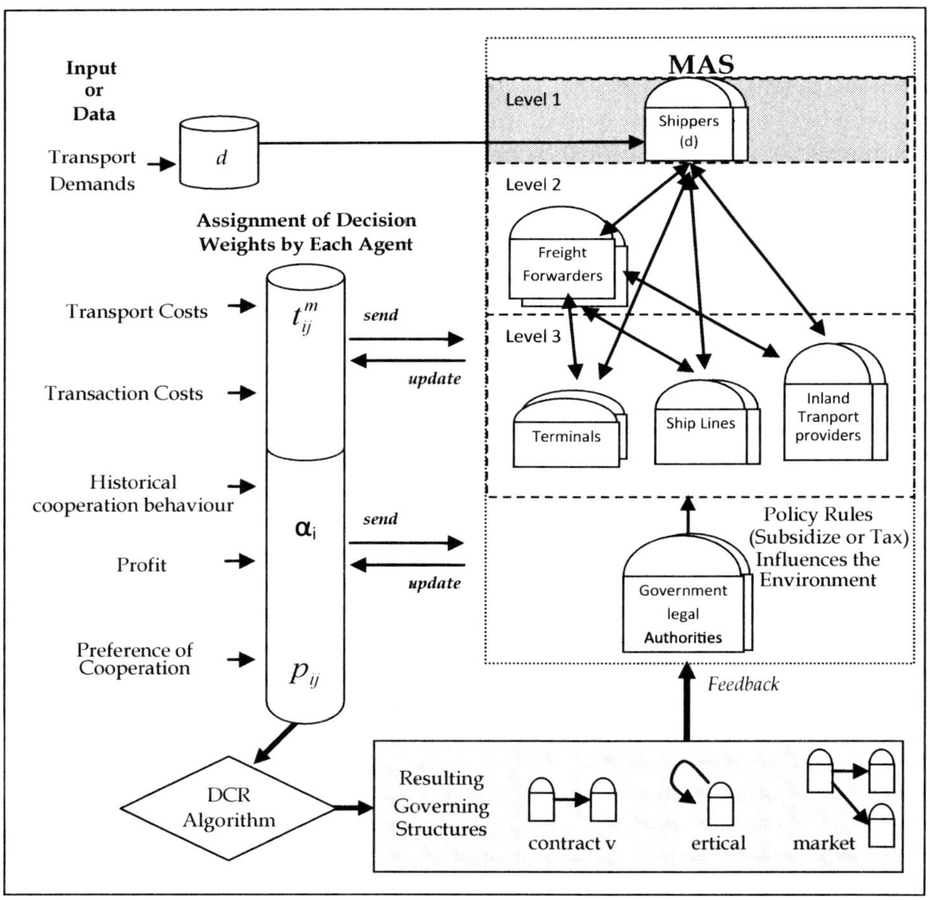

5.3 Conceptual Simulation Experiment

To illustrate the concepts from the above description, we will use a case example in which real actors are coordinating or contracting with each other along a set or well established Transport Corridor between Karlshamn, Sweden with Klaipeda, Lithuania (EastWest Transport Corridor, 2008). All Agents in the beginning of the time step will choose a set of preferences based on Transaction Costs, calculate transport costs for transporting, assign weights and identify (if possible) a preferred partner(s).

The Agents will be dynamically matching based on updated parameters from the MAS. As an example of matching inland transport providers with shippers, we illustrate the input parameters in Table 2. In preference rankings in which a negative value is scored indicates that the coordination between the Agents is 'unacceptable'. From the example in Table 2, AAK choose to cooperate with it's self to conduct the transport, which is considered a form of vertical integration. IKEA cooperates with DHL and this implies a contract form of organization. Volvo considers several partners in which Karlsham Xpress is the preferred choice. Since Volvo considers several partners it is an indicator of that a market form of organisation is suitable. Furthermore, if the Simulation result happens to show, that Volvo buys from different providers of transport from time to time, the result is another indicator of the suitability of a market structure. The choice of operators and cooperation with them has an influence on governance structure.

Software that is considered for the Simulation tool are: JACK Intelligent Agents™, MAGNET and TNG. The first software, JACK Intelligent Agents is a Multi-Agent based system environment for building, running and integrating Agents using a component-based approach (cf., (AOS:JACK Intelligent Agents, 2008)). The JACK software extends Java programming language by employing the following agent-oriented concepts: Agents, capabilities, events, plans and resource management. The Multi-Agent Negotiation Test bed system (MAGNET) (Collins, et al., 2002) is a framework for self-interested Agents, which are either suppliers, customers, or may posses both traits in conducting commerce among themselves via negotiation of contracts for tasks. The Agents may exhibit behaviour that is cooperative, competitive, and may possibly display tendencies that are both but not at the same time. The formation of a virtual organisation can be viewed for understanding the market infrastructure. Trade Network Game (TNG) that is located in Testfatsion (2008) which combines evolutionary

Table 2. Example of operator selection of transport users and providers using weights. AAK (user 1) ranks itself as the best choice. IKEA (user 2) ranks DHL (provider 6) as the best choice. Volvo (user 3) ranks provider Karlshamn Xpress (provider 2) as first choice.

Provider of Transport						
User of Transport	1 (AAK)	2 (Karlshamn Xpress)	3 (Schenker)	4 (DFDS)	5 (Food Tankers)	6 (DHL)
1 (AAK)	2,3	-0,5	-0,4	-0,5	-0,3	-0,2
2 (IKEA)	-0,3	-0,5	-0,2	-0,4	-0,3	4,3
3 (Volvo)	-0,2	5,1	2,4	1,1	3,1	3,3
4 (DHL)	1,4	-0,2	-0,1	-0,2	-0,5	1
5 (Electrolux)	3,1	-0,4	4,2	2,4	1,3	2,3
6 (SAAB)	-0,5	-0,3	-0,3	-0,5	-0,4	1

game play with preferential partner selection is suggested for evaluating alternative specifications for market structure, trade partner matching, trading, expectation formation, and trade strategy evolution. The evolutionary implications of these specifications can later be studied at three different levels: individual trader attributes; trade network formation; and social welfare, c.f. Agent-based Computational Economics website (ACE) (Tesfatsion, 2008).

6. CONCLUSION AND FUTURE WORK

In this chapter a conceptual model is proposed for simulating three Transaction Costs (switching, handling and searching), which are associated in determining the organisational structure in a Transport Corridor. The model introduces and describes an approach using MABS, which seems to provide additional means in understanding decisions on choice of cooperation in the Transport Corridor as well as other decisions effecting freight movements. By providing a framework model that integrates Transaction Costs and transport costs, MABS seems to be a suitable approach. The organisation forms can be analysed in the context of Transaction Costs and cooperation between actors in the Transport Corridor. A Simulation model could be developed by further extending the work on TCE in (Klos, 2001) by including additional variables, such as utility, income, number of firms such as in (Axtell, 1999) with a study of negotiation protocols that is discussed by Resnschein and Zlotkin in (Wooldridge, 2002).

Economic theory has provided many contributions to resource sharing and decision making, such as, compute optimum allocation introduced through economic models. The use of computer Simulation utilizing MABS introduces a novel approach to analyzing Transport Corridors and transportation systems where Transaction Costs are considered. This chapter has conceptually demonstrated how transaction cost theory could provide a useful base for models and tools to be further developed in assisting choice of Transport Corridors.

Further work with software such as JACK Intelligent Agents, MAGNET and TNG is required. Additional information and data collection from companies, i.e. questionnaire or interviews, would benefit the development of the Agents in the Transport Corridor. Modelling the actual contractual transactions, i.e., the "buying" and "selling" of transport services for satisfying transport demands could be considered in the Simulation.

A couple of situations that could be experimented are: opportunism with the Agents in the system and more evaluation on the organisation structure that is best fitted for the actors in the Transport Corridor, i.e. vertical integration, market, or contract. The switching costs of one agent to another, e.g. the use of a road hauler to a rail road can be better identified or possibly measured in sums of money. The coordination practices of the Agents in the system could be further analyzed and tested on how coalitions are formed. Some examples of agent coordination, which are considered deliberative are; cooperative planning, behaviour based decision making and negotiation based on either worth-oriented domains or task-oriented domains. The development of the suggested simulator in studying Transport Corridors in European Union financed project, such as in the *BalticGateway* (2008) or *EASTWEST* (2008) could be attractive, however such a simulator is vision to be applicable to other geographical areas.

ACKNOWLEDGMENT

We wish to thank Prof. Dr. Wayne Talley and Prof. Dr Photis M. Panayides for comments on an earlier draft. We thank the municipality of Karlshamn in Sweden for their financial support and the EAST-WEST Transport Corridor project.

REFERENCES

AOS: JACK Intelligent Agents, The Agent Oriented Software Group (AOS). Retrieved February 11, 2008 from http://www.agent-software.com

Axtell, R. (1999). *The Emergence of Firms in a Population of Agents: Local Increasing Returns, Unstable Nash Equilibria, And Power Law Size Distributions.* Washington D.C., US: Center on Social and Economic Dynamics, Brookings Institution.

Baltic Gateway. (2008). Retrieved December 20, 2007 from http://www.balticgateway.se

Coase, R. (1937). The nature of the firm. *Economica, 1*(4), 386-405.

Collins, J., Gini, M., & Mobasher, B. (2002). *Multi-Agent negotiation using combinatorial auctions with precedence constraints.* Paper presented at the Fifth International Conference on Autonomous Agents (AGENTS´01). University of Minnesota, Minneapolis, Minnesota.

Downing, T. E., Scott, M., & Pahl-Wostl, C., (2000). Understanding Climate Policy Using Participatory Agent-Based Social Simulation. In S. Moss & P. Davidsson (Ed.), *Multi-Agent Based Simulation: Proceedings of the Second International Workshop*, (1979). (pp. 198-213). Berlin: Springer-Verlag.

EastWest Transport Corridor. (2008). Retrieved December 20, 2007 from http:// www.eastwesttc.org.

Gale, D., & Shapley L. S. (1962). College admissions and stability of marriage. *American Mathematical Monthly, 69*(January), 9-15.

Henesey, L., & Törnquist, J. (2002). Enemy at the Gates: Introduction of Multi-Agent s in a Terminal Information Community. *Third International Conference on Maritime Engineering and Ports.* Rhodes, Greece: Wessex Institute of Technology, UK.

Henesey, L., Notteboom, T., & Davidsson, P. (2003). Agent-based Simulation of stakeholders relations: An approach to sustainable port and terminal management. *The International Association of Maritime Economists Annual Conference*, (IAME 2003). Busan, Korea.

Klos T. (2001). Agent-based computational transaction cost economics. *Journal of Economic Dynamics & Control, 25*(3-4), 503-526.

Klos, T. B. (2000). *Agent-based Computational Transaction Cost Economics.* Published doctoral dissertation, University of Gronigen, Groningen, The Netherlands.

Kondratowicz, L. (1992). Generating logistical chains scenarios for maritime transport policy making. In N. Wijnolst, C. Peters, & P. Liebman (Eds.), *European Short Sea Shipping: proceedings from the second European Research Roundtable Conference* (pp. 379-402). London, UK: Lloyds of London Press Ltd.

Kylaheiko, K., Cisic, D., & Komadina, P. (2000). A*pplication of Transaction Costs to Choice of Transport Corridors. Economics Working Paper* Archive at WUSTL, 2000. Retrieved January 12, 2008, from http://ideas.repec.org/p/wpa/wuwpit/0004001.html)

Ljungstrom, B. J. (1985). Changes in Transport Users' Motivations for Modal Choice: Freight Transport. *In ECMT, Round Table 69*. Paris, France.

Mivitrans, (1998). *Intermodal and Transportation conference*. Hamburg, Germany.

RAND. (2002). *Strategic Sourcing: Theory and Evidence from Economics and Business Management.* Retrieved January 12, 2008 from http://www.rand.org/publications/MR/MR865/MR865.chap2.pdf.

Tesfatsion, L. (2008). *Agent-based computational economics (ACE) website*. Retrieved January 9, 2008 from http://www.econ.iastate.edu/tesfatsi/ace.html.

Tesfatsion, L.S., (1997). A trade network game with endogeneous partner selection, In H. M. Amman, B. Ruesm, & A. B. Winston (Eds.), *Computational Approaches to Economic Problems.* (pp. 249-269). Dordecht,The Netherlands: Kluwer.

Williamson, O. (1979). Transaction cost economics: the governance of contractual relations. *Journal of Law and Economics,* (22), 233-262.

Williamson, O. (1991). Comparative Economic Organisation: The Analysis of Discrete Structural Alternatives. *Administrative Science Quarterly, 36*(2), 269-296.

Williamson, O. (1995). Hierarchies, Market, and Power in the Economy: An Economic Perspective. *Industrial and Corporate Change, 4*(1), 21-49.

Wooldridge, M. (2002). *An Introduction to Multi Agent Systems.* West Sussex, England: John Wiley and Sons Ltd.

Chapter XVIII
A Multi-Agent Simulation of Collaborative Air Traffic Flow Management

Shawn R. Wolfe
NASA Ames Research Center, USA

Peter A. Jarvis
NASA Ames Research Center, USA

Francis Y. Enomoto
NASA Ames Research Center, USA

Maarten Sierhuis
USRA-RIACS/Delft University of Technology, The Netherlands
NASA Ames Research Center, USA

Bart-Jan van Putten
USRA-RIACS/Utrecht University, The Netherlands
NASA Ames Research Center, USA

Kapil S. Sheth
NASA Ames Research Center, USA

ABSRACT

Today's air traffic management system is not expected to scale to the projected increase in traffic over the next two decades. Enhancing collaboration between the controllers and the users of the airspace could lessen the impact of the resulting air traffic flow problems. The authors summarize a new concept that has been proposed for collaborative air traffic flow management, the problems it is meant to address, and our approach to evaluating the concept. The authors present their initial simulation design and experimental results, using several simple route selection strategies and traffic flow management approaches. Though their model is still in an early stage of development, these results have revealed interesting properties of the proposed concept that will guide their continued development, refinement of the model, and possibly influence other studies of traffic management elsewhere. Finally, they conclude with the challenges of validating the proposed concept through simulation and future work.

INTRODUCTION

Air traffic in the United States of America (U.S.A.) is forecasted to double or triple by the year 2025 (Pearce, 2006). Recent simulations (Mukherjee, Grabbe, & Sridhar, 2008) of this increase in demand using current air traffic management techniques yielded an increase in average delay per flight from four minutes to over five hours – a clearly unacceptable situation. Accordingly, the National Aeronautics and Space Administration (NASA) is currently exploring several new concepts that may reduce or alleviate air traffic problems. One such concept is Collaborative Air Traffic Flow Management (CATFM), which seeks to lessen the impact on airspace user operations rather than eliminate the problem. Today in the U.S.A., the Federal Aviation Administration (FAA) makes the bulk of Air Traffic Flow Management (ATFM) decisions with only limited consultation with the airlines. In CATFM, the airspace users are given more opportunities to express their preferences, choose among options, and take proactive actions. It is presumed that this will result in decreased workload for the FAA, increased airline satisfaction, and more efficient traffic flow management (Odoni, 1987).

Several questions arise when evaluating if the CATFM concept will work in the future environment. Will the airlines take advantage of new opportunities for action, or will they be passive and let the FAA continue to solve traffic problems independently? Will increasing airline involvement decrease the FAA's workload? Will the options available to the airlines enable them to substantially increase the efficiency of their operations, in particular when many factors still remain out of their control? Will the uncoordinated actions of individual airlines increase the efficiency of the system as a whole, even though each airline is only concerned with their own operations (Waslander, Raffard, & Tomlin, 2008)? Might potential efficiency gains be offset by the actions of rogue operators, who purposely seek to interfere with the operations of a competitor (Hardin, 1968)?

Given that the CATFM concept involves many independent entities with their own beliefs and desires, we feel that the first step to answering some of these questions is through agent-based modeling and simulation. Our goal is to build a simulation of CATFM so that its strengths and weaknesses can be evaluated long before more costly human–in-the-loop simulations or limited field deployments are attempted (Wambsganss, 1996). Our simulation is in an early stage of development, but we have already found several interesting and important properties of CATFM (presented in our conclusions).

Though our study is certainly most relevant to air traffic, certain aspects are relevant to other forms of traffic as well. Our methodology can be applied to any concept of operations in these domains. Many of the basic concepts (e.g., choosing routes, traffic congestion, independent and uncoordinated agent actions) are the same and the overall structure is similar. Nonetheless, there are important differences. An aircraft's airborne speed must remain in a narrow range: significant speed increases are usually unachievable; slower speeds can produce stalls; and halting is impossible. This greatly constrains the actions that are available, and is further limited by the amount of fuel onboard (which is minimized to reduce operating costs).

ATFM generally has more centralized control than other forms of traffic management: In contrast, CATFM increases information sharing and distributes some elements of decision making. Finally, a significant portion of air traffic is comprised of fleets (i.e., airlines) – essentially allied pilots who are interested in cooperating for the common good of the company.

When viewed abstractly, systems developed and evaluated for CATFM could be generalized to other agent-based systems, particularly those that model people. Like many other real-word systems, the air traffic system involves a competition for limited and shared resources. The participants of this system are neither wholly cooperative (which is rarely realistic given self-interest), nor entirely competitive (which can lead to less efficient overall performance). Rather, there are two types of participants: a controlling entity, which seeks to maximize some global property such as system performance; and participating operators, which seek to maximize their own utility. The challenge is to design a robust system of constraints so that the actions of the participants work towards maximizing the desired global property. The utility functions of the participants are self-determined, may include antagonistic elements, and are generally unknowable, complicating matters. Yet, this situation occurs often not only in government-controlled systems, but also in any system with central authority, such as companies, organizational bodies, and games of many types.

We begin with a description of ATFM and related work. We describe the main features of the CATFM concept of operations and the observed operational problems it is meant to address. Our approach to developing a simulation of this concept of operations is presented, and we describe our simulation of the flight routing phase. We discuss the comparative results of different CATFM approaches and different airspace user strategies. We conclude with an analysis of these experiments, and present our goals for future development of the simulation.

BACKGROUND

Introduction to ATFM

Air traffic control (ATC), a superset of ATFM, provides safe, orderly, and efficient flow of aircraft operating within a given airspace (Nolan, 2003). Generally, an Air Traffic Service Provider (ATSP) is the authority responsible for providing air traffic management; the FAA is the ATSP for the U.S.A.'s National Airspace System (NAS). The FAA has four major types of facilities that participate in ATC. ATC towers manage the aircraft arriving, departing, and taxiing on the ground. Terminal radar approach control facilities control airspace within approximately thirty miles of a major airport. Air Route Traffic Control Centers (ARTCCs) are responsible for the remainder of controlled airspace in the NAS. There are twenty such ARTCCs in the continental United States, and each ARTCC is further subdivided into sectors. Finally, the Air Traffic Control System Command Center (ATCSCC) develops nation-wide strategic plans for traffic flow management throughout the NAS. It has final approval of all national flight restrictions and is responsible for resolving inter-facility issues. Our research has focused on ATFM at the ARTCC level, which consists mostly of "en route" traffic flying on instrument flight rules (meaning they rely on instrumentation and FAA guidance). The FAA usually assigns traffic to predefined air routes (essentially "sky highways") in order to increase the predictability of the traffic flow.

ATFM is a system-level function to manage the traffic flow based on capacity and demand. ATFM is the responsibility of a Traffic Management Unit (TMU) within each ARTCC and the ATCSCC for regional and national problems, respectively. The ATCSCC TMU develops strategic plans to ensure balanced flow throughout the NAS over a planning horizon of two to eight hours. The ARTCC TMUs develop tactical plans to manage air traffic within their local airspace over a planning horizon of up to two hours that are consistent with any relevant ATCSCC restrictions. The TMUs constantly monitor for potential conditions that could reduce airspace capacity such as adverse weather, and for excessive traffic demand that could overload a sector controller's ability to safely handle traffic (Adams, Kolitz, Milner, & Odoni, 1996). For example, a TMU may identify a Flow Constrained Area (an airspace region with a capacity-demand imbalance) due to anticipated severe convective weather. The TMU would then analyze which type of restriction should be invoked to alleviate the traffic imbalance. Since restrictions may affect adjacent centers, either directly or through ripple effects, ATCSCC approval is needed before invoking such a restriction. ATFM issues are reported during a bi-hourly planning teleconference, involving representatives from the ATCSCC, each ARTCC, and airspace users.

A variety of restrictions are available to the FAA, depending on the nature of the traffic flow problem (Sridhar, Chatterji, Grabbe, & Sheth, 2002); we describe some commonly used restrictions. A re-route procedure assigns a new route to an aircraft to avoid a problem area, such as a severe thunderstorm or congested airspace. (This is the only restriction we have implemented in our current simulation.) A Ground Delay Program (GDP) is used to delay aircraft at departure airports in order to manage the demand at an arrival airport. Flights are assigned delayed controlled departure times, thus changing their expected arrival time at the impacted airport. GDPs are implemented when capacity at an arrival airport has been reduced for a sustained period, due to weather or excessive demand. Miles-in-Trail (MIT) restrictions enforce an increased spatial separation between aircraft transiting through some point in the airspace, but may shift traffic problems upstream. Time-based metering provides dynamic sequence and schedule advisories to controllers to reduce delays for arrival aircraft approaching capacity-constrained airports.

Airlines manage their fleet of aircraft in an Airline Operations Center (AOC). Each AOC has a coordinator that monitors the restrictions and participates in the planning teleconference to make their concerns known to the FAA. A major thrust of the CATFM concept is to increase the role of the AOCs in ATFM.

Agent-Based ATFM Simulations

The Airspace Concept Evaluation System (ACES) (Sweet, Manikonda, Aronson, Roth, & Blake, 2002) is a distributed agent-based simulation of the entire NAS, including but not restricted to ATFM (Couluris, Hunter, Blake, Roth, Sweet, & Stassart, 2003). ACES uses a layered architecture to support several simulations at various levels of fidelity. Airspace participants, ranging from individuals to larger entities, are represented as agents. Given its broad coverage, ACES is able to perform cost-benefit evaluations on new concepts whose effects go beyond that of a particular element.

IMPACT (Intelligent agent-based Model for Policy Analysis of Collaborative Traffic flow management) is a swarm-based agent model of FAA agents and airline agents, used to evaluate three possible responses to capacity reductions: no advanced planning, GDPs without information sharing, and GDPs with shared airline schedules (Campbell, Cooper, Greenbaum, & Wojcik, 2000). In each scenario, the FAA agents decide whether or not to impose GDPs, based on predefined policies. The airline agents choose actions that minimize the estimated cost to their operations. As expected, their simulation measured the best performance when schedule information was shared, but found that GDPs without shared information (as occurs in today's operations) resulted in a *greater* average cost per flight than when no advanced planning occurred.

STEAM (Tambe, 1997) has been used to evaluate a collaborative system for real-time traffic synchronization (Nguyen-Duc, Briot, Drogoul, & Duong, 2003). Real-time traffic synchronization is the work of the individual sector controllers as they manage flights that run through multiple sectors. The airspace user agents do not participate in the collaboration: Only the sector controller agents and a few higher-level coordinating entities coordinate their problem-solving actions.

The Man-Machine Integrated Design and Analysis System (MIDAS) is an agent-based model of human performance when coupled with machine interfaces. MIDAS has been applied to ATFM (Corker, 1999), and emphasizes the capabilities and limitations of human cognitive ability instead of complex decision making.

ISSUES AND PROBLEMS

Characterizing Operations and Issues through Field Observations

To characterize current problems in air traffic flow management, field observations were conducted at several operational centers (Idris, Evans, Vivona, Krozel, & Bilimoria, 2006). A diverse set of facilities was included to provide a wide scope of operational characteristics and corresponding issues, including five ARTCCs, five AOCs, and the ATCSCC. The ARTCCs managed areas of varied geographical size with assorted weather characteristics and differing traffic patterns. The AOCs included both large and small carriers, with different operational models and customers. Finally, the ATCSCC provided a unique perspective of national air traffic flow management.

These field observations supported the development of the CATFM concept of operations in three ways. First, they made it possible to characterize the operational situations that result in air traffic flow constraints. These operational situations typically stem from two immediate causes: either from a decrease in airspace capacity (e.g., due to weather or airspace restrictions); or through an increase in demand (e.g., from pop-up traffic, overscheduling, or from traffic rerouted from another area). Second, once the flow constraint situations and their immediate causes were identified, the underlying operational issues that often lead to inefficient handling of these situations were identified. Finally, these observations provide a valuable record of *work practice*. By analyzing how the work is done, potential solutions were developed, and a corresponding agent-based model of ATFM operations was built.

Identified ATFM Issues

The primary finding from the field observations was that the current ATFM system limited the potential for collaborative problem solving. Primarily two factors cause these issues. First, the sharing of information between the FAA and airlines is limited. Thus, planning must be conducted without accurate information about the other entity's view of the current state, priorities and plans. These three elements correspond to the belief, desire and intention agent framework (Bratman, 1999). Second, the bulk of the problem solving activities falls upon the FAA, but their workload limits the solutions they can realistically pursue. We present a summary of these findings; the complete list can be found in (Idris, Vivona, Penny, Krozel, & Bilimoria, 2005).

Inaccurate Problem Assessment

Efficient management of traffic flow issues begins with an assessment of the problem. Incorrect assessments of either the demand or the capacity can lead to inaccurate problem assessments, including over- or underestimating the problem severity, missing a problem or incorrectly raising a non-existent problem. Factors that lead to inaccurate *demand* assessments include erroneous prediction of pop-up traffic, changes in departure times, flight plans or cancellations, and displacement of traffic from flow constraints elsewhere. Factors that lead to inaccurate *capacity* assessments include incorrect weather and airspace restriction predictions. These inaccuracies may lead to divergent assessments between the FAA and AOCs, resulting in inconsistent plans.

Differing Evaluations of Identified Problem

Once the traffic flow problem is identified, the FAA and the airlines regard the problem differently, for after safety, their concerns diverge. The FAA will seek to minimize the effect of the problem on the NAS and limit controller workload. The airlines are only concerned with the affect on their own flights and not the flights of competing airlines. Each airline seeks solutions that adhere to their business model, often with a goal of minimizing costs while limiting the negative effect on their customers. Moreover, different carriers will have different business models, therefore addressing cost, reliability and on-time service differently. Thus, even with a consensus on the traffic flow problem, different entities will often prefer different solutions.

Limited Mitigations

The ARTCC and ATCSCC TMUs have a limited set of restrictions available when choosing mitigations to a traffic flow management issue. These restrictions are typically coarse-grained and are applied uniformly to all airspace users. Often, the mitigations are overly restrictive, and because they are not selective, may disproportionately impact some airspace users.

High TMU Workload

Two factors contribute to a high TMU workload when the disruptions to the NAS grow severe. First, the reliance on direct synchronous communications such as teleconferences and phone calls increases the cost of communication, decreasing both the time available for such communications and for other activities. Secondly, actions targeting individual flights (such as rerouting) greatly increase the quantity of tasks that must be performed by the TMU. As a result, TMU workload becomes a limiting factor for the possible solutions.

Limited Coordination between FAA and Airlines

Due to the problems with communication and TMU workload, coordination between the FAA and the airspace users *decreases* as problems become more severe. Unfortunately, this means there is little or no coordination exactly at the times when it is needed most. The FAA and the AOC assess, evaluate and plan independently from one another outside of the planning teleconferences run by the ATCSCC. This is exacerbated by the relative unpredictability of both parties, potentially leading to a double penalty for either: The TMU may choose unnecessary mitigations and be unprepared for the actual problem, while the AOC may independently avoid one restriction only to be impacted by another, unanticipated restriction. Moreover, due to the decrease in communication caused by a high workload, the FAA may be late in notifying all interested parties that a restriction has been removed, resulting in some parties needlessly avoiding a problem that no longer exists.

SOLUTIONS AND RECOMMENDATIONS

CATFM Concept of Operations

The CATFM concept of operations recommends several changes to address these issues. Most changes fall under the following three categories, listed by order of increasing emphasis. First, automation must be used to reduce the workload of TMU personnel, reducing the need for the TMU planners to perform mundane tasks and lessening the cost of communication. Second, more information should be shared between the FAA and airspace users. By doing so, assessments can be made with more complete information, common assessments are possible, and actions are more predictable. Finally, and most importantly, when possible, the AOCs should be more involved in the traffic flow management process.

We summarize the four phases of the ATFM process in the CATFM Concept of Operations below; a more complete description can be found in (Idris, Vivona, Penny, Krozel, & Bilimoria, 2005).

Common Problem Identification

As described previously, ATFM problems are caused by situations where the demand for an airspace exceeds its capacity. Demand is best predicted by the airspace users who create it, whereas capacity is determined by the FAA, as it is an assessment of the FAA's ability to manage traffic in the affected area. This leads naturally to a collaborative situation where information is shared to produce a more accurate problem assessment, and to minimize the divergence of problem assessments.

Shared Impact Assessment

Various restrictions could address a given ATFM issue, each with a different impact on airline and FAA operations. By establishing a shared impact assessment, options can be evaluated more accurately and better contingency plans can be developed. Moreover, if early indications of probable TMU actions are provided, the AOCs may be able to adjust their plans to coincide with such actions, potentially reducing or eliminating the need for the proposed TMU action.

Traffic Flow Planning with AOC Input

Once a possible set of ATFM actions have been identified, along with their impact, a specific ATFM plan is instantiated to address the traffic flow problem. Instead of a planning decision being made unilaterally by the TMU (as occurs today), the AOCs can provide preferred solutions. These become additional inputs to the TMU's planning process, allowing for the accommodation of airspace user preferences when they do not violate other constraints. In addition, when the TMU workload allows it, the AOCs can suggest alternative plans that may result in an overall better solution.

Joint Plan Implementation

Once an ATFM plan with a set of actions has been chosen, it must be instantiated at the level of individual flights. In some cases, particularly with reroutes, choices must be made, such as which flights should be given the new route. When possible, the airlines should choose which of their flights are impacted by the ATFM action, according to their individual business plan. This reduces the workload of the TMU by shifting the burden of implementation to the AOC, and allows the airline to maximize their own benefit by directly choosing the most acceptable options.

Approach

We have built an initial agent-based simulation of CTFM with Brahms (Clancey, Sierhuis, Kaskiris, & Hoof, 2003). Brahms is a modeling and simulation environment for developing intelligent software agents, particularly to analyze work practice in organizations. Brahms can run in different simulation and runtime modes on distributed platforms, enabling flexible integration of people, hardware-software systems, and other simulations. Brahms was originally conceived as a business process modeling and simulation tool that incorporates the *social systems of work*, illuminating how formal process flow descriptions relate to people's actual situated activities in the workplace (Clancey, Sachs, Sierhuis, & Hoof, 1998). To simulate human behavior at the work practice level, one must model how people work together as individuals in organizations, performing both individual and teamwork activities. The Brahms language is unique in that it models not only individual agent and group behavior, but also systems and artifact behavior, as well as the interactions of people, systems, objects, and the environment. Most other multi-agent languages leave out artifacts and the interaction with the environment, making it difficult to develop a holistic model of real-world situations (Wooldridge & Jennings, 1995). Brahms is an agent language that operationalizes a theory for modeling work practice, allowing a researcher to develop models of human activity behavior that corresponds with how people actually behave in the real world (Sierhuis, 2001).

A methodology for designing and simulating future work systems has been developed and used with Brahms (Clancey, Sierhuis, Seah, Buckley, Reynolds, Hall, & Scott, 2007). The process begins with detailed observations of work practice, which is used to build a model of current operations. After model validation, a new concept of operations is developed, and a simulation of the future work system is created using validated components of the model of current operations whenever possible. After testing the concept in implementation, the process repeats. We have adapted this methodology to our circumstances, taking advantage of the pre-existing CATFM concept of operations and work practice observations. We are developing the model iteratively, building successively more accurate models from increasingly detailed sources of information. At every stage, we evaluate the concept of operations based on the findings of our simulation, modify the concept accordingly, and then increase model fidelity in the next stage.

So far we have built a rudimentary model of ATFM using second-hand sources of information such as the work practice observations described previously, other ATFM literature, and the concept of operations itself. In the next stage, we will interview subject matter experts and incorporate their conception of work practice into the model. This will allow us to fill in details not discernable from the recorded observations of work practice. To validate the model at this stage, historical situations will be simulated and the results will be compared with the historical outcomes. Likewise, historical data may also be used to infer behavior, either by intuition or through data mining techniques. In the third stage, we will perform new observations of work practice, enabling us to build a detailed model at the level of individual (rather than organizational) participants in the ATFM process. The model at this stage can also be validated by comparing the simulated behavior to the behavior observed in the actual system. Subsequent evaluation of the concept will require human subjects to participate in the CATFM process, with humans and agent proxies participating in a human-in-the-loop simulation.

Initial ATFM Simulation Design

We have created a simplified model of a subset of ATFM, including only the Joint Plan Implementation phase (see earlier section of same name) when flights are assigned routes. In order to simplify this selection process, we have redefined capacity to be a property of a route, rather than a sector, and assumed that the routes are independent. Route capacity, flight schedules and agent strategies are static throughout the simulation. In contrast, route demand changes dynamically throughout the simulation as the agents choose routes. We do not model runway constraints or temporal ordering, treating all flights as if they have the same departure time. Our simulation only deals with pre-flight planning and does not simulate the flights themselves.

Figure 1 provides an overview of our current agent architecture. We have built our initial model at the organizational level, with each organization (i.e., TMUs and AOCs) modeled as single agents. Each agent (TMU or AOC) has different responsibilities, with route selection performed by either the TMU agent or the AOC agents (see below). The AOC agents provide the TMU with their flight schedules and the value of each flight. The TMU agent informs the AOC agents of the current status of the airspace by aggregating the current demand on a given route, comparing this with the capacity, and broadcasting the route status (under capacity, at capacity, or oversubscribed) to the AOC agents. In the initial simulation, the TMU does not reroute flights or choose among AOC requests: It approves them all when the route is at or below capacity, and denies all requests when demand exceeds capacity (thus leaving the route unused). To be consistent with U.S.A. law against anti-competitive practices, no communication occurs between AOC agents in order to prevent coalitions or other AOC-AOC negotiations. We do not model communication issues, treating them as reliable, instantaneous, and clear.

For each origin-destination airport pair, we created three routes arbitrarily: a direct route and two alternate routes, 1.25 and 1.5 times the length of the direct route. The capacities of these routes vary, with typically the direct route having insufficient capacity for all scheduled traffic. Our fundamental question is: how will the CATFM concept perform in this simplified model? In order to answer this question, we created four ATFM approaches:

- **Blue Sky:** All capacities are infinite, so every flight takes the direct route. This is not a realistic approach but provides an upper bound on performance that we use as a baseline.
- **Current operations:** The TMU agent makes the route selection, putting flights on the best available routes (i.e., under capacity routes) in a random order without inspecting the flight value. This approach is closest to the current operations where the FAA makes route assignments with little input from the airlines.

Figure 1. Agent architecture

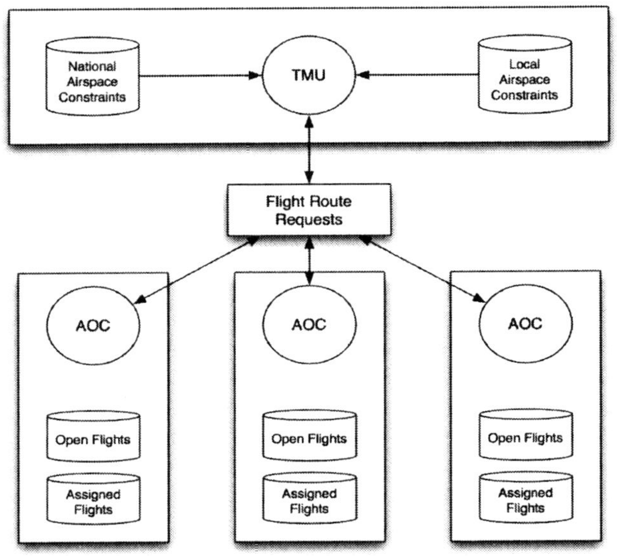

- **Global Optimum:** The TMU agent makes the route selection as in the Current Operations approach, but does so in order of greatest flight value. This greedy algorithm produces the best overall system performance, according to our metrics, but may give preferential route assignments to one airline over another due to differences in flight value distribution.
- **Airline Planning:** The AOC agents make the route selections, with each agent initially requesting the best route for every flight regardless of the strategy used. After the TMU agent broadcasts the status of all routes, the AOC agent may independently choose a new route for each flight. The process repeats iteratively (six iterations unless where noted otherwise) until the time for planning is exhausted. Within a simulation run, a given AOC agent will use the same strategy on each iteration (i.e., no changes in strategy during a run). We used the following simplified strategies:
    - **Aggressive:** An AOC agent with the Aggressive strategy will always request the best route for every flight at each iteration, regardless of the situation.
    - **Moderate:** An AOC agent with the Moderate strategy will request the next best route for some of its flights when faced with an overcapacity situation, repeating the prior request for the other flights.
    - **Conservative:** An AOC agent with the Conservative strategy will request the *worst* route for some of its flights when faced with an overcapacity situation, repeating the prior request for the other flights. The assumption is that the worst route is the least likely to fill up, so the conservative AOC agent attempts to forgo a chance at a better route assignment in exchange for a greater likelihood of finding an available route.

All approaches except Current Operations are deterministic.

Figure 2. Local traffic scenario involving seven airports

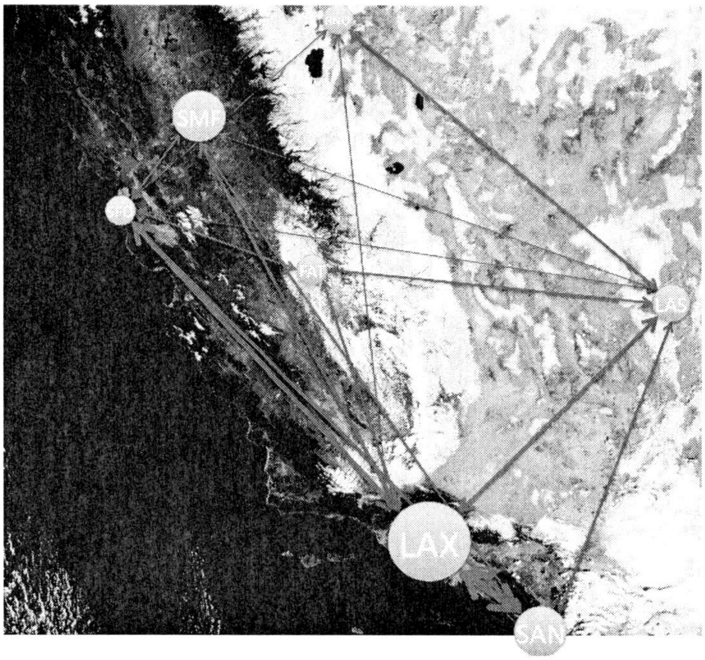

Experiment on a Local Traffic Scenario

We created a local traffic scenario (see Figure 2) that corresponds to traffic generated by three major carriers among several airports in the southwest of the U.S.A. The schedules and aircraft types were chosen based on our observations of the flight schedules of these carriers. Information on connecting crew, passengers, and route capacities were not available, however, so we used our best judgment based on nominal conditions, expected passenger behavior and operational patterns. In all cases, sufficient aggregate capacity was available among the three routes such that every flight could have *some* route assignment.

For a specific flight *F*, we define the following quantities:

p_c = passengers with connecting flights
p_u = passengers without connecting flights
c_c = onboard crew members with a connecting flight
t_a, = the actual flight time of *F*, in minutes
t_o, = the optimal flight time of *F* (from the Blue Sky simulation), in minutes

Each flight is assigned a flight value, which is a heuristic measure of the importance of the flight to the airline. We define v_F, the flight value of *F*, as

$$v_F = p_u + 3p_c + 5c_c \qquad (1)$$

When *F* is assigned a route, we calculate d_F, the delay for flight F, as follows:

$$d_F = t_a - t_o \tag{2}$$

When F is not assigned a route, we assume a standard sixty minutes of delay in a later stage that we do not simulate. Traffic demand naturally rises and falls throughout the day, so we assume that the level of demand falls significantly after our simulation ends. Other factors may also cause delays in practice but are not part of our model.

Finally, we seek to measure in our experiments the total passenger delay incurred by flight *F*, either through an immediate delay or through missed connections. We assume that when a passenger with a connecting flight is delayed, on average, that passenger will experience an additional two-hour delay. When connecting crew members are delayed, their personal delay is not counted (since they are not considered passengers in our simulation), but they are likely to delay the departure of their connecting flight, which in turn impacts many passengers. Therefore, we assume on average, any delay of a connecting crew member results in a total of five hours of passenger delay. Combining this with the above formulae, we calculate the total incurred passenger delay incurred by flight *F*, d_T, in minutes, as

$$d_T = (p_u * d_F) + (p_c * d_F) + 120p_c + 300c_c \text{ when } d_F > 0$$
$$d_T = 0 \text{ when } d_F = 0 \tag{3}$$

We ran the experiments once for each deterministic approach and fifty times for the randomized Current Operations approach, yielding some surprising results (Wolfe, Jarvis, Enomoto, & Sierhuis, 2007). The Airline Planning approach is highly sensitive to the strategies employed by the AOC agents and often performs poorly. Figure 3 shows an example with several strategies, where the light shaded bars indicate delay incurred by selecting a longer route, and the dark shaded bars indicate delay from failing to get an approved route assignment. Further examination of specific trials showed that the Aggressive strategy is disruptive to the system as a whole by pushing demand beyond capacity on the best routes. However, the best performing combination of airline strategies outperformed the Current Operations approach (see Figure 4), indicating the potential for improvement under the CATFM con-

Figure 3. Comparing ATFM approaches on the local scenario

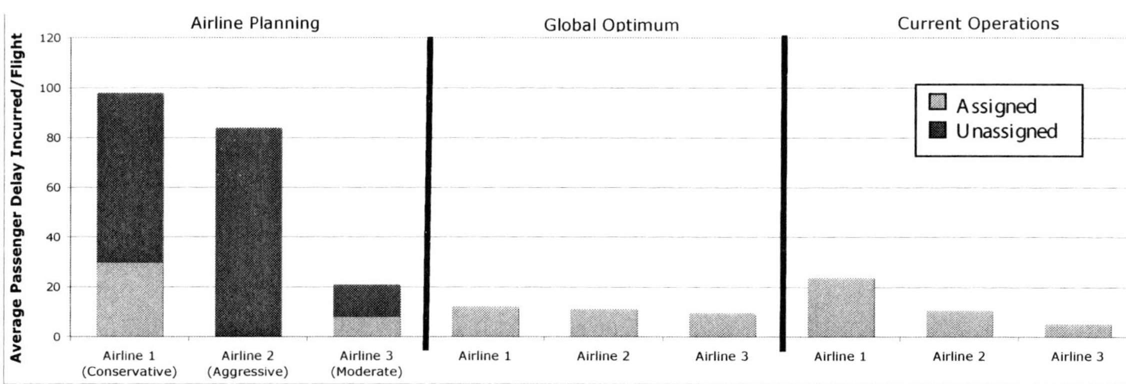

Figure 4. Best airline planning combination compared with Current Operations approach

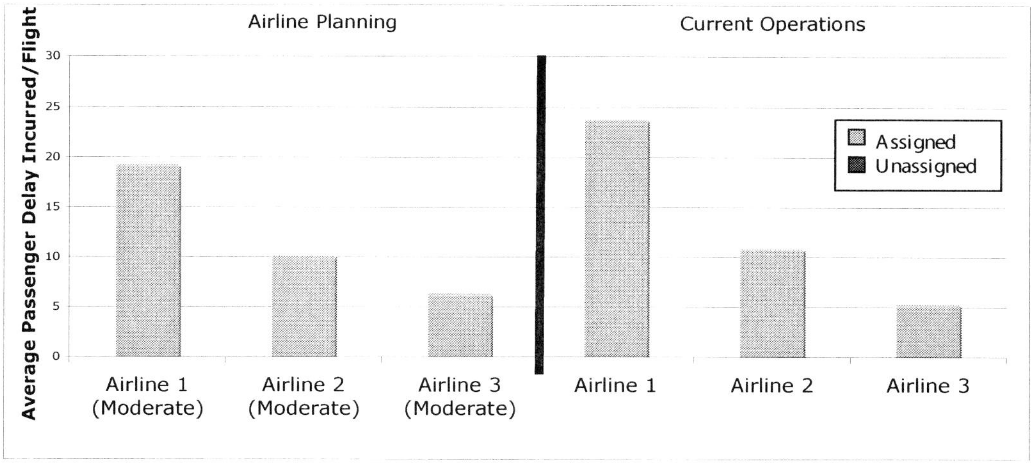

cept. The number of planning cycles can also affect solution quality in the Airline Planning approach, as shown in Figure 5.

Single Origin-Destination Experiment

In our previous experiment, a given AOC agent would use the same strategy on all origin-destination pairs, regardless of the situation. In reality, an airline is likely to use several strategies, matching them to the situation at hand. Since we aggregated the results over the origin-destination pairs, we could see how a strategy performed overall but could not isolate the specific situations where it performed well or poorly. We also wanted to evaluate new approaches that could address concerns that arose from our previous set of experiments, leading to the following additions:

- **Mixed:** This combines the Airline Planning and Optimal approaches. The airlines schedule their flights as before in the Airline Planning approach. Once the planning phase is over, however, the TMU agent will assign any unassigned flights using the Optimal approach. This ensures that any

Figure 5. Effect of additional planning cycles with Airline Planning approach

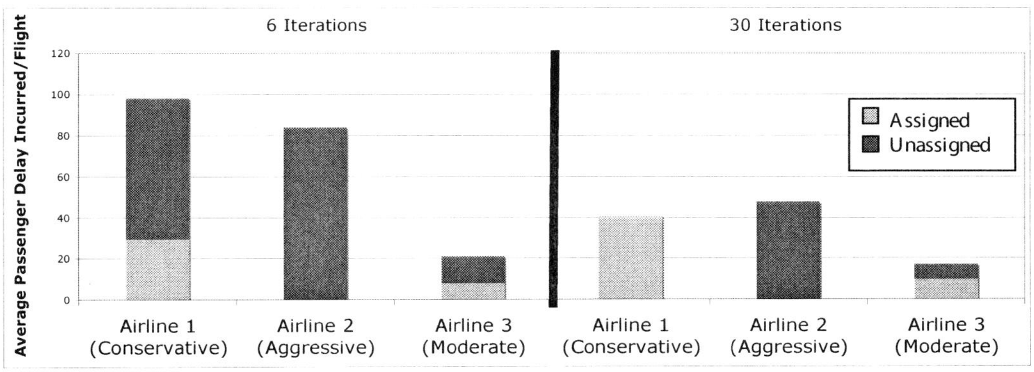

unused capacity will be utilized by flights for which the AOC agents failed to choose an acceptable route.

- **Equitable:** This is a variant of the Optimal approach. Each AOC agent gives a ranking of their flights but does not supply flight values. The TMU agent gives top priority to first-ranked flights, followed by second-ranked flights, and so on. This gives each airline an equal share of each route's capacity, regardless of the value of their flights.

We created three scenarios with the same origin-destination, with one primary route and two alternates as defined previously. In all three scenarios we had three AOC agents, each with four flights to schedule. The scenarios varied in the amount of capacity available:

- **Demand<Capacity:** each route can accommodate five flights.
- **Demand=Capacity:** each route can accommodate four flights.
- **Demand>Capacity**: each route can accommodate only three flights.

Therefore, all flights could be assigned a route on the Demand<Capacity and the Demand=Capacity scenarios, but this was not possible in the Demand>Capacity scenario.

We ran each scenario with all combinations of the three strategies for the three AOC agents, using both the Airline Planning and Mixed approaches, resulting in twenty-seven runs for each. Figure 6 and Figure 7 show the average performance (across all agents and competitor strategy combinations) for each strategy. Table 1 and Table 2 compare strategy alternatives by measuring whether an agent would do as well or better with a different strategy in the given situation, while keeping the competitor strategies constant. For instance, in the Demand<Capacity scenario under the Mixed approach (Table 2), an agent using the Aggressive (A) strategy would have performed as well or better with the Moderate (M) strategy in only 4% of the simulated situations– indicating that the Aggressive strategy was the better choice.

Several patterns emerge from this analysis. The Aggressive strategy is a poor choice when using the Airline Planning approach, consistent with earlier findings, because its insistence on the best route makes that route unusable, potentially leaving its flights unassigned. In contrast, the Aggressive strategy is a good choice when using the Mixed approach with adequate overall capacity. In such cases, the

Figure 6. Strategy performance averaged over all agents with Airline Planning approach

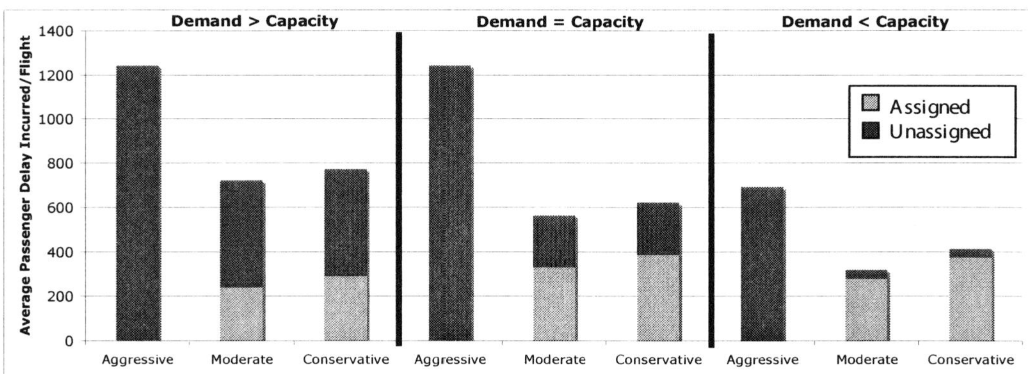

Figure 7. Strategy performance averaged over all agents with Mixed approach

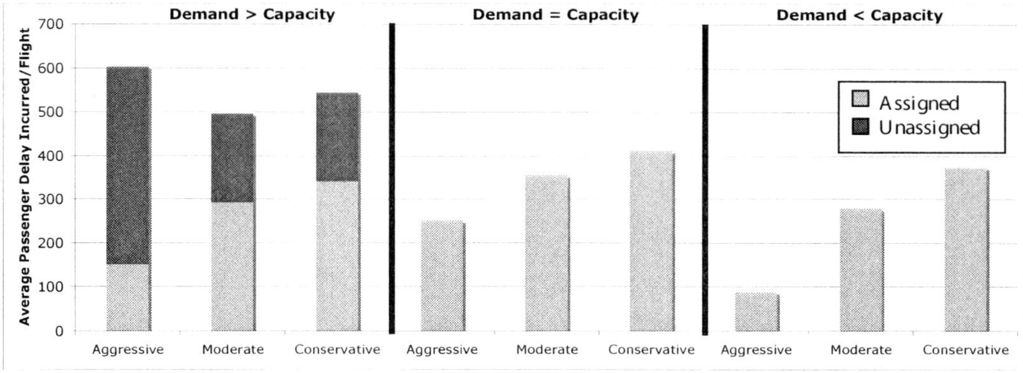

Table 1. Airline Planning approach: Cases equal or improved with alternate strategy

Demand > Capacity	Chosen Strategy A	M	C	Demand = Capacity	Chosen Strategy A	M	C	Demand < Capacity	Chosen Strategy A	M	C
Alternate Strategy A	-	0%	0%	Alternate Strategy A	-	0%	0%	Alternate Strategy A	-	44%	44%
M	100%	-	78%	M	100%	-	89%	M	56%	-	100%
C	100%	22%	-	C	100%	11%	-	C	56%	0%	-

Table 2. Mixed approach: Cases equal or improved with alternate strategy

Demand > Capacity	Chosen Strategy A	M	C	Demand = Capacity	Chosen Strategy A	M	C	Demand < Capacity	Chosen Strategy A	M	C
Alternate Strategy A	-	19%	33%	Alternate Strategy A	-	85%	100%	Alternate Strategy A	-	100%	100%
M	81%	-	78%	M	26%	-	93%	M	4%	-	100%
C	67%	22%	-	C	11%	7%	-	C	0%	0%	-

Aggressive strategy will either succeed in putting all of its flights onto the best route, or it will prevent all other airlines from using the best route. In the latter case, none of the Aggressive airline's flights will be scheduled, and the best route will be completely available when the TMU assigns the remaining flights, leading to a greater share of the best route. However, when there is not sufficient capacity, this strategy performs poorly because not all of its flights will be assigned.

As shown by the rows of Table 1 and Table 2, the best strategy cannot be determined only by the scenario and approach (with the sole exception of the Aggressive strategy in the Demand<Capacity scenario with the Mixed Approach). This is because the performance of a strategy is affected by the

Table 3. Total incurred passenger delay for three airlines with forty flights each

	Airline 1	Airline 2	Airline 3	Total
Current Operations	3552	4332	2939	10823
Optimal	3314	2806	3300	9420
Equitable	2969	3407	3073	9449

competitors' strategies: in particular, each strategy performed worse when a competitor used the same strategy. Therefore it was often preferable to use a unique but generally less attractive strategy than one used by a competitor.

Finally, we created a larger scenario with primary and secondary routes defined as before, but each with a capacity of forty flights, and three airlines with forty flights each. Table 3 shows the results of experiments on this scenario in terms of the total incurred passenger delay metric. In this case, the Equitable approach performed nearly as well as the Optimal approach; it is worth noting that the distributions of flight values were comparable among the three airlines.

CONCLUSION

We have described the design and methodology of a multi-agent simulation of ATFM, as well as experimental findings. At this time, our simulation is a coarse-grained model of operations, with agents corresponding to participating entities (i.e., TMUs and AOCs) rather than persons. Since we simplified other components that were not essential to the problem, actual performance in implementation may differ, but should produce similar conclusions under identical conditions, strategies and policy.

We evaluated several approaches to ATFM, and for the Airline Planning and Mixed approaches, also evaluated several simple route selection strategies. Of these, the Moderate strategy is intuitively the most appealing, and had the best overall performance in our experiments. In contrast, the Conservative strategy did not perform as well, but was usually preferable when it was different than all competitors' strategies. This theme was repeated throughout our experimental results; in nearly every case, the best strategy could not be chosen independently, as it was dependent on the strategies used by the other AOC agents. Finally, the Aggressive strategy worked very well with the Mixed approach when there was adequate capacity, casting doubt on the suitability of the Mixed approach. The Aggressive strategy also did well when the other AOC agents removed their flights from the best route, thus accommodating the aggressive AOC.

In our evaluation of the CATFM concept, we observed that nearly all the approaches that utilized our flight value metric (Equation 3) yielded better results than the Current Operations approach. This supports the claim that utilizing airspace user preferences in ATFM should lead to better solutions. However, this was not the case in all of our experimental results; certain combinations of strategies with the Airline Planning approach produced unacceptably poor results. Moreover, based on current experiments, we did not observe any indication that increasing AOC involvement would reduce FAA workload. In the Optimal and Equitable approaches, the TMU agent continued to perform route selection, and with additional criteria, so this represents an increase in workload. In the Airline Planning approach, the TMU did not perform route selection but the results were often unacceptable; in the

Mixed approach, the results were good, but often the TMU would still make many route selections and inadvertently rewarded aggressive behavior. Therefore, automation is most likely the key to reducing FAA workload. Finally, the AOC agents usually found better solutions when more planning cycles were available. This puts an emphasis on the earlier stages of the CATFM process, which we did not simulate – the earlier situational information is available, the better the likely solution.

In the end, the challenge of refining the CATFM concept will *not* be designing effective AOC agent strategies, as they will be determined by the airlines rather than the system designers. Each airline is likely to have a somewhat different strategy, geared towards their private business model and influenced by the people executing it. Nor is it reasonable to assume that these strategies would necessarily be optimal in all cases. Rather, the challenge is to design a *system* that rewards behavior yielding desirable system performance. In game-theoretical terms, this amounts to redesigning the game itself, rather than the player strategies. In our experiments, the Airline Planning approach was vulnerable to aggressive AOC agents; likewise, the Mixed approach often rewarded the Aggressive strategy. The Optimal approach is unlikely to be deployable in practice, as it would be difficult to create a single objective utility function (flight value in our experiments) over all airlines. Based on our experiments, the Equitable approach is the most promising, as it produced results on par with the Optimal approach (when airlines had comparable flights), but did so without relying on a universal flight evaluation.

FUTURE RESEARCH DIRECTIONS

We have completed the initial stage of development and will continue to expand the CATFM model. We have begun work on the next stage, expanding our model to capture the breadth of the CATFM concept of operations, covering all phases. Our current study simulated the instantiation of the ATFM plan (namely the selection of routes), which was necessary to evaluate the result of the process; however, as earlier phases produce inputs to later phases, it may be that the earlier phases have the greatest operational impact.

In addition to broader scope, a higher degree of fidelity would support stronger claims about the CATFM concept of operations. A more sophisticated flight model would eliminate many simplifying assumptions, such as simplified schedules, and route capacities in lieu of sector capacities. Modeling organizational roles and concentrating on interactions at the level of individual people would reveal the complexity of the proposed work practice and lead to more accurate characterizations of workload. Interviews with subject matter experts, case studies, and additional observations of work practice will yield insight as to how these processes work today.

The results from our initial experiments can be used to guide refinements to the concept of operations and develop policies that are more likely to be successful. Further experimentation with the Equitable approach in a wider array of situations is needed to evaluate its suitability. Additionally, more complex ATFM approaches and airline strategies may yield better overall solutions. Identifying likely airline strategies is of great importance, but difficult, due to their proprietary nature. Since the situations we are simulating are characteristic of *future* operations, rather than today's operations, airlines may not have developed appropriate strategies, and if they have, they may not be willing to share them.

Building a model of *future* operations is difficult at any stage of development. Our approach has been to build and validate a model of current operations, and then to modify that model to fit the future concept. Even validating the current model is a challenge, given the complexity of operations. Modifying a model of current operations to yield a model of future operations introduces uncertainty. We have dealt with this by simulating a variety of possible actions, essentially modeling several possibilities. Game theory can be utilized to develop likely strategies and to analyze properties of the system as a whole. Approaches to traffic management problems in other domains may translate to ATFM, and vice versa.

REFERENCES

Adams, M., Kolitz, S., Milner, J., & Odoni, A. (1996). Evolutionary Concepts for Decentralized Air Traffic Flow Management. *Air Traffic Control Quarterly, 4*(4), 281-306.

Bratman, M. E. (1999). *Intention, Plans, and Practical Reason.* Chicago, Illinois: University of Chicago Press.

Campbell, K. C., Cooper, W. W., Greenbaum, D. P., & Wojcik, L. A. (2000, June). *Modeling Distributed Human Decision-Making in Traffic Flow Management Operations.* Paper presented at the 3rd USA/Europe Air Traffic Management R&D Seminar, Napoli, Italy.

Clancey, W. J., Sachs, P., Sierhuis, M., & Hoof, R. v. (1998). Brahms: Simulating practice for work systems design. *International Journal on Human-Computer Studies, 49*, 831-865.

Clancey, W. J., Sierhuis, M., Kaskiris, C., & Hoof, R. v. (2003, May). *Advantages of Brahms for Specifying and Implementing a Multiagent Human-Robotic Exploration System.* Paper presented at the 16th International Florida Artificial Intelligence Research Society (FLAIRS) Conference, St. Augustine, Florida.

Clancey, W. J., Sierhuis, M., Seah, C., Buckley, C., Reynolds, F., Hall, T., & Scott, M. (2007, October). *Multi-Agent Simulation to Implementation: A Practical Engineering Methodology for Designing Space Flight Operations.* Paper presented at the Eighth Annual International Workshop "Engineering Societies in the Agents World" (ESAW 07), Athens, Greece.

Corker, K. M. (1999, December). *Human Performance Simulation in the Analysis of Advanced Air Traffic Management.* Paper presented at the 1999 Winter Simulation Conference, Phoenix, Arizona.

Couluris, G. J., Hunter, C. G., Blake, M., Roth, K., Sweet, D. N., & Stassart, P. A. (2003, August). *National Airspace System Simulation Capturing the Interactions of Air Traffic Management and Flight Trajectories.* Paper presented at the American Institute of Aeronautics and Astronautics (AIAA) Guidance, Navigation, and Control (GNC) Conference, Austin, Texas.

Hardin, G. (1968). The Tragedy of the Commons. *Science, 162*(3859), 1243-1248.

Idris, H., Evans, A., Vivona, R., Krozel, J., & Bilimoria, K. (2006, September). *Field Observations of Interactions Between Traffic Flow Management and Airline Operations.* Paper presented at the American Institute of Aeronautics and Astronautics (AIAA) 6th Aviation, Technology, Integration, and Operations Conference (ATIO), Wichita, Kansas.

Idris, H., Vivona, R., Penny, S., Krozel, J., & Bilimoria, K. (2005, September). *Operational Concept for Collaborative Traffic Flow Management based on Field Observations.* Paper presented at the American

Institute of Aeronautics and Astronautics (AIAA) 5th Aviation, Technology, Integration, and Operations Conference (ATIO), Arlington, Virginia.

Mukherjee, A., Grabbe, S., & Sridhar, B. (2008). *Alleviating Airspace Restriction through Strategic Control.* Paper presented at the American Institute of Aeronautics and Astronautics (AIAA) Guidance, Navigation, and Control Conference, Honolulu, Hawaii.

Nguyen-Duc, M., Briot, J.-P., Drogoul, A., & Duong, V. (2003, October). *An Application of Multi-Agent Coordination Techniques in Air Traffic Management.* Paper presented at the 2003 Institute of Electrical & Electronic Engineers/ Web Intelligence Consortium (IEEE/WIC) International Conference on Intelligent Agent Technology (IAT 2003), Halifax, Canada.

Nolan, M. S. (2003). *Fundamentals of Air Traffic Control* (4th ed.) Pacific Grove, California: Brookes/Cole Publishing.

Odoni, A. (1987). The Flow Management Problem in Air Traffic Control. *In Flow Control of Congested Networks* (pp. 269-288). Berlin, Germany: Springer-Verlag.

Pearce, R. A. (2006). The Next Generation Air Transportation System: Transformation Starts Now. *Journal of Air Traffic Control*, (pp. 7-10).

Sierhuis, M. (2001). *Modeling and Simulating Work Practice, BRAHMS: A multiagent modeling and simulation language for work system analysis and design* (Vol. 2001-10). Amsterdam, the Netherlands: University of Amsterdam.

Sridhar, B., Chatterji, G., Grabbe, S., & Sheth, K. (2002, August). *Integration of Traffic Flow Management Decisions.* Paper presented at the American Institute of Aeronautics and Astronautics (AIAA) Guidance, Navigation, and Control Conference, Monterey, California.

Sweet, D. N., Manikonda, V., Aronson, J. S., Roth, K., & Blake, M. (2002, August). *Fast-Time Simulation System for Analysis of Advanced Air Transportation Concepts.* Paper presented at the American Institute of Aeronautics and Astronautics (AIAA) Modeling and Simulation Technologies Conference and Exhibit, Monterey, California.

Tambe, M. (1997, July). *Agent architectures for flexible, practical teamwork.* Paper presented at the American Association for Artificial Intelligence Conference (AAAI-2007), Providence, Rhode Island.

Wambsganss, M. (1996). Collaborative Decision Making Through Dynamic Information Transfer. *Air Traffic Control Quarterly*, 4, 107-123.

Waslander, S. L., Raffard, R. L., & Tomlin, C. J. (2008). Market-Based Air Traffic Flow Control with Competing Airlines. *Journal of Guidance, Control and Dynamics*, 31(1), 148-161.

Wolfe, S. R., Jarvis, P. A., Enomoto, F. Y., & Sierhuis, M. (2007, November). *Comparing Route Selection Strategies in Collaborative Traffic Flow Management.* Paper presented at the 2007 Institute of Electrical & Electronic Engineers/ Web Intelligence Consortium/ Association for Computing Machinery (IEEE/WIC/ACM) International Conference on Intelligent Agent Technology (IAT 2007), Fremont, California.

Wooldridge, M., & Jennings, N. R. (1995). Intelligent Agents: Theory and Practice. *Knowledge Engineering Review, 10*(2), 115-152.

ADDITIONAL READING

The Brahms simulation environment has its own language (Hoof & Sierhuis, 2007), which is similar but distinct from other belief, desire, and intent frameworks (Sierhuis, 2007). This representation has been developed to support the simulation of work practice (Sierhuis & Clancey, 2002), a major application of Brahms technology. The theoretical basis of Brahms is related to that of situated cognition (Clancey, 2002). The Brahms tool set, simulation environment and additional information are publicly available from the Brahms website (Agent iSolutions).

Agent based modeling and simulation and agent-based techniques have been applied to various aspects of AOC operations. A simulation of the United Airlines AOC has been developed (Pujet, Feron, & Rakhit, 1998), where each AOC employee is modeled as a multi-class queueing server. This model was used to track task execution information, namely which entities performed which task at any given point in time, with the goal of supporting timely decision making. Castro and Oliveira have developed a multi-agent system to handle disruptions in operations by reallocating crew (Castro & Oliveira, 2007). Various agents compete using different methods problem-solving methods to find the best solution; in simulation, this approach produced better solutions than current human operators.

Agent-based solutions have been proposed to solve other areas of ATFM. Tumer and Agogino have developed a multi-agent algorithm for ATFM (Tumer & Agogino, 2007). They use a Monte-Carlo simulation to estimate the congestion within the NAS, based on agents' actions to speed up or slow down traffic. These agents use reinforcement learning to set the separation between airplanes in order to manage the congestion. OASIS is an agent-based system developed to maximize airport arrival throughput by managing aircraft arrival and runway utilization (Ljunberg & Lucas, 1992). Various functions of ATC Tower operations are managed by agents in OASIS, and are implemented in the Procedural Reasoning System (Ingrand, Georgeff, & Rao, 1992). Jonker, Meyer, and Dignum have also advocate the use of multi-agent systems in the ATC Tower operations (Jonker, Meyer, & Dignum, 2005). They describe a market-based control mechanism, and analyze its usage from a game-theoretical perspective.

Agent-based modeling and simulation has also been used to study the effect of increased volume and independent choice in other forms of traffic. A simulation of projected traffic in the seaport of Rotterdam estimated the effect of increased traffic in terms of delay (Ruit, Schuylenburg, & Ottjes, 1995). Automobile traffic has been simulated fairly extensively; of particular relevance to this book chapter are those focused on route selection. Klügl and Bazzan examined how individual drivers could learn to prefer certain routes and how forecasts of traffic influenced this ability (Klügl & Bazzan, 2004). Interestingly, their study showed that the best overall system performance was achieved when most, but not all, drivers had access to these traffic forecasts. Stark et al. (Stark, Helbing, Schönhof, & Holyst, 2006) investigated how cooperative strategies could be learned in a route selection context without any communication between drivers.

Several other relevant ATFM simulation environments are not agent-based. The Future ATM Concepts Evaluation Tool (FACET) (Bilimoria, Sridhar, Chatterji, Sheth, & Grabbe, 2000) is a NASA-developed tool for simulating air traffic flow that has been integrated into Flight Explorer, a commercial product used by nearly all major U.S. airlines. FACET contains modules that concentrate on trajectory modeling, weather modeling, and also contains a model of the airspace structure, including the ARTCC regions, sectors, and air routes. The Center-TRACON Automation System (CTAS) (Erzberger, 1994) is another NASA-developed simulation system, with a single ARTCC focus and a greater emphasis on human in the loop simulations. The Traffic Management Advisor, one of the CTAS suite of tools, is particularly relevant from an ATFM perspective, and has been extended to coordinate among multiple ARTCCs in the McTMA system (Hoang, 2004). The Linking Existing On Ground, Arrival and Departure project (LEONARDO) evaluated the feasibility of implementing Collaborative Decision Making (CDM) in airport processes, both through simulation and a limited deployments (European Commission, 2004). LEONARDO integrated decision support tools to promote information sharing among airport stakeholders, providing them with early and reliable planning updates. SKATE (Skills, Knowledge, and Attitudes for Teamwork), is a model for teamwork measurement developed and used in real-time simulations to validate the use of LEONARDO for CDM (EUROCONTROL, 2004).

The CATFM concept of operations has to the potential to enhance the Collaborative Decision Making (CDM) initiative (Ball, Hoffman, Chen, & Vossen, 2000; Federal Aviation Administration), a joint government and industry effort was established in the mid-1990s to enhance the interaction and collaboration between the ATSP and the users of airspace. CDM deals with improvement of ATFM through better information exchange among the participants of the aviation community. The goal of CDM is to create solutions for better utilization of airspace resources through technological and procedural solutions for traffic management problems that are encountered in the NAS, without compromising safety. The CDM group consists of several sub-groups, e.g., flow evaluation, future concepts, ground delay program enhancements, weather evaluation, etc., which deal with various aspects of the air traffic flow management problem. Several automation decision support tools have emerged as a result of the CDM effort over the years, including the Flight Schedule Monitor (Metron Aviation, 2006a) for managing arrival/departure times, the Collaborative Convective Forecast Product (National Oceanic and Atmospheric Administration, 2007) for a common assessment of convective weather, and the Post Operations Evaluation Tool (Metron Aviation, 2006b) for analysis support of NAS operations. Preliminary evaluation of CDM initiatives on elements such as GDP is promising (Ball, Hoffman, Knorr, Wetherly, & Wambsganss, 2001).

The Future Concepts Team is a sub-group of the CDM initiative. Over the past few years, the FCT group has focused their effort on future collaboration between the service provider and the airspace users to improve efficiency of operations in the NAS. The two main areas of interest are the Integrated Collaborative Routing (ICR) (Usmani, 2005) and the System Enhancements for Versatile Electronic Negotiation (SEVEN) (Gaertner, Klopfenstein, & Wilmouth, 2007). The ICR effort is geared towards better incorporation of airspace users' preferences for rerouting during events that cause congestion and weather related delays. The SEVEN concept is a longer-term initiative which aims to enhance the collaboration among the participants to a much higher level than what exists today through use of electronic data exchange and to explore the roles and responsibilities of participants, along with identification of associated issues and concerns. This enhanced collaboration encompasses all elements of the Flow Constrained Areas (for establishing areas of impacted traffic), the Ground Delay Programs and Airspace Flow Programs (for managing traffic during bad weather conditions) and Playbook routes (for specific rerouting strategies). The premise for Concept SEVEN is for the airspace users to provide prioritized flight lists and enabling them to update their options as the constraining events unfold.

Other concepts of operations have elements that are similar to the CATFM concept of operations. The Concept of Operations for the Next Generation Air Transportation System (Joint Planning and Development Office, 2007) defines how the air transportation system shall operate in the year 2025, forming a technological baseline to help stimulate the development of policy. The International Civil Aviation Organization has also developed requirements for an operational concept in 2025 (International Civil Aviation Organization, 2003), emphasizing collaborative decision making. It also provides a comprehensive view of operations, including airspace design, airport operations and collision avoidance, and describes potential benefits and a possible adoption strategy.

The FAA has developed useful training materials that explains terms, techniques, and programs associated with traffic flow management in the NAS (Federal Aviation Administration, 2007). Operational details of ATFM, including the ATFM roles and duties at the ATCSCC, ATFM tools, flight restriction guidelines, and overviews of the traffic patterns within each ARTCC are available from the FAA (Federal Aviation Administration, 2006). Finally, the Airline Handbook (Air Transport Association of America, 2007) provides a brief history of aviation and an overview of important aviation topics, including: the principles of flight, deregulation, the structure of the industry, airline economics, airports, air traffic control, safety, security and the environment, and a glossary of commonly used aviation terms.

ADDITIONAL READING REFERENCES

Agent iSolutions. *Brahms* [Electronic Version]. Retrieved December 15, 2007, from http://www.agentisolutions.com/

Air Transport Association of America. (2007). *The Airline Handbook*. Washington, D.C.: ATA Publications.

Ball, M., Hoffman, R., Chen, C.-Y., & Vossen, T. (2000). Collaborative Decision Making in Air Traffic Management: Current and Future Research Directions. In L. Bianco, P. Dell'Olmo & A. R. Odoni (Eds.), *New Concepts and Methods in Air Traffic Management* (pp. 17-30). New York, New York, USA: Springer.

Ball, M. O., Hoffman, R. L., Knorr, D., Wetherly, J., & Wambsganss, M. (2001). Assessing the Benefits of Collaborative Decision Making in Air Traffic Management. In G. L. Donohue & A. G. Zellweger (Eds.), *Air Transportation Systems Engineering, Progress in Astronautics and Aeronautics* (Vol. 193, pp. 239). Reston, Virginia: American Institute of Aeronautices and Astronautics.

Bilimoria, K., Sridhar, B., Chatterji, G., Sheth, K., & Grabbe, S. (2001). FACET: Future ATM Concepts Evaluation Tool. *Air Traffic Control Quarterly*, 9(1), 1-20.

Castro, A. J. M., & Oliveira, E. (2007, November). *Using Specialized Agents in a Distributed MAS to Solve Airline Operations Problems: a Case Study.* Paper presented at the Institute of Electrical & Electronic Engineers/ Web Intelligence Consortium/ Association for Computing Machinery (IEEE/WIC/ACM) International Conference on Intelligent Agent Technology (IAT 2007), Fremont, CA.

Clancey, W. J. (2002). Simulating activities: Relating motives, deliberation, and attentive coordination. *Cognitive Systems Research: Special Issue on Situated and Embodied Cognition*, 3(3), 471-499.

Erzberger, H. (1994, July). *Center-TRACON Automation System (CTAS).* Paper presented at the Capacity Technology Subcommittee, FAA Research and Development Advisory Committee, Washington, D.C.

EUROCONTROL. (2004). *A Measure to Assess the Impact of Automation on Teamwork* [Electronic Version]. Retrieved December 15, 2007, from http://www.eurocontrol.int/humanfactors/gallery/content/public/docs/DELIVERABLES/HF48-HRS-HSP-005-REP-07%20Released-withsig.pdf

European Commission. (2004). *LEONARDO Final Report* [Electronic Version]. Retrieved December 15, 2007, from http://ec.europa.eu/transport/air_portal/research/doc/rtd_5_leonardo.pdf

Federal Aviation Administration. (2007). *Collaborative Decision Making* [Electronic Version]. Retrieved December 15, 2007, from http://cdm.fly.faa.gov/

Federal Aviation Administration. (2006). *National System Strategy Team* [Electronic Version]. Retrieved December 15, 2007, from http://www.fly.faa.gov/Operations/NSST/nsst2006.pdf

Federal Aviation Administration. (2007). *Traffic Flow Management for Flight Operations Personnel* [Electronic Version]. Retrieved December 15, 2007. from http://www.fly.faa.gov/Products/Training/Traffic_Management_for_Pilots/TFM_NASv2.pdf

Gaertner, N., Klopfenstein, M., & Wilmouth, G. (2007). *Updated Operational Concept for System Enhancements for Versatile Electronic Negotiation (SEVEN)* [Electronic Version]. Retrieved December 15, 2007, from http://cdm.fly.faa.gov/Workgroups/ICEFM/2007/32F0907-005-R0%20Updated%20SEVEN%20Operational%20Concept.pdf

Hoang, T. (2004, September). *A Description of the Mutli-Center Traffic Management Advisor (McTMA) Simulation Environment.* Paper presented at the American Institute of Aeronautics and Astronautics (AIAA) Aircraft Technology, Integration and Operations (ATIO), Chicago, Illinois.

Hoof, R. v., & Sierhuis, M. (2007). *Brahms Language Specification, TM99-0008* [Electronic Version]. Retrieved December 15, 2007, from http://www.agentisolutions.com/documentation/language/ls_title.htm

Ingrand, F. F., Georgeff, M. P., & Rao, A. S. (1992). An architecture for real-time reasoning and system control. *Institute of Electrical & Electronic Engineers (IEEE) Expert* 7(6), 34-44.

International Civil Aviation Organization (2003, November). *ATM Operational Concept Document.* Paper presented at the 11th Air Navigation Conference, Montreal, Canada.

Joint Planning and Development Office. (2007). *Concept of Operations for the Next Generation Air Transportation System* [Electronic Version]. Retrieved December 15, 2007, from http://www.jpdo.gov/library/NextGen_v2.0.pdf

Jonker, G., Meyer, J.-J., & Dignum, F. (2005, December). *Towards a Market Mechanism for Airport Traffic Control.* Paper presented at the 12th Portuguese Conference on Artificial Intelligence (EPIA 2005), Covilha, Portugal.

Klügl, F., & Bazzan, A. L. C. (2004). Route Decision Behaviour in a Commuting Scenario: Simple Heuristics Adaptation and Effect of Traffic Forecast. *Journal of Artificial Societies and Social Simulation*, 7(1).

Ljunberg, M., & Lucas, A. (1992, September). *The OASIS Air Traffic Management System.* Paper presented at the Second Pacific Rim International Conference on Artificial Intelligence (PRICAI '92), Seoul, Korea.

Metron Aviation. (2006a). *Flight Schedule Monitor* [Electronic Version]. Retrieved December 15, 2007, from http://www.metronaviation.com/content/text/collaborativeDecisionMakingFSM_2.pdf.

Metron Aviation. (2006b). *Post Operations Evalution Tool* [Electronic Version]. Retrieved December 15, 2007, from http://www.metronaviation.com/content/text/collaborativeDecisionMakingPOET_2.pdf

National Oceanic and Atmospheric Administration. (2007). *Collaborative Convective Forecast Product: Product Description Document* [Electronic Version]. Retrieved December 15, 2007, from http://aviationweather.gov/products/ccfp/docs/pdd-ccfp.pdf

Pujet, N., Feron, E., & Rakhit, A. (1998, August). Modeling and identification of an Airline Operations Center as a multi-agent queueing system. Paper presented at the American Institute of Aeronautices and Astronautics (AIAA) Guidance, Navigation, and Control Conference (GNC 1998), Boston, Massachusetts.

Ruit, G. J. v. d., Schuylenburg, M. v., & Ottjes, J. A. (1995, June). Simulation of shipping traffic flow in the Maasvlakte port area of Rotterdam. Paper presented at the European Simulation Multiconference (ESM 1995), Prague, Czech Republic.

Sierhuis, M. (2007). It's not just goals all the way down - It's activities all the way down. In G. M. P. O'Hare, A. Ricci, M. J. O'Grady & O. Dikenelli (Eds.), *Engineering Societies in the Agents World VII (Vol LNAI 4457/2007). 7th International Workshop* (ESAW 2006) (pp. 1-24) Dublin, Ireland: Springer.

Sierhuis, M., & Clancey, W. J. (2002). Modeling and Simulating Work Practice: A human-centered method for work systems design. *Institute of Electrical & Electronic Engineers (IEEE) Intelligent Systems*, 17(5).

Stark, H.-U., Helbing, D., Schönhof, M., & Holyst, J. A. (2006). Alternating cooperation strategies in a route choice game: Theory, experiments, and effects of a learning scenario. In A. Innocenti & P. Sbriglia (Eds.), *Games, Rationality, and Behaviour* (pp. 256-273). Houndmills, England and New York, New York: Palgrave Macmillan.

Tumer, K., & Agogino, A. (2007, May). Distributed Agent-Based Air Traffic Flow Management. Paper presented at the 6th International Joint Conference on Autonomous Agents and Multiagent Systems (AAMAS 2007), Honolulu, Hawaii.

Usmani, A. (2005, November). *Upcoming TFM Enhancements: The Role of Simulation*. Paper presented at the FAA Eurocontrol Action Plan 9 (AP9) Traffic Flow Management (TFM) in Fast-Time Simulation Technical Interchange Meeting, Atlantic City, New Jersey.

KEY TERMS

Air Traffic Control (ATC): A service operated by the appropriate authority to promote the safe, orderly, and expeditious flow of air traffic.

Air Traffic Flow Management (ATFM): The regulation of air traffic in order to avoid exceeding airport or airspace capacity, and to ensure that available capacity is used efficiently.

Airline Operations Center (AOC): An airline unit responsible for dispatching flights and adjusting schedules in response to restrictions in the airspace system.

Brahms: A set of software tools to develop and simulate multi-agent models of human and machine behavior.

Collaborative Decision Making (CDM): Collaboration involving the system stakeholders in determining the best approach to a given situation. In the context of air transportation, it is the cooperative effort between the government and industry to exchange information for better decision-making.

Traffic Management Unit (TMU): A team of air traffic controllers who analyze the demand and external effects, such as weather, on the airspace system and implement initiatives to balance the demand with capacity.

Compilation of References

Adams, M., Kolitz, S., Milner, J., & Odoni, A. (1996). Evolutionary Concepts for Decentralized Air Traffic Flow Management. *Air Traffic Control Quarterly, 4*(4), 281-306.

Agogino, A., & Tumer, K. (2004). Efficient evaluation functions for multi-rover systems. In *The Genetic and Evolutionary Computation Conference*, (pp. 1–12), Seatle, WA.

Allen, G. L. (1999). Spatial abilities, cognitive maps, and wayfinding: Bases for individual differences in spatial cognition and behavior. In R. Golledge (Ed.), *Wayfinding behavior: Cognitive mapping and other spatial processes* (pp. 46-80). Baltimore: John Hopkins University Press.

Anderson, J. R. (1983). *The Architecture of Cognition.* Cambridge, Harvard University Press.

Andrighetto, G., Campennì, M., & Conte, R. (2007). *EMIL-M: Models of norms emergence, norms immergence and the 2-way dynamic* (Tech. Rep. No. 00507). Rome: The National Research Council (CNR), Laboratory of Agent Based Social Simulation (LABSS) at The Institute of Cognitive Science and Technology (ISTC).

Andrighetto, G., Conte, R., Turrini, P., & Paolucci, M. (2007). Emergence in the loop: Simulating the two way dynamics of norm innovation. In *Proceedings of the Dagstuhl Seminar on Normative Multi-agent Systems, March 2007.* Dagstuhl.

AOS: JACK Intelligent Agents, The Agent Oriented Software Group (AOS). Retrieved February 11, 2008 from http://www.agent-software.com

Arentze, T. A., & Timmermans, H. J. P. (2001). *Albatross: A Learning Based Transportation Oriented Simulation System.* EIRASS, Eindhoven, The Netherlands.

Arentze, T. A., & Timmermans, H. J. P. (2003). Measuring impacts of condition variables in rule-based models of space-time choice behavior: method and empirical illustration. *Geography Analysis, 35*, 24-45.

Arentze, T. A., & Timmermans, H. J. P. (2003). Modeling learning and adaptation processes in activity-travel choice. *Transportation, 30*, 37-62.

Arentze, T. A., & Timmermans, H. J. P. (2004). A learning-based transportation oriented simulation system. *Transportation Research B, 38*, 613-633.

Arentze, T. A., & Timmermans, H. J. P. (2004). *A theoretical framework for modelling activity-travel scheduling decisions in non-stationary environments under conditions of uncertainty and learning*. Paper presented at the Conference on Progress in Activity-Based Analysis, May 28 – 31, Maastricht, The Netherlands.

Arentze, T. A., and Timmermans, H. J. P. (2006). Social networks, social interactions and activity-travel behavior: a framework for micro-simulation. *In Proceedings of the 85th Annual Meeting of the Transportation Research Board*, Washington, D.C. (CD-ROM: 18 pp.). To appear in Environment and planning B.

Arentze, T. A., Timmermans, H. J. P., Janssens, D., & Wets, G. (2006). Modeling short-term dynamics in activity-travel patterns: From Aurora to Feathers. Paper presented at the Innovations in Travel Modeling Conference, May 21-23, Austin, Texas.

Arentze, T., Hofman, F., van Mourik, H., & Timmermans, H. (2000). *ALBATROSS: A multi-agent rule-based model of activity pattern decisions*. Paper 00-0022, Transportation Research Board Annual Meeting, Washington, D.C.

Arnott, R., & Rowse, J. (1999). Modeling parking. *Journal of Urban Economics, 45*(1), 97-124.

Arnott, R., Palma, A. D., & Lindsey, R. (1993). A structural model of peak-period congestion: A traffic bottleneck with elastic demand. *The American Economic Review, 83*(1), 161–179.

Arthur, P., & Passini, R. (1992). *Wayfinding: People, signs, and architecture.* New York: McGraw-Hill Book Co.

Arthur, W. B. (1994). Inductive reasoning and bounded rationality. *American Economic Review, 84*, 406–411.

Auto21 (2007). Retrieved July 17th, 2008, from: http://www.auto21.ca/

Avineri, E., & Prashker, J. (2003). Sensitivity to uncertainty: Need for paradigm shift. *Transportation Research Record, 1854*, 90–98.

Axelrod, R. (1997). *The complexity of cooperation.* Princeton, NJ: Princeton University Press.

Axhausen, K. (1988). *Eine ereignisorientierte Simulation von Aktivitätenketten zur Parkstandswahl*. Ph.D. thesis, Universität Karlsruhe, Germany.

Axhausen, K., & Herz, R. (1989). Simulating activity chains: German approach. *Journal of Transportation Engineering, 115*(3), 316–325.

Axhausen, K., Zimmermann, A., Schönfelder, S., Rindsfüser, G., & Haupt, T. (2002). Observing the rhythms of daily life: A six-week travel diary. *Transportation, 29*(2), 95–124.

Axtell, R. (1999). *The Emergence of Firms in a Population of Agents: Local Increasing Returns, Unstable Nash Equilibria, And Power Law Size Distributions.* Washington D.C., US: Center on Social and Economic Dynamics, Brookings Institution.

Axtell, R., Axelrod, R., Epstein, J., & Cohen, M. (1997). Replication of agent-based models, aligning simulation models: A case study and results. *The Complexity of Cooperation,* (pp. 183-205).

Balmer, M. (2007). *Travel demand modeling for multi-agent transport simulations: Algorithms and systems.* Ph.D. thesis, Swiss Federal Institute of Technology (ETH) Zürich, Switzerland.

Balmer, M., & Nagel, K. (2006). Shape morphing of intersection layouts using curb side oriented driver simulation. In J. Van Leeuwen, & H. Timmermans (Eds.), *Innovations in Design & Decision Support Systems in Architecture and Urban Planning* (pp. 167–183).

Balmer, M., Axhausen, K., & Nagel, K. (2006). Agent-based demand modeling framework for large scale micro-simulations. *Transportation Research Record, 1985*, 125–134.

Balmer, M., Cetin, N., Nagel, K., & Raney, B. (2004). Towards truly agent-based traffic and mobility simulations. In *Proceedings of the Third International Joint Conference on Autonomous Agents and Multi-Agent Systems*, (pp. 60–67), New York, NY.

Balmer, M., Nagel, K., & Raney, B. (2004). Large scale multi-agent simulations for transportation applications.

In Proceeding of Behavior Responses to ITC, Eindhoven (CD-Rom).

Baltic Gateway. (2008). Retrieved December 20, 2007 from http://www.balticgateway.se

Bana, S. V. (2001). *Coordinating Automated Vehicles via Communication*. PhD thesis, University of California at Berkeley, Berkeley, CA.

Bandini S., Federici, M. L., & Vizzari, G. (2007). Situated cellular agents approach to crowd modeling and simulation. *Cybernetics and Systems, 38*, 729.

Bandini, S., Manzoni, S., & Vizzari, G. (2004). Situated cellular agents: A model to simulate crowding dynamics. *IEICE Trans. Inf. & Syst., E87-D*, 726.

Banos, A., & Charpentier, A. (2007). Simulating pedestrian behavior in subway stations with agents. In F. Amblard (Ed.), *Fourth Conference of The European Social Simulation Association (ESSA)* (pp. 611-621). Toulouse: Institut de Recherche en Informatique de Toulouse (IRIT).

Barceló, J. (1991). Software environment for integrated RTI simulation systems. In Proceedings of the DRIVE Conference, *Advanced Telematics in Road Transport, 2,* 1095-1115. Amsterdam: Elsevier.

Barkowsky, T. (2002). Mental representation and processing of geographic knowledge - a computational approach. *Künstliche Intelligenz,* (4), 42.

Baxter, J. & Bartlett, P. L. (2001). Infinite-horizon policy-gradient estimation. *Journal of Artificial Intelligence Research, 15*, 319–350.

Bazzan, A. (1997). *An Evolutionary Game-Theoretic Approach for Coordination of Traffic Signal Agents*. PhD Thesis, University of Karlsruhe.

Bazzan, A. (2005). A Distributed Approach for Coordination of Traffic Signal Agents. *Autonomous Agents and Multi-Agent Systems, 10*(1), 131-164, Springer.

Bazzan, A. L. C., & Klügl, F. (2005). Case Studies on the Braess Paradox: Simulating Route Recommendation and Learning in Abstract and Microscopic Models. *Transportation Research, C 13*(4), 299-319.

Bazzan, A. L. C., Fehler, M., & Klügl, F. (2006). Learning To Coordinate In a Network of Social Drivers: The Role of Information. *Proceedings of International Workshop on Learning and Adaptation in MAS, 3898*, 115-128.

Bazzan, A. L. C., Oliveira, D., Klügl, F., & Nagel, K. (2008). Adapt or Not to Adapt – Consequences of Adapting Driver and Traffic Light Agents. *Conference proceedings of Adaptive Agents and Multi-Agent Systems III, 4865*, 1-14.

Bazzan, A. L., Bordini, R. H., Andrioti, G. K., & Vicari, R. M. (2000). Wayward agents in a commuting scenario - personalities in the minority game. *In Proceedings of the Fourth International Conference on MultiAgent Systems (ICMAS-2000)*, 55, Washington DC. IEEE Computer Society.

Bazzan, A. L., & Klügl, F. (2005). Case studies on the Braess paradox: Simulating route recommendation and learning in abstract and microscopic models. *Transportation Research C, 13*(4), 299–319.

Bazzan, A. L., Wahle, J., & Klügl, F. (1999). Agents in traffic modelling – from reactive to social behaviour. In *KI – Kunstliche Intelligenz*, (pp. 303–306).

Beltratti, A. (1996). *Models of economic growth with environmental assets* Norwell, MA: Kluwer Academic Publishers.

Ben-Akiva, M. E., & Boccara, B. (1995). Discrete choice models with latent choice-sets. *International Journal of Research in Marketing, 12*, 9-24.

Ben-Akiva, M., & Lerman, S. R. (1985). *Discrete choice analysis: theory and application to travel demand.* Cambridge: MIT Press.

Ben-Jacob, E. (1997). From snowflake formation to growth of bacterial colonies. Part II. Cooperative formation of complex colonial patterns. *Contemp. Phys., 38*, 205.

Bernstein, D. S., Givan, R., Immerman, N., & Zilberstein, S. (2002). The complexity of decentralized control of Markov Decision Processes. *Mathematics of Operations Research, 27*(4), 819–840.

Bernstein, D., Givan, R., Immerman, N., & Zilberstein, S. (2002). The complexity of decentralized control of markov decision processes. *Mathematics of Operations Research, 27*(4), 819–840.

Bertsekas, D. P. (2000). *Dynamic Programming and Optimal Control, Vols. 1 & 2*, 2nd ed., Nashua, NH: Athena Scientific.

Bhat, C., Guo, J., Srinivasan, S., & Sivakumar, A. (2004). A comprehensive econometric microsimulator for daily activity-travel patterns. *Transportation Research Record, 1894*, 57–66.

Bierlaire, M. (2001). The acceptance of modal innovation: The case of swissmetro. In *1st Swiss Transport Research Conference*, Monte Verità.

Bierlaire, M. (2003). Biogeme: a free package for the estimation of discrete choice models. In *Proceedings of the 3rd Swiss Transport Research Conference*, Monte Verita, Ascona.

Bishop, W. R. (2005). *Intelligent Vehicle Technology and Trends*. Norwood, MA: Artech House.

Blue, V. J., & Adler, J. L. (2000). Cellular automata microsimulation of bi-directional pedestrian flows. *J. Trans. Research Board, 1678*, 135–141.

Blue, V. J., & Adler, J. L. (2002). Flow capacities from cellular automata modeling of proportional splits of pedestrians by direction. In M. Schreckenberg & S. D. Sharma (Eds.), *Pedestrian and Evacuation Dynamics*. Berlin Heidelberg: Springer.

Blumer, A., Noonan, J., & Schmolze, J. G. (1995). *Knowledge based systems and learning methods for automated highway systems.* (Technical Report), Waltham, MA: Raytheon Corp.

Bonasso, R. P., Firby, R. J., Gat, E., Kortenkamp, D., Miller, D. P., & Slack, M. G. (1997). Experiences with an architecture for intelligent, reactive agents. *Journal of Experimental and Theoretical Artificial Intelligence,* 9(2-3), 237-256.

Bonneson, J. A., & McCoy, P. T. (1993). Estimation of safety at two-way stop–controlled intersections on rural highways. *Transportation Research Record, 1401*, 83–89.

Bosch (2008). *ACC for more room.* Retrieved August 4, 2008, from http://rb-k.bosch.de/en/safety_comfort/driving_comfort/driverassistancesystems/adaptivecruisecontrolacc/application_range/index.html

Bouchefra, K., Reynaud, R., & Maurin, T. (1995). IVHS viewed from a multiagent approach point of view. *In Proceedings of the IEEE Intelligent Vehicles Symposium*. Piscataway, NJ: IEEE. 113-117

Boutilier, C. (1996). Planning, learning and coordination in multiagent decision processes. In *Proceedings of the Sixth Conference on Theoretical Aspects of Rationality and Knowledge*, Holland.

Bovy, P. H. L., & Stern, E. (1990). *Route choice: Wayfinding in transport networks.* Dordrecht; Boston: Kluwer Academic Publishers.

Bowman, J., Bradley, M., Shiftan, Y., Lawton, T., & Ben-Akiva, M. (1999). Demonstration of an activity-based model for Portland. In *World Transport Research: Selected Proceedings of the 8th World Conference on Transport Research 1998*, *3*, 171–184. Elsevier, Oxford.

Bratman, M. E. (1999). *Intention, Plans, and Practical Reason.* Chicago, Illinois: University of Chicago Press.

Broggi, A., Bertozzi, M., Fascioli, A., Lo, C., & Piazzi, B. (1999). The argo autonomous vehicle's vision and control systems. International *Journal of Intelligent Control and Systems, 3*(4), 409–441.

Brooks, R. A. (1991). Intelligence without representation. *Artificial Intelligence Journal, 47*, 139–159.

Bürckert, H.-J., Fischer, K., & Vierke, G. (2000). Holonic transport scheduling with Teletruck. *Applied Artificial Intelligence, 14*(7), 697–725.

Burmeister, B., Haddadi, A., & Matylis, G. (1997). Application of multi-agent systems in traffic and transportation. *IEE Proceedings on Software Engineering, 144*(1), 51-60.

Burstedde, C., Kirchner, A., Klauck, K., Schadschneider, A., & Zittartz, J. (2002). Cellular automaton approach to pedestrian dynamics – applications. In M. Schreckenberg & S. D. Sharma (Eds.), *Pedestrian and Evacuation Dynamics* (pp. 87–98). Berlin Heidelberg: Springer.

Burstedde, C., Klauck, K., Schadschneider, A., & Zittartz, J. (2001). Simulation of pedestrian dynamics using a two-dimensional cellular automaton. *Physica A*, *295*, 507–525.

Caduff, D. (2007). *Assessing landmark salience for human navigation.* PhD thesis, University of Zurich, Zurich.

Campbell, K. C., Cooper, W. W., Greenbaum, D. P., & Wojcik, L. A. (2000, June). *Modeling Distributed Human Decision-Making in Traffic Flow Management Operations.* Paper presented at the 3rd USA/Europe Air Traffic Management R&D Seminar, Napoli, Italy.

Camponogara, E., & Kraus Jr., W. (2003). Distributed Learning Agents in Urban Traffic Control. In: 11th Portuguese Conference on Artificial Intelligence, EPIA. (pp. 324–335). Lecture Notes in Computer Science 2902. Berlin: Springer.

Cascetta, E., & Papola, A. (2001). Random utility models with implicit availability/perception of choice alternatives for simulation for travel demand. *Transportation Research, C, 9*, 249-263.

Castelfranchi, C. (1998). Simulating with cognitive agents: The importance of cognitive emergence. In J. Sichman, R. Conte, & N. Gilbert (Eds.), *Multi-Agent Systems and Agent-Based Simulation* (pp. 26-44). Berlin: Springer.

Cellular Automata - 7th International Conference on Cellular Automata for Research and Industry, acri 2006. (2006).

Cetin, N. (2005). *Large-Scale parallel graph-based simulations.* Ph.D. thesis, Swiss Federal Institute of Technology (ETH) Zürich, Switzerland.

Cetin, N., Burri, A., & Nagel, K. (2003). A large-scale agent-based traffic microsimulation based on queue model. In *Proceedings of Swiss Transport Research Conference (STRC).* Monte Verita, CH. See www.strc.ch. Earlier version, with inferior performance values: Transportation Research Board Annual Meeting 2003 paper number 03-4272.

Challet, D., & Zhang, Y. C. (1998). On the minority game: Analytical and numerical studies. *Physica A*, *256*, 514.

Challet, D., & Zhang, Y.-C. (1997). Emergence of cooperation and organization in an evolutionary game. *Physica A, 246*, 407–415.

Challet, D., Marsili, M., & Zhang, Y.-C. (2004). *Minority Games Interacting agents in financial markets.* Oxford, UK: Oxford University Press.

Charypar, D., & Nagel, K. (2003). Generating complete all-day activity plans with genetic algorithms. *In Proceeding of the 10th International Conference of Travel Behavior Research*, Lucerne, Switzerland (CD-Rom).

Charypar, D., & Nagel, K. (2005). Generating complete all-day activity plans with genetic algorithms. *Transportation*, *32*(4), 369–397.

Charypar, D., Axhausen, K., & Nagel, K. (2007). An event-driven parallel queue-based microsimulation for large scale traffic scenarios. In *Proceedings of the Word Conference on Transport Research.* Berkeley, CA.

Chatterjee, K., & McDonald, M (1999). Modelling the impacts of transport telematics: current limitations and future developments. *Transport Reviews*, *19*(1), 57-80.

Cheng, Y. (1998). Hybrid simulation for resolving resource conflicts in train traffic rescheduling. *Computers in Industry*, *35*(3), 233-246.

Chernick, M. R. (1999). *Bootstrap Methods A Practitioner's Guide.* New York: John Wiley & Sons Inc.

Chmura, T., Pitz, T., Möhring, M., & Troitzsch, K. G. (2005). Netsim. A software environment to study route choice behavior in laboratory experiments. In *Representing Social Reality* (pp. 339-344). Fölbach: European Social Simulation Association.

Chopard, B., & Droz, M. (1998). *Cellular Automata Modeling of Physical Systems*. Cambridge University Press.

Chowdhury, D., Nishinari, K., Santen, L., & Schadschneider, A. (2008). *Stochastic Transport in Complex Systems: From Molecules to Vehicles*. Elsevier.

Chowdhury, D., Santen, L., & Schadschneider, A. (2000). Statistical physics of vehicular traffic and some related systems. *Physics Reports*, *329*(4–6), 199–329.

Ciari, F., Balmer, M., & Axhausen, K. (2007). Mobility tool ownership and mode choice decision processes in multi-agent transportation simulation. In *Proceedings of Swiss Transport Research Conference (STRC)*. Monte Verita, CH. See www.strc.ch.

Clancey, W. J., Sachs, P., Sierhuis, M., & Hoof, R. v. (1998). Brahms: Simulating practice for work systems design. *International Journal on Human-Computer Studies*, *49*, 831-865.

Clancey, W. J., Sierhuis, M., Kaskiris, C., & Hoof, R. v. (2003, May). *Advantages of Brahms for Specifying and Implementing a Multiagent Human-Robotic Exploration System*. Paper presented at the 16th International Florida Artificial Intelligence Research Society (FLAIRS) Conference, St. Augustine, Florida.

Clancey, W. J., Sierhuis, M., Seah, C., Buckley, C., Reynolds, F., Hall, T., & Scott, M. (2007, October). *Multi-Agent Simulation to Implementation: A Practical Engineering Methodology for Designing Space Flight Operations*. Paper presented at the Eighth Annual International Workshop "Engineering Societies in the Agents World" (ESAW 07), Athens, Greece.

Clark, G. L. (1998). Stylized facts and close dialogue: Methodology in economic geography. *Annals of the Association of American Geographers, 88*(1), 73-87.

Cleveland, W. S. (1981). LOWESS: A program for smoothing scatterplots by robust locally weighted regression. *The American Statistician*, 35, 54.

Coase, R. (1937). The nature of the firm. *Economica, 1*(4), 386-405.

Collier, N., Howe, T., & North, M. (2003). Onward and upward: The transition to repast 2.0. In *Proceedings of the First Annual North American Association for Computational Social and Organizational Science Conference*, page 5, Pittsburgh, PA USA. June, Electronic Proceedings.

Collins, J., Gini, M., & Mobasher, B. (2002). *Multi-Agent negotiation using combinatorial auctions with precedence constraints.* Paper presented at the Fifth International Conference on Autonomous Agents (AGENTS'01). University of Minnesota, Minneapolis, Minnesota.

Cordeau, J.-F., Desaulniers, G., Desrosiers, J., Solomon, M. M., & Soumis, F. (2001). VRP with time windows. In P. Toth & D. Vigo (Eds.), *The Vehicle Routing Problem* (pp. 157–193). Philadelphia, PA: Society for Industrial and Applied Mathematics Monographs on Discrete Mathematics and Applications.

Corker, K. M. (1999, December). *Human Performance Simulation in the Analysis of Advanced Air Traffic Management*. Paper presented at the 1999 Winter Simulation Conference, Phoenix, Arizona.

Couluris, G. J., Hunter, C. G., Blake, M., Roth, K., Sweet, D. N., & Stassart, P. A. (2003, August). *National Airspace System Simulation Capturing the Interactions of Air Traffic Management and Flight Trajectories*. Paper presented at the American Institute of Aeronautics and Astronautics (AIAA) Guidance, Navigation, and Control (GNC) Conference, Austin, Texas.

County Surveyors' Society, & Department for Transport (2006). *Puffin crossings. Good practice guide* – Release 1. Retrieved August 4, 2008, from http://www.dft.gov.uk/pgr/roads/tss/gpg/puffingoodpracticeguide01

DARPA 2007. The DARPA urban challenge. http://www.darpa.mil/grandchallenge.

Dash, R. K., Jennings, N. R., & Parkes, D. C. (2003). Computational-mechanism design: A call to arms. *IEEE Intelligent Systems 18*(6), 40-47.

Davidsson, P., Henesey, L., Ramstedt, L., Törnquist, J., & Wernstedt, F. (2005). An analysis of agent-based approaches to transport logistics. *Transportation Research, 13C*(4), 255-271.

Davidsson, P., Henesey, L., Ramstedt, L., Törnquist, J., & Wernstedt, F. (2005). An analysis of agent-based aproaches to transport logistics. *Transportation Research Part C, 13*(4), 255–271.

Davison, A., & Hinkley, D. (1997). *Bootstrap Methods and their Application.* Cambridge Series in Statistical and Probabilistic Mathematics. Cambridge, UK: Cambridge University Press.

de Bruin, D., Kroon, J., van Klaveren, R., & Nelisse, M. (2004). Design and test of a cooperative adaptive cruise control system. *In Proceedings of IEEE Intelligent Vehicles Symposium* (pp. 392–396).

de Cara, M. A. R., Pla, O., & Guinea, F. (2000). Learning, competition and cooperation in simple games. *The European Physical Journal B, 13*, 413–416.

de Palma, A., & Marchal, F. (2002). Real case applications of the fully dynamic METROPOLIS tool-box: An advocacy for large-scale mesoscopic transportation systems. *Networks and Spatial Economics, 2*(4), 347–369.

Denis, M. (1997). The description of routes: A cognitive approach to the production of spatial discourse. *Current Psychology of Cognition, 16*, 409-458.

Department for Transport (2004). Values of time and operating costs. *Technical report, Transport Analysis Guidance Unit.* Retrieved August 4, 2008 from http://www.webtag.org.uk/webdocuments/3_Expert/5_Economy_Objective/3.5.6.htm#1_2b.

Derbyshire County Council (2003). Re: toll and bus fare for road user charge at the Upper Derwent valley. *e-mail and interview through phone, 2nd – 6th, June. Currently (Nov., 2003).*

Derrida, B. (1998). An exactly soluble non-equilibrium system: The asymmetric simple exclusion process. *Phys. Rep., 301*, 65.

Desjardins, C., Grégoire, P.-L., Laumônier, J., & Chaib-draa, B. (2007). Architecture and design of a multi-layered cooperative cruise control system. *In Proceedings of the Society of Automobile Engineering World Congress (SAE'07).*

Dia, H. (2002). An agent-based approach to modelling driver route choice behaviour under the influence of real-time information. *Transportation Research Part C: Emerging Technologies, 10*(5-6), 331–349.

Diakaki, C., Dinopoulou, V., Aboudolas, K., Papageorgiou, M., Benshabat, E., Seider, E., & Leibov, A. (2003). Extensions and New Applications of the Traffic Signal Control Strategy TUC. In: *82th Annual Meeting of the Transportation Research Board*, 2003. (pp. 12–16).

Dorer, K. & Calisti, M. (2005). An adaptive solution to dynamic transport optimization. *In Proc. of 4th Int. Joint Conf. on Autonomous Agents and Multiagent Systems (AAMAS 2005)*, (pp. 45–51). New York, NY: ACM Press.

Downing, T. E., Scott, M., & Pahl-Wostl, C., (2000). Understanding Climate Policy Using Participatory Agent-Based Social Simulation. In S. Moss & P. Davidsson (Ed.), *Multi-Agent Based Simulation: Proceedings of the Second International Workshop*, (1979). (pp. 198-213). Berlin: Springer-Verlag.

Drabkin, V., Friedman, R., Kliot, G., & Segal, M. (2007). Rapid: Reliable probabilistic dissemination in wireless ad-hoc networks. *In The 26th IEEE International Symposium on Reliable Distributed Systems*, Beijing, China.

Dresner, K. & Stone, P. (2004). *Multiagent traffic management: A protocol for defining intersection control policies UT-AI-TR-04-315*, The University of Texas at Austin, Department of Computer Sciences, AI Laboratory.

Dresner, K. & Stone, P. (2005). Multiagent traffic management: An improved intersection control mechanism. *In The Fourth International Joint Conference on Autonomous Agents and Multiagent Systems*, (pp. 471–477), Utrecht, The Netherlands.

Dresner, K. & Stone, P. (2007). Sharing the road: Autonomous vehicles meet human drivers In Proceedings of the Twentieth International Joint Conference on Artificial Intelligence, (pp. 1263–68), Hyderabad, India.

Dresner, K., & Stone, P. (2004). Multiagent traffic management: A reservation-based intersection control mechanism. In *Proceedings of the Third International Joint Conference on Autonomous Agents and Multiagent Systems* (pp. 530-537). Washington, DC, USA: IEEE Computer Society.

Dresner, K., & Stone, P. (2005). Multiagent traffic management: an improved intersection control mechanism. *Proceedings of the 4th International Joint Conference on Autonomous Agents and Multiagent Systems* (pp. 471-477). ACM Press.

Dresner, K., & Stone, P. (2006). Multiagent Traffic Management: Opportunities for Multiagent Learning. *Lecture Notes in Computer Science* (pp. 129-138), volume 3898. Springer-Verlag.

Dresner, K., & Stone, P. (2008). A multiagent approach to autonomous intersection management. *Journal of Artificial Intelligence Research, 31*, 591-656.

EastWest Transport Corridor. (2008). Retrieved December 20, 2007 from http:// www.eastwesttc.org.

Eckton, G. D. C. (2003). Road-user charging and the Lake District National Park. *Journal of Transport Geography, 11*(4), 307-317.

Edmonds, B. (1999). Modelling socially intelligent agents. *Applied Artificial Intelligence, 12*, 677–699. http://www.cpm.mmu.ac.uk/cpmrep26.html.

Edmonds, B., & Moss, S. J. (2005). *From KISS to KIDS: An ``anti-simplistic'' modeling approach* Manchester Metropolitan University Business School.

Efron, B. (1979). Bootstrap methods: Another look at the jackknife. *The Annals of Statistics, 7*, 1-2.

Ehlert, P. A. (2001). *The agent approach to tactical driving in autonomous vehicle and traffic simulation.* Master's thesis, Knowledge Based Systems Group, Delft University of Technology, Delft, The Netherlands.

Elias, B. (2003). Extracting Landmarks with Data Mining Methods. In W. Kuhn, M. Worboys, & S. Timpf (Eds.), *Spatial Information Theory: Foundations of Geographic Information Science* (375-389). Berlin: Springer Verlag.

Epstein, J., & Axtell, R. (1996). *Growing Artificial Societies: Social Science from the Bottom Up*. Brookings Institution Press.

FAA, F. A. A. (1990). *Emergency evacuation - cfr sec. 25.803* (Regulation No. CFR Sec. 25.803). : Federal Aviation Administration.

Fagiolo, G., Windrum, P., & Moneta, A. (2006). *Empirical validation of agent-based models: A critical survey.* LEM Working Paper Series.

Febbraro, A. D., & Sacco, N. (2004). On modelling urban transportation networks via hybrid petri nets. *Control Engineering Practice, 12*(10), 1225-1239.

Ferber, J. (1999). *Multi-agent systems*. Addison-Wesley.

Festinger, L. (1954). A theory of social comparison processes. *Human Relations, 7,* 117-140.

Fischer, K., Muller, J. P., Pischel, M., & Schier, D. (1995). A model for cooperative transportation scheduling. *In Proc. of the 1st Int. Conf. on Multiagent Systems,* (pp. 109–116). Menlo park, California: AAAI Press / MIT Press.

Floyd, S., & Jacobson, V. (1993). Random Early Detection gateways for Congestion Avoidance. *IEEE/ACM Transactions on Networking, 1*(4), 397-413.

Forbes, J. R. (2002). *Reinforcement Learning for Autonomous Vehicles*. PhD thesis, University of California at Berkeley, Berkeley, CA.

Forbes, J., Huang, T., Kanazawa, K., & Russell, S. J. (1995). The batmobile: towards a bayesian automated taxi. *In Proceedings of International Joint Conference on Artificial Intelligence* (pp. 1878–1885), Morgan Kaufmann.

Fowkes, A. S. (2000). Recent developments in stated preference techniques in transport research. In Ortézar, J., (Ed.), *Stated Preference Modelling Techniques, volume 4 of PTRC Perspectives,* (pp. 37–52). PTRC Education and Research Services Ltd, London.

Frank, A. U., Bittner, S., & Raubal, M. (2001). *Spatial and cognitive simulation with multi-agent systems.* Paper presented at the Spatial Information Theory - Foundations of Geographic Information Science (Int. Conference COSIT, September 2001), Morro Bay, U.S.A.

Franklin, S., & Graesser, A. (1996). Is it an agent or just a program? A taxonomy for autonomous agents. In Intelligent Agents III, *Agent Theories, Architectures, and Languages* (LNAI, No.1193). 21-35.

Freksa, C. (1991). Qualitative spatial reasoning. In D. M. Mark & A. U. Frank (Eds.), *Cognitive and linguistic aspects of geographic space* (pp. 361-372). Dordrecht, The Netherlands: Kluwer Academic Press.

Frenken, K. (2006). Technological innovation and complexity theory. *Economics of Innovation and New Technology, 15*(2), 137-155.

Fruin, J. J. (1971). *Pedestrian Planning and Design.* New York: Metropolitan Association of Urban Designers and Environmental Planners.

Fukui, M., & Ishibashi, Y. (1999). Jamming transition in cellular automaton models for pedestrians on passageway. *J. Phys. Soc. Jpn., 68*, 3738.

Fukui, M., & Ishibashi, Y. (1999). Self-organized phase transitions in cellular automaton models for pedestrians. *J. Phys. Soc. Jpn., 68*, 2861.

Fürstenberg, K. C., & Lages, U. (2005). New European approach for intersection safety - The EC-project INTERSAFE. In *2005 IEEE Intelligent Vehicles Symposium* (pp. 177-180). IEEE.

Furukawa, H., Shiraishi, Y., Inagaki, T., & Watanabe, T. (2003). Mode awareness of a dual-mode adaptive cruise control system. In *IEEE International Conference on Systems, Man, and Cybernetics, 2005, 1*, 832-837. IEEE.

Gale, D., & Shapley L. S. (1962). College admissions and stability of marriage. *American Mathematical Monthly, 69*(January), 9-15.

Galea, E. R. (Ed.). (2003). *Pedestrian and Evacuation Dynamics 2003.* London: CMS Press.

Garber, N. J., & Hoel, L. A. (1988). *Traffic and highway engineering.* St. Paul, USA: West Publishing Company.

Gärling, T., Böök, A., & Linderg, E. (1986). Spatial orientation and wayfinding in the designed environment - a conceptual analysis and some suggestions for post-occupancy evaluation. *Journal of Architectural and Planning Research, 3*, 55-64.

Gawron, C. (1998). An iterative algorithm to determine the dynamic user equilibrium in a traffic simulation model. *International Journal of Modern Physics C, 9*(3), 393–407.

Gazi, V. (2005). Swarm aggregations using artificial potentials and sliding-mode control. *IEEE Transactions on Robotics* , (pp. 1208–1214).

Gershenson, C. (2005). Self-Organizing Traffic Lights. *Complex Systems, 16*(1), 29–53. Champaign, IL: Complex Systems Publications, Inc.

GEVAS (2008). Travolution. Retrieved August 4, 2008 from http://www.gevas.eu/index.php?id=65&L=1

Gibson, J. J. (1986). *The ecological approach to visual perception.* Hillsdale, NJ: Lawrence Erlbaum.

Gilbert, C., Robertson, G., Le Maho, Y., Naito, Y., & Ancel, A. (2006). Huddling behavior in emperor penguins: Dynamics of huddling. *Physiology & Behavior, 88* (4-5), 479-488.

Gilbert, N. (2004). Open problems in using agent-based models in industrial and labor dynamics. In R. Leombruni, & M. Richiardi (Eds.), *Industry and labor dynamics: The agent-based computational approach* (pp. 401-405). Sigapore: World Scientific.

Gilbert, N., & Troitzsch, K. G. (1999). *Simulation for the Social Scientist.* Buckingham: Open University Press.

Gilbert, N., & Troitzsch, K. G. (2005). *Simulation for the social scientist* (2nd ed.). McGraw-Hill, Maidenhead: Open University Press.

Gipps, P. G. (1981). A behavioural car-following model for computer simulation. *Transportation Research,* 15B, 105-111.

Gipps, P. G. (1986). A model for the structure of lane-changing decisions. *Transportation Research, 20B*, 403-414.

Gipps, P. G., & Marksjö, B. (1985). A micro-simulation model for pedestrian flows. *Mathematics and Computers in Simulation, 27*, 95–105.

Godbole, D. N., & Lygeros, J. (1994). Longitudinal control of a lead car of a platoon. *IEEE Transaction on Vehicular Technology, 43*(4), 1125–1135.

Golledge, R. (1999). Human wayfinding and cognitive maps. In R. Golledge (Ed.), *Wayfinding behavior: Cognitive mapping and other spatial processes* (pp. 5-45). Baltimore: John Hopkins University Press.

Grazziotin, P. C., Turkienicz, B., Sclovsky, L., & Freitas, C. M. D. S. (2004). CityZoom – A Tool for the Visualization of the Impact of Urban Regulations. *In Proceedings of the 8th Iberoamerican Congress of Digital Graphics*. (pp. 216-220).

Greene, W. H. (2003). *Econometric analysis*. London: Prentice Hall International.

Greschner, J., & Gerland, H. E. (2000). Traffic signal priority: Tool to increase service quality and efficiency. In *Proceedings APTA Bus & Paratransit Conference* (pp. 138-143). American Public Transportation Association.

Gupta, N., Agogino, A., & Tumer, K. (2006). Efficient agent-based models for non-genomic evolution. In *Proceedings of the Fifth International Joint Conference on Autonomous Agents and Multi-Agent Systems*, Hakodate, Japan.

Hallé, S. & Chaib-draa, B. (2004). Collaborative driving system using teamwork for platoon formations. In *The third workshop on Agents in Traffic and Transportation*.

Hallé, S. & Chaib-draa, B. (2005). A Collaborative Driving System based on Multiagent Modelling and Simulations. *Journal of Transportation Research Part C (TRC-C): Emergent Technologies, 13*(4), 320–345.

Hallouzi, R., Verdult, V., Hellendorn, H., Morsink, P. L., & Ploeg, J. (2004). Communication based longitudinal vehicle control using an extended kalman filter. *In Proceedings of International Federation of Automatic Control Symposium on Advances in Automotive Control* (pp. 745-750).

Hanson, S., & Burnett, K. (1982). The analysis of travel as an example of complex human behaviour in spatially-constraint situation: Definition and measurement issues. *Transportation Research Part A: Policy and Practice, 16*(2), 87–102.

Hardin, G. (1968). The Tragedy of the Commons. *Science, 162*(3859), 1243-1248.

Harel, D. (1987). Statecharts: A visual formalism for complex systems. In *Science of Computer Programming* (pp. 231–274). North-Holland: Elsevier Science Publishers.

Harwood, D. W., Bauer, K. M., Potts, I. B., Torbic, D. J., Richard, K. R., Rabbani, E. R. K., Hauer, E., Elefteriadou, L., & Griffith, M. S. (2003). Safety effectiveness of intersection left- and right-turn lanes. *Transportation Research Record, 1840*, 131–139.

Hebert, T., & Valavanis, K. (1998). Navigation of an autonomous vehicle using an electrostatic potential field. *In Proceedings of the 1998 IEEE International Conference on Control Applications , 2*, 1328–1332.

Helbing, D. (1992). A fluid-dynamic model for the movement of pedestrians. *Complex Systems, 6*, 391–415.

Helbing, D., & Johansson, A. (2007). Quantitative agent-based modeling of human interactions in space and time. In F. Amblard (Ed.), *Fourth Conference of The European Social Simulation Association* (ESSA) (pp. 623–637). Toulouse: Institut de Recherche en Informatique de Toulouse (IRIT).

Helbing, D., & Molnár, P. (1995). Social force model for pedestrian dynamics. *Phys. Rev. E, 51*, 4282-4286.

Helbing, D., Buzna, L., Johansson, A., & Werner, T. (2005). Self-Organized Pedestrian Crowd Dynamics:

Experiments, Simulations, and Design Solutions. *Transportation Science , 39*, 1-24.

Helbing, D., Buzna, L., Johansson, A., & Werner, T. (2005). Self-organized pedestrian crowd dynamics: Experiments, simulations, and design solutions. *Transportation Science, 39*, 1–24.

Helbing, D., Farkás, I. J., Molnár, P., & Vicsek, T. (2002). Simulation of pedestrian crowds in normal and evacuation situations. In M. Schreckenberg, & S. D. Sharma (Eds.), *Pedestrian and Evacuation Dynamic* (pp. 21–58). Berlin, Germany: Springer.

Helbing, D., Farkas, I., & Vicsek, T. (2000). Freezing by heating in a driven mesoscopic system. *Phys. Rev. Let., 84*, 1240–1243.

Helbing, D., Farkas, I., & Vicsek, T. (2000). Simulating dynamical features of escape panic. *Nature, 407*, 487–490.

Helbing, D., Farkas, I., Molnár, P., & Vicsek, T. (2002). Simulation of pedestrian crowds in normal and evacuation situations. In M. Schreckenberg & S. D. Sharma (Eds.), *Pedestrian and Evacuation Dynamics* (pp. 21–58). Berlin Heidelberg: Springer.

Helbing, D., Johannson, A., & Al-Abideen, H. (2007). Crowd turbulence: the physics of crowd disasters. In *The Fifth International Conference on Nonlinear Mechanics (ICMN-V)* (pp. 967-969). Shanghai.

Helbing, D., Johansson, A., & Al-Abideen, H. Z. (2007). The dynamics of crowd disasters: An empirical study. *Phys. Rev. E, 75*, 046109.

Henderson, L. F. (1974). On the fluid mechanics of human crowd motion. *Transpn. Res., 8*, 509–515.

Henesey, L., & Törnquist, J. (2002). Enemy at the Gates: Introduction of Multi-Agent s in a Terminal Information Community. *Third International Conference on Maritime Engineering and Ports.* Rhodes, Greece: Wessex Institute of Technology, UK.

Henesey, L., Notteboom, T., & Davidsson, P. (2003). Agent-based Simulation of stakeholders relations: An approach to sustainable port and terminal management. *The International Association of Maritime Economists Annual Conference*, (IAME 2003). Busan, Korea.

Hensher, D., & Puckett, S. (2005). Road user charging: The global relevance of recent developments in the United Kingdom. *Transport Policy, 12*(5), 377–383.

Hertkort, G., & Wagner, P. (2005). Adaptation of time use patterns to simulated travel times in a travel demand model. In H. J. P. Timmermans (Ed.), *Progress in Activity-Based Analysis,* Amsterdam, 161-174.

Hess, S., Bierlaire, M., & Polak, J. W. (2005). Estimation of value of travel-time savings using mixed logit models. *Transportation Research Part A: Policy and Practice, 39*(2-3), 221-236.

Heye, C. (2002). *Deskriptive Beschreibung der Ergebnisse der Internetbefragung: Umsteigen an Haltestellen der VBZ (Technical Report No. KoMoNa02-03).* University of Zurich: Department of Geography, Zurich, Switzerland.

Heye, C., & Timpf, S. (2003). *Factors influencing the physical complexity of routes in public transportation networks.* Paper presented at the 10th International Conference on Travel Behaviour Research, Lucerne.

HHLA (2008). HHLA Container Terminals GmbH: A division of Hamburger Hafen und Logistik AG. Retrieved August 4, 2008, from http://www.hhla.de/fileadmin/download/HHLA_Container_Broschuere_ENG.pdf

Hidas, P. (2005). Modelling Vehicle Interactions in Microscopic Simulation of Merging and Weaving. *Transportation Research, 10C*, 37-62.

Hinkley, D. V. (1988). Bootstrap methods. *Journal of the Royal Statistical Society. Series B, 50*(3), 321-337.

Hirtle, S. C., & Heidorn, P. B. (1993). The structure of cognitive maps: Representations and processes. In T. Gärling & R. G. Golledge (Eds.), *Behavior and environment: Psychological and geographical approaches* (pp. 1-29, Chapter 27).

Hoen, P. J., & La Poutré, J. A. (2003). A decommitment strategy in a competitive multi-agent transportation setting. *In Proc. of 2nd Int. Joint Conf. on Autonomous*

Agents and Multiagent Systems (AAMAS 2003), (pp. 1010–1011), New York, NY: ACM Press.

Holland, J. (1992). *Adaptation in Natural and Artificial Systems*. Bradford Books. Reprint edition.

Hoogendoorn, S. P., & Bovy, P. (2003). Simulation of pedestrian flows by optimal control and differential games. *Optim. Control Appl. Meth., 24*, 153.

Hoogendoorn, S. P., & Daamen, W. (2005). Pedestrian behavior at bottlenecks. *Transportation Science, 39 2*, 0147-0159.

Hoogendoorn, S. P., Daamen, W., & Bovy, P. H. L. (2003). Microscopic pedestrian traffic data collection and analysis by walking experiments: Behaviour at bottlenecks. In E. R. Galea (Ed.), *Pedestrian and Evacuation Dynamics '03* (pp. 89–100). CMS Press, London.

Hopcroft, J., & Ullman, J. (1979). *Introduction to automata theory, languages and computation* (1st ed.). Boston, MA: Addison-Wesley.

Horiguchi, R., Katakura, M., Akahane, H., & Kuwahara, M. (1994). A Development of A Traffic Simulator for Urban Road networks: AVENUE. *Conference proceedings of Vehicle Navigation and Information Systems Conference*, (pp. 245-250).

Huberman, B. A., & Hogg, T. (1988). The behavior of computational ecologies. In *The Ecology of Computation*, (pp. 77–115). North-Holland.

Hughes, R. L. (2000). The flow of large crowds of pedestrians. *Mathematics and Computers in Simulation, 53*, 367–370.

Hughes, R. L. (2002). A continuum theory for the flow of pedestrians. *Transportation Research Part B, 36*, 507–535.

Hunt, J., Johnston, R., Abraham, J., Rodier, C., Garry, G., Putman, S., & de la Barra, T. (2000). Comparisons from Sacramento mode test bed. *Transportation Research Record, 1780*, 53–63.

Idris, H., Evans, A., Vivona, R., Krozel, J., & Bilimoria, K. (2006, September). *Field Observations of Interactions Between Traffic Flow Management and Airline Operations*. Paper presented at the American Institute of Aeronautics and Astronautics (AIAA) 6th Aviation, Technology, Integration, and Operations Conference (ATIO), Wichita, Kansas.

Idris, H., Vivona, R., Penny, S., Krozel, J., & Bilimoria, K. (2005, September). *Operational Concept for Collaborative Traffic Flow Management based on Field Observations*. Paper presented at the American Institute of Aeronautics and Astronautics (AIAA) 5th Aviation, Technology, Integration, and Operations Conference (ATIO), Arlington, Virginia.

Ihaka, R. & Gentleman, R. (1996). A language for data analysis and graphics. *Journal of Computational and Graphical Statistics, 5*, 299-314.

ILOG, Inc. (1992). *Using the CPLEX Callable Library and CPLEX Mixed Integer Library.*

Infopolis2. (1999). *Needs of Travellers: An Analysis Based on the Study of Their Tasks and Activities* (pp. 62). Brussels: Commission of the European Communities - DG XIII. WP3.2/DEL3/1999.

Inoue, M. (2004). Current Overview of ITS in Japan. *Conference proceedings of The 11th World Congress on Intelligent Transport Systems (CD-ROM).*

Ioannou, P. A., & Chien, C. C. (1993). Autonomous intelligent cruise control. *IEEE Transactions on Vehicular Technology, 42*(4), 657-672.

Isaacs, R. (1999). *Differential Games: A Mathematical Theory with Applications to Warfare and Pursuit, Control and Optimization.* New York: Dover Publications.

Itoh, M., Sakami, D., & Tanaka, K. (2000). Dependence of human adaptation and risk compensation on modification in level of automation for system safety. In *IEEE International Conference on Systems, Man, and Cybernetics, 2000, 2*, 1295-1300. IEEE.

Jaganthan, S., Clarke, T. L., Kaup, D. J., Koshti, J., Malone, L., & Oleson, R. (2007). Intelligent Agents: Incorporating Personality into Crowd Simulation. *I/ITSEC Interservice/Industry Training, Simulation & Education Conference.* Orlando.

Jain, S., & Neal, R. M. (2004). A Split-Merge Markov Chain Monte Carlo Procedure for the Dirichlet Process Mixture Model, *Journal of Computational and Graphical Statistics*, *13*, 158-182.

Janssens, D., Wets, G., Timmermans, H. J. P., & Arentze, T. A. (2007). Modeling short-term dynamics in activity-travel patterns: the Feathers model. *In Proceedings of WCTR Conference*, Berkeley, (CD-ROM: 24 pp.).

Jefferies, P., Hart, M. L., & Johnson, N. F. (2002). Deterministic dynamics in the minority game. *Physical Review E*, 65 (016105).

Jennings, N., & Bussmann, S. (2003). Agent-based control systems: Why are they suited to engineering complex systems? *Control Systems Magazine, IEEE*, *23*(3), 61–73.

Joh, C.-H., Arentze, T. A., & Timmermans, H. J. P. (2006). Measuring and predicting adaptation behavior in multidimensional activity-travel patterns. *Transportmetrica*, *2*, 153-173.

Johnson, N. R. (1987, Oct). Panic at "The Who Concert Stampede": An Empirical Assessment. *Social Problems*, *34*(4), 362–373.

Johnson-Laird, P. N. (1992). Mental models. In S. C. Shapiro (Ed.), *Encyclopedia of artificial intelligence* (2nd ed.) (pp. 932-939). New York: John Wiley & Sons.

Kaptelinin, V., Nardi, B., & Macaulay, C. (1999). The activity checklist: A tool for representing the 'space' of context. *Interactions*(july/august), 29-39.

Kaup, D. J., Clarke, T. L., Malone, L., & Oleson, R. (2006). Crowd Dynamics Simulation Research. *Summer Simulation Multiconference*. Calgary, Canada.

Kaup, D. J., Clarke, T. L., Malone, L., Jentsch, F., & Oleson, R. (2007). Introducing Age-Based Parameters into Simulations of Crowd Dynamics. *American Sociological Association's 102nd Annual Meeting*. New York.

Keating, J. P. (1982). The myth of panic. *Fire Journal*, *May*, 57-62.

Kendall E. A., Malkoun, M. T., & Jiang, C. H. (1997). The layered agent pattern language. In *Proceedings of the Conference on Pattern Languages of Programs (PLoP'97)*. Monticello, IL.

Kerner, B. S., & Rehborn, H. (1996). Experimental properties of complexity in traffic flow. *Physical Review E*, *53*(5), R4275–4278.

Khan, M. A., Turgut, D., & Bölöni, L. (2008). A study of collaborative influence mechanisms for highway convoy driving. 5th Workshop on AGENTS IN TRAFFIC AND TRANSPORTATION. Estoril, Portugal.

Khatib, O. (1985). Real-time obstacle avoidance for manipulators and mobile robots. *In IEEE International Conference on Robotics and Automation*, *2*, 500–505.

Kiencke, U., & Nielsen, L. (2000). *Automotive control systems: for engine, driveline and vehicle*. Berlin, Germany: Springer-Verlag.

Kirchner, A., & Schadschneider, A. (2002). Simulation of evacuation processes using a bionics-inspired cellular automaton model for pedestrian dynamics. *Physica A*, *312*, 260.

Kirchner, A., Klüpfel, H., Nishinari, K., Schadschneider, A., & Schreckenberg, M. (2003). Simulation of competitive egress behavior: Comparison with aircraft evacuation data. *Physica A*, *324*, 689.

Kirchner, A., Klüpfel, H., Nishinari, K., Schadschneider, A., & Schreckenberg, M. (2004). Discretization effects and the influence of walking speed in cellular automata models for pedestrian dynamics. *J. Stat. Mech.*, *10*, P10011.

Kirchner, A., Namazi, A., Nishinari, K., & Schadschneider, A. (2003). Role of conflicts in the floor field cellular automaton model for pedestrian dynamics. In E. R. Galea (Ed.), (p. 51). London: CMS Press.

Kirchner, A., Nishinari, K., & Schadschneider, A. (2003). Friction effects and clogging in a cellular automaton model for pedestrian dynamics. *Phys. Rev. E*, *67*, 056122.

Klos T. (2001). Agent-based computational transaction cost economics. *Journal of Economic Dynamics & Control, 25*(3-4), 503-526.

Klos, T. B. (2000). *Agent-based Computational Transaction Cost Economics*. Published doctoral dissertation, University of Gronigen, Groningen, The Netherlands.

Klügl, F., & Bazzan, A. L. C. (2004). Route decision behaviour in a commuting scenario: Simple heuristics adaptation and effect of traffic forecast. *Journal of Artificial Societies and Social Simulation, 7*(1).

Klügl, F., & Rindsfüser, G. (2007). Large-scale agent-based pedestrian simulation. *Lect. Notes Comp. Sc., 4687*, 145.

Klügl, F., Bazzan, A. L. C., & Wahle, J. (2003). Selection of Information Types Based on Personal Utility: A Testbed for Traffic Information Markets. *Conference proceedings of the Second International Joint Conference on Autonomous Agents and Multiagent Systems*, (pp. 377-384).

Klügl, F., Bazzan, A., & Ossowski, S., editors (2005). *Applications of Agent Technology in Traffic and Transportation*. Springer.

Klügl, F., Herrler, R., & Fehler, M. (2006). *Sesam: Implementation of agent-based simulation using visual programming*. Paper presented at the AAMAS 2006, Hakodate.

Klügl, F., Wahle, J., Bazzan, A. L. C., & Schreckenberg, M. (2000). Towards anticipatory traffic forecast – modelling of route choice behaviour. In *Proceeding of the workshop "Agents in Traffic Modelling" at the Autonomous Agents 2000*. Barcelona.

Klüpfel, H. (2006). The simulation of crowds at very large events. In A. Schadschneider, T. Pöschel, R. Kühne, M. Schreckenberg, & D. Wolf (Eds.), *Traffic and Granular Flow '05* (p. 341). Berlin: Springer.

Klüpfel, H. (2007). The simulation of crowd dynamics at very large events — calibration, empirical data, and validation. In N. Waldau, P. Gattermann, H. Knoflacher, & M. Schreckenberg (Eds.), (p. 285). Berlin: Springer.

Klüpfel, H., & Meyer-König, T. (2003). Models for crowd movement and egress simulation. In S. F. et al. (Ed.), *Traffic and Granular Flow '03* (pp. 357–372). Berlin: Springer.

Klüpfel, H., & Meyer-König, T. (2003). Simulation of the evacuation of a football stadium. In S. F. et al. (Ed.), *Traffic and Granular Flow '03* (pp. 423–430). Berlin: Springer.

Klüpfel, H., Meyer-König, T., Wahle, J., & Schreckenberg, M. (2000). Microscopic simulation of evacuation processes on passenger ships. In S. Bandini & T. Worsch (Eds.), *Theory and Practical Issues on Cellular Automata*. Berlin Heidelberg: Springer.

Knoblauch, R. L., Pietrucha, M. T., & Nitzburg, M. (1996). Field studies of pedestrian walking speed and start-up time. *Transportation Research Record*. No. 1538 Pedestrian and Bicycle Research.

Kohout, R., & Erol, K. (1999). In-time agent-based vehicle routing with a stochastic improvement heuristic. *In Proc. of the 16th national conference on Artificial Intelligence and the 11th on Innovative Applications of Artificial Intelligence (AAAI '99/IAAI '99)*, (pp. 864–869). Menlo Park, CA: American Association for Artificial Intelligence.

Kolodko, J. & Vlacic, L. (2003). Cooperative autonomous driving at the intelligent control systems laboratory. *IEEE Intelligent Systems, 18*(4), 8–11.

Kondratowicz, L. (1992). Generating logistical chains scenarios for maritime transport policy making. In N. Wijnolst, C. Peters, & P. Liebman (Eds.), *European Short Sea Shipping: proceedings from the second European Research Roundtable Conference* (pp. 379-402). London, UK: Lloyds of London Press Ltd.

Krajzewicz, D., Bonert, M., & Wagner, P. (2006). The open source traffic simulation package sumo. In *Proceedings of RoboCup 2006 Infrastructure Simulation Competition*. Bremen.

Kretz, T., & Schreckenberg, M. (2006). Moore and more and symmetry. In N. Waldau, P. Gattermann,

H. Knoflacher, & M. Schreckenberg (Eds.), (pp. 317–328). Berlin: Springer.

Kretz, T., Grünebohm, A., & Schreckenberg, M. (2006). Experimental study of pedestrian flow through a bottleneck. *J. Stat. Mech.*, P10014.

Kretz, T., Grünebohm, A., Kaufman, M., Mazur, F., & Schreckenberg, M. (2006a). Experimental study of pedestrian counterflow in a corridor. *J. Stat. Mech.*, P10001.

Krieg-Brückner, B., Frese, U., Lüttich, K., Mandel, C., Mossakowski, T., & Ross, R. J. (2005). Specification of an ontology for route graphs. In C. Freksa, M. Knauff, B. Krieg-Brückner, B. Nebel & T. Barkowsky (Eds.), *Spatial cognition IV. Reasoning, action, and interaction* (Vol. LNAI 3343, pp. 390-412). Berlin: Springer.

Krishna, V. (2002). *Auction theory*. London: Academic Press.

Kuipers, B. (1982). The 'map in the head' metaphor. *Environment and Behaviour, 14*(2), 202-220.

Kuipers, B. (2000). The spatial semantic hierarchy. *Artificial Intelligence, 119*, 191-233.

Kuipers, B., Tecuci, D. G., & Stankiewicz, B. J. (2003). The skeleton in the cognitive map: A computational and empirical exploration. *Environment and Behavior, 35*(1), 81-106.

Kurumatani, K. (2003). Social Coordination with Architecture for Ubiquitous Agents: CONSORTS. *Conference proceedings of International Conference on Intelligent Agents, Web Technologies and Internet Commerce 2003* (CD-ROM).

Kurumatani, K. (2004). Mass User Support by Social Coordination among Citizens in a Real Environment. *Multiagent for Mass User Support*, (pp. 1-19). LNAI 3012, Springer.

Kylaheiko, K., Cisic, D., & Komadina, P. (2000). *Application of Transaction Costs to Choice of Transport Corridors. Economics Working Paper* Archive at WUSTL, 2000. Retrieved January 12, 2008, from http://ideas.repec.org/p/wpa/wuwpit/0004001.html)

Lakoba, T. I., Kaup, D. J., & Finkelstein, N. M. (2005). Modifications of the Helbing-Molnár-Farkas-Vicsek social force model for pedestrian evolution. *Simulation, 81 5*, 339–352.

Larsen, A., Madsen, O., & Solomon, M. (2002). Partially dynamic vehicle routing-models and algorithms. *Journal of the Operational Research Society, 53*, 637–646.

Laumônier, J., & Chaib-draa, B. (2006). Partial local FriendQ multiagent learning: application to team automobile coordination problem. In L. Lamontagne and M. Marchand (Ed.), *Canadian AI, Lecture Notes in Artificial Intelligence* (pp. 361–372). Berlin, Germany: Springer-Verlag.

Lazar, A. A., Orda, A., & Pendarakis, D. E. (1997). Capacity allocation under noncooperative routing. *IEEE Transactions on Networking*, 5(6), 861–871.

Lebacque, J.-P. (2003). Intersection modeling, application to macroscopic network traffic flow models and traffic management. In S. P. Hoogendoorn, S. Luding, P. V. L. Bovy, M. Schreckenberg, & D. E. Wolf (Eds.), *Traffic and Granular Flow 2003* (pp. 261–278). Berlin: Springer.

Lee, J., Huang, R., Vaughn, A., Xiao, X., Hedrick, K., Zennaro, M., et al. (2003). Strategies of path-Planning for a UAV to track a ground vehicle. *AINS Conference*.

Pineda, J. (1988). A parallel algorithm for polygon rasterization. *SIGGRAPH '88: Proceedings of the 15th annual conference on Computer graphics and interactive techniques*, (pp. 17-20).

Lee, K., Hui, P. M., Wang, B.-H., & Johnson, N. F. (2001). Effects of announcing global information in a two-route traffic flow model. *Journal of the Physical Society of Japan, 70*(12), 3507–3510.

Lefebvre, N., & Balmer, M. (2007). Fast shortest path computation in time-dependent traffic networks. In *Proceedings of Swiss Transport Research Conference (STRC)*. Monte Verita, CH. See www.strc.ch.

Leiser, D., & A. Zilbershatz (1989). The Traveller—A Computational Model of Spatial Network Learning. *Environment and Behavior 21*(4), 435-463.

Leong, H. W., & Liu, M. (2006). A multi-agent algorithm for vehicle routing problems with time window. *In Proc. of the ACM Symposium on Applied Computing (SAC 2006),* (pp. 106–111). New York, NY: ACM Press.

Lewin, K. (1951). *Field Theory in Social Science.* Harper.

Lindner, C. C., & Rodger, C. A. (1997). *Design Theory.* Boca Raton, FL: CRC Press.

Littman, M. (2001). Friend-or-Foe Q-learning in General-Sum Games. In C.E. Brodley and A. P. Danyluk (Ed.), *Proceedings of the Eighteenth International Conference on Machine Learning* (pp. 322–328), San Francisco, CA: Morgan Kaufmann.

Littman, M. L. (1994). Markov games as a framework for multi-agent reinforcement learning. In *Proceedings of the 11th International Conference on Machine Learning,* (pp. 157–163).

Liu, X., Liang, X., & Tang, B. (2004). Minority game and anomalies in financial markets. *Physica A: Statistical and Theoretical Physics, 333,* 343–352.

Ljungstrom, B. J. (1985). Changes in Transport Users' Motivations for Modal Choice: Freight Transport. *In ECMT, Round Table 69.* Paris, France.

Lotzmann, U., Möhring, M., & Troitzsch, K. G. (2008). Simulating norm formation in a traffic scenario. In *Proceedings of the Fifth Annual Conference of the European Social Simulation Association (ESSA).* Brescia.

Lu, X.-Y., Tan, H.-S., Empey, D., Shladover, S. E., & Hedrick, J. K. (2000). *Nonlinear longitudinal controller development and real-time implementation* (Technical Report UCB-ITS-PRR-2000-15). Los Angeles, CA: University of Southern California, California Partners for Advanced Transit and Highways (PATH).

Lynch, K. (1960). *The image of the city.* Cambridge: MIT Press.

Ma, R., & Kaber, D. B. (2005). Situation awareness and workload in driving while using adaptive cruise control and a cell phone. *International Journal of Industrial Ergonomics, 35*(10), 939-953.

Mahmassani, H. S., & Jayakrishnan, R. (1991). System Performance and User Response Under Real-Time Information in a Congested Traffic Corridor. *Transportation Research 25A*(5), 293-307.

Maniccam, S. (2003). Traffic jamming on hexagonal lattice. *Physica, A321,* 653.

Maniccam, S. (2005). Effects of back step and update rule on congestion of mobile objects. *Physica A, 346,* 631.

Marconi, S., & Chopard, B. (2002). A multiparticle lattice gas automata model for a crowd. *Lecture Notes in Computer Science, 2493,* 231.

Marsden, G., McDonald, M., & Brackstone, M. (2001). Towards an understanding of adaptive cruise control. *Transportation Research Part C: Emerging Technologies 9*(1), 33-51.

Max-Neef, M. (1992). Development and human needs. In P. Ekins, & M. Max-Neef (Eds.), *Real-life economics: Understanding wealth creation.* London, New York: Routledge.

McDermott, D., & Davis, E. (1984). Planning Routes through Uncertain Territory. *Artificial Intelligence, 22*(1), 107-156.

McFadden, D. (1974). Conditional logit analysis of qualitative choice behaviour. In Zarembka, P., (Ed.), *Frontiers in Econometrics.* New York: Academic Press.

McNally, M. G. (2000). The four step model. In Handbook of Transport Modelling (pp. 35-52). Oxford: Pergamon Press.

Meier, E. (1997). Verkehrsnachfragemodelle: Hilf- oder Allheilmittel? In A. Müller (Ed.), *Wege und Umwege in der Verkehrsplanung.* Zürich: vdf Hochschulverlag.

Meister, K. (2004). *Erzeugung kompletter Aktivitätenpläne für Haushalte mit genetischen Algorithmen.* Master's thesis, IVT, ETH Zürich. See www.ivt.ethz.ch/docs/students/dip44.pdf.

Meister, K., Rieser, M., Ciari, F., Horni, A., Balmer, M., & Axhausen, K. (2008). Anwendung eines agentenbasierten Modells der Verkehrsnachfrage auf die Schweiz. In *Proceedings of Heureka '08*. Stuttgart, Germany.

Mes, M., van der Heijden, M., & van Harten, A. (2007). Comparison of agent-based scheduling to look-ahead heuristics for real-time transportation problems. *European Journal of Operational Research, 181*(1), 59–75.

Miller, H. J. (1992). Human Wayfinding, Environment-Behavior Relationships, and Artificial Intelligence. *Journal of Planning Literature, 7*(2), 139-150.

Mivitrans, (1998). *Intermodal and Transportation conference*. Hamburg, Germany.

Mokhtarian, P. L., & Salomon, I. (2001). How derived is the demand for travel? some conceptual and measurement considerations. *Transportation Research Part A: Policy and Practice, 35*(8), 695-719.

Montello, D. R. (1993). Scale and multiple psychologies of space. In A. U. Frank & I. Campari (Eds.), *Spatial information theory: Theoretical basis for gis, 716*, 312-321. Heidelberg-Berlin: Springer Verlag.

Montello, D. R. (2005). Navigation. In P. Shah & A. Miyake (Eds.), *Handbook of visuospatial cognition.* (pp. 257-294): Cambridge, Cambridge University Press.

Moriarty, D., & Langley, P. (1998). Learning cooperative lane selection strategies for highways. *In Proceedings of the Fifteenth National Conference on Artificial Intelligence* (pp. 684–691), Menlo Park, CA: AAAI Press.

Muir, H. C., Bottomley, D. M., & Marrison, C. (1996). Effects of motivation and cabin configuration on emergency aircraft evacuation behavior and rates of egress. *Intern. J. Aviat. Psych., 6*(1), 57–77.

Mukherjee, A., Grabbe, S., & Sridhar, B. (2008). *Alleviating Airspace Restriction through Strategic Control.* Paper presented at the American Institute of Aeronautics and Astronautics (AIAA) Guidance, Navigation, and Control Conference, Honolulu, Hawaii.

Müller, K. (1981). *Zur Gestaltung und Bemessung von Fluchtwegen für die Evakuierung von Personen aus Bauwerken auf der Grundlage von Modellversuchen.* Dissertation, Technische Hochschule Magdeburg, in German.

Muramatsu, M., & Nagatani, T. (2000). Jamming transition in two-dimensional pedestrian traffic. *Physica A, 275*, 281–291.

Muramatsu, M., & Nagatani, T. (2000). Jamming transition of pedestrian traffic at crossing with open boundary conditions. *Physica A, 286*, 377–390.

Muramatsu, M., Irie, T., & Nagatani, T. (1999). Jamming transition in pedestrian counter flow. *Physica A, 267*, 487–498.

Nagai, R., & Nagatani, T. (2006). Jamming transition in counter flow of slender particles on square lattice. *Physica A, 366*, 503.

Nagai, R., Fukamachi, M., & Nagatani, T. (2006). Evacuation of crawlers and walkers from corridor through an exit. *Physica A, 367*, 449–460.

Nagel, K. (1997). *Experiences with iterated traffic microsimulations in dallas.* pre-print adap-org/9712001.

Nagel, K. (2001). Multi-modal traffic in TRANSIMS. In *Pedestrian and Evacuation Dynamics*, (pp. 161–172). Springer, Berlin.

Nagel, K. (2002). Traffic networks. In S. Bornholdt & H. G. Schuster (Eds.), *Handbook of graphs and networks* (pp. 248-272): Wiley.

Nagel, K., & Rickert, M. (2001). Parallel implementation of the TRANSIMS micro-simulation. *Parallel Computing, 27*(12), 1611–1639.

Nagel, K., & Schleicher, A. (1994). Microscopic traffic modeling on parallel high performance computers. *Parallel Computing, 20*, 125–146.

Nagel, K., & Schreckenberg, M. (1992). A cellular automaton model for freeway traffic. *Journal de Physique I, 2*, 2221–2229.

NAHSC (1998). *Technical Feasibility Demonstration Summary Report (Technical Report).* Troy, MI, USA: National Automated Highway System Consortium.

Nakashima, H. (2003). Grounding to the Real World - Architecture for Ubiquitous Computing -. *Foundations of Intelligent Systems*, (pp. 7-11). Lecture Notes in Computer Science, 2871, Springer Berlin.

Nakayama, A., Hasebe, K., & Sugiyama, Y. (2005). Instability of pedestrian flow and phase structure in a two-dimensional optimal velocity model. *Phys. Rev. E, 71*, 036121.

Naranjo, J., Gonzalez, C., Reviejo, J., Garcia, R., & de Pedro, T. (2003). Adaptive fuzzy control for inter-vehicle gap keeping. *IEEE Transactions on Intelligent Transportation Systems, 4*(3), 132–142.

Nardi, B. (Ed.). (1996). *Context and consciousness - activity theory and human-computer interaction.* Cambridge, MA: The MIT Press.

Nash, C. (2003). Marginal cost pricing and other pricing principles for user charging in transport: a comment. *Transport Policy, 10*, 345–348.

National Highway Traffic Safety Administration (2002). *Economic impact of U.S. motor vehicle crashes reaches $230.6 billion*, new NHTSA study shows NHTSA Press Release 38-02. http://www.nhtsa.dot.gov.

Naumann, R. & Rasche, R. (1997). Intersection collision avoidance by means of decentralized security and communication management of autonomous vehicles. *In Proceedings of the 30th ISATA - ATT/IST Conference.*

Navin, P. D., & Wheeler, R. J. (1969). Pedestrian flow characteristics. *Traffic Engineering, 39*, 31–36.

Nelson, H. E., & Mowrer, F. W. (2002). Emergency movement. In P. J. DiNenno (Ed.), *SFPE Handbook of Fire Protection Engineering* (Third ed., p. 367). Quincy MA: National Fire Protection Association.

Nguyen-Duc, M., Briot, J.-P., Drogoul, A., & Duong, V. (2003, October). *An Application of Multi-Agent Coordination Techniques in Air Traffic Management.* Paper presented at the 2003 Institute of Electrical & Electronic Engineers/ Web Intelligence Consortium (IEEE/WIC) International Conference on Intelligent Agent Technology (IAT 2003), Halifax, Canada.

Noda, I., Ohta, M., Shinoda, K., Kumada, Y., & Nakashima, H. (2003). Evaluation of Usability of Dial-a-Ride Systems by Social Simulation. *Proceedings of Multi-Agent-Based Simulation III (4th International Workshop),* (pp. 167-181).

Nolan, M. S. (2003). *Fundamentals of Air Traffic Control* (4th ed.) Pacific Grove, California: Brookes/Cole Publishing.

Norman, D. A. (1988). *The design of everyday things.* Doubleday.

Norris, G. A., & Jager, W. (2004). Household-level modeling for sustainable consumption. In *Proceedings of the Third International Workshop on Sustainable Consumption.* Tokyo.

Norris, J. (1997). *Markov chains.* Cambridge: Cambridge University Press.

North, M. J., Howe, T. R., Collier, N. T., & Vos, R. J. (2005). The repast simphony runtime system. In *Proceedings of Agent 2005 Conference on Generative Social Processes, Models and Mechanisms.* Argonne, IL: Argonne National Laboratory.

Odell, J., Parunak, H. V. D., & Fleischer, M. (2003). Modeling Agents and their Environment: the communication environment. *Journal of Object Technology, 2*(3), 39-52.

Odoni, A. (1987). The Flow Management Problem in Air Traffic Control. *In Flow Control of Congested Networks* (pp. 269-288). Berlin, Germany: Springer-Verlag.

Oeding, D. (1963). *Verkehrsbelastung und Dimensionierung von Gehwegen und anderen Anlagen des Fußgängerverkehrs* (Forschungsbericht No. 22). : Technische Hochschule Braunschweig, in German.

Older, S. J. (1968). Movement of pedestrians on footways in shopping streets. *Traffic Engineering and Control, 10*, 160-163.

Oliveira, D., & Bazzan, A. L. C. (2006). Traffic Lights Control with Adaptive Group Formation Based on Swarm Intelligence. In: *The Fifth International Workshop on Ant Colony Optimization and Swarm Intelligence,* ANTS 2006, 2006. (pp. 520–521). Lecture Notes in Computer Science. Berlin: Springer.

Oliveira, D., Bazzan, A. L. C., & Lesser, V. (2005). Using Cooperative Mediation to Coordinate Traffic Lights: a case study. In: *The Fourth International Joint Conference on Autonomous Agents and Multiagent System*, 2005. New York: IEEE Computer Society. (pp. 463–470).

Oliveira, D., Bazzan, A. L. C., Silva, B. C., Basso, E. W., Nunes, L., Rossetti, R. J. F., Oliveira, E. C., Silva, R., & Lamb, L. C. (2006). Reinforcement learning based control of traffic lights in non-stationary environments: a case study in a microscopic simulator. In: *Fourth European Workshop On Multi-Agent Systems*, (EUMAS06). (pp. 31–42).

Oliveira, E., & Duarte, N. (2005). Making way for emergency vehicles. In the European Simulation and Modelling Conference (pp. 128-135). Ghent: EUROSIS-ETI.

Oota, J. (1995). Introduction to petri net. *Technical report, Faculty of Information Science and Technology*, Aichi prefectural university.

Ortuzar, J. D. D., & Willumsen, L. G. (2001). *Modeling transport*. Chichester, England: John Wiley & Sons.

Otakeguchi, K., & Horiuchi, T. (2004). Conditions and Analysis of the Up-Link Information Gathered from Infrared Beacons in Japan. *Conference proceedings of The 11th World Congress on Intelligent Transport Systems (CD-ROM)*.

Palmer, R., Arthur, W. B., Holland, J. H., LeBaron, B., & Tayler, P. (1994). Artificial economic life: a simple model of a stockmarket. *Physica D, 75*, 264–274.

Panait, L., & Luke, S. (2005) Cooperative Multi-Agent Learning: the state of the art. *Autonomous Agents and Multi-Agent Systems, 11*(3), 387–434, Hingham, MA, USA.

Parkes, D. C. (2001). *Iterative Combinatorial Auctions: Theory and Practice*. PhD thesis, University of Pennsylvania.

Parunak, V., Savit, R., & Riolo, R. (1998). Agent-based modeling vs. equation-based modeling: A case study and users guide. *Proceedings of Workshop on Multi-agent systems and Agent-based Simulation (MABS'98)*, 10–25.

Payne, J. W., Laughunn, D. J., & Crum, R. (1980). Translation of gambles and aspiration level effects on risky choice behavior. *Management Science, 26*(10), 1039-1060.

Pearce, R. A. (2006). The Next Generation Air Transportation System: Transformation Starts Now. *Journal of Air Traffic Control*, (pp. 7-10).

Peeta, S., Zhang, P., & Zhou, W. (2005). Behavior-based analysis of freeway car-truck interactions and related mitigation strategies. *Transportation Research Part B: Methodological, 39*(5), 417–451.

Pellegrini, P. A., Fotheringham, S., & Lin, G. (1997). An empirical evaluation of parameter sensitivity to choice set definition in shopping destination choice models. *Papers in Regional Science, 76*, 257-284.

Pendrith, M. D. (2000). Distributed reinforcement learning for a traffic engineering application. C. Sierra, M. Geni and J.S. Rosenschein (Ed.), *Proceedings of the Fourth International Conference on Autonomous Agents* (pp. 404-411), New York, NY: ACM Press.

Pendyala, R. (2004). *Phased Implementation of a Multimodal Activity-Based Travel Demand Modeling System in Florida. Volume II: FAMOS Users Guide*. Research report, Florida Department of Transportation, Tallahassee. See www.eng.usf.edu/~pendyala/publications.

Peng, H., bin Zhang, W., Arai, A., Lin, Y., Hessburg, T., Devlin, P., Tomizuka, M., & Shladover, S. (1992). *Experimental automatic lateral control system for an automobile (Technical Report UCB-ITS-PRR-92-11)*, Los Angeles, CA: University of Southern California, California Partners for Advanced Transit and Highways (PATH).

Persaud, B. N., Retting, R. A., Gardner, P. E., & Lord, D. (2001). Safety effect of roundabout conversions in the united states: Empirical bayes observational before-after study. *Transportation Research Record, 1751*, 1–8.

Persson, J. A., Davidsson, P., Johansson, S. J., & Wernstedt, F. (2005). Combining agent-based approaches and classical optimization techniques. *In Proc. of the European workshop on Multi-Agent Systems (EUMAS 2005)*, (pp. 260–269). Koninklijke Vlaamse Academie van Belie voor Wetenschappen en Kunsten.

Persson, M., Botling, F., Hesslow, E., & Johansson, R. (1999). Stop & go controller for adaptive cruise control. In *Proceedings of the 1999 IEEE International Conference on Control Applications* (pp. 1692-1697), IEEE.

Perugini, D., Lambert, D., Sterling, L., & Pearce, A. (2003). A distributed agent approach to global transportation scheduling. *In IEEE/WIC Int. Conf. on Intelligent Agent Technology (IAT 2003),* (pp. 18–24).

Perumalla, K. S., & Bhaduri, B. L. (2006). On accounting for the interplay of kinetic and non-kinetic aspects in population mobility models. In A. G. Bruzzone, A. Guasch, M. A. Piera, & J. Rozenblit (Eds.), *International Mediterranean Modeling Multiconference (I3M)* (pp. 201-206). Barcelona.

Peterson, J. L. (1981). *Petri net theory and the modeling of systems.* Englewood Cliffs, N.J: Prentice-Hall.

Pfeiffer, P. E., & Schum, D. A. (1973). *Introduction to Applied Probability Theory.* New York: Academic Press.

Pomerleau, D. (1995). Neural network vision for robot driving. In S. Nayar and T. Poggio (Eds.), *Early Visual Learning* (pp. 161-181), New York, NY: Oxford University Press.

Pontikakis, E. (2006). *Wayfinding in GIS: Formalization of basic needs of a passenger when using public transportation.* PhD thesis, TU Vienna, Vienna.

Portugali, J. (1996). *The Construction of Cognitive Maps.* Kluwer Publishers.

Predtechenskii, V. M., & Milinskii, A. I. (1978). *Planing for Foot Traffic Flow in Buildings.* Amerind Publishing, New Dehli. (Translation of: Proekttirovanie Zhdanii s Uchetom Organizatsii Dvizheniya Lyuddskikh Potokov, Stroiizdat Publishers, Moscow, 1969)

Presson, C. C., & Montello, D. R. (1988). Points of reference in spatial cognition: Stalking the elusive landmark. *British Journal of Developmental Psychology, 6*, 378-381.

PTV (accessed 2008). Traffic Mobility Logistics. See www.ptv.de.

RAND. (2002). *Strategic Sourcing: Theory and Evidence from Economics and Business Management.* Retrieved January 12, 2008 from http://www.rand.org/publications/MR/MR865/MR865.chap2.pdf

Raney, B., & Nagel, K. (2004). Iterative route planning for large-scale modular transportation simulations. *Future Generation Computer Systems, 20*(7), 1101–1118.

Raney, B., & Nagel, K. (2006). An improved framework for large-scale multi-agent simulations of travel behaviour. In P. Rietveld, B. Jourquin, & K. Westin (Eds.), *Towards better performing European Transportation Systems* (p. 42). London: Routledge.

Raney, B., & Nagel, K. (in press). An improved framework for large-scale multi-agent simulations of travel behavior. In P. Rietveld, B. Jourquin & K. Westin (Eds.), *Towards better performing european transportation systems.*

Raney, B., Cetin, N., Völlmy, A., & Nagel, K. (2002). *Large scale multi-agent transportation simulations.* Paper presented at the 42nd ERSA Congress (European Regional Science Association), Dortmund.

Raney, B., Cetin, N., Völlmy, A., Vrtic, M., Axhausen, K., & Nagel, K. (2002). *Towards an activity-based microscopic simulation of all of Switzerland.* Zürich: Institut für Verkehrsplanung, Transporttechnik, Strassen- und Eisenbahnbau (IVT), ETHZ.

Raubal, M. (2001). Human wayfinding in unfamiliar buildings: A simulation with a cognizing agent. *Cognitive Processing, 2-3*, 363-388.

Raubal, M., & Worboys, M. (1999). *A formal model of the process of wayfinding in built environments.* Paper presented at the Spatial Information Theory - cognitive and computational foundations of geographic information science, Stade, Germany.

Raza, H., & Ioannou, P. (1997). *Vehicle following control design for automated highway systems (Technical Report UCB-ITS-PRR-97-2)*, Los Angeles, CA: University of Southern California, California Partners for Advanced Transit and Highways (PATH).

Redmond, L. S., & Mokhtarian, P. L. (2001). The positive utility of the commute: modeling ideal commute time and relative desired commute amount. *Transportation, 28*(2), 179-205.

Reif, J., & Wang, H. (1999). Social potential fields: A distributed behavioral control for autonomous robots. *Robotics and Autonomous Systems, 27*(3), 171-194).

Repast Organization for Architecture and Development (2005). Repast 3.0. Retrieved August 4, 2008 from http://repast.sourceforge.net.

Resnik, M. D. (1987). *Choices: An introduction to decision theory* Minnesota: University of Minnesota Press.

Reynolds, C. W. (1987). Flocks, herds and schools: A distributed behavioral model. *SIGGRAPH Computer Graphics , 21*, 25-34.

Reynolds, C. W. (1993). An Evolved, Vision-Based Behavioral Model of Coordinated Group Motion. *From Animals to Animats, Proc. 2nd International Conf. on Simulation of Adaptive Behavior.* Cambridge, MA: MIT Press.

Reynolds, C. W. (1999). Steering behaviors for autonomous characters. In AYu (Ed.), *Proceedings of the 1999 Game Developer's Conference* (pp. 763–782). San Francisco, CA: Miller Freeman.

Rieser, M., & Nagel, K. (2008). Network breakdown "at the edge of chaos" in multi-agent traffic simulations. *European Journal of Physics.* Doi 10.1140/epjb/e2008-00153-6.

Rindsfuser, G., & Klugl, F. (2005). The scheduling agent using Sesam to implement a generator of activity programs. In H. J. P. Timmermans (Ed.), *Progress in Activity-Based Analysis, Amsterdam*, (pp. 115-137).

Ripley, B. D. (1987). *Stochastic simulation.* New York: John Wiley & Sons Inc.

Rizzo, M., McGehee, D. V., Dawson, J. D., & Anderson, S. N. (2001). Simulated car crashes at intersections in drivers with Alzheimer disease. *Alzheimer Disease and Associated Disorders, 15*(1), 10–20.

Rogsch, C., Klingsch, W., Seyfried, A., & Weigel, H. (2007). How reliable are commercial software-tools for evacuation calculation? In *Interflam 2007 - Conference Proceedings* (pp. 235–245).

Roozemond, D. A. (1999). Using autonomous intelligent agents for urban traffic control systems. *In Proceedings of the 6th World Congress on Intelligent Transport Systems.*

Rosenblatt, J. K. (1995). DAMN: A distributed architecture for mobile navigation. In H. Hexmoor and D. Kortenkamp (Eds.), *Proceedings of the American Association of Artificial Intelligence Spring Symposium on Lessons Learned from Implemented Software Architectures for Physical Agents*, Menlo Park, CA: AAAI Press.

Rosetti, R. J. F., & Liu, R. (2004). A dynamic network simulation model based on multi-agent systems. *In Proceeding of the 3rd Workshop of Agents in Traffic and Transportation*, (pp. 88-93), AAMAS, New York.

Ross, S. M. (1997). *Simulation.* New York: Academic Press.

Rossetti, R. J. F., & Bampi, S. (1999). A Software Environment to Integrate Urban Traffic Simulation Tasks. *Journal of Geographic Information and Decision Analysis, 3*(1), 56-63.

Rossetti, R. J. F., Bampi, S., Liu, R., Van Vliet, D., & Cybis, H. B. B. (2000). An agent-based framework for the assessment of drivers' decision-making. *In Proceedings of the IEEE Conference on Intelligent Transportation Systems* (pp. 387-392). Piscataway, NJ: IEEE.

Rossetti, R. J. F., Bordini, R. H., Bazzan, A. L. C., Bampi, S., Liu, R., & Van Vliet, D. (2002). Using BDI agents to improve driver modelling in a commuter scenario. *Transportation Research, 10C*, 373-398.

Rothengatter, W. (2003). How good is first best? marginal cost and other pricing principles for user charging in transport. *Transport Policy, 10*, 121–130.

Rothman, D. H., & Zaleski, S. (1994). Lattice-gas models of phase separation: Interfaces, phase transitions, and multiphase flow. *Rev. Mod. Phys., 66*, 1417.

Rothman, D., & Zaleski, S. (1997). *Lattice-gas Cellular Automata*. Cambridge University Press.

Rüetschi, U. J., & Timpf, S. (2004). Modelling wayfinding in public transport: Network space and scene space. In C. Freksa, M. Knauff, B. Krieg-Brückner, B. Nebel & T. Barkowsky (Eds.), *Spatial cognition iv: Reasoning, action, interaction; international conference frauenchiemsee* (Vol. LNCS 3343, pp. 24-41). Heidelberg, Berlin: Springer.

Rüetschi, U.-J. (2007). *Wayfinding in science space - modelling transfers in public transport*. PhD thesis, University of Zurich, Zurich.

Russell, S. J., & Norvig, P. (2002). *Artificial Intelligence: A Modern Approach*. 2nd ed. Upper Saddle River, NJ: Prentice Hall.

Saloma, C. (2006). Herding in real escape panic. In N. Waldau, P. Gattermann, H. Knoflacher, & M. Schreckenberg (Eds.), *Pedestrian and Evacuation Dynamics 2006*. Berlin: Springer.

Salvini, P., & Miller, E. (2005). ILUTE: An operational prototype of a comprehensive microsimulation model of urban systems. *Network and Spatial Economics*, *5*(2), 217–234.

Sandholm, T. W. & Lesser, V. R. (2001). Leveled commitment contracts and strategic breach. *Games and Economic Behaviour*, *35*, 212–270.

Sandholm, T., & Crites, R. (1995). Multiagent reinforcement learning in the iterated prisoner's dilemma. *Biosystems*, *37*, 147–166.

Sayed, T. & Zein, S. (1999). Traffic conflict standards for intersections. *Transportation Planning and Technology*, *22*(4), 309–323.

Schadschneider, A. (2002). Cellular automaton approach to pedestrian dynamics – theory. In M. Schreckenberg & S. D. Sharma (Eds.), *Pedestrian and Evacuation Dynamics* (pp. 75–86). Berlin Heidelberg: Springer.

Schadschneider, A., Klingsch, W., Klüpfel, H., Kretz, T., Rogsch, C., & Seyfried, A. (2009). Evacuation dynamics: Empirical results, modeling and applications. In B. Meyers (Ed.), *Encyclopedia of Complexity and System Science*. Springer.

Schadschneider, A., Pöschel, T., Kühne, R., Schreckenberg, M., & Wolf, D. (Eds.). (2006). *Traffic and Granular Flow '05*. Berlin: Springer.

Schepperle, H., & Böhm, K. (2007). Agent-based traffic control using auctions. In *Cooperative Information Agents XI* (pp. 119-133). Berlin/Heidelberg, Germany: Springer.

Schepperle, H., & Böhm, K. (2008). Auction-based traffic management: Towards effective concurrent utilization of road intersections. In *The 10th IEEE Conference on E-Commerce Technology and the 5th International Conference on Enterprise Computing, E-Commerce and E-Services* (pp. 105-112). IEEE.

Schepperle, H., Barz, C., Böhm, K., Kunze, J., Laborde, C. M., Seifert, S., & Stockmar, K. (2006). *Auction mechanisms for traffic management*. In *Group Decision and Negotiation (GDN) 2006* (pp. 214-217). Universitätsverlag Karlsruhe.

Schepperle, H., Böhm, K., & Forster, S. (2008). Traffic management based on negotiations between vehicles – a feasibility demonstration using agents. In *Agent-Mediated Electronic Commerce IX/Trading Agent Design and Analysis*, Springer, forthcoming.

Schillo, M., Kray, C., & Fischer, K. (2002). The eager bidder problem: a fundamental problem of DAI and selected solutions. *In Proc. of 1st Int. Joint Conf. on Autonomous Agents and Multiagent Systems (AAMAS 2002)*, (pp. 599–606), New York, NY: ACM Press.

Schnittger, S., & Zumkeller, D. (2004). Longitudinal microsimulation as a tool to merge transport planning and traffic engineering models - the MobiTopp model. In *Proceedings of the European Transport Conference*. Strasbourg.

Schoggen, P. (1989). *Behavior Settings*. Stanford, CA: Stanford University Press.

Schönfelder, S., Axhausen, K., & Antille, M., N.and Bierlaire (2002). Exploring the potentials of automatically collected GPS data for travel behaviour analysis – a Swedish data source. In J. Mölthen, & A. Wytzisk (Eds.), *GI-Technologien für Verkehr und Logistik – IfGI, 13*, 155–179. Münster, Germany: Institut für Geoinformatik.

Schreckenberg, M., & Sharma, S. D. (Eds.). (2002). *Pedestrian and Evacuation Dynamics.* Berlin Heidelberg: Springer.

Schulze, M. (2007). *Summary of chauffeur project.* Retrieved July 17th, 2008, from http://cordis.europa.eu/telematics/tap_transport/research/projects/chauffeur.html

Schütz, G. M. (2001). Exactly solvable models for many-body systems. In C. Domb & J. L. Lebowitz (Eds.), *Phase Transitions and Critical Phenomena, Vol. 19.* Academic Press.

Seghers, B. H. (1974). Schooling Behavior in the Guppy (Poecilia reticulata): An Evolutionary Response to Predation. *Evolution , 28*, 486-489.

Sengupta, R., Rezaei, S., Shladover, S. E., Cody, D., Dickey, S., & Krishnan, H. (2007). Cooperative collision warning systems: Concept definition and experimental implementation. *Journal of Intelligent Transportation Systems, 11*(3), 143–155.

Seyfried, A., Rupprecht, T., Passon, O., Steffen, B., Klingsch, W., & Boltes, M. (2007). Capacity estimation for emergency exits and bootlenecks. In *Interflam 2007 - Conference Proceedings.*

Seyfried, A., Steffen, B., & Lippert, T. (2006). Basics of modelling the pedestrian flow. *Physica A, 368*, 232-238.

Seyfried, A., Steffen, B., Klingsch, W., & Boltes, M. (2005). The fundamental diagram of pedestrian movement revisited. *J. Stat. Mech.*, P10002.

SFSO (2000). Eidgenössische Volkszählung. Swiss Federal Statistical Office, Neuchatel.

SFSO (2001). Eidgenössische Betriebszählung 2001 - Sektoren 2 und 3. Swiss Federal Statistical Office, Neuchatel.

SFSO (2006). Ergebnisse des Mikrozensus 2005 zum Verkehrs. Swiss Federal Statistical Office, Neuchatel.

Sheikholeslam, S. & Desoer, C. A. (1990). Longitudinal control of a platoon of vehicles. *In Proceedings of the American Control Conference, 1,* 291–297).

Shiose, T., Onitsuka, T., & Taura, T. (2001). Effective Information Provision for Relieving Traffic congestion. *Conference proceedings of The 4th International Conference on Intelligence and Multimedia Applications,* (pp. 138-142).

Shladover, S. E. (2007). Path at 20 – history and major milestones. *IEEE Transactions on Intelligent Transportation Systems, 8*(4), 584–592.

Shocker, A. D., Ben-Akiva, M. E., Boccara, B., & Nedugadi, P. (1991). Consideration set influences on consumer decision-making and choice: issues, models and suggestions. *Marketing Letters, 2*, 181-197.

Siegel, A. W., & White, S. H. (1975). The development of spatial representations of large-scale environments. In H. W. Reese (Ed.), *Advances in child development and behavior, 10*, 9-55. New York: Academic Press.

Sierhuis, M. (2001). *Modeling and Simulating Work Practice, BRAHMS: A multiagent modeling and simulation language for work system analysis and design* (Vol. 2001-10). Amsterdam, the Netherlands: University of Amsterdam.

Silva, B. C., Junges, R., Oliveira, D., & Bazzan, A. L. C.. (2006). *ITSUMO: an Intelligent Transportation System for Urban Mobility.* In: *The 5th International Joint Conference on Autonomous Agents and Multiagent Systems* (AAMAS 2006) - Demonstration Track. (pp. 1471-1472).

Smith, R. (1980). The contract net protocol: High-level communication and control in a distributed problem solver. *IEEE Transactions on Computers, C-29*(12), 1104–1113.

Sridhar, B., Chatterji, G., Grabbe, S., & Sheth, K. (2002, August). *Integration of Traffic Flow Management Decisions.* Paper presented at the American Institute of Aeronautics and Astronautics (AIAA) Guidance, Navigation, and Control Conference, Monterey, California.

Steiner, T. J., & Bristow, A. L. (2000). Road pricing in national parks: a case study in the Yorkshire Dales National Parks. *Transport Policy, 7*, 93–103.

Stevens, C. F., & Zador, A. M. (1998). Input synchrony and the irregular firing of cortical neurons. *Nature Neuroscience*, 1, 210-217.

Stone, P., & Veloso, M. (2000). Multiagent systems: A survey from a machine learning perspective. *Autonomous Robots, 8*(3), 345–383.

Stopher, P. R. (2004). Reducing road congestion: a reality check. *Transport Policy, 11*(2), 117–131.

Sukthankar, R., Baluja, S., & Hancock, J. (1998). Multiple adaptive agents for tactical driving. *Applied Intelligence, 9*(1), 7–23.

Sutton, R. S., & Barto, A. G. (1998). *Reinforcement Learning: An Introduction*. London: MIT Press.

Swait, J. (2001). Choice set generation with the generalized extreme value family of discrete choice models. *Transportation Research, B, 35*, 643-666.

Swait, J., & Ben-Akiva, M. E. (1987). Incorporating random constraints in discrete models of choice set generation. *Transportation Research, B, 21*, 91-102.

Sweet, D. N., Manikonda, V., Aronson, J. S., Roth, K., & Blake, M. (2002, August). *Fast-Time Simulation System for Analysis of Advanced Air Transportation Concepts*. Paper presented at the American Institute of Aeronautics and Astronautics (AIAA) Modeling and Simulation Technologies Conference and Exhibit, Monterey, California.

Tajima, Y., & Nagatani, T. (2002). Clogging transition of pedestrian flow in T-shaped channel. *Physica A, 303*, 239–250.

Tambe, M. (1997, July). *Agent architectures for flexible, practical teamwork*. Paper presented at the American Association for Artificial Intelligence Conference (AAAI-2007), Providence, Rhode Island.

Tambe, M., & Zhang, W. (2000). Toward flexible teamwork in persistent teams: extended report. *Journal of Autonomous Agents and Multi-agents Systems, 3*, 159–183.

Tan, M. (1993). Multi-agent reinforcement learning: Independent vs. cooperative agents. In *Proceedings of the Tenth International Conference on Machine Learning (ICML 1993)*, (pp. 330–337). Morgan Kaufmann.

Tanahashi, I., Kitaoka, H., Baba, M., H. Mori, H., Terada, S., & Teramoto, E. (2002). NETSTREAM, a Traffic Simulator for Large-scale Road networks. *R & D Review of Toyota CRDL, 37*(2), 47-53.

Tesfatsion, L. (2008). *Agent-based computational economics (ACE) website*. Retrieved January 9, 2008 from http://www.econ.iastate.edu/tesfatsi/ace.html.

Tesfatsion, L.S., (1997). A trade network game with endogeneous partner selection, In H. M. Amman, B. Ruesm, & A. B. Winston (Eds.), *Computational Approaches to Economic Problems.* (pp. 249-269). Dordecht,The Netherlands: Kluwer.

Thill, J. C., & Horowitz, J. L. (1997). Modeling non-work destination choices with choice sets defined by travel-time constraints. In M. M. Fisher & A. Getis (Eds.), *Recent Developments in Spatial Analysis-, -Spatial Statistics, Behaviour Modelling and Neuro-computing* (pp. 186-208). Heidelberg: Springer.

Thomas, R., & Donikian, S. (2008). A spatial cognitive map and a human-like memory model dedicated to pedestrian navigation in virtual urban environments. In T. Barkowsky, M. Knauff, G. Ligozat & D. R. Montello (Eds.), *Spatial cognition V: reasoning, action, interaction, LNCS*, 421-438. Berlin, Heidelberg: Springer.

Thompson, P. & Psaraftis, H. (1993). Cyclic transfer algorithms for multivehicle routing and scheduling problems. *Operations Research, 41*(5), 935-946.

Thompson, P. A., & Marchant, E. W. (1994). Simulex; developing new computer modelling techniques for evaluation. In *Fire Safety Science – Proceedings of the Fourth International Symposium* (pp. 613–624).

Thompson, R. G., & Richardson, A. J. (1998). A parking search model. *Transportation Research Part A: Policy and Practice*, *32*(3), 159–170.

Timmermans, H. J. P., & Golledge, R. G. (1990). Applications of behavioral research on spatial problems II: preference and choice. *Progress in Human Geography, 14,* 311-354.

Timmermans, H. J. P., Arentze, T. A., & Joh, C. H. (2002). Analyzing space-time behavior: new approaches to old problems. *Progress in Human Geography, 26*(2), 175-190.

Timpf, S. (2002). Ontologies of Wayfinding: a traveler's perspective. *Networks and Spatial Economics, 2*(1), 9-33.

Timpf, S. (2005). Wayfinding with mobile devices: Decision support for the mobile citizen. In S. Rana & J. Sharma (Eds.), *Frontiers of geographic information technology*. London, Berlin: Springer.

Timpf, S., & Kuhn, W. (2003). Granularity transformations in wayfinding. In C. Freksa, W. Brauer, C. Habel & K. F. Wender (Eds.), *Spatial cognition iii* (pp. 77-88). Berlin: Springer.

Timpf, S., Volta, G. S., Pollock, D. W., & Egenhofer, M. J. (1992). A conceptual model of wayfinding using multiple levels of abstractions. In A. U. Frank, I. Campari & U. Formentini (Eds.), *Theories and methods of spatio-temporal reasoning in geographic space*, *639*, 348-367. Heidelberg-Berlin: Springer Verlag.

Togawa, K. (1955). *Study on Fire Escapes Basing on the Observation of Multitude Currents* (Report of the Building Research Institute).: Ministry of Construction, Japan.

Torrent-Moreno, M., Killat, M., & Hartenstein, H. (2005). The challenges of robust inter-vehicle communications. In *2005 IEEE 62nd Vehicular Technology Conference, 1*, 319-323. IEEE.

Toth, P. & Vigo, D., editors (2002). The Vehicle Routing Problem. *SIAM Monographs on Discrete Mathematics and Applications*. Society for Industrial and Applied Mathematics.

TRANSIMS (accessed 2008). TRansportation ANalysis and SIMulation System. See transims.tsasa.lanl.gov.

Troitzsch, K. G. (2008). Simulating Collaborative Writing: Software Agents Produce a Wikipedia. In *Proceedings of the Fifth Annual Conference of the European Social Simulation Association (ESSA)*. Brescia.

Tsamboulas, D. A. (2001). Parking fare thresholds: a policy tool. *Transport Policy*, *8*(2), 115-124.

Tsugawa, S. (2005). Issues and recent trends in vehicle safety communication systems. *International Association of Traffic and Safety Sciences Research, 29*, 7–15.

Tsugawa, S., Kato, S., Matsui, T., Naganawa, H., & Fujii, H. (2000). An architecture for cooperative driving of automated vehicles. *In Proceedings of IEEE Intelligent Transportation Systems* (pp. 422–427).

Tumer, K. (2005). Designing agent utilities for coordinated, scalable and robust multi-agent systems. In P. Scerri, R. Mailler, & R. Vincent, (Eds.), *Challenges in the Coordination of Large Scale Multiagent Systems*. Springer.

Tumer, K., & Wolpert, D. (Eds.) (2004). *Collectives and the Design of Complex Systems*. Springer, New York.

Tumer, K., & Wolpert, D. H. (2000). Collective intelligence and Braess' paradox. In *Proceedings of the Seventeenth National Conference on Artificial Intelligence*, (pp. 104–109), Austin, TX.

Tumer, K., Welch, Z., & Agogino, A. (2008). Aligning social welfare and agent preferences to alleviate traffic congestion. In *Proceedings of the Seventh International Joint Conference on Autonomous Agents and Multi-Agent Systems*, Estoril, Portugal.

Tummolini, L., Castelfranchi, C., Omicini, A., Ricci, A., & Viroli, M. (2004). Exhibitionists and Voyeurs do it better: a Shared Environment for Flexible Coordination with Tacit Messages. *In 1st International Workshop on Environments for Multiagent Systems (LNAI, No.3374).* (pp. 215-231).

Tversky, B. (1993). Cognitive maps, cognitive collages, and spatial mental models. In A. U. Frank & I. Campari

(Eds.), *Spatial information theory: A theoretical basis for GIS, 716*, 14-24. Heidelberg-Berlin: Springer Verlag.

Ueyama, T., & Fukuda, T. (1993). Self-organization of cellular robots using random walk with simple rules. *In Proceedings of 1993 IEEE International Conference on Robotics and Automation, 3*, 595-600.

UNECE (2006). Convention on road traffic done at Vienna on 8 November 1968 (2006 consolidated version). Retrieved August 4, 2008, from http://www.unece.org/trans/conventn/Conv_road_traffic_EN.pdf

Ünsal, C., Kachroo, P., & Bay, J. S. (1999). Simulation study of multiple intelligent vehicle control using stochastic learning automata. *IEEE Transactions on Systems, Man and Cybernetics - Part A: Systems and Humans, 29*(1), 120–128.

Vehicle Information and Communication System Center (1995). http://www.VICS.or.jp/english/index.html

Vickrey, W. (1961). Counterspeculation, auctions, and competitive sealed tenders. *Journal of Finance, 16*, 8–37.

Vickrey, W. S. (1969). Congestion theory and transport investment. *The American Economic Review, 59*(2), 251–260.

Vose, D. (2000). *Risk analysis: a quantitative guide*. Chichester: John Wiley.

Vovsha, P., Petersen, E., & Donnelly, R. (2002). Microsimulation in travel demand modeling: lessons learned from the New York best practice model. *Transportation Research Record, 1805*, 68–77.

Vrtic, M., Froehlich, P., & Axhausen, K. (2003). Schweizerische Netzmodelle für Strassen- und Schienenverkehr. In T. Bieger, C. Lässer, & R. Maggi (Eds.), *Jahrbuch 2002/2003 Schweizerische Verkehrswirtschaft* (pp. 119–140). St. Gallen: SVWG Schweizerische Verkehrswissenschaftliche Gesellschaft.

W3C (2006). *eXtensible Markup Language (XML)*. World Wide Web Consortium (W3C). See www.w3.org/XML.

Waddell, P., Borning, A., Noth, M., Freier, N., Becke, M., & Ulfarsson, G. (2003). Microsimulation of urban development and location choices: Design and implementation of UrbanSim. *Networks and Spatial Economics, 3*(1), 43–67.

Waerden, P. J .H. J., Borgers, A. W. J., & Timmermans, H. J. P. (2003). The influence of key events and critical incidents on transportation mode choice switching behavior: a descriptive analysis. *In Proceedings of the IATBR Conference,* Lucerne, (CD-Rom: 24pp.).

Waldau, N., Gattermann, P., Knoflacher, H., & Schrekkenberg, M. (Eds.). (2007). *Pedestrian and Evacuation Dynamics 2005.* Berlin: Springer.

Waldock, A., & Nicholson, D. (2007). Cooperative Decentralised Data Fusion Using Probability Collectives. *Proceedings of the 1st International Workshop on Agent Technology for Sensor Networks (ATSN-07), held at the 6th International Joint Conference on Autonomous Agents and Multiagent Systems.*

Wambsganss, M. (1996). Collaborative Decision Making Through Dynamic Information Transfer. *Air Traffic Control Quarterly*, 4, 107-123.

Ward, N. J. (2000). Automation of task processes: An example of intelligent transportation systems. *Human Factors and Ergonomics in Manufacturing, 10*(4), 395-408.

Wardrop, J. (1952). Some theoretical aspects of road traffic research. *Proceedings of the Institute of Civil Engineers, 1*, 325–378.

Waslander, S. L., Raffard, R. L., & Tomlin, C. J. (2008). Market-Based Air Traffic Flow Control with Competing Airlines. *Journal of Guidance, Control and Dynamics*, 31(1), 148-161.

Watkins, C., & Dayan, P. (1992). Q-learning. *Machine Learning, 8*(3/4), 279–292.

Weidlich, W., & Haag, G. (1983). *Concepts and models of a quantitative sociology: The dynamics of interacting populations* (Series in Synergetics, 14). Berlin: Springer.

Weidmann, U. (1992). Transporttechnik der Fussgänger. *Zürich: Institutfür Verkehrsplanung.*

Weidmann, U. (1993). *Transporttechnik der Fußgänger - Transporttechnische Eigenschaften des Fußgängerverkehrs (Literaturauswertung)* (Schriftenreihe des IVT No. 90). : ETH Zürich. (Second Edition, in German)

Weisman, J. (1981). Evaluating architectural legibility: Wayfinding in the built environment. *Environment and Behavior, 13*(2), 189-204.

Werner, S., Krieg-Brückner, B., & Herrmann, T. (2000). Modelling navigational knowledge by route graphs. In C. Freksa, W. Brauer, C. Habel & K. F. Wender (Eds.), *Spatial cognition II, LNCS 1849*, 295-316) Berlin: Springer.

Werner, T., & Helbing, D. (2003). The social force pedestrian model applied to real life scenarios. In E. R. Galea (Ed.), (p. 17). London: CMS Press.

West, P., & Broniarczyk, S. (1998). Integrating multiple opinions: the role of aspiration level on consumer response to critic consensus. *Journal of Consumer Research, 25*(1), 38-51.

Weyns, D., Parunak, H., Michel, F., Holvoet, T., & Ferber, J. (2005). Environments for multiagent systems, State-of-the-art and research challenges. *In 1st International Workshop on Environments for Multiagent Systems (LNAI, No.3374).* (pp. 1-47).

Weyns, D., Schumacher, M., Ricci, A., Viroli, M., & Holvoet, T. (2005). Environments in multiagent systems. *The Knowledge Engineering Review, 20*(2), 127-141.

Widmer, P., & Axhausen, K. W. (2001). *Aktivitätenorientierte Personenverkehrsmodelle: Vorstudie*. Bern, Schweiz: Eidgenössisches Departement für Umwelt, Verkehr, Energie und Kommunikation / Bundesamt für Strassen.

Wiedemann, R. (1974). *Simulation des Straßenverkehrsflusses*. Schriftenreihe Heft 8, Institute for Transportation Science, University of Karlsruhe, Germany.

Wiener, J. M., & Franz, G. (2005). Isovists as a means to predict spatial experience and behavior. *Lecture notes in artificial intelligence, 3343*, 42-57.

Wierwille, W. W., Hanowski, R. J., Hankey, J. M., Kieliszewski, C. A., Lee, S. E., Medina, A., Keisler, A. S., & Dingus, T. A. (2002). *Identification and evaluation of driver errors: Overview and recommendations FHWA-RD-02-003*, Virginia Tech Transportation Institute, Blacksburg, Virginia, USA. Sponsored by the Federal Highway Administration.

Williams, R. J. (1992). Simple statistical gradient-following algorithms for connectionist reinforcement learning. *Machine Learning, 8*, 229–256.

Williamson, O. (1979). Transaction cost economics: the governance of contractual relations. *Journal of Law and Economics,* (22), 233-262.

Williamson, O. (1991). Comparative Economic Organisation: The Analysis of Discrete Structural Alternatives. *Administrative Science Quarterly, 36*(2), 269-296.

Williamson, O. (1995). Hierarchies, Market, and Power in the Economy: An Economic Perspective. *Industrial and Corporate Change, 4*(1), 21-49.

Winter, S. (2003). Route Adaptive Selection of Salient Features. In W. Kuhn, M. Worboys, & S. Timpf (Eds.), *Spatial Information Theory: Foundations of Geographic Information Science* (pp. 349-361). Berlin: Springer Verlag.

Wolf, D., & Grassberger, P. (Eds.). (1996). *Friction, Arching, Contact Dynamics.* Singapore: World Scientific.

Wolf, J., Schönfelder, S., Samaga, U., Oliveira, M., & Axhausen, K. (2004). Eighty weeks of Global Positioning System traces. *Transportation Research Record, 1870*, 46–54.

Wolfe, S. R., Jarvis, P. A., Enomoto, F. Y., & Sierhuis, M. (2007, November). *Comparing Route Selection Strategies in Collaborative Traffic Flow Management.* Paper presented at the 2007 Institute of Electrical & Electronic Engineers/ Web Intelligence Consortium/ Association for Computing Machinery (IEEE/WIC/ACM) International Conference on Intelligent Agent Technology (IAT 2007), Fremont, California.

Wolpert, D. (2004). Information theory - the bridge connecting bounded rational game theory and statistical physics. *Understanding Complex Systems* (pp. 262-290). Springer-Verlag.

Wolpert, D. H., & Tumer, K. (2001). Optimal payoff functions for members of collectives. *Advances in Complex Systems, 4*(2/3), 265–279.

Wolpert, D. H., Tumer, K., & Frank, J. (1999). Using collective intelligence to route internet traffic. In *Advances in Neural Information Processing Systems - 11*, (pp. 952–958). MIT Press.

Wolpert, D., & Tumer, K. (1999). An introduction to COllective INtelligence. *Technical Report NASA-ARC-IC-99-63.* NASA Ames Research Center.

Wolpert, D., & Tumer, K. (2001). Optimal payoff functions for members of collectives. *Advances in Complex Systems, 4*(2/3), 265-279 .

Wolpert, D., Wheeler, K., & Tumer, K. (1999). General principles of learning-based multi-agent systems. *Proceedings of the 3rd Annual Conference on Autonomous Agents* (pp. 77-83). ACM Press.

Wooldridge, M. (1999). Intelligent Agents. In Weiss, G. (Ed.), *Multiagent Systems: A modern approach to distributed artificial intelligence.* Cambridge, MA: The MIT Press.

Wooldridge, M. (2002). *An Introduction to Multi Agent Systems.* West Sussex, England: John Wiley and Sons Ltd.

Wooldridge, M., & Jennings, N. R. (1995). Intelligent Agents: Theory and Practice. *Knowledge Engineering Review, 10*(2), 115-152.

Yamori, K. (1998). Going with the flow: Micro-macro dynamics in the macrobehavioral patterns of pedestrian crowds. *Psychological Review, 105*(3), 530–557.

Yang, J., Jaillet, P., & Mahmassani, H. (1999). On-line algorithms for truck fleet assignment and scheduling under real-time information. *Transportation Research Record, 1667*, 107–113.

Yoshii, T., Akahane, H., & Kuwahara, M. (1996). Impacts of the Accuracy of Traffic Information in Dynamic Route Guidance Systems. *Conference proceedings of The 3rd Annual World Congress on Intelligent Transport Systems (CD-ROM).*

Yu, W. J., Chen, R., Dong, L. Y., & Dai, S. Q. (2005). Centrifugal force model for pedestrian dynamics. *Phys. Rev. E, 72*, 026112.

About the Contributors

Ana Bazzan received her PhD in 1997 from the University of Karlsruhe, Germany, and an MSc in computer science from the Institute of Informatics at the University of Rio Grande do Sul (UFRGS) in Porto Alegre, Brazil. From 1997 to 1998, she had a postdoc research associate position in the Multi-Agent Systems Laboratory at the University of Massachusetts in Amherst, under the supervision of prof. Victor Lesser. In 1999 she joined the Institute for Informatics at UFRGS as a professor and got tenure three years later. During 2006 and 2007 she had a fellowship from the Alexander von Humboldt Foundation at the University of Würzburg in Germany. She is affiliated with the research groups on Artificial Intelligence and Multi-Agent Systems at UFRGS. Her research interests include: game-theoretic paradigms for coordination of agents, multiagent learning, coordination and cooperation in MAS, agent-based simulation, RoboCup Rescue, and traffic simulation and control. Other professional activities: associate editor of the journal *Advances in Complex Systems*, chair of program committee for the 17th Braz. Symp. on Artificial Intelligence (2004), and co-organizer of workshop series on Agents in Traffic and Transportation.

Franziska Klügl is Universitetslektor at the Örebro University since September 2008. She is also responsible for agent-based Modelling at the Research Centre for Modelling and Simulation at the Campus Alfred Nobel in Karlskoga. She received her PhD in computer science from the University of Würzburg in 2000. From 2000 to 2008 she worked as an assistant professor at the University of Würzburg and headed there the multi-agent simulation group. She is co-organizer of the workshop series on "Agents in Traffic and Transportation" and was involved in the organization of several scientific events like PC Chair of the European Workshop of Multi-Agent Systems in 2008. Her research interests are in the area of methodologies, applications and tools for multi-agent simulation ranging from learning and adaptive agents to visual programming for modelling multi-agent models. The main application areas of her research are all forms of traffic simulation.

* * *

Adrian Agogino is a researcher at the University of California Affiliated Research Center at NASA Ames Research Center. His primary interests include complexity, multi-agent coordination, reinforcement learning and visualization. He has authored over thirty peer reviewed publications in these areas. His professional background also includes graphics, user interface design and hardware digital electronic design for systems such as satellite decoding, security and cryptography.

Theo Arentze received a PhD in decision support systems for urban planning from the Eindhoven University of Technology. He is now an associate professor at the Urban Planning Group at the same university. His main fields of expertise and current research interests are activity-based modeling, discrete choice modeling, knowledge discovery and learning-based systems, and decision support systems with applications in urban and transport planning.

Michael Balmer is project leader of ongoing micro-simulation projects at the transport planning group of prof. Kay W. Axhausen at the Eidgenössische Technische Hochschule Zürich (ETH Zurich). He holds a PhD in civil engineering from the ETH Zurich since Summer 2007 and an master in computer science from the ETH Zurich since spring 2003. His work is embedded in the research project MATSim (Multi-Agent Transport Simulation; www.matsim.org), while the focus lies into the system development, software engineering and performance of the MATSim toolkit. The outcome of his dissertation builds the software basis of MATSim's modular approach.

Klemens Böhm is full professor for computer science at Universität Karlsruhe (TH). Before joining Karlsruhe University in 2004, he has been a professor at Magdeburg University. Prior to that, he has been affiliated with ETH Zurich and GMD Darmstadt. His research interests are distributed information systems, e.g., peer-to-peer systems and grid infrastructures, data management in ubiquitous environments, and data warehousing. Klemens puts much effort in interdisciplinary research and application-oriented projects, currently ranging from biosystematics to traffic-data management.

Ladislau Boloni is an associate professor at the School of Electrical Engineering and Computer Science of University of Central Florida. He received a PhD degree from the Computer Sciences Department of Purdue University in May 2000. He is a senior member of IEEE, member of the ACM, AAAI and the Upsilon Pi Epsilon honorary society. His research interests include autonomous agents, grid computing and wireless networking.

Brahim Chaib-draa received the diploma in computer engineering from the Ecole Supérieure d'Electricité (SUPELEC) de Paris, Paris, France, in 1978 and the PhD degree in computer science from the Université du Hainaut-Cambrésis, Valenciennes, France, in 1990. In 1990, he joined the Computer Science and Software Engineering Department of Laval University, Quebec, QC, Canada, where he is a professor and group leader of the Decision and Actions for Multi-Agent Systems Group. His research interests include agent and multiagent technologies, natural language for interaction, formal systems for agents and multiagent systems, distributed practical reasoning, and real-time and distributed systems. He is the author of several technical publications in these areas. He is on the editorial boards of *Concurrent Engineering Research and Applications (CERA), The International Journal of Grids and Multiagent Systems*, and *Applied intelligence*. Dr. Chaib-draa is a member of the ACM, the IEEE Computer Society, and the AAAI.

David Charypar studied Computer Science at ETH Zurich from 1998 to 2003. During a semester project he got involved with travel behavior research. After his Master thesis on real time fluid simulation in the field of computer graphics he started his PhD studies in Kai Nagels Group at ETH Zurch, later changing the the Institute for Transport Planning and Systems (IVT). His work in the MATSim team includes developing the concept behind the agent's activity planning process as well as creating the current parallel event-driven traffic flow microsimulation which was recently extended by support for ignalled intersection. Currently, he is working on within-day replanning processes for the simulated agents.

Thomas L. Clarke is principal mathematician at the Institute for Simulation and Training of the University of Central Florida and is also associate professor in the Modeling and Simulation program. He has a PhD in applied mathematics from the University of Miami and worked at NOAA before coming to UCF. He has a diverse background in applying mathematics to simulation and has investigated areas such as the application of catastrophe theory and nonlinear dynamics.

Charles Desjardins has obtained in 2006 a BEng degree in software engineering from Laval University in Quebec City, QC, Canada. Afterwards, he has pursued a master's degree in computer science, again at Laval University. Apart from his interest in various software engineerings topics, his studies have focused on reinforcement learning, and on applying these algorithms to learn efficient behaviors in complex environments.

Kurt Dresner is a 6th-year PhD student in the Learning Agents Research Group at the University of Texas at Austin. He did his undergraduate work at Harvey Mudd College, in Claremont, California. Kurt was motivated to start working on autonomous intersection control when he found himself stuck at a red light at 2 a.m. after a late night in the lab. In addition to autonomous vehicles, he is interested in broader applications of multiagent systems as well as machine learning.

Francis Enomoto is a computer engineer in the Intelligent Systems division at NASA Ames Research Center. His current research encompasses air traffic management, decision support systems, and unmanned aerial vehicles. Previously, he led research projects in experimental and computational aerodynamics, computer graphics, and computer aided design. He holds a M.S. in Aeronautics and Astronautics from Stanford University and a B.S. in Mechanical Engineering from the University of Hawaii.

Edgar F. Esteves holds a BEng (Hons) in informatics engineering and is currently finishing his MSc in informatics engineering at University of Porto. He has been researching artificial intelligence techniques applied to microscopic traffic modelling for about one year, and has recently gained interest in pedestrian modelling and simulation. In both streams of research, his work has focused on the conceptualization and implementation of modelling and simulation frameworks, interactive editors, environment representation and visualisation (both 2D and 3D) tools, and algorithms for locomotion and visualization. His master's dissertation will report on his recent achievements in the field of pedestrian modelling and simulation frameworks.

Paulo A. F. Ferreira holds a BEng (Hons) and a MSc in informatics engineering from the University of Porto, Portugal, obtained in 2007 and 2008, respectively. He is currently a research assistant at the Artificial Intelligence and Computer Science Lab (LIACC), working on the application of several AI techniques to the simulation of urban traffic and transportation networks. He accumulates now more than one year of experience in the field. He has been one of the developers of the microscopic traffic simulator underlying the MAS-T^{2}er Lab platform, under development at LIACC/FEUP. Before his current research interests he worked for a company named CentralCasa, where he developed features for a service -oriented platform responsible for the remote management and control of demotic, security and surveillance devices via web applications.

Qi Han is a post doctorate researcher of Urban Planning Group at Eindhoven University of Technology, The Netherlands. Her current research interests are in developing models and experiments for dynamic and interactive agents' behavior and choice processes in marketing, retailing, tourism, transportation and management. She has published her research in several international journals, such as *Tourism Analysis*, *Transportation Research Record*, *Transportmetrica* and *Geojournal*.

Lawrence Henesey is employed at the Department of Systems and Software Engineering, Blekinge Institute of Technology, Karlshamn, Sweden. Dr. Henesey researches the application of techniques from Artificial Intelligence in Logistics. Concurrently with his research in improving the performance of container terminals using artificial intelligence technology. In addition, Dr. Henesey divides his work 50% with TTS Port Equipment AB in Gothenburg, Sweden on automated guided vehicle systems for port projects. Dr. Henesey graduated Cum-Laude with a MSc in transport and maritime management from the University of Antwerp, Belgium in 1999 and has a BS in international political economy from Old Dominion University, Norfolk, Virginia, USA, 1992.

Davy Janssens is a member of the Transportation Research Institute (IMOB), Hasselt University. His research interests include the use of data mining procedures for personalization, customer satisfaction studies, classification based on association rules and modeling activity-travel behavior.

Peter Jarvis is a computer scientist at NASA Ames Research Center specializing in applying Artificial Intelligence Technologies to problems in Space and Aeronautics. Before joining NASA in 2004 Peter worked for SRI International applying the same technologies in the military application domain. Peter holds a PhD in computer science from the University of Brighton (UK) and held a post doctoral position at the University of Edinburgh (UK).

David J. Kaup is Provost distinguished research professor in mathematics at the University of Central Florida with a joint appointment at the Institute for Simulation and Training. He has a PhD in physics from the University of Maryland. He was a professor at Clarkson University before coming to UCF. His research interests are in the area of non-linear waves, in particular solitons, and he has been recently applying this background to the area of modeling the dynamics of crowds.

Hubert Klüpfel studied physics at the universities of Würzburg and Stony Brook, NY, and obtained his PhD in 2003 under the supervision of prof. Schreckenberg at Duisburg University. In 2001 he co-founded TraffGo GmbH as a university spin-off and is now executive director of TraffGo HT GmbH, an SME specialized in the simulation and optimization of pedestrian flows and evacuation processes.

Tobias Kretz finished his PhD thesis "Pedestrian Traffic - Simulation and Experiments" in February 2007 in the group "Physics of Transport and Traffic" of prof. Schreckenberg at Duisburg University. In July 2007 he joined PTV AG company as product manager for the simulation of pedestrian crowds in the microscopic multi-modal traffic simulation software VISSIM.

Koichi Kurumatani is multi-agent group leader at Information Technology Research Institute (ITRI), National Institute of Advanced Industrial Science and Technology (AIST) since July 2004. He was a researcher in 1989-1991, a senior researcher in 1991-1999, Emergent Global Dynamics Lab. Leader in 1999-2001 at Electrotechnical Laboratory (ETL), Multi-Agent Team Leader at Cyber Assist Research Center (CARC) in 2001-2004. He received M.S. and Ph.D. degrees from the University of Tokyo in 1986 and 1989, respectively. His research interests include ubiquitous/pervasive computing, sensor network, multi-agent, social simulation, complex networks.

Julien Laumonier received the diploma in computer engineering from the École Polytechnique de l'Université de Nantes (France) and the master degree from Lumière University in Lyon (France) in 2002. He received a PhD in computer science from Université Laval in Quebec City, Canada in June 2008. His research interests include multiagent technology, reinforcement learning and intelligent transportation systems.

Nicolas Lefebvre studied Computer Science at ETH Zurich, with study exchanges at the Swiss Federal Institute of Technology in Lausanne and at the Univerisdad Politécnica de Madrid. After finishing his master thesis, he joined the MATSim team at IVT, ETH Zurich, to work on improving the routing algorithms used by MATSim. This lead to the implementations of the fast Landmarks-A* algorithm currently used by MATSim. Currently, he is working as head of software engineering for digitec AG in Zurich, a large provider of IT products and services in Switzerland.

Ulf Lotzmann obtained his diploma degree in computer science from the University of Koblenz-Landau in 2006. Already during his studies he has participated in development of several simulation tools. Since 2005 he has specialized in agent-based systems mainly within the domain of traffic simulation and is the creator of the TRASS system. He is also involved in several other recent projects of the research group, primarily the "Emergence in the loop: simulating the two way dynamics of norm innovation (EMIL)" project, funded by the New and Emerging Science and Technology programme of the European Commission (2006-2009).

Tamás Máhr did his master's in technical engineering at the Budapest University of Technology and Economics in 2001. After that he engaged in PhD research at the department of Telecommunication and Media Informatics, where he studied distributed resource allocation in differentiated services networks. In 2004, Tamás transferred to the Delft University of Technology in The Netherlands where he researches distributed multi-agent systems applied in the logistics domain. His focus is on analyzing the ability of different multi-agent routing solutions to handle unexpected events that occur during the execution of transportation plans. He is working towards defining a robustness measure that can characterize planning methods in terms of their ability to withstand unforeseen changes.

Linda C. Malone is a professor in the Industrial Engineering Department at University of Central Florida. She got her PhD degree in statistics from Virginia Tech after having gotten her BS and MS degrees in mathematics (at Emory and Henry College and University of Tennessee respectively). Her primary research interests include Response Surface Analysis and Quality. She was an associate editor of the *Journal of Statistical Computation and Simulation* for over 25 years and was a founding coeditor of the *STATS Magazine*. She was awarded the honor of fellow of the American Statistical Association.

Konrad Meister has been working as a research associate at the Institute of Transport Planning and Systems (IVT) at ETH Zurich since 2004. His background is computer and systems science. In the PhD project related to MATSim-T, he is particularly concerned with optimization of the agents' travel choices sich as activity timing or mode choice. The efficiency of the optimization is directly relevant to the quick computation of the equilibration problem in the multi-agent simulation system.

Mark Van Middlesworth is an undergraduate at Harvard University in Boston, MA. His interest in artificial intelligence and autonomous agents began in high school, when he met Peter Stone at the University of Texas at Austin. Dr. Stone invited Mark to join his team for the Trading Agent Competition in Supply Chain Management, providing an early introduction to the world of academic research. He hopes to continue his research through college, and ultimately pursue a graduate degree in Computer Science. When not in school, he spends his time in Austin, Texas, where he enjoys rock climbing, mountain biking, and catching up on the local music scene.

After studying physics and meteorology in Cologne and Paris, **Kai Nagel** got his PhD in computer science at the University of Cologne about "fast microscopic traffic simulations". From 1995 to 1999 he was at Los Alamos National Laboratory as part of the "TRANSIMS" team. 1999-2004 he was assistant professor for Computer Science at ETH Zurich at the Institute for Scientific Computing. Since 2004 he is full professor for "Transport systems planning and transport telematics" at the Technical University of Berlin. His research interests include: large transportation simulations, modeling and simulation of socio-economic systems, multi-agent simulations.

Rex Oleson is a student at the University of Central Florida working on a PhD in modeling and simulation. He has a Masters in mathematics from UCF and a BS in physics and mathematics from Susquehanna University. His research interests are in the areas of agent based simulation and particularly the application in the area of modeling human crowd movement.

Denise de Oliveira received her MSc (2005) and BSc (2002), in computer science, from the Institute of Informatics at the University of Rio Grande do Sul (UFRGS) in Porto Alegre, Brazil. Since 2005 she is a PhD student, with full Scholarship, first from the Brazilian National Counsel of Technological and Scientific Development (CNPq). For two semesters (Summer 2006 and Winter 2006/2007) she was enrolled as a PhD student and assistant at the Chair of Artificial Intelligence and Applied Informatics, University of Würzburg, Germany, supported by CAPES Foundation (ProBral Program). She is affiliated with the research groups on Artificial Intelligence and Multi-Agent Systems at UFRGS. Her main research interests are: multiagent learning, urban search and rescue simulation (RoboCup Rescue), coordination and cooperation in MAS, traffic simulation and control.

Eugénio C. Oliveira is a full professor and the coordinator of LIACC (Artificial Intelligence and Computer Science Laboratory) at the University of Porto. professor Oliveira is also the director of the PhD programme on informatics engineering at the University of Porto and has been responsible for 10 past successful PhD students' theses, and other four that are currently in progress. He is a member of the editorial boards of journals such as *Agents and Multi-Agents Systems International Journal*, *Agent Oriented Software Engineering*, and *Intelligent Decision Technologies*. He is also a co-founder of AGENTLink European Network of Excellence. He's got his PhD in artificial intelligence from the New

University of Lisbon, in 1984, and was awarded with the Gulbenkian Prize for Science and Technology in 1983. From 1984 to 1985, he was a "Guest Academic" at IBM/IEC in Brussels, and in 2008 the Area Chair for Agents in the European Conference on Artificial Intelligence.

Sascha Ossowski is the director of the Centre for Intelligent Information Technologies (CETINIA) at University Rey Juan Carlos in Madrid. Formerly, he was an HCM/TMR research fellow at the AI Department of Technical University of Madrid. He obtained his MSc degree in informatics from the University of Oldenburg (Germany) in 1993, and received a PhD in artificial intelligence from UPM in 1997. Prof. Ossowski is holding several research grants in the field of advanced software systems, funded by the European Commission and the Spanish Government. He has authored more than 100 research papers, focusing on the application of artificial intelligence techniques to real world problems such as transportation management, m-health, or e-commerce. Recently, he has been particularly active in the field of co-ordination mechanisms for agents and services, as well as models of trust and regulation in virtual organisations. He is co-editor of more than 20 books, proceedings, and special issues of international journals. He is a general chair of the ACM Annual Symposium on Applied Computing (SAC), chairs the Steering Committee of the European workshop series on Multiagent Systems (EUMAS), serves as a member of the editorial board for several international journals, and acts as programme committee member for numerous international conferences and workshops

Jan A. Persson is employed at the Department of Systems and Software Engineering, Blekinge Institute of Technology, Karlshamn, Sweden. Dr. Persson researches in optimization and simulation methods for decision support in the area of intelligent logistics and transportation systems.

Bart-Jan van Putten was an intern at NASA Ames Research Center (California, USA), where he completed a research project on agent-based modeling and simulation. Previously he had completed internships at Philips Research (NL) and OcŽ Research (NL), as well as teaching assistant duties in decision support systems, e-learning, and knowledge modeling for Utrecht University (NL). Bart-Jan holds a MSc (cum laude) in content and knowledge engineering from Utrecht University (NL) and a BSc in industrial design engineering from Eindhoven University of Technology (NL).

While studying Computer Science from 2000 to 2005 at ETH Zurich, **Marcel Rieser** got involved with transportation planning by taking several courses on that topic. After finishing his master thesis, he started to work within the MATSim team at the Technical University of Berlin in 2006. After a lot of system integration work, leading to the current modular design of MATSim, he works now on extending the traffic flow simulation of MATSim with public transportation features. He is also involved in the coordination of the further development of MATSim between Zurich and Berlin.

Rosaldo J. F. Rossetti holds a BEng(Hons) in civil engineering from UFC University (1995), and a MSc and a PhD in computer science from UFRGS University (1998 and 2002, respectively), Brazil. He did though most of his PhD research studies as a research student at Leeds University's Institute for Transport Studies, UK, within the Network Modelling Group (where SATURN e DRACULA tools were developed). From 2002 to 2006, Dr. Rossetti was the director of the BSc(Hons) programme in systems management and computing, at Altântica University, in Portugal. There he was also a co-founder and head of the Systems Management and Computing Lab, a R&D Unit. In 2006, Dr. Rossetti joined the

University of Porto, where he is an assistant professor in the Department of Informatics Engineering. Dr. Rossetti is a research fellow at LIACC/FEUP and a member of IEEE, ACM and the Portuguese Association for AI.

Christian Rogsch studied safety engineering at Wuppertal University, Germany. Since 2005 he is PhD student at the "Institute for Building Material Technology and Fire Safety Science" of prof. Klingsch in Wuppertal. His main interests are different kinds of simulations, especially fire, smoke and evacuation.

Andreas Schadschneider is professor for theoretical physics at Cologne University. He has obtained his PhD in 1991 in the field of solid state physics. Since more than 15 years he is working on problems of non-equilibrium physics. Here his focus are transport problems with interdisciplinary applications, e.g. in traffic engineering, biology and social dynamics. He has organized several international conferences and is author is various review articles on the application of methods from physics to traffic and transport problems.

Heiko Schepperle is a researcher in computer science at the Universität Karlsruhe (TH) since 2003, working towards a PhD degree. He holds a diploma degree in computer science from the same university. His research interests are negotiation mechanisms in multi-agent systems and agent-based driver-assistance systems in traffic applications.

Armin Seyfried is research assistant at the Jülich Supercomputing Centre. Computational physics was the focus of his diploma and PhD thesis (1998). He started to work in the field of pedestrian and evacuation dynamics during five years working at an engineering consultant for fire safety. Since 2004 he leads a research group at the Forschungszentrum Jülich and the University of Wuppertal concentrating on modelling and experimental studies of pedestrian dynamics.

Kapil Sheth received the BTech degree in aeronautical engineering from IIT Kharagpur, India, and the MS and PhD in engineering sciences from UC San Diego. Kapil has been working in the Air Traffic Management arena since 1996 and is a co-founder of the Future ATM Concepts Evaluation Tool (FACET). He was a member of the team that received NASAÕs Software of the Year Award, Turning Goals Into Reality Award and RaytheonÕs Corporate Level Excellence in Technology Award. Currently, Kapil is employed as an aerospace engineer at NASA Ames Research Center in the Automation Concepts Research Branch and specializes in air traffic management and weather impact research. He serves on the Air Transportation Systems Technical Committee of the AIAA.

Maarten Sierhuis is a senior research scientist and lead of the Collaborative Assistant Systems Group at RIACS/USRA, located at NASA Ames Research Center. He is a co-principal investigator for the Brahms project, working in the Work Systems Design & Evaluation group in the collaborative and assistant systems (CAS) area within the Intelligent Sciences Division at NASA Ames Research Center. Previously, he worked at NYNEX Science & Technology, the former R&D organization for the former NYNEX Corporation (now Verizon) in White Plains, NY. He received a PhD in social science informatics, from the University of Amsterdam, The Netherlands and holds an engineering degree in Informatics, from the Polytechnic University in The Hague, The Netherlands.

Jordan Srour is a third year PhD candidate at the Rotterdam School of Management of Erasmus University. She pretty much loves anything that has to do with the vehicle routing problem – especially as applied to problems in intermodal freight transport. Prior to becoming a PhD candidate at RSM (and subsequent to receiving her MSc in transportation engineering at The University of Texas at Austin), Jordan worked as a transportation engineer at Science Applications International Corporation (SAIC) where she performed evaluations of new technologies designed to improve the efficiency, safety, and security of freight operations. Jordan's most significant work in the realm of agents is a comparative study of on-line optimization and agent-based solution techniques for a truckload pick-up and delivery problem with time windows.

Peter Stone is an Alfred P. Sloan research fellow and associate professor in the Department of Computer Sciences at the University of Texas at Austin. He received his PhD in computer science in 1998 from Carnegie Mellon University. From 1999 to 2002 he was a senior technical staff member in the Artificial Intelligence Principles Research Department at AT&T Labs - Research. Peter's research interests include machine learning, multiagent systems, robotics, and e-commerce. In 2003, he won a CAREER award from the National Science Foundation for his research on learning agents in dynamic, collaborative, and adversarial multiagent environments. In 2004, he was named an ONR Young Investigator for his research on machine learning on physical robots. Most recently, he was awarded the prestigious IJCAI 2007 Computers and Thought Award.

Takeshi Takama is working as a research fellow at Stockholm Environment Institute /Oxford office after he completed his DPhil at Transport Studies Unit / Centre for Environment (OUCE) of University of Oxford as an Oxford Kobe scholar. He researches various projects related to environment and development, climate change, human behaviours, transportation, agent-based modelling, decision making under uncertainties. He leads work on both large and small scale projects funded by the EU, SIDA, UN, etc. He recently completed an EU funded CAVES project to design and operate the validation activities of agent-based modelling. He is also a fellow of OUCE at University of Oxford and Nagoya Sangyo University in Japan. He supervises post-graduate students at University of Oxford and presents his work locally at the university and internationally at conferences and meetings.

Harry Timmermans is professor of Urban Planning at the Eindhoven University of Technology. He has research interest in modeling consumer choice behavior and the integration of such models in design and decision support systems. Application domains include transportation, housing, retailing and leisure. He has published in many international journals and services on the editorial board of journals in urban planning, transportation, tourism, retailing and management.

Sabine Timpf holds a doctorate in the technical sciences from the Technical University of Vienna, a Diplomingenieur from the University of Hannover (Germany) and Master of Science from the University of Maine (USA). Her main research interests are in geographic information science, spatial cognition, human navigation and geosimulation. After her doctorate she did research and taught at the University of Zurich in Switzerland. While in Zürich, she repeatedly visited the Cognitive Systems Research Group at the University of Bremen. From 2006 to 2007 she did research at the University of Würzburg in Germany. She now holds a position as professor of Geoinformatics at the University of Augsburg in Germany.

Kagan Tumer is an associate professor at Oregon State University. Prior to joining OSU in 2006, he was a senior research scientist at NASA Ames Research Center. Dr. Tumer's research interests are learning, control and optimization in large distributed systems with a particular emphasis on multiagent coordination. Applications of his work include coordinating multiple robots, controlling unmanned aerial vehicles, reducing traffic congestion and managing air traffic. His work has led to over one hundred publications, including a book titled "Collectives and the Design of Complex Systems", and a best paper award at the 2007 Conference on Autonomous Agents and Multiagent Systems. Dr. Tumer received his PhD from the Electrical and Computer Engineering Department at the University of Texas, Austin in 1996. He is the section editor for computer science for "Advances in Complex Systems" (Elsevier), a member of the board of editors for the "Complex Systems and Inter-Disciplinary Science" book series (World Scientific) and was an associate editor of "Pattern Recognition Letters" (Elsevier). He holds one patent, has chaired multiple workshops/symposia, and is on the program committee of numerous international conferences (AAMAS, AAAI, ICML, IJCNN, GECCO, ICPR and ANNIE). Dr. Tumer is a member of AAAI and a senior member of IEEE.

Matteo Vasirani is a PhD student and a member of the Artifical Intelligence Research Group at the University Rey Juan Carlos in Madrid (Spain). He obtained his MSc in informatics from the University of Modena (Italy) in 2004. Between 2003 and 2004 he worked in the Agent and Pervasive Computing Research Group at the University of Modena. His research interests span from coordination and learning in multiagent systems, to intelligent transportation systems. He is author of 12 publications in journals, books and international conferences, and he participated in 4 research projects, funded by the European Commission and the Spanish Government.

Mathijs de Weerdt completed his master's in computer science at the Utrecht University in The Netherlands. After that he received his PhD on "Plan Merging in Multiagent Systems" at the Delft University of Technology. In his thesis Mathijs shows how his work relates to dial-a-ride planning problems in personal transportation. Since 2004 he has served as an assistant professor in the Algorithmics group in Delft. Mathijs obtained a VENI (personal) grant to study the interaction of efficient planning and task allocation algorithms with coordination mechanisms for self- interested agents leading to a number of publications. In his current work he combines game theory with algorithms for multiagent coordination. The results of this work are mainly applied in the transportation domain.

Since 2007, **Zachary T. Welch** has been working toward a PhD under the tutelage of Dr. Kagan Tumer, bringing over fifteen years of professional software engineer engineering experience to help seed the Adaptive Agents and Distributed Intelligence research group at Oregon State University. He has been active in the recent development of several new multi-agent system simulations, ranging from learning problems in rover and traffic domains to the aesthetically rewarding swarms of Boids. He spends his free time mastering mandolin, banjo, dobro, fiddle, guitar, and bass in order to bootstrap entirely new Bluegrass bands. He hopes the future brings opportunities to mix his passions for MAS and music, if only to help demonstrate solutions for similar coordination problems.

Geert Wets received a degree as commercial engineer in business informatics from the Catholic University of Leuven (Belgium) in 1991 and a PhD from Eindhoven University of Technology (The Netherlands) in 1998. Currently, he is a full professor at the faculty of Applied Economics at Hasselt

University (Belgium) where he is director of the Transportation Research Institute (IMOB) and the program coordinator of the Bachelor/Master in Transportation Sciences. His current research entails transportation modeling and traffic safety modeling. He has published his research in several international journals such as *Journal of the Royal Statistical Society, Accident Analysis and Prevention, Environment and Planning, Geographical Analysis, Knowledge Discovery and Data Mining, Transportation Research Record and Information Systems.*

Shawn Wolfe is a computer scientist at NASA Ames Research Center and the lead of the agent-based Collaborative Traffic Flow Management simulation team. Currently he is also developing methods to identify weather-related deviations in aircraft flight track data. He has been developing software to support aviation and space research since 1992, including knowledge management, information retrieval, automated software engineering, expert systems, and data mining applications. Shawn is currently a PhD student at the University of California, Santa Cruz and holds a MS in computer science from the University of Oregon and a BS in computer science from Iowa State University.

Tomohisa Yamashita is a researcher at Multi-Agent Group, Information Technology Research Institute (ITRI), National Institute of Advanced Industrial Science and Technology (AIST) since 2005. He was Japan Society for the Promotion of Science (JSPS) Research Fellowship for Young Scientists from 2000-2003, a visiting research fellow at the Brookings Institution from 2002-2003, and a researcher at Cyber Assist Research Center (CARC) from 2003-2004. He received PhD degrees from the University of Hokkaido in 2002. His research interests include social simulation, multi-agent, ubiquitous computing, and game theory.

Rob Zuidwijk works as an associate professor at RSM Erasmus University and is interested in the use of information and information technology in supply chains, in particular in closed loop supply chains and intermodal transport. His research interests focus on the development of quantitative models that assess the value of information in aforementioned contexts. He has published in journals such as *SIAM Journal Mathematical Analysis*, *Communications of the ACM*, *California Management Review*, *European Journal of Operational Research*, and *Production and Operations Management*.

Index